New Insights into the Structure of Matter

The First Decade of Science at Jefferson Lab

Journal of Physics: Conference Series Volume 299

Held over various years.

ISBN: 978-1-62276-676-5

New Insights into the Structure of Matter

The First Decade of Science at Jefferson Lab

Journal of Physics: Conference Series Volume 299

Held over various years.

ISBN: 978-1-62276-676-5
ISSN: 1742-6588

IOP Institute of Physics

Printed from e-media with permission by:

Curran Associates, Inc.
57 Morehouse Lane
Red Hook, NY 12571

Some format issues inherent in the e-media version may also appear in this print version.

Printed by Curran Associates, Inc. (2013)

For permission requests, please contact the Institute of Physics
at the address below.

Institute of Physics
Dirac House, Temple Back
Bristol BS1 6BE UK

Phone: 44 1 17 929 7481
Fax: 44 1 17 920 0979

techtracking@iop.org

Additional copies of this publication are available from:

Curran Associates, Inc.
57 Morehouse Lane
Red Hook, NY 12571 USA
Phone: 845-758-0400
Fax: 845-758-2634
Email: curran@proceedings.com
Web: www.proceedings.com

TABLE OF CONTENTS

CHAPTER 11

CHAPTER 12

CHAPTER 13

CHAPTER 14

CHAPTER 15

FOREWORD

Jefferson Lab was created in 1984 and started operating in about 1996. 2011 is an appropriate time to try to take a look at the results that have appeared, what has been learned, and what has been exciting for our scientific community.

Rather than attempt to construct a coherent view with a single author or at least a small number, we have, instead, invited small groups of people who have been intimately involved in the work itself to make contributions. These people are accelerator experts, experimentalists and theorists, staff and users. We have, in the main, sought reviews of the actual sub-fields. The primary exception is the first paper, which sets the scene as it was, in one person's view, at the beginning of Jefferson Lab.

In reviewing the material as it appeared, I was impressed by the breadth of the material. Major advances are documented from form factors to structure functions, from spectroscopy to physics beyond the standard model of nuclear and particle physics. Recognition of the part played by spin, the helicities of the beams, the polarizations of the targets, and the polarizations of final state particles, is inescapable. Access to the weak interaction amplitudes through measurements of the parity violating asymmetries has led to quantification of the strange content of the nucleon and the neutron radius of lead, and to measurements of the electroweak mixing angle.

Lattice QCD calculations flourished and are setting the platform for understanding of the spectroscopy of baryons and mesons. But the star of the game was the accelerator. Its performance enabled the physics and also the use of the technology to generate a powerful free electron laser. These important pieces of Jefferson Lab physics are given their place.

As the third Director of Jefferson Lab, and on behalf of the other physicists and others presently associated with the lab, I would like to express my admiration and gratitude for the efforts of the directors, chief scientists, associate directors, physicists, engineers, technicians and administrators who made it all possible. In sum, we should celebrate the science that Jefferson Lab has realized in this, its first long decade of physics.

Hugh Montgomery, Director

Making the Case for Jefferson Lab

Franz Gross
College of William and Mary, Williamsburg, Virginia 23185, USA and
Jefferson Lab, Newport News, Virginia 23606, USA

Abstract. This chapter is a personal account of the initial planning and competition for a new laboratory, which eventually became known as the Thomas Jefferson National Accelerator Facility, with the official nickname "Jefferson Lab." The period covered starts as far back as 1964, with the introduction of quarks, and extends up to the late 1980s after the initial team was assembled, the superconducting design was in place, and construction was well underway. I describe some of the major experiments that were proposed to justify the laboratory, reflect on the present status of those initially proposed experiments, and very briefly outline some of the new ideas that emerged after the laboratory was constructed. The science is presented in a simple manner intended for a lay audience, with some of the ideas illustrated by cartoons that were often used in popular lectures given during this period.

1. Washington shoot-out

It was the morning of February 17, 1983, and the auditorium in Washington, DC was filled with so many people it was hard to see who was there. Everyone was tense; five proposals for new electron accelerators, each with a different design, were to be presented and discussed over the next two days. The evaluation panel, chaired by D. Allan Bromley of Yale University, would listen, ask questions, and ultimately make recommendations to the Nuclear Science Advisory Committee (NSAC), which advised the Department of Energy (DOE) and the National Science Foundation (NSF).

The proposal from the newly formed Southeastern Universities Research Association (SURA) was to be presented first, followed by the presentations from the University of Illinois and the National Bureau of Standards (now NIST). Neither Illinois nor NIST wanted to build a high energy accelerator; they were there to present proposals for a new low energy accelerator should the nation decide to go in that direction. The front runners in the competition, scheduled to make their presentations on the second day, were the Massachusetts Institute of Technology (MIT) and the Argonne National Laboratory (ANL).

The performance of the five teams over these two days would largely determine who would be awarded the most important scientific project in a decade, and set the direction for research in nuclear physics for many years. Because all could hear and comment on each of the proposals in an open and unpredictable atmosphere, the participants came to call this the "Washington shoot-out."

SURA believed that the accelerator could not accomplish its mission unless its energy was at least 4 GeV, while MIT favored a less expensive accelerator with a lower energy of 2 GeV. Since this was controversial, SURA had devoted a considerable fraction of its allotted time explaining why the higher energy was necessary. When SURA had completed its presentation, Bromley began the discussion of the SURA proposal by asking the audience if "anyone" could tell him

Journal of Physics: Conference Series **299** (2011) 012001 doi:10.1088/1742-6596/299/1/012001

why a 4 GeV accelerator was needed. Would the spokesmen for MIT begin the argument for 2 GeV at this time? Should SURA answer the question again?

2. Hadronic physics from 1960 to 1980

In order to appreciate the excitement at the "shoot-out," it is necessary to go back to the early 1960s. A revolution in physics was underway. In 1964 Murray Gell-Mann (1969 Nobel Prize) [1] and independently George Zweig [2] proposed that many properties of hadronic matter (matter that interacts through the so-called strong forces) could be understood if the world were made of "quarks." Later it was realized that these quarks come in three identical copies, now called "colors." The forces between quarks are a consequence of this property of color, but the physical world cannot distinguish *individual* colors – it is "color-blind" (in mathematical language, an exact symmetry of nature). Initially it was unclear whether or not the quarks were real or just mathematical objects helpful to understand the many observed states of hadronic matter. But experiments completed at the Stanford Linear Accelerator (SLAC) by the team headed by Frideman, Kendall and Taylor (1990 Nobel Prize) [3, 4, 5] gave direct evidence for the physical existence of quarks, and then, in the early 1970s, the theory of the strong (colored) interactions, called Quantum Chromodynamics or QCD, was developed [6]. It was later shown that the forces between quarks become very small at large energies, a property of QCD referred to as "asymptotic freedom" (2004 Nobel Prize) [7, 8, 9]. This lead indirectly to a mathematical understanding of why quarks are confined – why they can never be completely separated from one another. By the late 1970s physicists had both the experimental evidence and a theory of quarks that explained why they could not be observed in isolation, so that even the skeptics began to believe quarks were real.

Because of these major discoveries, nuclear physics faced a crisis in the late 1970s. It had initially been believed that the nuclear force might eventually be understood in terms of the exchange of mesons between neutrons and protons (referred to collectively as "nucleons"), and that the size and shape of the nucleon was likely a result of a meson cloud surrounding a bare nucleon core. But now nucleons and mesons were known to be composite systems of quarks. Why weren't the effects of the underlying quark structure immediately visible? Could new phenomena be discovered that were a direct consequence of QCD and our new understanding of nuclear theory?

To provide an experimental answer to these questions, the Friedlander panel (1976) [10] and the Livingston panel (1977) [11] both recommended that a new high energy, continuous beam, electron accelerator be built for nuclear physics research. Soon after (1979) the first Nuclear Science Advisory Committee (NSAC) [12] included such an accelerator in its first Long Range Plan. The scientific community was on record urging the government to invest in a new accelerator, and these recommendations led directly to the five proposals under consideration at the Washington shoot-out.

3. Specifications for the new accelerator

These advisory panels recommended an electron accelerator (instead of a proton accelerator) because it was believed that electrons were a "clean" well-understood probe that could penetrate deep inside the atomic nucleus and directly feel the presence of the quarks. Figure 1 shows a slide that was often presented at colloquia to explain this property of electrons. It illustrates how electrons interact by exchanging a single photon (represented by the wavy line connecting the electron to one of the three quarks inside a nucleon). The photon exchange mechanism is so weak that most of the electrons pass through matter without interacting at all, and if an electron does interact, the probability that a second photon will be exchanged at the same time is less than 1/100. This simplicity makes it comparatively easy to calculate the interaction theoretically, one of the great advantages of the electron as a probe. James McCarthy, from the

University of Virginia (UVA) and leader of the SURA team, was fond of saying that electrons were like a "sharp knife" that could cut cleanly through hadronic matter revealing its inner structure.

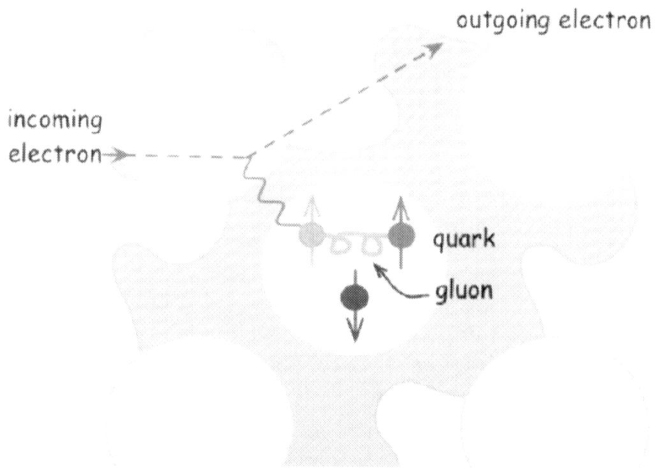

Figure 1. The electron shown as a "clean" probe of the partonic nature of hadronic matter.

High energy electron accelerators, like SLAC, had already played a crucial role in the *discovery* of quarks, but these accelerators could not make the new measurements required for a detailed study of the *behavior* of quarks in nuclei. Currents required to make the magnetic fields needed to accelerate and control high energy electrons are so large that even the best copper conductors and cavities would melt if they were operated continuously. To get around this limitation, the accelerators at the time were powered for very short time intervals so that the accelerator could cool between pulses. The time between pulses for the SLAC accelerator was usually about 8/1000 of a second and the length of the pulse was only about a microsecond. A convenient term which describes the operation of a pulsed accelerator is the *duty factor*, the ratio of the pulse time to the time between pulses. For SLAC the duty factor was less than 1/10,000, or 10^{-4} [13]. Other accelerators had larger duty factors, but did not have either sufficient energy or beam current. To perform the needed measurements an accelerator with a high beam current and a *continuous* beam (or a duty factor close to 1) was needed, and hence a completely new accelerator had to be constructed.

To understand why a continuous beam is needed, it is first necessary to realize that any new phenomena related to the underlying quark structure of matter would most likely be observed only through a *coincidence* measurement (in which the scattered electron is observed in coincidence with one of the nuclear fragments emitted at the same time), and then to understand why coincidence measurements require a high duty factor.

The first point is illustrated in Fig. 2. The figure shows a simple nucleus (the helium nucleus, ^{4}He, sometimes called the alpha particle) containing four nucleons: two neutrons and two protons. It is known that most of the time the motion of the quarks is confined into four spherical clusters (the nucleons), but there is a small probability that they might arrange themselves into unusual configurations, and the figure shows an example of such a (very rare) configuration. If only the scattered electron is observed, the cross section is proportional to the *sum* of all possibilities, so rare events cannot be distinguished from common ones. However, when the quarks fluctuate into some rare configuration, the result of the scattering will probably lead to some unusual configuration of hadrons in the final state, which will serve as a "fingerprint" of

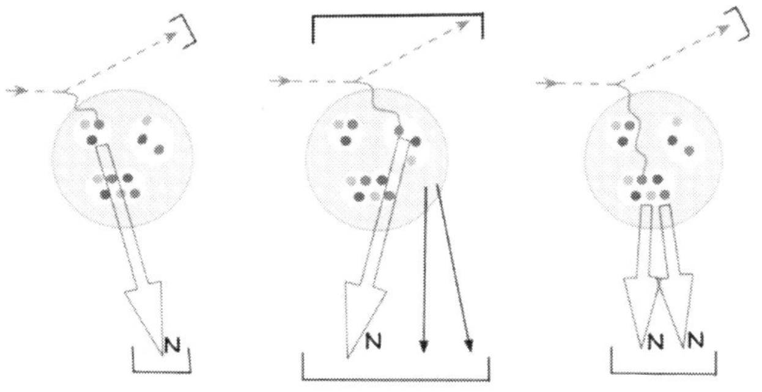

Figure 2. Measurement of hadronic fragments in *coincidence* with the scattered electron can reveal rare structures inside of hadronic matter.

the initial configuration. For example, if a nucleon changes its size in the nuclear medium, as suggested in the leftmost panel of Fig. 2, detection of the single nucleons in coincidence with the scattered electron can reveal this change. If the nucleon is excited inside the nucleus (as illustrated by the elongated nucleon depicted in the center panel, a very rare event) it is most likely that many hadronic fragments will accompany the collision, and if two nucleons overlap and form the unique 6-quark state predicted by QCD (also a rare event illustrated in the right panel) features of this state might be revealed in the signature of two-nucleon emissions.

The second point, that a continuous beam is required in order to do coincidence experiments, is illustrated in Fig. 3. Any detector has a resolution time $\Delta\tau$ within which individual events cannot be distinguished. This time can be converted into a resolution *length*, $c\Delta\tau$, by multiplying by the velocity of light, c. With a pulsed beam (upper panels of Fig. 3) so many electrons arrive at the target in the time interval $\Delta\tau$ that it is impossible to distinguish real coincidences from accidental coincidences. If the beam is continuous (lower panels), the electrons are spread out in time, and any coincidence is unlikely to be "accidental." In brief, a continuous beam accelerator

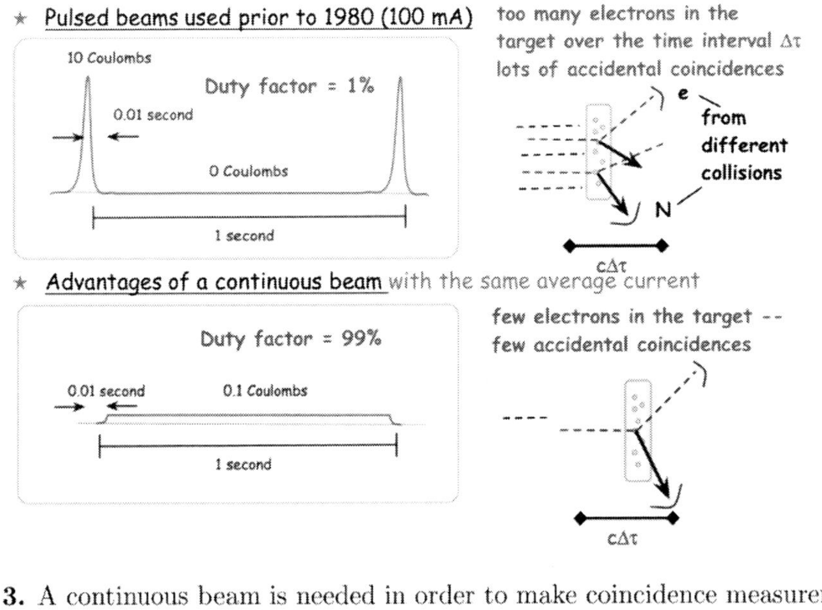

Figure 3. A continuous beam is needed in order to make coincidence measurements.

delivers its electrons spread out evenly over time, so that very few are in the target region at the same time, and accidental coincidences are very rare (they still happen, but can be separated from the real ones using experimental techniques).

4. Building on scientific recommendations (1979 – 1982)

The early recommendations of the Friedlander and Livingston panels which led to the NSAC report of 1979 were only a start. In order to justify the expenditure of $500 million of public money, a detailed scientific justification with practical accelerator designs was needed. McCarthy, well known for his experimental research in electron scattering and a member of the Livingston panel, wanted to build the next accelerator. He organized a conference at the University of Virginia (UVA) in January, 1979 [14], which examined various design possibilities. Although SURA would not be organized until mid 1980, this conference marks the beginning of the involvement of the Southeastern Universities. McCarthy assembled his team of young designers, including Blaine Norum (who became professor of physics at UVA) and Richard York (who became Associate Director for Accelerators, NSCL, Michigan State University), and John Sheppard (who became head of the Facility for Advanced Accelerator Experimental Tests at the Stanford Linear Accelerator Center).

On the national scene, the Users Group for the Bates Accelerator (run by MIT) invited participants to attend a planning meeting at MIT on January 1980, and both McCarthy and I attended. The Bates Users Group prepared a book (known as the "Blue Book"), released in the summer of 1981 [15], that described the new science, outlined the design specification for an accelerator capable of carrying out the necessary studies, and presented 26 short but detailed descriptions of specific experiments of very high scientific merit that could only be done with the new accelerator. These short "mini-proposals" were an important part of the scientific justification, filling in many details with estimates of the time required for each experiment and the errors expected.

Meanwhile, in the spring of 1980 McCarthy visited The College of William and Mary (W&M) and enlisted the support of physicists Hans von Baeyer (who later became the Secretary of SURA), Robert Siegel (Director of the Space Radiation Effects Laboratory, destined to become the site of Jefferson Lab), Robert Welsh (member of the Livingston panel), and me. Siegel suggested that other universities should be asked to join UVA and W&M, and this idea lead to the formation of SURA. Harry Holmgren (University of Maryland) and Tom Clegg (University of North Carolina) became the President and Treasurer of SURA, respectively, with McCarthy as Vice President. McCarthy and I became SURA scientific spokesmen, and to strengthen the project proposal being prepared by the SURA team, I convened meetings of SURA physicists that produced several new "mini-proposals" not previously included in the Blue Book.

The SURA team did not know when the project would be awarded and, fearing that they might not be ready in time, rushed to prepare their first proposal at the end of 1980. This was too early, but was very helpful in getting off to a quick start and helping SURA establish a national reputation. When DOE called for submissions by the end of 1982, SURA was well prepared with a second proposal [16].

5. The great energy debate

By early 1982, a year before the Washington "shoot-out," if was clear that the scientific community was in disagreement about whether the energy of the new accelerator should be 2 GeV or 4 GeV. Most of the MIT scientists favored the lower energy because (they argued) (i) it would be better for the study of complex nuclei, (ii) 4 GeV was too low to study the emergence of quark degrees of freedom anyway (some argued that 10 GeV would be necessary – too high for a practical high duty factor, high current machine at the time but, as the 12 GeV upgrade currently under construction has taught us, necessary for a more complete study), and

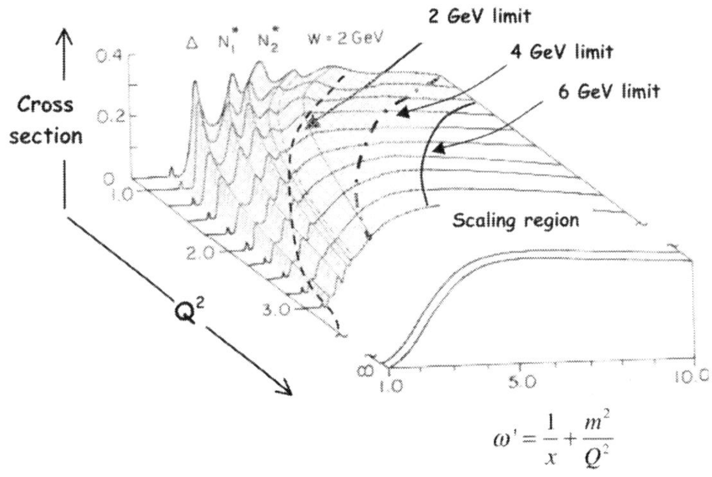

Figure 4. (Adapted from the Barnes panel report [17].) The total cross section for inelastic electron scattering from a free proton depends on two variables: Q^2 and ω' (or x).

(iii) a 4 GeV accelerator was so expensive that the funding might not be forthcoming. They also wanted to use their Bates accelerator as a first stage, and this would not be possible if the energy was as high as 4 GeV. SURA physicists and several physicists from SLAC, including Stan Brodsky, a leader in the application of QCD to nuclear physics, believed that 4 GeV was necessary because it was just high enough to reach the scaling region where quarks had already been "seen."

A panel of distinguished physicists was convened by NSAC to study the energy issue. It was chaired by Peter Barnes (Carnegie Mellon University and later to become the director of the proton accelerator at Los Alamos) and included Brodsky among its members. After many contentious meetings the Barnes panel finally recommended that the energy be 4 GeV [17]. While no simple argument can adequately summarize all of the discussion, the graph in Fig. 4 gives a rough picture of the committee's reason for selecting the higher energy. The figure shows how the cross section depends on Q^2 (the negative of the square of the four-momentum transferred by the scattered electron), and the dimensionless scaling variable x,

$$x = \frac{Q^2}{2m\nu}, \tag{1}$$

where m is the nucleon mass and ν the energy of the virtual photon in the rest system of the nucleon. When Q^2 is large, and x is not too close to unity, the cross section for inelastic electron scattering from a proton flattens out into a broad plateau (referred to as the "scaling region") where it is (approximately) independent of Q^2 and depends only on the scaling variable x. The existence of the scaling region is strong evidence that the electrons are scattering from point-like quarks inside the proton. The shaded regions in the upper left half of the surface show the parts of the surface that can be studied with 2 GeV, 4 GeV, and 6 GeV accelerators. This shows that 4 GeV provides just enough energy to barely reach the scaling region, and that this region *cannot be reached* with only 2 GeV.

The decision to go to 4 GeV seemed to be a significant victory for SURA, but in early 1983 the SURA team was dismayed to learn that the charge to the Bromley review panel included a reconsideration of the energy of the accelerator. The issue was still open and the SURA team had to prepare to justify the choice of 4 GeV once again.

6. The Washington shoot-out and its aftermath (1983-1985)

So Bromley's question on the morning of February 17 was expected, but the way it was raised seemed to ignore the previous recommendations of the Barnes panel and the presentation SURA had just given. Bromley asked this question several times during the first day, but no one came forward to justify the choice of 2 GeV. The spokesmen for the MIT proposal did not attend the Washington meeting until the *second* day, and by then their arguments for 2 GeV were late and seemed unconvincing. Later, during questioning by Hermann Grunder, the chair of Bromley's technical subpanel (and later to become the first Jefferson Lab director), MIT conceded that it did not have a credible proposal for a 4 GeV accelerator.

In the weeks that followed the Bromley panel decided to support the recommendations of the Barnes panel, and in this way the final choice was between the SURA and Argonne proposals. In its final recommendations to NSAC [18], which were accepted and forwarded to DOE and NSF on April 29, 1983, the SURA proposal was selected for three significant reasons:

- SURA had promised to create at least 35 new faculty positions in nuclear physics to support the science being studied by the new accelerator. (By 1996, 127 such positions had been created.)
- The SURA design could readily be extended to 6 GeV, while the Argonne design was frozen at 4 GeV.
- The SURA design was more conservative (if downgraded to half of the design current) while there were significant concerns about "potential beam loss" if the Argonne design did not meet specifications. In short, the Argonne design could fail completely and produce no beam (the "hard" failure mode) while the SURA design, even if it failed, would always produce a beam (the "soft" failure mode).

The Bromley panel voiced some concern about the relative inexperience of the SURA team, and urged them to hire an experienced management and construction team.

Unfortunately, Argonne protested, and DOE's decision became embroiled in some controversy, which reached as far as the halls of Congress where Senators Mark Hatfield (Oregon) and Bennett Johnston (Louisiana) asked [19] "is there a scientific need for such a machine? Is it the most cost-effective facility for nuclear science?" They went on to ask "Is its justification to satisfy a political or geographical constituency, because so many states and universities are involved and the Southeast is without a major accelerator?" It looked like the 7 years of careful scientific review and preparation would be undone, and it was certainly an irony that the efforts of SURA to generate enthusiasm among the Southeastern universities might turn from an advantage (35 promised faculty positions) into a disadvantage.

In partial response to this criticism, DOE asked for another review, and in Sep. 1984 an NSAC subcommittee chaired by Eric Vogt (director of TRIUMF) reported that it "reaffirms a 4 GeV CW electron accelerator as the first major construction project for nuclear physics" [20]. Soon after, funding became available and the SURA team was able to hire Hermann Grunder as its first director. Grunder arrived in Virginia in May 1985, and brought an experienced team with him, including Christoph Leeman as Associate Director for Accelerator Physics. A year later Grunder hired J. Dirk Walecka, a member of the original Friedlander panel and a distinguished professor of physics from Stanford University, to be the laboratory's first Scientific Director. Leeman later became Deputy Director in 2000 and served as Director from 2002 until his retirement in 2008.

7. Decision to build a superconducting accelerator (1985)

The original design for the SURA machine used a pulse storage ring, as shown in Fig. 5 [21]. Here the linear accelerator (linac) produces a pulse of electrons which is then injected into the

Journal of Physics: Conference Series **299** (2011) 012001 doi:10.1088/1742-6596/299/1/012001

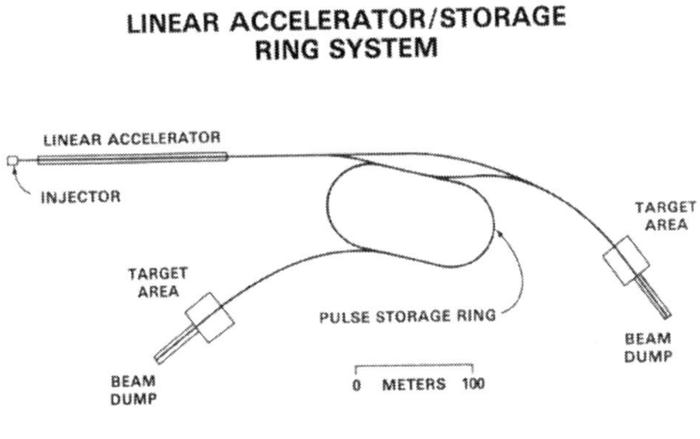

Figure 5. Sketch of the original pulse stretcher ring design for the SURA accelerator.

pulse storage ring. The ring is designed so that the length of the pulse matches the length of the ring; each injected pulse just fills the ring. Then, between successive pulses, the electrons are slowly drained out of the ring, converting the pulse into a continuous current. All of these components operated at room temperature and used conventional copper cavities.

While there were many technical problems to be solved in order to achieve the goals of the SURA design, the pulse storage ring was considered a comparatively "safe" design because it was based on extensions of known technology and, if it did not achieve its goals, it would still produce a high energy beam [18]. Its principal disadvantages were that it could not produce an extremely sharp, well collimated beam and, if it produced multiple beams, they must all have the same energy.

Grunder believed that superconducting technology might be used to build a better accelerator. With this technology the accelerator could be constructed of metal cavities (niobium as it turned out) that become superconducting when cooled to liquid helium temperatures (*i.e.* its resistance to the flow of current drops to almost zero, allowing a large current to flow through the accelerator without heating its components) [22]. Under these conditions the linear accelerator could be operated continuously without overheating, and because the beam in the accelerator would deliver its electrons continuously, the number of electrons in the accelerator at any give time would always be much less that the large number crammed into the pulses of a conventional accelerator, reducing the repulsive forces between them and giving an extremely well focused beam.

While superconducting cavities had already been developed and used for lower energy accelerators, in 1985 no one had designed radio-frequency (RF) accelerator cavities with the high field gradients needed for a 4 GeV machine. Furthermore, RF cavities with high gradients had been promised and expected for many years, so it was easy to overlook the steady progress being made. Grunder asked and received permission from DOE for the new laboratory to review the state of superconducting technology. The review led to a new conceptual design for the accelerator, with many advantages, and in December of 1985, less than eight months after Grunder had taken over leadership of the project, DOE approved the concept of a superconducting linac for CEBAF. This was a remarkable development, clearing the way for Jefferson Lab to eventually become a leading world-wide center for superconducting RF cavity design.

The actual design adopted is illustrated in Fig. 6. The accelerator is basically a linear accelerator (linac), wrapped around five times. As long as the recirculation arcs have a large radius so that the radiation of photons by electrons traveling in a circle is negligible, and

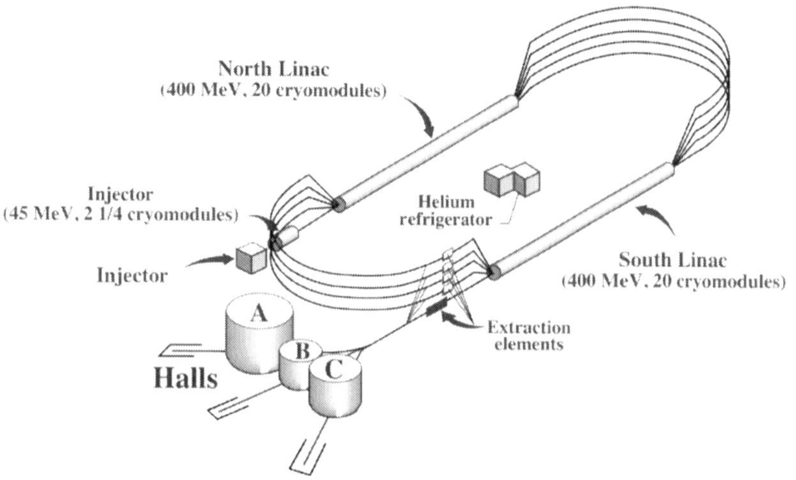

Figure 6. Sketch of the final design for the CEBAF accelerator. The beam travels once through the North and South linacs with each recirculation, when part of it may be extracted an injected to any of the three Halls. The linac energies shown are for operation at 4 GeV; at 6 GeV each linac operates at 600 MeV.

because the electrons are ultrarelativistic and travel with essentially the same speed through their journey, the multiple linacs along each side (not shown in the figure) can be combined into a single linac, saving money and real estate. The number of recirculations is a balance between the lower costs of fewer (or shorter) sections of linac and the higher costs and greater complexity of the arcs with more recirculations. The initial design called for 4 recirculations. It was eventually decided that a 5th recirculation would save money without significantly increasing the risk.

It was recognized from the beginning [23] that the experimental program needed three experimental halls (or "end stations") with different capabilities, as illustrated in Fig. 6. Hall A houses two high resolution spectrometers well matched to the excellent quality of the CEBAF beam, and is ideal for making accurate measurements of two particle final states. Hall B has the CEBAF Large Acceptance Spectrometer (CLAS) that can detect and identify the several particles expected from the multi-particle reactions illustrated in the middle panel of Fig. 2. Hall C has a medium resolution spectrometer and space for experiments requiring special equipment.

The superconducting design insured several advantages for the scientific program with the simultaneous use of multiple end stations:

- Beam quality was very high; all of the electrons travel in nearly the same direction with a very small angular spread and their momentum spread is also very small. This makes it possible to do very high resolution experiments.
- Beams injected into the three end stations could each have different energies, as long as they were a multiple of the energy of each recirculation. For example, at the full energy of 6 GeV, the beams directed to each hall could be any multiple of 1.2 GeV up to 6 GeV.
- The microstructure of the main beam made it possible to send beams with very different currents into each hall. For example, the detectors in the CLAS spectrometer are exposed and will be blinded unless the current on target is very low (the large acceptance of CLAS compensates for the low current required), while the spectrometers in Hall A have a very small acceptance and require high current in order to give reasonable counting rates. The ability to independently adjust the currents in each of these halls makes it possible to do

experiments in all of them simultaneously.

- If the electron injector produces a polarized beam, this polarization will not be seriously compromised by the comparatively smooth trip around the recirculating linac. Today beams at Jefferson Lab are produced with very high polarization.

8. The overall Jefferson Lab 6 GeV nuclear physics program: then and now

The remaining sections are an introduction to the overall 6 GeV physics program being concluded in 2012, and described in more detail in the rest of this volume. For information about the new 12 GeV physics program now being developed, see the Jefferson Lab web site at `www.jlab.org/12GeV/`.

The discussion of the overall 6 GeV program is organized into two parts, each consisting of smaller programs focused on specific issues:

- **programs featured in the 1985 Preconceptual Design Report (PCDR)**
  - charge distribution in the neutron and deuteron
  - coincidence measurements of single particle emission
  - two-body correlations in nuclei
  - production of excited nucleons
  - strangeness in nuclei
- **programs that emerged after 1985**
  - numerical solutions of QCD
  - parity-violating asymmetries
  - structure functions for deep inelastic scattering
  - generalized parton distributions (GPDs).

The five experimental programs featured in the PCDR [23] were extracted and distilled from the 26 short mini-proposals included in the proposal submitted by SURA in 1982. The last four programs describe exciting new developments unanticipated (or not emphasized) in 1985.

Recall that colored quarks are the fundamental particles of QCD. Because quarks are confined, states of free quarks do not exist; the only states that can exist in nature are colorless states which can be build out of combinations of two different types of "building blocks": either (i) three-quarks with one quark of each color (there are only three) combined into a colorless state, or (ii) quark-antiquark pairs (where the "anticolor" of the antiquark renders the pair "colorless"). A nucleon is made of one three-quark system and a "sea" of quark-antiquark pairs. A nucleus with mass number A is therefor a complex color neutral composite of A three-quark systems and a sea of quark-antiquark pairs, somehow clustered into A nucleons. The extent to which the nucleons retain their identity in an environment of other nucleons in close proximity is one of the puzzles to be unraveled, and the existence of the quark substructure opens up many possibilities: are there unique structures of quarks, such as 6-quark bags, present in a nucleus?

9. Programs featured in 1985

Five programs featured in 1985 address these questions in two different ways. Some of them study the structure of the nucleon itself, in order to unravel the distribution of quarks inside of free nucleons. Others study nuclei, with the goal of understanding how nucleons (and hence quarks) are distributed inside of bound systems with more than one nucleon.

9.1. Charge distribution in the neutron and deuteron

In 1985, the charge distributions inside the neutron and the deuteron were very poorly known, and measurement of these distributions was important for similar reasons.

Journal of Physics: Conference Series **299** (2011) 012001 doi:10.1088/1742-6596/299/1/012001

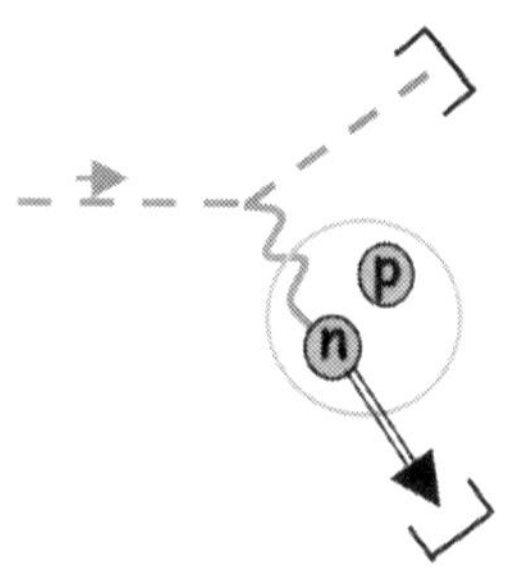

Figure 7. A polarized electron (dashed line with arrow) scatters from the neutron in the deuteron (the simplest nuclear target containing a neutron). Both the scattered electron and the neutron are detected in spectrometers.

The neutron is composed of two *down* quarks with charge $-\frac{1}{3}$ and one *up* quark with charge $+\frac{2}{3}$, making it electrically neutral. If the *up* and *down* quarks had exactly the same spatial distribution within the neutron, the neutron charge would be identically zero everywhere. Measurement of this distribution is therefore an important indicator of how the quarks behave inside the neutron.

This charge distribution can be measured by scattering electrons from a neutron, but unfortunately the neutron is unstable, so the scattering must be done from a deuteron (the *only* stable bound state of one neutron and one proton) or some other simple nucleus (often ^{3}He). If the recoiling neutron is detected in coincidence with the scattered electron, as shown in Fig. 7, events where the electron scatters from the neutron can be distinguished from those where it scatters from the proton (with a known correction with a small error). A second complication is that the neutron has both a charge and magnetic moment distribution, and the magnetic moment distribution is both less sensitive to the quark distribution and also dominates electron scattering, swamping out the effect of the charge and making it very difficult to "see" how the quarks are distributed inside of the neutron. A good way to separate out the charge from the magnetic moment distribution is to measure the transfer of spin from an electron to the scattered neutron, and this also requires that the neutron be measured in coincidence with the scattered electron. For these two reasons (use of a deuteron target and need to measure the polarization transfer) a coincidence measurement is essential, and requires the new accelerator.

On more than one occasion I was told by someone that he "might support the construction of the accelerator if I could name *one* important experiment" that it could do and that could not be done by another (existing) accelerator. When confronted with such a challenge, I always chose to describe the measurement of the neutron charge distribution; it was hard for anyone to deny the simplicity and importance of this experiment and all could see why it required a new accelerator to measure it.

What happened? The neutron charge measurements produced very accurate results. We now know that the *up* and *down* quarks have different distributions, with a greater density of *up* quarks at the center and *down* quarks at the periphery. This knowledge, important in its own right, also can be used to more reliably interpret the meaning of electron scattering experiments from all nuclei.

Similar measurements were made on the proton and those measurements lead to a big surprise! The proton charge distribution is much easier to measure and had been measured previously by another technique (referred to as a Rosenbluth separation) at SLAC. As part of the Jefferson Lab program, the proton charge distribution was measured again using the new polarization transfer technique, and the *results disagreed with the previous SLAC measurements*. A careful study of the source of this disagreement lead to an interesting discovery: it now seems that the two-photon exchange contribution (illustrated in Fig. 8(b)), previously thought to be very small and never observed, strongly affects the extraction of the charge from a Rosenbluth separation

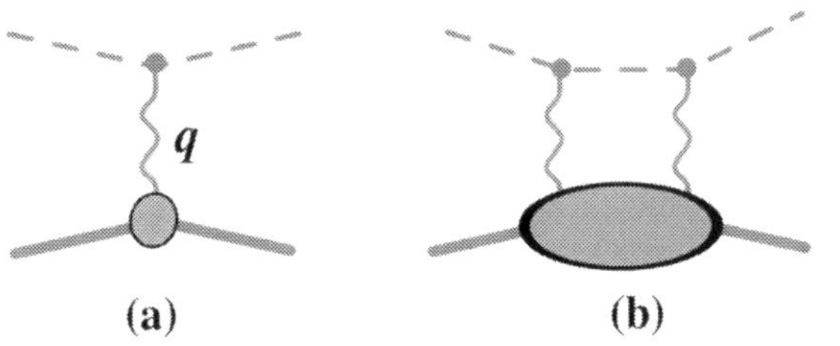

Figure 8. Feynman diagrams showing (a) the usual one-photon exchange approximation to the elastic scattering of electrons from nucleons, and (b) the two-photon exchange contribution.

but has only a little effect on its extraction from a polarization transfer measurement. An extensive program to confirm this picture is currently underway.

The deuteron is the only bound state of two nucleons and is therefore the simplest nucleus. Measurement of its charge distribution gives important information about the nuclear force – in this case the effective force between two nucleon three-quark clusters (labeled the NN force, for short) that accounts for the formation of all nuclei and is responsible for many aspects of the structure of stars.

While this charge distribution can be measured in electron scattering, detailed measurements of the distribution at short distances, where the effects of the NN force might be studied, are masked by effects from charge quadrupole scattering, a large effect insensitive to the details of the NN force. While looking at the deuteron in coincidence with the scattered electron, it is possible to control the scattering so that the (smaller) charge distribution can be disentangled from the larger (and less interesting) quadrupole scattering.

What happened? At Jefferson Lab measurements of the charge distribution of the deuteron were separated from the larger quadrupole contributions and measured to $Q^2 \sim 1.8$ GeV^2. The charge distribution has a zero around $Q^2 \simeq 0.7$ to 0.8 GeV^2. One of the surprising things learned from this experiment, completed during first years of the Jefferson Lab program, is that these experimental results can be understood using relativistic state of the art models based on nuclear and meson degrees of freedom, with no need to consider the underlying quark degrees of freedom explicitly.

9.2. Coincidence measurements of single particle emission

The measurement of a single nucleon emitted from a nucleus with A nucleons, in coincidence with a scattered electron, is illustrated in Fig. 9, which shows the two important contributions to the process. In the left diagram, the struck nucleon is emitted without rescattering from any of the remaining $A-1$ nucleons in the system, while in the right hand diagram the struck nucleon scatters from parts of the $A-1$ system (it undergoes "final state interactions"). If the rescattering (right figure) is very small (it is never exactly zero), the process depends on the product of (i) the nucleon charge and magnetic distributions, or "form factor" $F(Q)$ (represented by the small grey circle also shown in the one-photon exchange diagram of Fig. 8(a)), and (ii) the momentum distribution $\rho(k)$ of the nucleon just before it scatters from the electron. In this case the momentum, k, of the nucleon in the nucleus *before* it was struck (referred to as the "missing momentum") can be deduced from the momenta of the outgoing nucleon and the virtual photon, both of which are determined by the measurement.

The rescattering process also depends on the charge distribution of the nucleon in the nuclear medium, but the rest of this amplitude no longer depends simply on $\rho(k)$, but on a more complicated function. Because of this, a good theory is needed to unfold both $\rho(k)$ and $F(Q)$ from this experiment, so neither $\rho(k)$ nor $F(Q)$ are *directly* observable. It is also useful to find situations where the final-state interactions are very small, or at least very well understood. If both $\rho(k)$ and $F(Q)$ can be extracted from this measurement, everything about the shape, size,

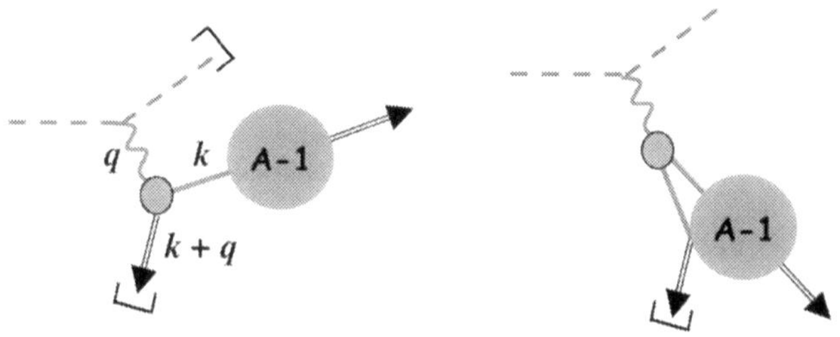

Figure 9. Cartoon showing two of the processes that take place when the emission of a single particle is measured in coincidence with the scattered electron.

and distribution of three-quark clusters in a nucleus will be revealed.

What happened? The question of whether or not the nucleon changes its size in a nuclear medium has been addressed in coincidence experiments where the proton was knocked out of ^{4}He, leaving the simple ^{3}H nucleus behind. There is a clear reduction in the ratio of polarization transfers for scattering from ^{4}He when compared with scattering from the proton, which can be explained by assuming the proton is modified in ^{4}He, but a firm conclusion is difficult to draw at this time because an alternative calculation of the final state interactions can also explain the results without assuming any modification of the nucleons. Stay tuned for further developments.

Data for proton knockout from ^{3}He has been separated into two-body (*pd*) and three-body (*ppn*) final states, giving clean information about the size of these two processes out to large missing momenta. These data can also be reasonably well explained by models based on nuclear and meson degrees of freedom. Other measurements from few body targets (including the deuteron) and under different kinematic conditions are being analyzed and a more complete picture will emerge soon.

In one particularly interesting study, *real photons* (radiated from electrons) may be absorbed by the deuteron producing a neutron and proton final state. At the high photon energies produced by the accelerator, the total cross sections for this deuteron photodisintegration process exhibit another type of "scaling" (only loosely related to the scaling observed in the inelastic scattering studies sketched in Fig. 4). This scaling prediction, which can be derived from QCD at *very* high energies, referred to as "perturbative QCD" or simply pQCD, is that any differential cross section $d\sigma/dt$ at fixed angle and *very* high energies should behave as

$$\frac{d\sigma}{dt} \sim \left(\frac{1}{s}\right)^{n-2}$$

where s is the square of the c.m. energy and n is the *minimum number of point-like particles* involved in the process. Applied to deuteron photodisintegration, $n = 1 + 6 + 3 + 3 = 13$ (1 photon, 6 quarks in the deuteron, and 3 quarks in each of the nucleons in the final state) leading to the prediction that s^{11} times the differential cross section should go like a constant as $s \to \infty$. The recent data taken at Jefferson Lab exhibit this behavior (which sets in most rapidly for scattering at 90° c.m.). On the face of it, this would seem to be a brilliant confirmation of the predictive power of pQCD and assurance that this theory can be used to understand deuteron photodisintegration with (comparatively low energy) 4–6 GeV electrons; however, attempts to calculate the *strength* of the s^{-11} behavior have failed, and other predictions (such as polarization observables) do not always agree with the data. We are left with a puzzle: why does pQCD scaling behavior work to such low energies when other pQCD predictions (strength and polarization dependence) fail?

9.3. Two-body correlations in nuclei

The very simplest theoretical picture of a large nucleus treats it as a gas of non-interacting nucleons that move freely through the nuclear volume. Ironically, this overly simple picture is useful because the strong short range repulsion between nucleons keeps them apart, preventing the nucleus from collapsing, guaranteeing that many body forces (involving three or more nucleons) must be small, and clearing the way for the longer range forces to be averaged so that they can be treated as a mean field. From this point of view, any correlations between the nucleons are either a reflection of the strong short range forces, or a signal of the departure from the mean field. In either case, direct observation of correlations is extremely interesting and tells us how the NN force is modified by the nuclear medium, or whether or not there are highly correlated 6-quark states present in nuclei. While this was believed to be an important program, in 1985 it was thought that correlations might show up as the knockout of a pair of nucleons, such as might emerge from a 6-quark bag as suggested by the right-most cartoon in Fig. 2.

What happened? The observations of correlations is one of the most exciting developments to emerge from the Jefferson Lab program but the "best" observation of correlations did not involve a coincidence experiment at all! This was an "inclusive" measurement in which *only* the scattered electron was observed. As shown in Fig. 4, for large Q^2 this cross section depends only on the scaling variable x, and it is easy to show that scattering from a single, free nucleon is possible only if $x \leq 1$. The most direct way to see correlations is therefore to look at processes in which $x > 1$. Under this condition, scattering is *impossible* unless the nucleon is bound in the nucleus. The ratio of the scattering cross section from different nuclei for $x > 1$ will therefore depend in part on the size of the two-body correlations in these nuclei (and, by extension, the ratios for $x > n - 1$ depend on n-body correlations). This fact was predicted before 1985, but sufficiently precise data were unavailable and the accurate observations at Jefferson Lab have opened up this area of study.

As anticipated in 1985, correlations have also been observed in one and two nucleon knock-out experiments. The recent observation of the large ratio of np to pp pairs seen in knockout from ^{12}C can be explained by the dominant role of the tensor part of the NN force.

9.4. Study of excited states of the nucleon and search for "missing" states

The nucleon is a complicated system composed of at least three quarks confined in a small volume by the colored forces that interact between quarks. When struck by a pion (the pion, π, is the lowest mass state of quark-antiquark pairs and beams of pions have been used to study the structure of nucleons and nuclei) or a photon, it can be excited into resonance states (most with a very short lifetime). The mass and structure of these baryonic resonance states, together with the way in which their excitation by photon and pion beams differs, give very direct information about the QCD forces that bind the quarks.

In the early 1980s these states had been primarily studied using pion beams (the counting rates for production of these states by pion accelerators is much larger than the rates using electrons and photons) and there was a puzzle: many states predicted by quark models were not seen. Nathan Isgur suggested that a reason for this was that the probability that a pion might excite some of these states was very small, and showed that these "missing states" might be more readily seen in the process $\gamma^* + N \to N^* \to N + \pi + \pi$ (where γ^* is a real or virtual photon produced by the electron). These measurements required the large acceptance spectrometer that could measure the momenta of the outgoing nucleon and two pions (and many other multiple particle states) and these requirements eventually lead to the construction of the CLAS detector in Hall B.

What happened? Study of the excited states of the nucleon has been a major part of the Hall B program from the earliest days and is described in this volume. The Hall B CLAS detector can detect an impressive number of final states, including nucleons, photons, and multiple pions and kaons, and their decay products. It has become increasingly evident that this program must be supported by a major theoretical effort, including the coupled channel calculations being carried out by the Excited Baryon Analysis Center (EBAC), with the goal of isolating resonance states from non-resonant background which must be dynamically calculated at the same time. Form factors of the lowest lying excited states of the nucleon, some very poorly known and others completely unknown in 1985, have now been measured, and while there are intriguing hints of missing states, final conclusions must await further experimental and theoretical analysis of the data.

9.5. Study of strangeness in nuclei

The strange baryons, especially the neutral Λ composed of at least one *up*, one *down*, and one *strange* quark, is unstable but can form stable nuclear states (referred to as hypernuclei), usually by substituting one Λ for one of the nucleons through one of the elementary reactions $\gamma^* + p \to K^+ + \Lambda$ or $\gamma^* + n \to K^0 + \Lambda$, where γ^* is a real or virtual photon produced through the (e, e') scattering process, and the K^+ and K^0 are strange mesons containing one *strange* antiquark. The K's interact weakly with the rest of the nucleus, passing through the nuclear medium and providing a clean "tag" for the conversion of a nucleon to a Λ. These states tell us both about the familiar NN force as well as the new ΛN force between the Λ and the nucleon, related through the underlying quark structure to the NN force. The ability to explain *both* the ΛN and NN forces together is a definitive test of our understanding of how two three-quark clusters interact.

However, to accurately unravel the ΛN force from the energy levels of the hypernuclear states requires that the energies of these states be measured very accurately, and that individual states which sometimes lie very close to each other, be separated and identified. This in turn requires measurements with good energy resolution. In the 1980s, most of the information about these states had been learned from K^- (or π^+) beams, which produce hypernuclei through the reaction $K^- + n \to \Lambda + \pi^-$ (or $\pi^+ + n \to K^+ + \Lambda$), and the resolution was limited to about 1.5 MeV. To make important contributions to this study, resolutions below 500 keV were sought.

What happened? The hypernuclear program was supported from the very beginning by a strong group of Japanese and Italian collaborators. Complementary programs in Hall C (which required construction of special spectrometers and has gone through three distinct phases) and Hall A have been established, and resolutions in the range of 500–600 keV have been achieved. Data from a variety of hypernuclei have been obtained and a number of new states identified and studied. Over the next decade Jefferson Lab data will complement and be combined with data from the new facility at J-PARC (the Japan proton accelerator research complex).

10. Programs that emerged after 1985

Several programs featured in this volume emerged after 1985. The theoretical program to solve QCD numerically was anticipated in the first proposal for a Jefferson Lab theory group [24] but not funded until much later. The parity violation program was also anticipated but was not featured. Deep inelastic scattering using *unpolarized* electrons, already advanced in 1985, did not require coincidence measurements, and measurements of the *polarization* dependent structure functions, which require an intense beam with high polarization, only emerged as a strong program later when it was realized that the new CEBAF accelerator had an excellent figure-of-merit for these measurements. Finally, the program to extract generalized parton distributions (GPDs) from deeply virtual Compton scattering, only developed later.

These programs, among the most productive and exciting being undertaken at Jefferson Lab, are examples of the fact that a laboratory will often develop in ways not anticipated during the initial planning.

Journal of Physics: Conference Series **299** (2011) 012001 doi:10.1088/1742-6596/299/1/012001

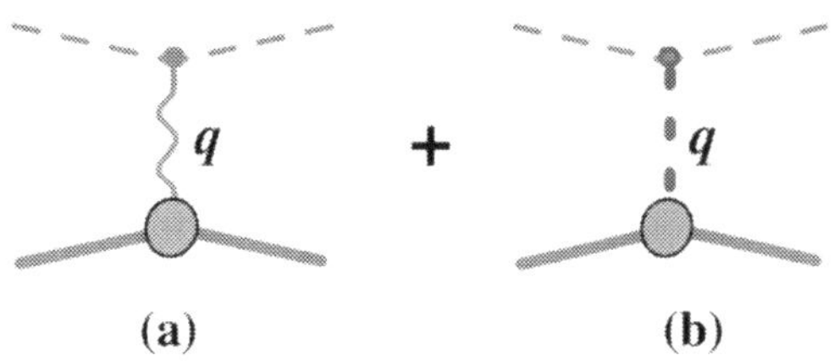

Figure 10. Feynman diagrams showing (a) the exchange of the virtual photon previously discussed in Fig. 8 (denoted M_γ), and (b) the exchange of a neutral Z boson (denoted $M_Z = M_Z^0 + M_Z^p$), which carries the weak force.

10.1. Numerical solutions of QCD (LQCD)

At very high energies, QCD can be solved analytically and the solutions compared with high energy data. The agreement is very impressive, leading to the expectation that QCD is the correct theory of the strong interactions. But very little is known about the solutions at the lower energies involved in normal nuclear reactions or at the scales needed to understand the structure of nucleons and nuclei. Does QCD really explain low energy phenomena as well, or does it break down and is a new theory needed in this region? Solving QCD at lower energy scales and comparing the solutions with experiment is of the highest priority; it is the central question in this field.

At low energies, QCD can be solved numerically on a lattice of space-time points, and the program to obtain such solutions is referred to as lattice QCD, or LQCD. Solutions have given a mass spectrum for mesons and baryons in good agreement with data, giving confidence that QCD also explains low energy phenomena. Numerical results have shown how quarks are confined, confirming what was previously only a conjecture.

To obtain LQCD solutions requires large and very fast clusters of computers, so the accuracy of the solutions we obtain is always limited. For the most part these uncertainties can be controlled and understood (LQCD solutions come with error bars, just like experimental data, and are therefore sometimes called "data"). To solve LQCD for the very small masses of the *up* and *down* quarks is currently near the limits of our computer capability, so many solutions have been obtained for (unrealistically) large quark masses, and are extrapolated to small masses (with associated uncertainties). Furthermore, the lattice used does not have the full rotational symmetry of nature, and certain tricks must be used to extract some quantities correctly.

It turns out that LQCD solutions can only be done in Euclidean space-time (related to the physical space-time by making time an imaginary number). For some quantities, such as particle masses, this is not a restriction. Other quantities cannot be calculated in Euclidean space, so LQCD cannot, in principal, calculate all of the things we want to know. Since LQCD cannot solve all of the problems that interest us, nor give an understanding of all of the physics involved in the solution, it must be supported by theoretical models that can connect numerical solutions to physical pictures and extend our calculations into areas not accessible to LQCD. In this respect LQCD can be an important guide; it can sometimes calculate physical quantities that cannot be directly measured, but which play an important role in the construction of realistic physical models.

This is a rapidly moving field and we expect still better solutions using more and more creative techniques in the years ahead.

10.2. Parity-violating asymmetries

The "Standard Model" is the modest name given to the theory that includes QCD and the combined theory of the electromagnetic and weak interactions (known as the "electroweak" theory). According to the Standard Model, an electron scatters from a nucleon by exchanging not only a virtual photon (the carrier of the electromagnetic force) but also by exchanging a neutral Z boson, one of the carriers of the weak force. The total scattering is then the sum of these two exchanges, as illustrated in Fig. 10. The strength of the Z boson exchange is about 100,000 times smaller than the photon exchange, so this term can normally be ignored (and, in fact, is more that 1000 times smaller that the two photon exchange mechanism shown in Fig. 8). How can one hope to see such a small effect and why should we care?

The answer to the first part of this question lies in a remarkable property of the weak interactions: they violate parity. Loosely speaking, parity is the symmetry that insures that fundamental laws cannot distinguish left from right. While macroscopic systems often show left-right differences, fundamental laws do not – at least they were not known to until its violation was discovered to be a feature of the weak interactions. The discovery, in the late 1950s, that the weak interactions do not conserve parity came as a great surprise; as far as we know this is not a feature of any of the other fundamental forces of nature.

In the context of electron scattering, a violation of parity will show up as a difference between the scattering of a right-handed electron (one with its spin pointing in the direction of its motion) from that of a left-handed electron (with its spin pointing opposite to its direction of motion) when *no other particles in the reaction are polarized*. The scattering cross section is proportional to the square of the two terms in Fig. 10. The exchange of a neutron Z boson contains a term (M_Z^p) that will change sign when the spin of the electron is switched, giving an *asymmetry* defined to be the difference divided by the sum of the two possible contributions

$$A = \frac{\sigma_R - \sigma_L}{\sigma_R + \sigma_L} = \frac{2(M_\gamma + M_Z^0)M_Z^p}{(M_\gamma + M_Z^0)^2 + (M_Z^p)^2} \sim \frac{2M_Z^p}{M_\gamma} \sim 10^{-4}Q^2,$$

where Q^2 is in GeV^2. Measurement of such small asymmetries is a challenge, and is done by picking the small asymmetry out of the much larger background by flipping the spin from left to right and synchronizing the small variation of the cross section with the spin direction. The excellent beam quality of the CEBAF superconducting accelerator design has helped to make such a program possible.

The study of the weak interactions contains lots of new information that we *cannot* learn from the study of the electromagnetic interaction only. This was recognized in the initial SURA proposal, but it was not featured in the PCDR prepared in 1985. J. Dirk Walecka, Jefferson Lab's first Scientific Director, recognized the importance of this program, and after his arrival in 1986 sought to build a culture that would ensure that these difficult parity-violating measurements would become a regular part of the program.

Study of the weak interaction, Fig. 10(b), is extremely interesting for two, somewhat distinct, reasons. First, a full description of the coupling of the nucleon of either the photon or the Z requires summing the contributions from all three quarks in the nucleon: the u, d, and s. The photon couplings to the proton and the neutron give two observables linearly dependent on all three of these distributions, but only by measuring the coupling of the Z boson to the proton (for example) can we obtain the third linear combination needed to *separately* determine the contribution coming from strange quarks. This search has been a major effort at Jefferson Lab. Early expectations suggested that this contribution might be large, but results obtained so far show that it is very small and this places significant limits on the types of models that can be constructed to explain the structure of the nucleon.

The second direction for the parity-violating program is to use it to test the validity of the Standard Model itself. The Standard Model seems to describe all known phenomena in the realm of nuclear physics; any discovery of a failure would be of very great significance. The primary focus of tests carried out at Jefferson Lab are precision measurements of the strength of the weak interaction. This strength depends on only two parameters, the overall strength of the weak interaction and a mixing angle that can be related to the mass of the Z boson. Once

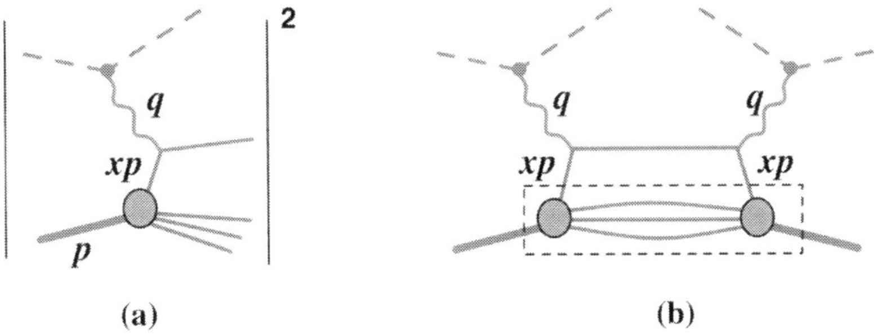

Figure 11. (a) Feynman diagram showing the DIS amplitude at large Q^2 ($Q^2 = -q^2$) where the leading term comes from the virtual photon being absorbed by a single quark, and (b) the cross section, proportional to the absolute value squared of diagram (a).

these parameters have been measured, a number of relations and predictions follow, and any violation in these predictions will signal the breakdown of the theory.

10.3. Structure functions for deep inelastic scattering
As mentioned briefly in the discussion surrounding Fig. 4, the existence of point-like quarks inside the nucleon was inferred first from the study of deep inelastic scattering (DIS) from the proton. At large $Q^2 > 1$ GeV^2 and for x not too close to unity, the structure functions that define the DIS cross section begin to scale; they depend primarily on the variable x defined in Eq. (1), with only a weak dependence on Q^2 that can be calculated theoretically. The Feynman diagram that describes DIS is shown in more detail in Fig. 11(a). The DIS cross section is proportional to the square of this diagram, shown in Fig. 11(b). It turns out that the variable x can be interpreted as the fraction of the total momentum p of the nucleon carried by the struck quark, as displayed in the figure.

To fully describe the parity-conserving (electromagnetic) part of the DIS cross section, four structure functions are needed, which are usually denoted $F_1(x)$, $F_2(x)$, $g_1(x)$ and $g_2(x)$. These are extracted from the part of the cross section enclosed in the dashed rectangle shown in Fig. 11(b). The first two (the F's) are the only ones that contribute when the electron beam is unpolarized; The other two (the g's) arise when the beam and target are polarized, and have been a particular focus of an intense study at Jefferson Lab. Data for $g_2(x)$, the smallest and most difficult to measure, has been the last to be obtained.

This program of measurements has contributed to efforts to resolve the long standing "spin" puzzle: polarization data show that the spin carried by quarks is only a small fraction of the total proton spin ($\frac{1}{2}$). This is a surprising result; it requires that other contributions to the spin, which must come from the orbital angular momentum of the quarks, possible contributions from strange quark-antiquark pairs, or the spin and angular momentum of the "gluons" (the carriers of the force between the quarks), must be larger than naively expected.

The excellent quality of the CEBAF beam, which can be highly polarized, has made these studies possible. When all of these functions are well known, the distributions in space, spin, and angular momentum of the quarks inside of the nucleon will be tightly constrained. More measurements will be carried out when the Jefferson Lab accelerator reaches 12 GeV.

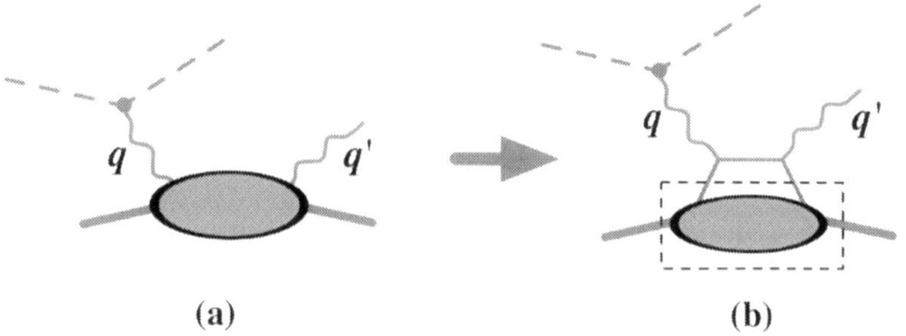

Figure 12. Feynman diagrams showing (a) virtual Compton scattering, and (b) the simplification possible if the virtual photon has a very large Q^2. The part of the right hand diagram enclosed in the dashed rectangle is the GPD discussed in the text.

10.4. Generalized parton distributions (GPDs)
In the mid 1990s it was realized that measurements of the virtual Compton scattering cross section would open a new window into the study of the structure of the nucleon. Virtual Compton scattering occurs when a virtual photon (produced by the electron) is scattered from a nucleon, producing a final state with a real photon and a nucleon, as shown in Fig. 12. When Q^2 is very large, the full diagram Fig. 12(a) may be "factorized" as shown in Fig. 12(b). In this diagram one of the quarks in the nucleon absorbs the high energy virtual photon and then radiates a real photon which emerges in the final state. Since the energy of the two photons and the interacting quark are all very large, this part of the diagram can be calculated using pQCD. The low energy part of the interaction is confined to the lower part of the diagram (which defines the generalized parton distributions, or GPDs, enclosed in the dashed rectangle).

The GPDs are generalizations of the parton distributions illustrated in Fig. 11. The parton distributions depend only on the scaling variable x, interpreted as the fraction of the target momentum carried by the quark, and therefore give us a picture of the distribution of quark momentum in the direction of motion of the nucleon. The GPDs depend on three variables: x, an asymmetry parameter, ξ, and the momentum transferred to the target, t. Together these additional variables can be analyzed to tell about the structure of the nucleon transverse to its direction of motion, which will ultimately allow us to construct a full three-dimensional image of the proton. They can also give detailed information about the angular momentum carried by the u and d quarks, shedding light on the spin puzzle.

The GPDs can also provide a description of the deeply virtual production of mesons. Here the cross section depends not only on the GPDs, but also on the structure of the meson itself. The 12 GeV upgrade is essential to the development of this program.

11. Conclusions

The Jefferson Lab of today is the end result of a long process of scientific planning, open competition, and scientific peer review, starting in the late 1970s and culminating with first beam on target on July 25, 1994, and the physics program beginning a year later. Without the open competition championed by the DOE, the laboratory would have taken a different form and it is unlikely that more than a hundred new faculty positions would have been created to support its scientific program.

The results obtained so far include some surprises, and have provided us with a detailed picture of the quark structure of the nucleon and a much better understanding of when we can see the quark degrees of freedom hidden in a nucleus. Analysis of data already obtained and new theoretical work stimulated by these findings continues, so the full impact of this program is yet to be felt. The superconducting accelerator, with its continuous, tightly focused beam, is now being upgraded to 12 GeV. We expect this new facility to provide many new insights, and some surprises, in the coming years.

Acknowledgment
For more complete accounts of the early history of Jefferson Lab, see the excellent articles by Catherine Westfall [25, 26]. Special thanks to Blaine Norum for providing details about the SLAC accelerator.

It was a great privilege to be a part of this unique endeavor and to work with so many fine people. A few (not already mentioned) are Bill Bertozzi, Carl Carlson, Tony Chargin, John Domingo, Bill Donnelly, Bernard Frois, Bernhard Mecking, Ingo Sick, John Tjon, Steve Wallace, Roy Whitney, and Rolf Winter. This work is supported by Jefferson Science Associates, LLC under U.S. DOE Contract No. DE-AC05-06OR23177.

References

[1] M. Gell-Mann, Phys. Lett. **8**, 214 (1964).
[2] G. Zweig, *An SU_3 Model for Strong Interaction Symmetry and its Breaking*, CERN Report 8419/TH.412 (1964).
[3] J. I. Friedman, Rev. Mod. Phys. **63**, 615 (1991).
[4] H. W. Kendall, Rev. Mod. Phys. **63**, 597 (1991).
[5] R. E. Taylor, Rev. Mod. Phys. **63**, 573 (1991).
[6] S. Weinberg, Rev. Mod. Phys. **52**, 515 (1980).
[7] D. J. Gross, Proc. Nat. Acad. Sci. **102**, 9099 (2005).
[8] H. D. Politzer, Proc. Nat. Acad. Sci. **102**, 7789 (2005).
[9] F. Wilczek, Proc. Nat. Acad. Sci. **102**, 8403 (2005).
[10] Friedlander Panel *Future of Nuclear Science, NAS 1977* (National Research Council, National Academy of Science) (1977).
[11] Livingston Panel *The Role of Electron Accelerators in US Medium Energy Nuclear Science* (ORNL/PPA-77/4) (1977).
[12] *Long Range Plan for Nuclear Science* (DOE/NSF Nuclear Science Advisory Committee) (1979).
[13] R. Neal, *The Stanford Two-Mile Accelerator* (W. A. Benjamin, Inc.) (1968).
[14] *Conference on the Future Possibilities for Electron Accelerators* (Charlotesville, VA) January 1979.
[15] *Future Directions in Electromagnetic Nuclear Physics* (Troy: Rensselaer Polytechnic Institute) (1981).
[16] *Proposal for a National Accelerator Laboratory* (Southeastern Universities Research Association) (1982).
[17] Barnes Panel *The Role of Electromagnetic Interactions in Nuclear Science* (DOE/NSF Nuclear Science Advisory Committee) (1982).
[18] Bromley panel 1983 *A Report of the Panel on Electron Accelerator Facilities* (DOE/NSF Nuclear Science Advisory Committee) (1983).
[19] I. Goodwin, Phys. Today **37**(9), 55 (1984).
[20] *Report of the NSAC Ad Hoc Subcommittee on a 4 GeV CW Electron Accelerator for Nuclear Physics* (DOE/NSF Nuclear Science Advisory Committee) (1984).
[21] B. E. Norum, J. S. McCarthy and R. C. York, Nucl. Instrum. Meth. B **10/11**, 337 (1985).
[22] C. Leemann CEBAF: Design Overview and Project Status *Proceedings of The Third Workshop on RF Superconductivity, Argonne National Laboratory, Illinois, USA* p 141 (1988).
[23] December, 1985 *CEBAF Preconceptual Design Report* (Continuous Electron Beam Accelerator Facility, Newport News, VA). The primary purpose of this report, prepared at the request of the DOE in December 1985, was to describe the new superconducting design, but Chapter 2 included an overview of the scientific justification. December, 1985.
[24] *Proposal for a Theory Group for the Continuous Electron Beam Accelerator Facility* (Southeastern Universities Research Association) (1983).
[25] C. Westfall, *The Founding of CEBAF, 1979 to 1987* (The Continuous Electron Beam Accelerator Facility) (1994).
[26] C. Westfall, Hist. Stud. Phys. Biol. Sci. **32**, 369 (2002).

Journal of Physics: Conference Series **299** (2011) 012002 doi:10.1088/1742-6596/299/1/012002

Nucleon Form Factors – A Jefferson Lab Perspective

John Arrington[1], Kees de Jager[2] and Charles F. Perdrisat[3]

[1] Argonne National Laboratory, Argonne, Illinois 60439, USA
[2] Jefferson Lab, Newport News, Virginia 23606, USA
[3] College of William and Mary, Williamsburg, Virginia 23187, USA

Abstract. The charge and magnetization distributions of the proton and neutron are encoded in their elastic electromagnetic form factors, which can be measured in elastic electron–nucleon scattering. By measuring the form factors, we probe the spatial distribution of the proton charge and magnetization, providing the most direct connection to the spatial distribution of quarks inside the proton. For decades, the form factors were probed through measurements of unpolarized elastic electron scattering, but by the 1980s, progress slowed dramatically due to the intrinsic limitations of the unpolarized measurements. Early measurements at several laboratories demonstrated the feasibility and power of measurements using polarization degrees of freedom to probe the spatial structure of the nucleon. A program of polarization measurements at Jefferson Lab led to a renaissance in the field of study, and significant new insight into the structure of matter.

1. Introduction

The electromagnetic form factors encode the spatial distributions of charge and magnetization in the nucleon. In the simple non-relativisitic picture, the two form factors of a spin-1/2 object, G_E and G_M, are the Fourier transforms of the spatial distribution of the charge and magnetization inside the object. In the nucleon, the quarks are the carriers of charge, and so these observables are directly connected to the spatial distribution of quarks in the nucleon, as well as a probe of the underlying dynamics.

In the 1950s electron scattering became the tool of choice for measuring the nucleon form factors. An active program mapped out the form factors as well as possible, but the ability to extract the form factors using unpolarized cross section measurements was limited. By the 1980s most of the experiments provided only incremental improvements on the precision or on the Q^2 range of existing measurements, or were early proof-of-principle tests of new techniques.

The advent of electron beams with high luminosity and polarization, combined with new polarized targets, recoil polarimeters, and large-acceptance detectors, led to a revolution in the study of nucleon form factors. In the last 10 years measurements at Jefferson Lab have rewritten the textbook on the proton and neutron form factors. The techniques that have allowed this dramatic resurgence of interest in the form factors have also opened up other possibilities, allowing us to isolate the contribution of strangeness in the nucleon and making cleaner and more precise measurements of the impact of the nuclear environment on the internal structure of the proton and neutron.

2. Historical context

In the Born approximation, where the interaction occurs *via* the exchange of a single virtual photon, the unpolarized e–N elastic cross section can be written in terms of the Sachs form factors, G_E and G_M, as

$$\frac{d\sigma}{d\Omega} = \frac{\sigma_{Mott}}{\varepsilon(1+\tau)} \left[\tau G_M^2(Q^2) + \varepsilon G_E^2(Q^2) \right] , \tag{1}$$

where $\tau = Q^2/4M^2$, $-Q^2$ the square of the four-momentum transfer, M the nucleon mass, and $\varepsilon = 1/[1 + 2(1+\tau)\tan^2(\frac{\theta_e}{2})]$ is the linear polarization parameter of the virtual photon. The value of ε depends on the scattering angle θ_e, with $\varepsilon \to 1$ in the limit of forward scattering, and $\varepsilon = 0$ for 180° scattering. The term σ_{Mott} denotes the cross section for the scattering of two spin-1/2 point-like objects. The quantity in brackets is referred to as the reduced cross section which, at fixed Q^2, depends only on the values of G_E, G_M, and ε. The electric and magnetic form factors can also be written in terms of the Dirac and Pauli form factors: $G_E = F_1 - \tau F_2$, $G_M = F_1 + F_2$. Writing the cross section in terms of the Sachs form factors (Eq. (1)) yields a simpler expression, without the cross terms that appear when using F_1 and F_2. By performing a Rosenbluth separation — making cross section measurements at a fixed Q^2 value but two or more ε values — one can separate the values of G_E^2 and G_M^2.

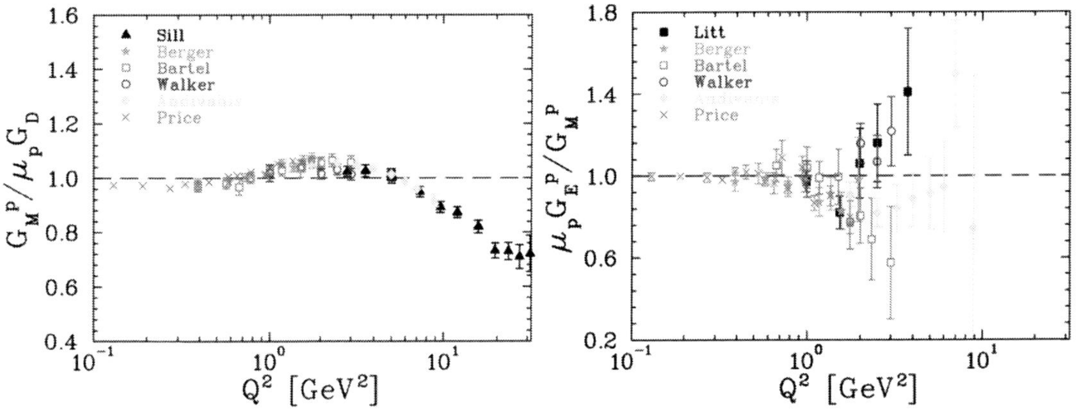

Figure 1. Database for $G_M^p/\mu_p G_D$ (left) and $\mu_p G_E^p/G_M^p$ (right) obtained by the Rosenbluth method as of the mid-1990s.

This technique is limited in its ability to make a clean separation of the form factors. One can see from the form of the reduced cross section that there is very little sensitivity to G_E for $Q^2 \gg 1$ GeV^2 — or for $(G_E/G_M)^2 \ll 1$ — and little sensitivity to G_M for $Q^2 \ll 1$, except for $\theta_e \to 180°$. Nonetheless, it was possible to make measurements of G_E^p, and especially G_M^p, over a wide range in Q^2, as seen in Fig. 1, which shows the status of proton form factor measurements in the mid-1990s (see Refs. [1, 2, 3] for details of the measurements). The Q^2 dependence of the proton magnetic form factor is well approximated by the dipole form up to 10 GeV^2 ($G_M^p/\mu_p \approx G_D = (1 + Q^2/0.71)^{-2}$, with μ_p the proton magnetic moment), and falls $\sim 30\%$ below the dipole form at $Q^2 \approx 30$ GeV^2. While G_E^p was also reasonably well approximated by the dipole form, systematic variations between the results of different experiments were much larger than in G_M^p. However, the general conclusion was that the data were consistent with form factor scaling, *i.e.* $\mu_p G_E^p/G_M^p$ was independent of Q^2, up to at least 5 GeV^2. This was consistent with simple nonrelativistic quark models, as well as the perturbative QCD expectation at large Q^2.

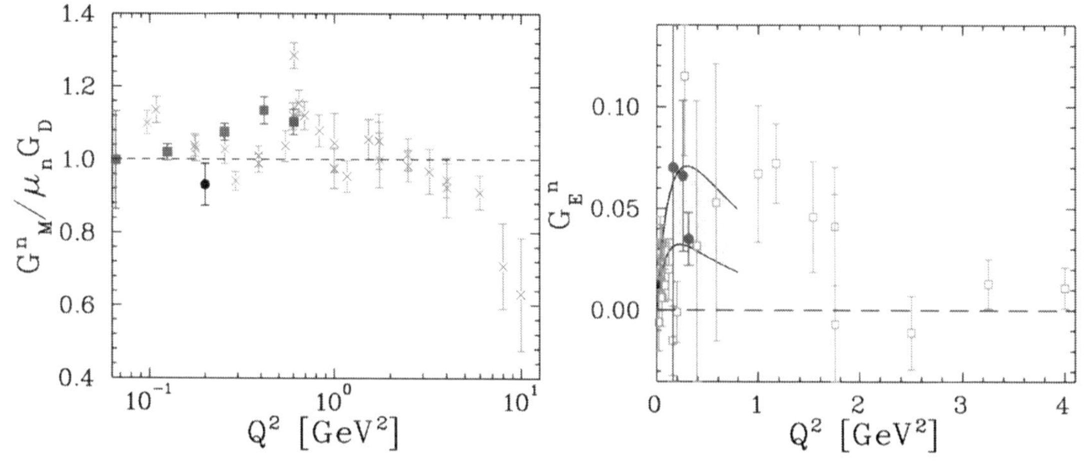

Figure 2. Database for $G_M^n/\mu_n G_D$ (left) and G_E^n (right) before the Jefferson Lab turn-on. For G_M^n, the crosses are from quasi-elastic scattering (inclusive and coincidence), the filled squares are from ratio measurements, and the filled circle from a polarization experiment. For G_E^n, the filled circles are from polarization measurements, and the open squares from quasi-elastic scattering data. The shaded band shows the range of results extracted from a model-dependent analysis of elastic e–d scattering [4].

For the neutron, measurements using the Rosenbluth technique were even more difficult. The need to use light nuclei as "effective" neutron targets necessitated large corrections in extracting the e–n cross section, thus limiting the precision and kinematic coverage for extractions of G_M^n. The extraction of G_E^n was even more problematic, as G_E^n is much smaller than the other electromagnetic form factors. Therefore, there were no precise extractions of G_E^n, and for $Q^2 \gtrsim 2$ GeV^2 only upper limits could be set. In the limit $Q^2 \to 0$, G_E^n must approach zero (the charge of the neutron), while at finite Q^2, any non-zero value must come from a difference in the spatial distribution of up and down quarks in the neutron.

Figure 2 shows the status of neutron form factor measurements around the time of the Jefferson Lab turn-on in 1996. For G_E^n, the data are mainly from quasi-elastic scattering from the deuteron and from elastic e–d scattering, which has large model-dependent uncertainties, indicated by the shaded band. Only a few proof-of-principle polarization measurements had been performed. For G_M^n, most of the results are from inclusive quasi-elastic scattering measurements from the deuteron, and the large uncertainties show the impact of the proton subtraction and the nuclear corrections. A few data points had been measured using the so-called ratio technique which, along with the polarization measurements, will be described in the following sections.

It had long been known [5, 6, 7, 8] that, in principle, including measurements of polarization observables would allow a much improved determination of the form factors. Using a polarized electron beam, one can measure either the cross section asymmetry from a polarized target or the polarization transferred to an unpolarized nucleon. In both cases, the measured asymmetry depends only on the ratio G_E/G_M, thus providing sensitivity to the electric form factor even if its contribution to the unpolarized cross section is extremely small. However, it required the development of high intensity, highly polarized electron beams and polarized targets or recoil polarimeters with high figures of merit to apply this theory to advance our knowledge of the form factors.

Already at the third Program Advisory Committee (PAC) meeting of CEBAF in February of 1989 three letters of intent (LOI) had been presented to measure nucleon form factors with

Table 1. List of Jefferson Lab experiments related to nucleon form factors. Experiments marked with "†" were focused on two-photon exchange contributions. Proposals are available from http://www.jlab.org/exp_prog/generated/approved.html.

Expt.(Hall)	Reaction	Goal, Q^2 range	Expt.(Hall)	Reaction	Goal, Q^2 range
E93-027(A)	$p(\vec{e},e'\vec{p})$	G_E^p, 0.5–3.5	E95-001(A)	$^3\vec{\mathrm{He}}(\vec{e},e')$	G_M^n, 0.1–0.6
E99-007(A)	$p(\vec{e},e'\vec{p})$	G_E^p, 3.5–5.6	E94-017(B)	$^2\mathrm{H}(e,e'N)$	G_M^n, 1–4.8
E04-108(C)	$p(\vec{e},e'\vec{p})$	G_E^p, 2.5–8.5	E93-026(C)	$^2\vec{\mathrm{H}}(\vec{e},e'n)$	G_E^n, 0.5–1.0
E04-019(C)	$p(\vec{e},e'\vec{p})^\dagger$	G_E^p, 2.5	E93-038(C)	$^2\mathrm{H}(\vec{e},e'\vec{n})$	G_E^n, 0.4–1.5
E01-001(A)	$p(e,e'p)^\dagger$	G_E^p, 2.6–4.1	E02-013(A)	$^3\vec{\mathrm{He}}(\vec{e},e'n)$	G_E^n, 1.4–3.4
E05-017(C)	$p(e,e'p)^\dagger$	G_E^p, 0.4–5.8	E04-110(C)	$^2\mathrm{H}(\vec{e},e'\vec{n})$	G_E^n, 4.3
E08-007(A)	$p(\vec{e},e'\vec{p})$	G_E^p, 0.25–0.7	E07-005(B)	$p(e^\pm,e'p)^\dagger$	TPE, 0.5–3.0
E08-007(A)	$\vec{p}(\vec{e},e'p)$	G_E^p, 0.015–0.4			

the new techniques being developed; two to measure G_E^n and one to measure G_E^p. These LOIs were then developed into full-blown proposals, resulting in five fully approved proposals in 1994, shortly before the start of operations. The first experiments provided much improved measurements of proton and neutron form factors, as well as some unexpected and exciting results. This led to a variety of proposals, summarized in Table 1, aimed at fully developing and exploiting these new techniques to extend the kinematic coverage and precision of the data. Within a decade of the first measurement, this program would drastically transform our state of understanding of nucleon form factors. Related measurements examining parity-violating elastic scattering as well as meson and transition form factors are discussed in other reviews in this volume.

3. Proton form factors

The original motivation for the new measurements of the proton electric form factor was the internal inconsistency of the database of the time, which showed a rapid increase in the uncertainty and scatter of the results with Q^2, starting at ~1 GeV^2; see Fig. 1. The initial experiment in Hall A yielded unexpected results, which created significant excitement in the field. A great deal of effort went into trying to understand the implications of the new data on our picture of the proton, and to extend the polarization measurements to higher Q^2 values. There was also significant activity aimed at understanding the discrepancy with previous measurements, and making significant improvements in precision at low Q^2 values.

First, we discuss the new techniques that made such measurements feasible, then we present the results of the initial high-Q^2 experiments, as well as the later studies that were an offshoot of these new techniques.

3.1. Techniques

Polarization transfer experiments measure the polarization of the recoiling proton by rescattering it in an appropriate material and determining the resulting azimuthal asymmetry distribution, thus providing a measurement of the two components of the polarization in a plane perpendicular to the proton momentum. After being struck by a polarized electron, the proton has in-plane polarization components parallel (P_ℓ) and perpendicular (P_t) to the proton momentum. The component normal to the scattering plane is zero in the Born approximation. To measure both P_ℓ and P_t simultaneously requires precession of P_ℓ into a normal component, using a dipole magnet. The precession in the dispersive plane of a dipole is given by $\chi = \gamma \cdot (\theta_B + \theta_{target} - \theta_{fpp}) \cdot \kappa_p$, where γ, θ_B, θ_{target}, θ_{fpp} and $\kappa_p = \mu_p - 1$ are the proton's relativistic factor, the mean bending angle of the dipole, the entrance and exit angle of an individual proton trajectory, and the anomalous proton magnetic moment, respectively. The resulting distribution in the azimuthal angle φ is

$$N(\vartheta, \varphi) = N_0 \cdot [1 + h(P_b A_y(\vartheta) P_n^{fpp} \sin\varphi - P_b A_y(\vartheta) P_t^{fpp} \cos\varphi)], \quad (2)$$

where N_0 is the average number of events in a given interval of θ and φ, P_b is the electron beam longitudinal polarization and $h = \pm 1$ the helicity state of the beam, and $P_n^{fpp} \approx P_\ell \sin\chi$ and $P_t^{fpp} \approx P_t$ are the polarization components in the polarimeter; $A_y(\vartheta)$ is the analyzing power at a given polar scattering angle ϑ. The relative beam-helicity difference distribution in a given ϑ interval is then:

$$\frac{N^{h=+1}(\varphi) - N^{h=-1}(\varphi)}{2N_0} = P_b A_y (P_n^{fpp} \sin\varphi - P_t^{fpp} \cos\varphi). \quad (3)$$

Taking only the spin precession in the dispersive plane, χ, into account, the ratio $P_t^{fpp} \sin\chi / P_n^{fpp}$ is directly related to the ratio P_t/P_ℓ at the target, which in turn is a measure of G_E^p/G_M^p:

$$\frac{G_E^p}{G_M^p} = -\frac{P_t}{P_\ell}\frac{(E_e + E_{e'})}{2M}\tan(\theta_e/2) = -\frac{P_t}{P_\ell}\sqrt{\frac{\tau(1+\varepsilon)}{2\varepsilon}}. \quad (4)$$

As G_E^p/G_M^p is defined by the ratio of two polarization components, knowledge of the beam polarization and polarimeter analyzing power is not necessary. The remaining source of systematic uncertainty comes from the accuracy of the spin transport, *i.e.* the calculation of the components of the polarization at the target from the asymmetry in the focal plane. As Q^2 increases, the precession in the non-dispersive plane due to focusing elements in the spectrometer becomes significant and has to be taken into account. This residual systematic uncertainty can be evaluated on the basis of optical studies [9].

In almost all recoil polarization experiments at Jefferson Lab, both final state particles were detected to reduce the inelastic contamination. The first focal-plane polarimeter (FPP) built in Hall A was used for the GEp(I) experiment, E93-027 [9, 10]. Several changes were required to utilize the full beam energy of the accelerator and extend the measurements to higher Q^2. For a fixed beam energy, the cross section scales as Q^{-12} and the analyzing power decreases with increasing Q^2. Furthermore, the electron solid angle matching the proton acceptance increases with Q^2. To compensate for these factors in GEp(II) (E99-007), the electron was detected in a large solid-angle electromagnetic calorimeter, and the polarimeter was reconfigured with 95 g/cm^2 of CH_2 instead of graphite to increase the effective analyzing power.

As the maximum momentum of the proton spectrometer in Hall A limited the range of possible Q^2, GEp(III) (E04-018) used the HMS spectrometer in Hall C, which can detect proton momenta up to 7.5 GeV/c. To increase the figure of merit at high Q^2, a new FPP was built, consisting of two polarimeters in succession, each containing a slab of polyethylene, 50 g/cm^2 thick, followed by a set of drift chambers. A larger calorimeter ("BigCal") was needed to match the proton acceptance at these large Q^2 values. An entirely new calorimeter was built, consisting of 1744 lead glass bars 4×4 cm^2 in cross section, and 40-45 cm long, with a frontal area of 2.6 m^2. A trigger signal from the calorimeter was required for the definition of an $(e, e'p)$ or an $(e, \gamma p)$ event.

3.2. High-Q^2 regime

GEp(I) ran in mid-1998 and measured the G_E^p/G_M^p ratio up to $Q^2 = 3.5$ GeV^2 [9]. The results, shown in Fig. 3, revealed an unexpected decrease of G_E^p/G_M^p with increasing Q^2, in disagreement with scaling. When GEp(II) extended the measurement of G_E^p/G_M^p to $Q^2 = 5.6$ GeV^2, the ratio was found to continue to decrease linearly with Q^2, to a value of $\mu_p G_E^p/G_M^p$=0.28±0.09 at

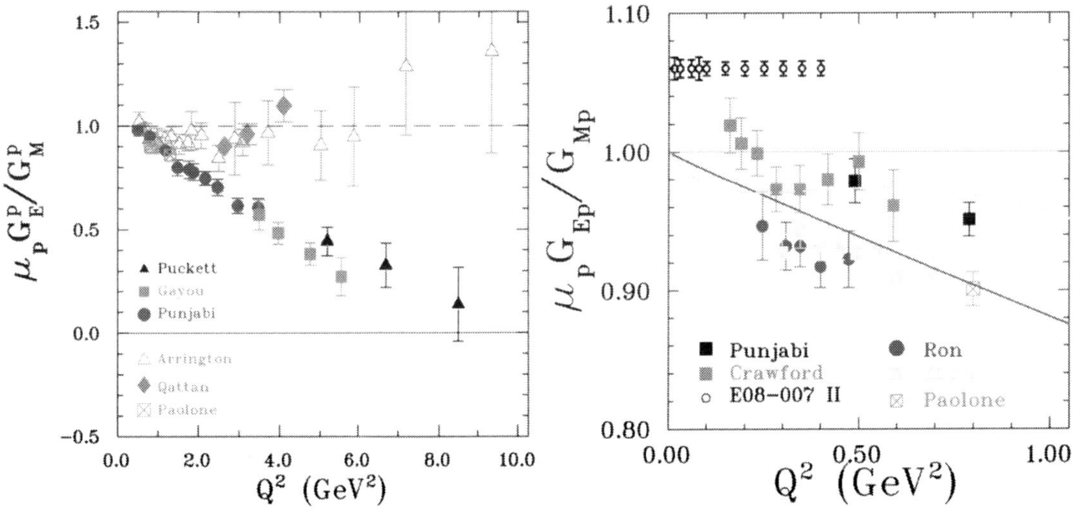

Figure 3. Left: The ratio $\mu_p G_E^p / G_M^p$ from Jefferson Lab and MIT-Bates (BLAST) polarization experiments, a global analysis of cross section measurements [13] (open triangles), and the "Super-Rosenbluth" results [14] (filled diamonds). Right: Focus on the low-Q^2 results from polarization measurements [9, 15, 16, 17, 18], and projections for the polarized target measurement of E08-007. The dashed line is the fit of Kelly [19], made before most of these data were available, and an updated [17] version of the fit from Ref. [20].

$Q^2 = 5.6$ GeV^2 [11]. The GEp(III) measurement extended the measurements up to 8.5 GeV^2 [12], with the data suggesting a slower decrease of the ratio G_E^p / G_M^p above $Q^2 = 5$–6 GeV^2.

The results from polarization transfer and Rosenbluth extraction of G_E^p / G_M^p are compared in Fig. 3. The discrepancy between the two results is significant over a wide range of Q^2. A detailed reanalysis of the world's cross section data [21] showed that the apparent discrepancy between different extractions of G_E^p (Fig. 1) was the result of neglecting normalization uncertainties when combining data from different measurements, but that these uncertainties could not account for the systematic difference between Rosenbluth and polarization transfer results.

Following the unexpected results of the first polarization transfer experiment, a new Rosenbluth separation of unprecedented accuracy was prepared and the experiment, E01-001, obtained data in 2002. The unique feature of this "Super-Rosenbluth" experiment was the detection of the proton, (e, p), yielding fixed proton momentum for a given Q^2, minimal variation of rate as a function of angle, and smaller ε-dependent corrections than in (e, e'). This, combined with smaller radiative corrections, led to significantly reduced uncertainties in the ratio G_E^p / G_M^p. The results [14] (filled diamonds in Fig. 3) cover a Q^2 range from 2.6 to 4.1 GeV^2, and agree with the previous Rosenbluth extractions. At the same time, traditional Rosenbluth measurements in Hall C [22] provided additional measurements, albeit with much larger uncertainties (these data are included in the global analysis). At that point it was clear that there was a systematic difference between Rosenbluth and recoil polarization data.

Intensive discussion of various possible explanations followed. With the exception of two-photon exchange (TPE) contributions, radiative corrections were believed to be well understood for the unpolarized case, and small in the relevant polarization observables [23]. Historically, only the infrared (IR)-divergent component of the TPE corrections had been taken into account. The remaining TPE terms had been evaluated in a soft photon approximation [24], and found to be small but potentially important at large Q^2 values. While the TPE contributions are small, the contribution from G_E^p decreases rapidly as Q^2 increases, and so the TPE corrections can be important in understanding the discrepancy. It was later shown that a ε-dependent correction of $\sim 5\%$ could bring the Rosenbluth results in agreement with the polarization measurements [13, 25], while having little impact on the polarization transfer results. The status of these studies and of new experiments designed to measure two-photon exchange contributions are discussed in Sec. 3.4.

3.3. Low-Q^2 regime

The Q^2 region below 1 GeV^2 is of interest because it covers the range over which the pion cloud is believed to make a significant contribution to the electromagnetic structure of the proton and the neutron. While this yields a significant part of the neutron's electric form factor at small Q^2, it is also expected to be important for the form factors of the proton, as discussed for example by Friedrich and Walcher [26]. Extremely low values of Q^2 are interesting because the form factor behavior as $Q^2 \to 0$ is connected to the nucleon charge and magnetic radii.

While the neutron was the focus of early measurements at low Q^2, the demonstration that extremely precise measurements of G_E^p/G_M^p were possible led to improved low-Q^2 measurements of the proton form factor. In recent years careful studies of the optical properties of the Hall A HRS spectrometers have resulted in a significant reduction of the systematic errors in polarization transfer data, thus allowing measurements of G_E^p/G_M^p below $Q^2 \sim 1$ GeV^2 with a total error of 1% or less. New data from Bates [15], along with updated results [18] from the Jefferson Lab 'LEDEX' experiment E05-013 [27] and preliminary results from the dedicated E08-007 measurements [17], are shown in the right panel of Fig. 3. These high precision data do not show any indication of structure in the low-Q^2 "pion cloud" region, although there is at present a systematic disagreement between the polarized target data from Bates [15] and the high precision Jefferson Lab polarization data [18, 17]. New results from an extensive set of cross section measurements at low Q^2 from Mainz [28] also see a reduction in $\mu_p G_E/G_M$ in this region, consistent with the new Jefferson Lab data, although they only provide a fit to the form factors, rather than direct extractions of G_E and G_M. Phase II of E08-007 will make extremely high precision measurements using a polarized target. This will provide another comparison between the polarized target and recoil polarization measurements, and will extend the low-Q^2 data down to $Q^2 = 0.015$ GeV^2. This will allow for significantly improved extractions of the magnetic form factor at very low Q^2, where the cross section has greatly reduced sensitivity.

Knowledge of the form factors is also important in the analysis of a range of other experiments. Good knowledge of the e–N scattering cross section is crucial to the interpretation of high-precision quasi-elastic scattering measurements aimed at understanding nucleons in nuclei. They are also necessary input to the analysis of parity-violating electron scattering, where the contributions from strange quarks can be isolated given sufficient precision on the electromagnetic form factors and the parity-violating asymmetry. In addition, a better determination of the charge and magnetic radii provides input for the hadronic corrections to the hyperfine structure of hydrogen.

3.4. Two-photon exchange

After the observation of a large discrepancy between Rosenbluth and polarization measurements of $\mu_p G_E/G_M$, two-photon exchange corrections received a great deal of attention as a possible explanation. In 2003, two papers, appearing back-to-back in PRL, shed significant light on the issue. The first paper [25] provided a general formalism for scattering beyond the Born approximation, while the second [29] provided a direct calculation of the TPE correction, including the case of two hard photons, calculated in a hadronic basis.

The analysis of Guichon and Vanderhaeghen [25] not only provided the general formalism, but also demonstrated that relatively small TPE contributions might bring the Rosenbluth results into agreement with the new polarization data. They were able to resolve the discrepancy with TPE amplitudes at the 2–3% level, consistent with the expectation of order α_{EM} corrections for higher-order electromagnetic diagrams. These TPE corrections, derived based on a set of very simple assumptions, were, however, at odds with the measured cross section ratio of positron-proton and electron-proton scattering [30]. A later analysis [31] used modified assumptions to extract a set of TPE amplitudes of similar magnitude that could explain the discrepancy between Rosenbluth and polarization transfer measurements while being consistent with the comparison of electron and positron data. Because the impact of the TPE corrections is much smaller for polarization transfer measurements, the impact of TPE on the neutron form factors is also small. Rosenbluth extractions of G_M^n were not very precise, and all precision extractions of G_E^n were based on polarization measurements. The recent measurement of the ε dependence of the polarization observables [32] allows a more detailed extraction of the TPE amplitudes [33, 34]. There is still model dependence in the ε dependence of the TPE amplitudes that is not fully constrained, but the inclusion of the polarization observables provides additional important constraints.

Following the initial calculation of TPE corrections in a hadronic model, several other theoretical calculations of the two hard-photon exchange contribution were undertaken (see [35]). A difficulty arises from the fact that in the intermediate state inside the box diagram, the proton can be any excited baryon compatible with angular momentum and parity conservation. One approach, expected to be reliable at lower Q^2, has been to start with a calculation for an unexcited hadron in the intermediate state [29, 36], and then estimate the effect of higher resonance states [37]. Another approach focuses on the high-energy region [38, 39], assuming that the virtual photon interacts with one valence quark and that the residual system of quarks and gluons is accurately described by generalized parton distributions (GPDs). These two approaches yield qualitatively similar behavior: small corrections to polarization observables and a small change in the ε dependence of the cross section. This change in the ε dependence results in a large correction to the Rosenbluth extraction of G_E^p at high Q^2, where the initial ε dependence is extremely small.

While much of the focus has been on improving the calculation of TPE diagrams, there have also been efforts to improve the treatment of higher-order terms in the radiative corrections [24, 40, 41, 42]. For example, hard Bremsstrahlung terms yield an additional $\sim 1\%$ difference between high and low ε values [42] that is not present in the soft-photon approximation. However, as with most of these comparisons, the result is compared to Mo and Tsai radiative corrections [43]. While this is the general approach adopted in nearly all experimental extractions of the cross sections, many analyses have applied improved corrections, *e.g.* for multi-photon bremsstrahlung, and further work is required to determine to what extent the corrections actually applied to the data differ from the new calculations.

While the reduced cross section must depend linearly on ε in the Born approximation, TPE contributions will introduce a curvature in the ε dependence of the cross section (in the Rosenbluth procedure). A global analysis of Rosenbluth measurements set tight limits on the nonlinear contributions to the ε dependence of the cross section [44], but these limits are not tight enough to rule out the nonlinear contributions predicted by most TPE calculations. New measurements of the kinematical dependence of the G_E^p/G_M^p ratio, obtained in recoil polarization at a constant Q^2 of 2.48 GeV^2 in Hall C (E04-019), show no ε dependence within the statistical uncertainty [45], supporting the calculations that conclude that polarization measurements are not significantly affected by TPE. The most direct test of TPE contributions, the comparison of positron and electron scattering, yields evidence for a non-zero TPE contribution at the 3σ level [30]. At the present time there is no definitive experimental evidence that TPE contributions are the major cause of the difference between Rosenbluth and recoil polarization extractions of the proton form factor, although preliminary results from positron-electron comparisons at Novosibirsk [46] yield an excess positron cross section consistent with TPE calculations, and data taking has begun for E07-005, a much more extensive set of comparisons in Hall B.

There is an active program at Jefferson Lab to determine if TPE corrections do in fact fully explain the discrepancy, and to map out their impact on both the cross section and polarization observables. In Hall C, E05-017 extended the high-precision "Super-Rosenbluth" measurements, covering a large range in Q^2 and ε to better map out the Rosenbluth–polarization difference and to improve the limits on nonlinear contributions by a factor of two or more over a wide range in Q^2. The discrepancy has motivated experiments to compare positron and electron scattering at Novosibirsk, DESY, and Jefferson Lab [47]. Data taking has begun for E07-005, which will provide a new comparison of the e^+–p and e^-–p cross sections over a broad kinematic range. This experiment is running in Hall B using a tertiary beam of both positrons and electrons over a broad range of energies. This will allow a quantitative study of TPE contributions over a range of ε for Q^2 up to 2–3 GeV^2, and will directly determine whether TPE corrections fully explain the discrepancy between Rosenbluth and polarization measurements in this region. While G_E^p is a unique case where these small corrections have a large impact on the interpretation of the results, TPE effects contribute to all electron scattering measurements. A stringent test of the calculations is important to ensure that these corrections are well understood for future measurements that aim for extremely high precision in a variety of reactions.

4. Neutron form factors

For the neutron, form factor extractions using the Rosenbluth technique were limited by the small size of the electric form factor and the need to measure e–n scattering using quasi-elastic scattering from deuteron targets. As discussed in Sec. 2, it was known that G_E^n was positive, but there was a factor of two uncertainty in the values derived from elastic electron–deuteron scattering. While the situation for G_M^n was better, both the precision and Q^2 range were limited compared to measurements on the proton.

The availability of high-polarization beams, effective polarized neutron targets, and large neutron detectors and recoil polarimeters made it possible to dramatically improve our knowledge of the neutron form factors. Experiments at other labs had demonstrated the feasibility of such measurements, and begun the process of validating the techniques through comparisons of measurements utilizing different techniques or different target nuclei. The Jefferson Lab program added to these studies by expanding precision measurements of G_M^n above Q^2=1 GeV^2 and by providing essentially all of the direct measurements of G_E^n above 0.8 GeV^2.

4.1. Neutron magnetic form factor

Figure 4 shows the present status of measurements of the neutron magnetic form factor. In the high-Q^2 regime, E94-017 [51] completed a study of G_M^n at Q^2 up to ~ 5 GeV^2 in 2000 by measuring the exclusive neutron/proton cross section ratio from the deuteron with the CLAS detector in Hall B. The ratio of the $^2\mathrm{H}(e,e'n)$ and $^2\mathrm{H}(e,e'p)$ reactions in quasi-elastic kinematics is approximately equal to the ratio of elastic scattering off a free neutron and proton, respectively:

$$R_D = \frac{\frac{d\sigma}{d\Omega}[^2\mathrm{H}(e,e'n)_{QE}]}{\frac{d\sigma}{d\Omega}[^2\mathrm{H}(e,e'p)_{QE}]} = a \cdot R_{free} = a \cdot \frac{\tau(G_M^n)^2 + \varepsilon(G_E^n)^2}{\tau(G_M^p)^2 + \varepsilon(G_E^p)^2}. \quad (5)$$

The correction factor a is close to unity for quasi-elastic kinematics and larger Q^2 values and can be accurately calculated [52] as a function of Q^2 and θ_{pq}, the angle between the momentum transfer and the knocked-out nucleon in the center of mass, using standard models for the deuteron. The value of G_M^n is then determined from the measured value of R_D and the known

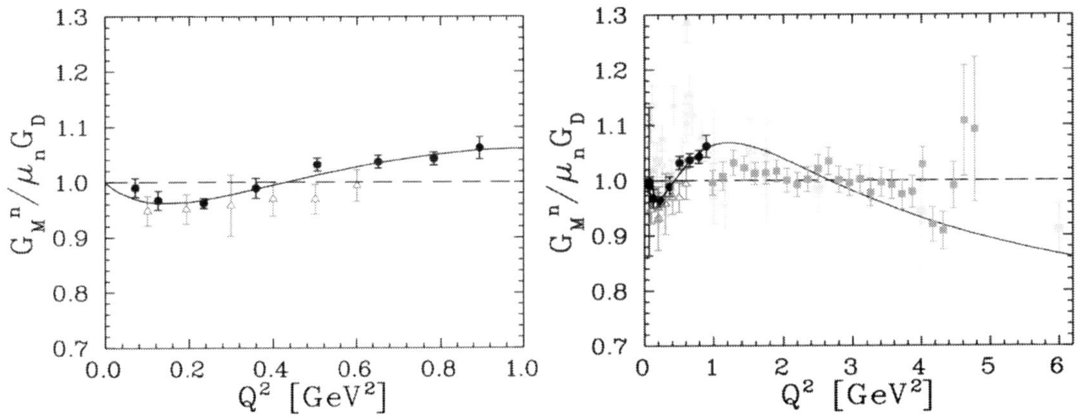

Figure 4. Current status of extractions of $G_M^n/\mu_n/G_D$. The left figure shows only the high precision, low Q^2 Jefferson Lab polarization [48] (open triangles) and Mainz ratio [49, 50] (filled circles) measurements, while the right figure shows the full data set, including the CLAS ratio [51] measurements (filled squares).

values of G_M^p and G_E^p, with very small corrections due to the lack of knowledge of G_E^n. The cross sections for the ^{2}H$(e,e'n)$ and ^{2}H$(e,e'p)$ reactions were measured simultaneously. Measuring the ratio R_D reduces or completely eliminates several systematic uncertainties, such as those from final state interactions in the deuteron, from knowledge of the luminosity and from radiative corrections. The remaining major uncertainty was due to the neutron detector efficiency, which was measured through pion electroproduction off a hydrogen target. These efficiency measurements were performed simultaneously with the primary deuteron measurement, by positioning a hydrogen target upstream of the deuterium target. The experiment took data at two different beam energies, and for each energy independently analyzed neutrons detected in time-of-flight scintillators and neutrons detected in the calorimeter, yielding four essentially independent but overlapping measurements [51].

Experiment E95-001 [53, 54, 48] determined G_M^n in 1999 at Q^2 values smaller than 1 GeV^2 by measuring the beam asymmetry in inclusive quasi-elastic scattering of electrons off a polarized ^{3}He target in Hall A. The polarized ^{3}He target system in Hall A utilizes the spin exchange between optically pumped alkali-metal vapor and noble-gas nuclei to produce an ensemble of spin polarized ^{3}He nuclei. Polarized ^{3}He nuclei serve as an effective polarized neutron target, because their ground state is dominated by a spatially symmetric S-state in which the spins of the protons cancel. The central feature of the target system is a sealed glass cell, that contains an admixture of ^{3}He gas at a pressure of 0.7 MPa and an alkali-metal vapor. These cells have two chambers, a heated upper chamber in which the spin exchange takes place and a lower one, through which the electron beam passes. Lower cells with lengths from 20 to 40 cm have been used, corresponding to a target thickness of up to $1 \cdot 10^{22}$ nuclei/cm^2. In earlier versions of the target only rubidium was used as the spin-exchange medium, resulting in an in-beam polarization of up to ~40% at a beam current of up to ~10 μA. A detailed description of the polarized ^{3}He target system can be found in Ref. [55].

The beam-target asymmetry is:

$$A = -\frac{(\cos\theta^* v_{T'} R_{T'} + 2\sin\theta^* \cos\phi^* v_{TL'} R_{TL'})}{v_L R_L + v_T R_T}, \tag{6}$$

where θ^* and ϕ^* are the polar and azimuthal target polarization angles with respect to $\vec{q}$, R_i denote various nucleon response functions, and v_i the corresponding kinematic factors. By orienting the target polarization parallel to $\vec{q}$, one measures the ratio of $R_{T'}$ to the unpolarized cross section. In quasi-elastic kinematics $R_{T'}$ is dominantly sensitive to $(G_M^n)^2$:

$$R_{T'} \propto P_n(G_M^n)^2 + P_p(G_M^p)^2, \tag{7}$$

where P_n and P_p denote the effective polarizations of the neutron and the proton, respectively. The extraction of G_M^n requires an iterative process since the asymmetry depends on both $R_{T'}$ and the unpolarized cross section, which also depends strongly on G_M^n. In addition, corrections for the nuclear medium [56] are necessary to take into account effects of final state interactions and meson exchange currents. Such corrections are calculable nonrelativistically at low Q^2 [53]. At intermediate Q^2, relativistic effects have to be taken into account, making calculations much more difficult. However, there the size of the corrections is expected to be small and they have been neglected in the analysis [54]. Both measurements are in good agreement in the overlap region (Fig. 4) and with the CLAS data in the larger Q^2 range, but disagree significantly from earlier Mainz measurements [49, 50], that used the same ratio method as the CLAS measurement, but with an off-site neutron detector calibration.

4.2. Neutron charge form factor

The same techniques that have been used to measure G_E^p/G_M^p can also be applied to measure G_E^n/G_M^n, except that one measures scattering from neutrons in ^{2}H or ^{3}He. In the past decade, a series of Jefferson Lab beam-asymmetry measurements of neutron knock-out from a polarized target or studies of polarization transfer have provided accurate data on G_E^n. The first such measurements at Jefferson Lab were carried out in Hall C.

Arnold, Carlson and Gross [8] were the first to show that the measurement of the up-down asymmetry in a neutron polarimeter after the spin of the knocked-out neutron has been precessed by a vertical dipole field yields access to the ratio G_E^n/G_M^n. This was the technique used by experiment E93-038, which in 2000 used a large scintillator neutron polarimeter to determine G_E^n at Q^2 values of 0.45, 1.13 and 1.45 GeV^2 [57]. The neutron polarimeter consisted of a large dipole magnet with a vertically oriented field, an active analyzer (preceded and followed by a veto/tagger), and a top and bottom array of scintillators. The dipole magnet precesses the spin of the neutrons through an angle χ in the horizontal plane and sweeps protons and other charged particles out of the acceptance. The active analyzer consists of twenty $100\times10\times10$ cm^3 scintillators. The long axes were oriented horizontally, perpendicular to the central flight path, stacked vertically into four layers of five detectors. The up/down rear arrays of scintillators each consisted of 12 detectors stacked in three horizontal layers with the long axes of the scintillators oriented parallel to the flight path of the neutrons. The up-down scattering asymmetry measured in this rear array is proportional to the projection of a recoil polarization on to a horizontally-oriented sideways axis. Additional elements of this polarimeter were a lead curtain, veto taggers and extensive steel and concrete shielding surrounding the scintillator detectors [58].

The precession through an angle χ results in a scattering asymmetry $\xi(\chi)$

$$\xi(\chi) = A_y P^n \sin(\chi - \delta). \tag{8}$$

The ratio G_E^n/G_M^n can be determined from the value of the precession angle where the polarization asymmetry is observed to be zero:

$$\frac{G_E^n}{G_M^n} = -\sqrt{\frac{\tau(1+\varepsilon)}{2\varepsilon}}\tan\delta. \tag{9}$$

By determining a so-called cross-ratio of the up-down asymmetry in the polarimeter for both values of the beam helicity the result becomes independent of both the target luminosity and the polarimeter efficiency. Corrections for charge-exchange reactions in the lead curtain were determined to be $\approx 3\%$ based on measurements with a liquid hydrogen target and detailed Monte Carlo simulations.

E93-026 ran in 2001 and used a deuterated ammonia (ND_3) target as an effective polarized neutron target to measure G_E^n at Q^2 of 0.5 and 1.0 GeV^2 [59]. The solid polarized target used the dynamical nuclear polarization technique [60, 61] to reach an in-beam polarization of $\sim$24% at electron beam intensities of up to 100 nA. Ammonia ND_3 granules, doped by radiation damage with a small concentration of free radicals, were immersed in liquid helium. Because the occupation of the magnetic substates in the radicals follow a Boltzmann distribution, the free electrons are polarized to more than 99% in a $\sim$5 T field generated by a pair of superconducting coils and at a $\sim$1 K temperature. A radiofrequency field is then applied to induce transitions to states with a preferred nuclear spin orientation. Because the relaxation time of the electrons is much shorter than that of the nuclei, polarized nuclei are accumulated. The electron beam had to be rastered uniformly in a 1 cm radius to minimize local heating and depolarization. A two-magnet chicane compensated for the deflection of the electron beam by the target field. The scattered electrons were detected in the HMS detector and the knocked-out neutrons in a 160$\times$160 cm^2 large scintillator detector, six planes thick, preceded by two planes of thin scintillators serving as veto detectors of charged particles, shielded by a 2.5 cm thick lead curtain from direct gamma rays originating in the target. The measured beam-target asymmetry A_{en}^V, with the target polarization vector in the scattering plane and perpendicular to the momentum transfer vector, can be directly related to the ratio G_E^n/G_M^n

$$A_{en}^V = -\frac{2\sqrt{\tau(\tau+1)}\tan(\theta_e/2)G_E^n/G_M^n}{(G_E^n/G_M^n)^2 + \tau/\epsilon}. \tag{10}$$

Corrections for charge-exchange reactions in the target material and in the lead curtain were determined by a Monte Carlo simulation.

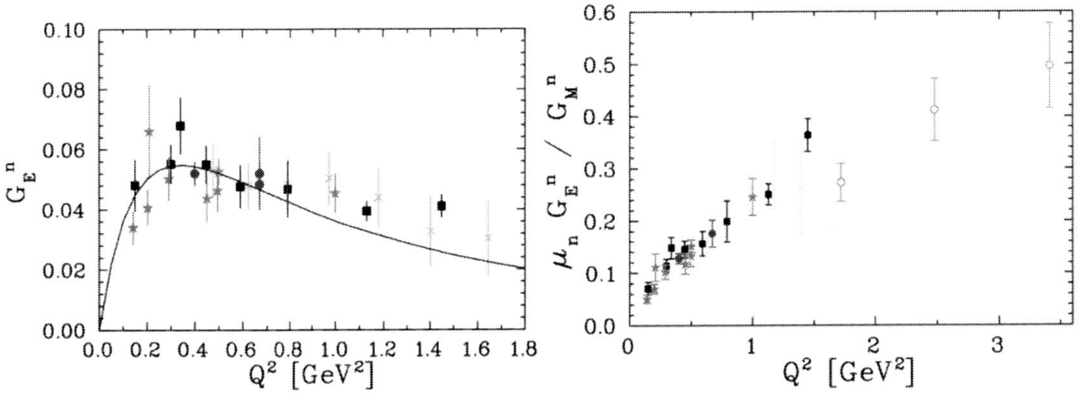

Figure 5. Current status of extractions of G_E^n. Squares are extractions from recoil polarization measurements on ^{2}H, stars (circles) are from cross section asymmetries from polarized ^{2}H (^{3}He) targets. The crosses are from a modern analysis of electron-deuteron elastic scattering [62], and the curve represents the Galster fit to G_E^n. Data sets are the same as shown in Ref. [3] with the addition of new results from BLAST [63] (stars) and the high-Q^2 results from E02-013 (open circles).

Figure 5 shows the combined results from all of the high-precision extractions of G_E^n. Results from recoil polarization measurements on ^{2}H and cross section asymmetry measurements from polarized ^{2}H and ^{3}He targets are in good agreement. The limited figure of merit of the polarized ^{2}H targets, due to restrictions in the polarization, the dilution factor, and the beam current, inhibited its use at higher Q^2 values.

Experiment E02-013 ran in Hall A in 2006 with a polarized ^{3}He target, a large acceptance neutron detector and a large acceptance electron spectrometer ("BigBite"). The target utilized hybrid optical pumping in which a mixture of rubidium and potassium is used to enhance the spin exchange, resulting in an in-beam polarization of $\sim 50\%$. The BigBite spectrometer consisted of a large dipole magnet with an angular acceptance of close to 100 msr and a momentum acceptance of 90%. Its detector package contained two planes of drift chambers and a lead-glass shower calorimeter. A special target holding field magnet was designed that also provided the required magnetic shielding of the target cell from the fringe field of BigBite. The neutron detector was the largest dedicated neutron detector ever built. It had an active area of 1.6×5 m^2, consisting of 250 scintillators stacked in 7 planes that were interspersed with a 2.5 cm thick iron conversion plane and preceded by 7.5 cm of lead and iron shielding and 2 veto planes made from 200 thin scintillator bars. The use of these three novel devices at a luminosity of $3 \cdot 10^{37}$ cm^{-2}s^{-1} made it possible to extend the G_E^n data set to $Q^2 = 3.4$ GeV^2, as shown in the right panel of Fig. 5. Charge-exchange corrections were determined by taking data, in addition to ^{3}He, on a number of targets with different N/Z ratios, such as hydrogen, deuterium and ^{12}C. Nuclear medium effects, including pion production, were determined through Glauber-type calculations.

5. Impact of the Jefferson Lab program

The dramatically improved data set that has become available in the last decade, has had a transformative effect on the study of nucleon form factors. While the issue of two-photon exchange corrections led to a brief period of uncertainty, it soon became clear that the surprising new results on G_E^p/G_M^p were correct and that our textbook picture of the proton form factors would have to be revised. The impact of these new experimental results was magnified by the parallel developments on the theory side, in particular several attempts to learn more about the internal substructure of the nucleon within the framework of generalized parton distributions.

The most dramatic new result was the fall-off of the ratio G_E/G_M for the proton. One question raised by this striking behavior was whether G_E^p may have a zero crossing at high Q^2. This possibility was considered rather exotic when the high-Q^2 polarization results were first available, as the idea that G_E^p and G_M^p both followed the dipole form and decreased monotonically towards zero had become common wisdom based on the earlier Rosenbluth measurements. However, there is nothing unusual about a zero crossing in G_E^p. In fact, Dombey stated in a 1969 review article [7] that "As $G_E = F_1 - \tau F_2$, it is *a priori* quite likely that G_E becomes negative for large values of $[Q^2]$". The recent GEp(III) experiment [12] shows a decrease in the fall of G_E^p/G_M^p with Q^2, suggesting that a zero crossing, if it occurs, is at higher Q^2 than suggested by the earlier measurements. Results from the new G_E^n experiment [64] suggest that G_E^n falls less rapidly, roughly following the dipole form for Q^2 from 1.5 to 3.4 GeV^2.

The expectation for high Q^2 was that the form factors would behave according to the leading term in perturbative QCD (pQCD). This led to the expectation that the ratio G_E/G_M would become independent of Q^2 at large Q^2 values (corresponding to $F_1/F_2 \propto Q^2$), as suggested by the older Rosenbluth results. The pQCD predictions were re-examined, and while the leading power behavior yields a constant ratio, there are terms that give a logarithmic Q^2 dependence that can be large even at the highest Q^2 values of existing data [65]. Figure 6 shows scaling for the proton and neutron, along with a selection of calculations.

Miller's calculation [66] is based on an extension of the cloudy bag model, in which three relativistically moving constituent quarks are surrounded by a pion cloud. Roberts and collaborators [67] solve a Poincare covariant Faddeev equation for dressed quarks in which correlations between those are expressed *via* diquarks. The other three calculations use different

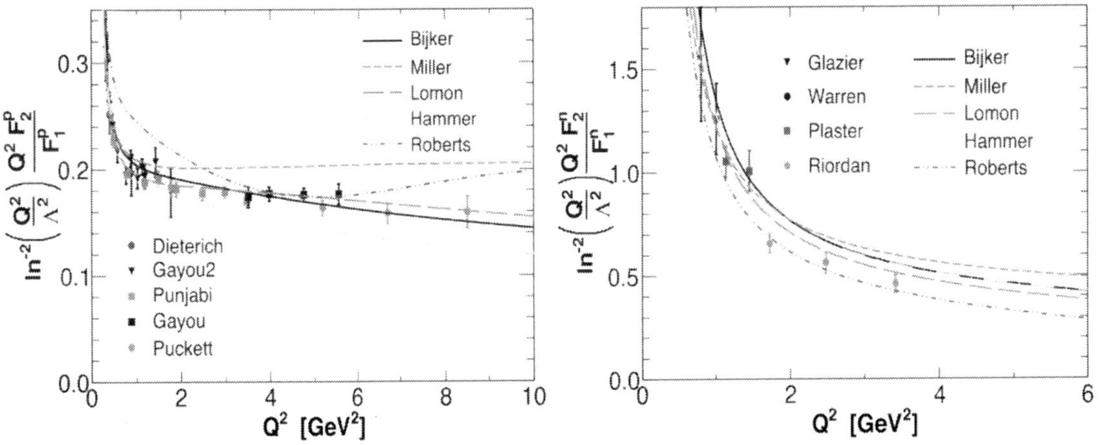

Figure 6. The F_2/F_1 ratio for the proton and neutron, after applying a logarithmic scaling correction from [65], with $\Lambda = 300$ MeV. The proton (left) is consistent with this logarithmic scaling above 1–2 GeV^2, while the preliminary neutron results, extending up to $Q^2 = 3.4$ GeV^2, do not yet show the same scaling behavior.

extensions of the Vector Meson Dominance (VMD) model where the scattering amplitude is expressed in a bare nucleon form factor, multiplied by the amplitude of the photon interaction with a vector meson. In Bijker and Iachello's model [68], the virtual photon is assumed to couple to the assumed intrinsic structure of the quarks and one of three vector mesons (ρ, ω and ϕ). By adding more parameters, such as the width of the ρ meson and the masses of heavier vector mesons, Lomon [69, 70] succeeded in describing all electromagnetic form factor data. Hammer and Meissner [71] included the isovector $\pi\pi$ channel through dispersion relations. Because the VMD models require a significant number of parameters to provide good fits to the data, they are not expected to have significant predictive power.

While the proton data are consistent with the modified scaling of Ref. [65], the neutron data do not show this scaling, and it has been suggested [3] that the nonperturbative mass scale required for the proton indicates that the perturbative prediction is not applicable in this Q^2 range. These logarithmic terms are connected to spin-flip contributions, which for nearly massless quarks must come from the orbital angular momentum of the quarks. Quark orbital angular momentum is an important feature in many of the nucleon models that show G_E^p/G_M^p decreasing with Q^2.

In the Born approximation, the form factors are Fourier transforms of the charge and magnetization distributions in the Breit (center-of-momentum) frame. The extraction of those distributions in the rest frame requires relativistic boost corrections, which scale as Q^2/m^2. At low Q^2, the mass of the constituent quarks has evolved to ~400 MeV, so the corrections are expected to be relatively small for very low Q^2 values. Kelly [72] used a simple model for these boost corrections in obtaining the results shown in Fig. 7 for the charge and magnetization distributions of the proton and the neutron. For the proton, the central magnetization density is 50% larger than the central charge density, as a result of its sharper drop-off. For the neutron, there is a positive central charge distribution with an extended negative tail, strongly supporting the picture that the neutron has a (p, π^-) component in its wave function.

In parallel with the improvements in the experimental techniques, new tools were also being developed to allow for the extraction of additional information about the structure of the nucleon. The development of the framework of GPDs led to new approaches that go beyond the traditional

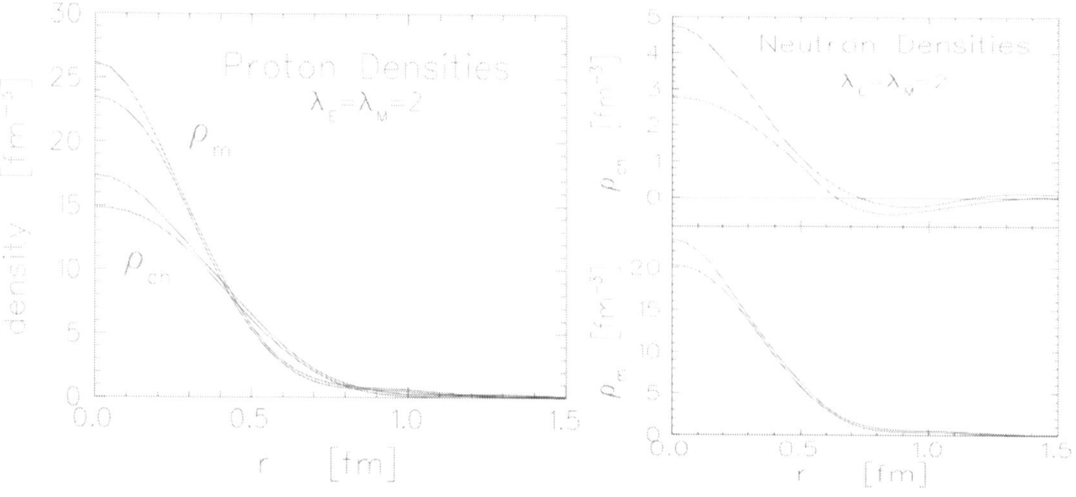

Figure 7. Charge and magnetization densities for the proton and neutron. Figures taken from Ref. [72].

one-dimensional pictures of the spatial charge distributions and look into correlations in the momentum, space, and spin structure of the quarks in the nucleon. Several groups looked at isolating the spatial distribution for low and high momentum quarks, or for the quarks with spins parallel, anti-parallel, or transverse to the nucleon spin [73, 74], as illustrated in Fig. 8. More recently, both the charge and magnetization densities, along with their uncertainties, have been extracted [75]. The impact of these data on constraining GPDs, and broader discussion of the interpretation of nucleons in terms of the GPDs, is included in a later article in this volume [76].

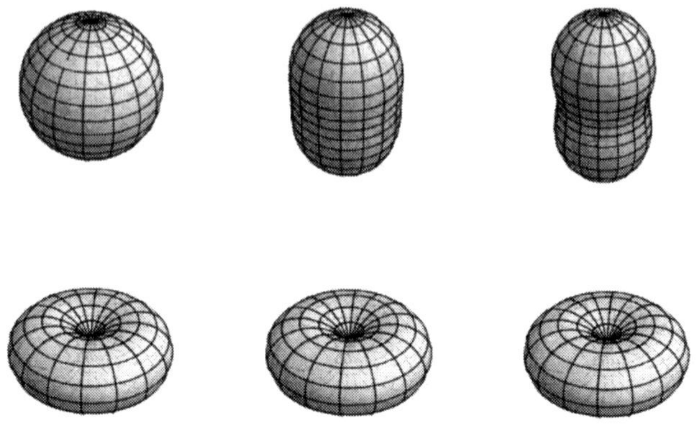

Figure 8. Visualizations of the spatial quark distributions in the proton [73]. The top (bottom) distributions are for quark spin parallel (anti-parallel) to the proton spin. Contours of constant density are shown for quark momenta 0, 1, and 2 GeV/c (from left to right).

The approaches built on GPDs provided more detailed information on nucleon substructure, but significant modeling is required to build GPDs from the constraints provided mainly by form factors and structure functions. Related studies were performed that led to the development of a model-independent procedure to extract information on the spatial distributions that depend only on the form factors as input. Miller [77] showed that in the infinite momentum frame (IMF) the transverse charge distribution as a function of the impact parameter is simply the two-dimensional Fourier transform of the Dirac form factor $F_1(Q^2)$. This provides a model-independent extraction of the transverse spatial distribution of charge for a nucleon in the IMF, which is more closely connected to the quark parton momentum distributions and GPDs. While the proton's transverse charge distribution was consistent with expectations, the neutron yielded a small, negative region of charge in the very center. This novel feature was not consistent with expectations based on simply treating the IMF distribution as the transverse spatial distribution in the rest frame.

The Drell-Yan-West relation links the x dependence of a parton distribution function $q(x)$ at large x to the Q^2 dependence of a form factor $F(Q^2)$:

$$q(x) \propto (1-x)^{(\nu-1)} \longleftrightarrow F(Q^2) \propto Q^{-\nu}. \tag{11}$$

The preliminary G_E^n data at large Q^2 show a slower fall-off with Q^2 than the G_E^p data, indicating a dominance of d quarks over u quarks at large x. This observation, first suggested by Kroll [78], is in agreement with global fits of parton distribution functions, such as CTEQ6M [79]. Using models of neutron GPDs, the correlation between the transverse spatial distributions of the nucleon in the IMF and the quark momenta was examined, and it was shown that this negative central charge in the neutron was the result of this strong d-quark dominance at high x [80]. In the IMF, the center of momentum of the nucleon is the longitudinal momentum weighted average of the transverse quark positions. In the limit where one quark has $x \to 1$, the position of that single quark provides the dominant contribution in defining the transverse center of the nucleon, and thus contributions from large x become localized near the center of the nucleon. Figure 9 shows the charge density of the neutron coming from different x regions. At low x, the density of up and down quarks is similar, and there is a net positive charge with a broad spatial distribution. At large x, the density of down quarks is more than twice that of up quarks, so the net charge is negative and the distribution becomes more localized, yielding the small negative core in the neutron.

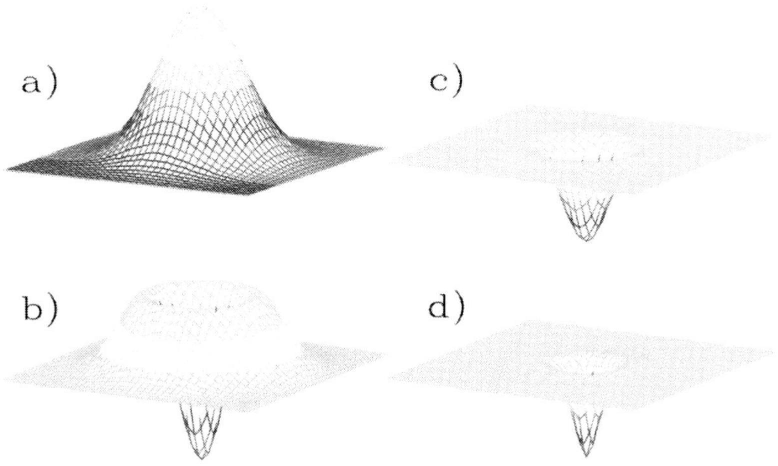

Figure 9. Contributions to the infinite momentum frame transverse charge density of the neutron coming from low (a) to high (d) momentum quarks [80].

Such visualizations of the nucleon charge distributions on the light front have been expanded to the case of nucleons polarized transversely to the light front [81]. Both the (proton and neutron) unpolarized distributions are shifted along the polarization axis due to the nucleon's large anomalous magnetic moment, which results in an induced electric dipole moment perpendicular to the polarization axis.

The new form factor data provide information on much more than just the spatial structure of the nucleon. The new high-Q^2 measurements, combined with the new, complete set of low-Q^2 form factor data, provide powerful constraints on models of nucleon structure. Figure 6 shows a small selection of calculations and it is clear that there is a significant scatter even between modern calculations. Prior to the Jefferson Lab program, only G_M^p was known with high precision over a large Q^2 range; the other form factor extractions were limited in both precision and kinematic coverage. Many theoretical approaches were used to model nucleon structure and with parameters adjusted to reproduce G_M^p, there were only limited constraints from the other form factors. The improved measurements of G_E^p and G_M^n provided real challenges for calculations that had, until this time, been minimally constrained. High-precision measurements of G_E^n have been more difficult to incorporate in evaluating models of nucleon structure. Pion cloud contributions are neglected in many calculations, but play an important role in the behavior of G_E^n at low Q^2. The newest G_E^n measurement extends the data to $Q^2 = 3.4$ GeV^2, providing a complete set of form factor extractions in the Q^2 region where pion cloud contributions should be small and the form factor is expected to be sensitive to the quark core of the neutron.

Early experiments on the proton focused on the high-Q^2 region, but precision measurements at low Q^2 also became an area of renewed interest. Had the up and down quark distributions been identical, the neutron would have been neutral everywhere yielding $G_E^n = 0$ at all Q^2 values. The non-zero values of G_E^n at low Q^2 demonstrate that there is a net positive charge in the core of the neutron and a negative charge distribution on the outside, as expected from virtual fluctuations of the neutron into a proton and a negative pion. An intriguing analysis [26] suggested the possibility that similar contributions may be present in all of the form factors. For the proton, the pion cloud is a small contribution on top of the quark core, and precision data are necessary to test models of pion contributions for all four form factors. New measurements [15, 27] at low Q^2, shown in Fig. 3, showed hints of structure in the ratio $\mu_p G_E^p/G_M^p$, motivating experiment E08-007, which will cover this Q^2 range with significantly higher precision.

Having a precise and complete data set at low Q^2 also provides the opportunity to study the contributions of the different quark flavors. Neglecting strange and heavier quarks, the proton and neutron form factors consist of contributions from the up and down quarks. Since many nucleon properties appear to have significant contributions from strangeness, neglecting strange quarks may not be well justified, and additional information is required to constrain the impact of strangeness on the nucleon form factors. Measurements of parity-violating electron scattering can provide information on the coupling of a Z-boson to the proton. This provides a complete set of observables that allows for a separation of the up, down, and strange quark contributions to the nucleon form factors, as discussed in detail in the following article [82]. Data on both proton and neutron are also needed to extract the isovector form factors (proton minus neutron). This is the combination that can be most reliably extracted from lattice QCD calculations, where taking the difference between proton and neutron removes the influence of disconnected diagrams which are difficult to calculate.

Finally, data at even lower Q^2 values, below 0.1 GeV^2, can also have an important impact. Extraction of the proton charge radius depends on having reliable measurements of the form factors at extremely low Q^2, and what we have learned about two-photon exchange corrections has led to an updated extraction of the proton charge radius [83, 84]. Ultra-high precision atomic physics measurements, *e.g.* hyperfine splitting in hydrogen and muonic hydrogen, are extremely sensitive to proton structure corrections, including terms directly calculated from the low-Q^2 proton form factors [81]. Future experiments (Table 1) will extend proton form factor extractions down below 0.02 GeV^2 with significantly improved precision over older Rosenbluth experiments.

6. Future plans and outlook

State of the art polarized electron beams, coupled with high figure-of-merit polarized targets and recoil polarimeters, have enabled a program of measurements at Jefferson Lab that has dramatically modified our picture of nucleon form factors. Individual experiments, in particular the surprising results of the proton electric form factor measurements, have led to a re-examination of long held pictures of nucleon structure. Taken together, these experiments have provided data of dramatically improved quality for most form factors over large kinematic ranges, and have led to a resurgence in efforts to evaluate nucleon models against a complete set of form factor measurements. With neutron measurements beginning to cover a significant portion of the Q^2 range for which precise proton data already exist, we can begin to extract model-independent information about the difference between the up and down quark distributions, and the high-Q^2 extractions of G_E^n show that there is a noticeable difference between the up and down quark spatial distributions in the neutron [64].

These new data, as well as the techniques that made these measurements possible, will continue to have impact on other experimental investigations. Precise knowledge of the electromagnetic form factors is necessary to probe the strange quark contributions to the nucleon using parity-violating lepton scattering [82]. It is also important input to high-precision nuclear structure measurements that utilize quasi-elastic scattering from the nucleus, and measurements of hyperfine splitting in hydrogen. In addition, the measurement of nucleon form factors utilizing polarization measurements allows for much more sensitive investigations into the effect of the nuclear medium on the nucleon form factors. The comparison of free to bound proton form factors, using polarization transfer in quasi-elastic scattering from ^{4}He [85, 86], yields smaller corrections than similar measurements that relied on Rosenbluth separations, although the impact of these corrections on the interpretation is a topic of great interest [87].

Soon, the last of the measurements from Table 1 will be completed as the 6 GeV program is brought to a close. The increased electron energies available after the 12 GeV upgrade, coupled with further improvements in the experimental equipment, will enable a dramatic extension of the program presented here, doubling or tripling the Q^2 range of most of these measurements. This will provide valuable constraints on generalized parton distributions, an important focus of the upgrade, at very high Q^2 values. In addition to the benefits gained by extending the Q^2 range for the individual form factors, this will also provide a complete set of form factor measurements at large Q^2, where the pion cloud contributions are expected to be small and the measurements can be directly compared to calculations of the quark core of the nucleon. This will make evaluation of nucleon models more reliable, as pion cloud contributions are typically difficult to include in a self-consistent fashion.

Acknowledgments

This work was supported by the Department of Energy, Office of Nuclear Physics, contract nos. DE-AC05-84ER40150 and DE-AC02-06CH11357 and by the National Science Foundation, grant PHY-0753777. The authors thank those of their colleagues who assisted in the preparation of this overview and apologies for omissions made necessary by the constraints of length and time.

References

[1] C. E. Hyde-Wright and K. de Jager, Ann. Rev. Nucl. Part. Sci. **54**, 217 (2004).
[2] C. F. Perdrisat, V. Punjabi and M. Vanderhaeghen, Prog. Part. Nucl. Phys. **59**, 694 (2007).
[3] J. Arrington, C. D. Roberts and J. M. Zanotti, J. Phys. G **34**, S23 (2007).
[4] S. Platchkov *et al.*, Nucl. Phys. **A510**, 740 (1990).
[5] A. I. Akhiezer, L. N. Rozentsweig and I. M. Shmuskevich, Sov. Phys. JETP **6**, 588 (1958).
[6] A. I. Akhiezer and M. P. Rekalo, Sov. Phys. Dokl. **13**, 572 (1968).
[7] N. Dombey, Rev. Mod. Phys. **41**, 236 (1969).
[8] R. G. Arnold, C. E. Carlson and F. Gross, Phys. Rev. C **23**, 363 (1981).
[9] V. Punjabi *et al.*, Phys. Rev. C **71**, 055202 (2005). Erratum-ibid. C **71**, 069902 (2005).
[10] M. K. Jones *et al.*, Phys. Rev. Lett. **84**, 1398 (2000).
[11] O. Gayou *et al.*, Phys. Rev. Lett. **88**, 092301 (2002).
[12] A. J. R. Puckett *et al.*, Phys. Rev. Lett. **104**, 242301 (2010).
[13] J. Arrington, Phys. Rev. C **69**, 022201(R) (2004).
[14] I. A. Qattan *et al.*, Phys. Rev. Lett. **94**, 142301 (2005).
[15] C. B. Crawford *et al.*, Phys. Rev. Lett. **98**, 052301 (2007).
[16] M. Paolone *et al.*, Phys.Rev.Lett. **105**, 072001 (2010).
[17] X. Zhan *et al.*, arXiv:nucl-ex/1102.0318.
[18] G. Ron *et al.*, to be submitted to Phys. Rev. C (2011).
[19] J. J. Kelly, Phys. Rev. C **70**, 068202 (2004).
[20] J. Arrington, W. Melnitchouk and J. A. Tjon, Phys. Rev. C **76**, 035205 (2007).
[21] J. Arrington, Phys. Rev. C **68**, 034325 (2003).
[22] M. E. Christy *et al.*, Phys. Rev. C **70**, 015206 (2004).
[23] A. Afanasev, I. Akushevich and N. Merenkov, Phys. Rev. D **64**, 113009 (2001).
[24] L. C. Maximon and J. A. Tjon, Phys. Rev. C **62**, 054320 (2000).
[25] P. A. M. Guichon and M. Vanderhaeghen, Phys. Rev. Lett. **91**, 142303 (2003).
[26] J. Friedrich and T. Walcher, Eur. Phys. J. **A17**, 607 (2003).
[27] G. Ron *et al.*, Phys. Rev. Lett. **99**, 202002 (2007).
[28] J. Bernauer *et al.*, Phys. Rev. Lett. **105**, 242001 (2010).
[29] P. G. Blunden, W. Melnitchouk and J. A. Tjon, Phys. Rev. Lett. **91**, 142304 (2003).
[30] J. Arrington, Phys. Rev. C **69**, 032201(R) (2004).
[31] J. Arrington, Phys. Rev. C **71**, 015202 (2005).
[32] M. Meziane *et al.*, arXiv:nucl-ex/1012.0339.
[33] J. Guttmann, N. Kivel, M. Meziane and M. Vanderhaeghen, arXiv:hep-ph/1012.0564.
[34] D. Borisyuk and A. Kobushkin, arXiv:hep-ph/1012.3746.
[35] C. E. Carlson and M. Vanderhaeghen, Ann. Rev. Nucl. Part. Sci. **57**, 171 (2007).
[36] P. G. Blunden, W. Melnitchouk and J. A. Tjon, Phys. Rev. C **72**, 034612 (2005).
[37] S. Kondratyuk and P. G. Blunden, Phys. Rev. C **75**, 038201 (2007).
[38] Y. C. Chen, A. Afanasev, S. J. Brodsky, C. E. Carlson and M. Vanderhaeghen, Phys. Rev. Lett. **93**, 122301 (2004).
[39] A. V. Afanasev, S. J. Brodsky, C. E. Carlson, Y.-C. Chen and M. Vanderhaeghen, Phys. Rev. D **72**, 013008 (2005).
[40] A. V. Afanasev, I. Akushevich and N. P. Merenkov, Phys. Rev. D **65**, 013006 (2002).
[41] Y. M. Bystritskiy, E. A. Kuraev and E. Tomasi-Gustafsson, Phys. Rev. C **75**, 015207 (2007).
[42] A. V. Afanasev, arXiv:hep-ph/0711.3065.
[43] L. W. Mo and Y.-S. Tsai, Rev. Mod. Phys. **41**, 205 (1969).
[44] V. Tvaskis *et al.*, Phys. Rev. C **73**, 025206 (2006).
[45] L. Pentchev, AIP Conf. Proc. **1056**, 357 (2008).
[46] J. Arrington *et al.*, nucl-ex/0408020.
[47] J. Arrington, AIP Conf. Proc. **1160**, 13 (2009).
[48] B. Anderson *et al.*, Phys. Rev. C **75**, 034003 (2007).
[49] H. Anklin *et al.*, Phys. Lett. B **428**, 248 (1998).
[50] G. Kubon *et al.*, Phys. Lett. B **524**, 26 (2002).
[51] J. Lachniet *et al.*, Phys. Rev. Lett. **102**, 192001 (2009).
[52] S. Jeschonnek and J. W. Van Orden, Phys. Rev. C **62**, 044613 (2000).
[53] W. Xu *et al.*, Phys. Rev. Lett. **85**, 2900 (2000).
[54] W. Xu *et al.*, Phys. Rev. C **67**, 012201 (2003).
[55] J. Alcorn *et al.*, Nucl. Instrum. Meth. A **522**, 294 (2004).
[56] J. Golak, G. Ziemer, H. Kamada, H. Witala and W. Gloeckle, Phys. Rev. C **63**, 034006 (2001).

[57] R. Madey *et al.*, Phys. Rev. Lett. **91**, 122002 (2003).
[58] B. Plaster *et al.*, Phys. Rev. C **73**, 025205 (2006).
[59] G. Warren *et al.*, Phys. Rev. Lett. **92**, 042301 (2004).
[60] D. G. Crabb and D. B. Day, Nucl. Instrum. Meth. A **356**, 9 (1995).
[61] T. D. Averett *et al.*, Nucl. Instrum. Meth. A **427**, 440 (1999).
[62] R. Schiavilla and I. Sick, Phys. Rev. C **64**, 041002(R) (2001).
[63] E. Geis *et al.*, Phys. Rev. Lett. **101**, 042501 (2008).
[64] S. Riordan *et al.*, Phys. Rev. Lett. **105**, 262302 (2010).
[65] A. V. Belitsky, X.-d. Ji and F. Yuan, Phys. Rev. Lett. **91**, 092003 (2003).
[66] G. A. Miller, Phys. Rev. C **66**, 032201 (2002).
[67] I. C. Cloet, G. Eichmann, B. El-Bennich, T. Klahn and C. D. Roberts, Few Body Syst. **46**, 1 (2009).
[68] R. Bijker and F. Iachello, Phys. Rev. C **69**, 068201 (2004).
[69] E. L. Lomon, Phys. Rev. C **66**, 045501 (2002).
[70] E. L. Lomon, nucl-th/0609020.
[71] H. W. Hammer and U.-G. Meissner, Eur. Phys. J. **A20**, 469 (2004).
[72] J. J. Kelly, Phys. Rev. C **66**, 065203 (2002).
[73] G. A. Miller, Phys. Rev. C **68**, 022201 (2003).
[74] A. V. Belitsky, X.-d. Ji and F. Yuan, Phys. Rev. D **69**, 074014 (2004).
[75] S. Venkat, J. Arrington, G. A. Miller and X. Zhan, Phys. Rev. C **83**, 015203 (2011).
[76] C. E. Hyde, M. Guidal and A. V. Radyushkin, *Deep Virtual Exclusive Processes and Generalized Parton Distributions*, this volume.
[77] G. A. Miller, Phys. Rev. Lett. **99**, 112001 (2007).
[78] P. Kroll, hep-ph/0710.2771.
[79] J. Pumplin *et al.*, JHEP **07**, 012 (2002).
[80] G. A. Miller and J. Arrington, Phys. Rev. C **78**, 032201 (2008).
[81] C. E. Carlson, V. Nazaryan and K. Griffioen, arXiv:0805.2603.
[82] K. Paschke, A. Thomas, R. Michaels and D. Armstrong, *Strange Vector Form-Factors from Parity-Violating Electron Scattering*, this volume.
[83] R. Rosenfelder, Phys. Lett. B **479**, 381 (2000).
[84] P. G. Blunden and I. Sick, Phys. Rev. C **72**, 057601 (2005).
[85] S. Strauch *et al.*, Phys. Rev. Lett. **91**, 052301 (2003).
[86] W. K. Brooks, S. Strauch and K. Tsushima, *Medium Modifications of Hadron Properties and Partonic Processes*, this volume.
[87] R. Schiavilla, O. Benhar, A. Kievsky, L. E. Marcucci and M. Viviani, Phys. Rev. Lett. **94**, 072303 (2005).

New Insights into the Structure of Matter: The First Decade of Science at Jefferson Lab IOP Publishing
Journal of Physics: Conference Series **299** (2011) 012003 doi:10.1088/1742-6596/299/1/012003

Strange Vector Form Factors from Parity-Violating Electron Scattering

Kent Paschke[1], **Anthony Thomas**[2], **Robert Michaels**[3], **and David Armstrong**[4]

[1] University of Virginia, Charlottesville, Virginia 22903, USA
[2] University of Adelaide, South Australia 5005, Australia
[3] Jefferson Lab, Newport News, Virginia 23606, USA
[4] College of William and Mary, Williamsburg, Virginia 23187, USA

Abstract. The simplest models might describe the nucleon as 3 light quarks, but this description would be incomplete without inclusion of the sea of glue and $q\bar{q}$ pairs which binds it. Early indications of a particularly large contribution from strange quarks in this sea to the spin and mass of the nucleon motivated an experimental program examining the role of these strange quarks in the nucleon vector form factors. The strangeness form factors can be extracted from the well-studied electromagnetic structure of the nucleon using parity-violation in electron-nuclear scattering to isolate the effect of the weak interaction. With high luminosity and polarization, and a very stable beam due to its superconducting RF cavities, CEBAF at Jefferson Lab is a precision instrument uniquely well suited to the challenge of measurements of the small parity-violating asymmetries. The techniques and results of the two major Jefferson Lab experimental efforts in parity-violation studies, HAPPEX and G0, as well as efforts to describe the strange form factors in QCD, will be reviewed.

1. Introduction

The achievement which clearly identified Quantum Electrodynamics (QED) as the correct theory of the electromagnetic interaction was the quantitative explanation of the measurement of the Lamb shift in hydrogen. For the strong force we have Quantum Chromodynamics (QCD), a local gauge theory built on color, which has been verified with considerable precision in the high energy, "asymptotically free" regime. However, in the nonperturbative regime, where QCD is truly strong, it is still being tested, as our capacity for calculation using lattice QCD grows. By analogy with the Lamb shift in QED, a fundamental test of QCD in the nonperturbative regime is its capacity to accurately explain vacuum polarization. Since there are no strange valence quarks in the proton, the strangeness form factors of the nucleon present the ideal testing ground.

Historically, the enormous interest in the strangeness form factors arose from two things. The discovery of the EMC spin crisis, with its very small fraction of the proton spin carried by quarks, was widely interpreted in terms of a large negative fraction of the proton spin carried by strange quarks – $\Delta s = -0.16 \pm 0.08$ [1]. At about the same time, analysis of the octet baryon masses in terms of SU(3) symmetry suggested values of the strangeness sigma commutator, $\sigma_s = \langle p|m_s\bar{s}s|p\rangle$, as large as 330 MeV [2]. Since $\sigma_s = 330$ MeV represents a contribution to the nucleon mass of more than one third from a single sea quark flavor, this was an astonishing result. Both of these indications suggested a far greater role for the strange quark in proton structure than had been anticipated before. Modern lattice estimates of the strange sigma commutator suggest a value almost an order of magnitude smaller [3]. In addition, it seems likely that SU(3) breaking leads to a considerably smaller value of Δs [4]. Nevertheless, in the late 1980's the older values were a powerful motivation for the experimental program.

In this environment, a number of authors estimated the strange quark charge radius and the strange magnetic moment of the proton in various models. Indeed, almost the entire arsenal of hadronic models have been brought to bear on the question of the strange form factors, including, but not limited to, vector meson dominance, the Skyrme model, kaon loops, the chiral quark model, dispersion relations, the NJL model, the quark-meson coupling model, the chiral bag model, heavy-baryon chiral perturbation theory, *etc.*; reviews of these are available elsewhere [5, 6]. Many reported large values. For example, Jaffe [7] found $\langle r_s^2 \rangle = 0.16 \pm 0.06$ fm^2 and $\mu_s = -0.31 \pm 0.09\mu_N$, while Jaffe and Manohar observed [1] that until "such time as these experiments are perfected, the prediction of $F_2^{(0)}$ remains an excellent goal for theorists who think they have understood the flavor structure of the proton."

This excited theoretical discussion inspired novel experimental suggestions as to how one might actually measure these elusive quantities and it became clear that parity-violating electron scattering was the tool of choice [8, 9]. Thus began a rigorous and demanding program lasting more than 20 years, which as we report, has experimentally defined the strange electric and magnetic form factors of the proton. This program would not have been possible without the long term support of funding agencies and major laboratories, including MIT Bates, Mainz and, of course, Jefferson Lab.

Ignoring the tremendous technological and experimental challenges, the principle of the measurement of the strange vector form factors is simple. There are three vector operators whose matrix elements must be determined, $\bar{q}\gamma^\mu q$ with $q = u$, d and s. Using charge symmetry, which is respected at better than 1% by the strong force, one obtains two constraints from the proton and neutron electromagnetic form factors. For the third constraint, the fact that the weak vector charges differ from the electromagnetic charges means that the measurement of parity-violating electron scattering from the proton is sufficient. Indeed, a measurement of the left-right asymmetry, $A_{\rm LR}$, in the scattering of longitudinally polarized electrons from an unpolarized proton target yields the strange form factors through

$$A_{\rm LR} = -\frac{G_F Q^2}{4\pi\alpha\sqrt{2}} \left\{ (1-4\sin^2\theta_W) - \frac{\epsilon G_E^p G_E^n + \tau G_M^p G_M^n}{\epsilon (G_E^p)^2 + \tau (G_M^p)^2} - \frac{\epsilon G_E^p G_E^s + \tau G_M^p G_M^s}{\epsilon (G_E^p)^2 + \tau (G_M^p)^2} \right.$$
$$\left. - \frac{(1-4\sin^2\theta_W)\epsilon' G_M^p G_A^p}{\epsilon (G_E^p)^2 + \tau (G_M^p)^2} \right\}. \qquad (1)$$

Here $\tau = Q^2/4M_p^2$, $\epsilon = \left(1 + 2(1+\tau)\tan^2\frac{\theta}{2}\right)^{-1}$, and $\epsilon' = \sqrt{\tau(1+\tau)(1-\epsilon^2)}$. $G_E^s(Q^2)$ and $G_M^s(Q^2)$ are the strange electric and strange magnetic form factors, respectively, and parameterize the strange quark contribution to the vector structure of the nucleon. Also appearing in this expression is the axial form factor G_A^p, which becomes significant at backward scattering angles where ϵ' is larger. To lowest order, this form factor is the same as is measured in charged-current neutrino scattering. However, for electron scattering, large radiative corrections are expected to modify G_A^p. Decomposing G_A^p into isoscalar $G_A^e(T=0)$, and $G_A^e(T=1)$ pieces, Zhu *et al.* [10] found, in a model-dependent analysis, that the radiative corrections for the isoscalar piece are small and under reasonable theoretical control, while the isovector part is large ($\sim 30\%$) and less-well constrained theoretically. This latter term is interesting in its own right, as it includes the effective parity-violating coupling of the photon to the nucleon, the so-called "anapole" term [11], and, as we will describe, the parity-violating electron scattering program described here can also shed some light on this topic.

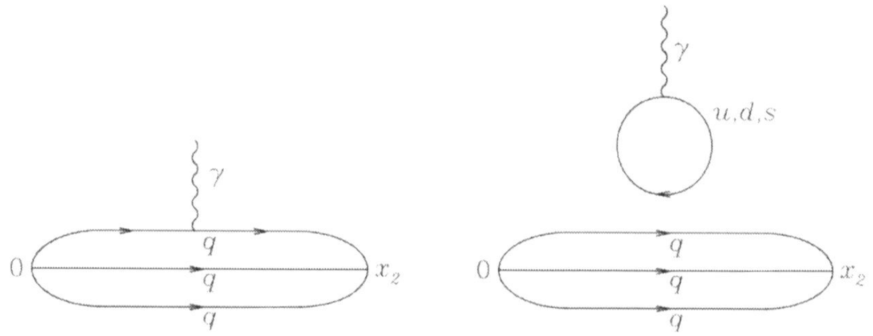

Figure 1. Diagrams illustrating the two topologically different insertions of the current within the framework of lattice QCD.

In this chapter, we first discuss the prediction of strange form factors from modern theory, focusing on lattice QCD. We then describe the experimental aspects of this program at Jefferson Lab. Then we present briefly the global analysis which yields the strange form factors, compare the extracted form factors with theory, and provide an outlook for the future.

2. Calculation of strange form factors within QCD

While the theoretical study of strange form factors within QCD cannot lay claim to two decades of effort, it is not so far off. Indeed, the initial studies of the electromagnetic form factors of the octet baryons began in the early 1990's, with work by Leinweber, Woloshyn and Draper [12]. It was these studies which gave direct information on the contribution to the baryon form factors from individual valence quarks that eventually led, some 14 years later, to remarkably accurate predictions [13, 14] of the strange quark form factors. The technique used in this work was indirect [15], so that for the strange magnetic moment it was necessary to combine lattice QCD calculations of the valence form factors, under the assumption of charge symmetry, with experimental information on the octet magnetic moments. This work required control of the extrapolation of the valence moments as a function of quark mass in both quenched and full QCD [16]. Only in the last two years (so again two decades from the first calculations) has it been possible to make a direct calculation in full QCD, with the Kentucky group reporting very accurate values of the strange magnetic moment of the proton in very good agreement with the earlier indirect calculations and with experiment.

2.1. Indirect method

As illustrated in Fig. 1, the three point function required to extract a magnetic moment in lattice QCD involves two topologically distinct processes. (Of course, in full QCD these diagrams incorporate an arbitrary number of gluons and quark loops.) The left-hand diagram illustrates the connected insertion of the current to one of the "valence" quarks of the baryon. In the right-hand diagram the external field couples to a quark loop. The latter process, where the loop involves an s-quark, is entirely responsible for the strange quark contribution to the nucleon form factor.

Under the assumption of charge symmetry [17], the magnetic moments of the octet baryons satisfy [15]:

$$\begin{aligned} p &= e_u\, u^p + e_d\, d^p + O_N \; ; \; n = e_d\, u^p + e_u\, d^p + O_N \, , \\ \Sigma^+ &= e_u\, u^\Sigma + e_s\, s^\Sigma + O_\Sigma \; ; \; \Sigma^- = e_d\, u^\Sigma + e_s\, s^\Sigma + O_\Sigma \, , \\ \Xi^0 &= e_s\, s^\Xi + e_u\, u^\Xi + O_\Xi \; ; \; \Xi^- = e_s\, s^\Xi + e_d\, u^\Xi + O_\Xi \, . \end{aligned} \tag{2}$$

Here, p and Ξ^- are the physical magnetic moments of the proton and Ξ^-, and similarly for the other baryons. The valence u-quark sector magnetic moment in the proton, corresponding to the left hand side of Fig. 1, is denoted u^p. Charge symmetry has been used to replace the d-quark contribution in the neutron by u^p, d in the Σ^- by u in the Σ^+ (u^Σ), and so on. The labels on quark magnetic moments allow for the environment sensitivity implicit in the three-point function. That is, the naive expectations of the constituent quark model, namely $u^p/u^\Sigma = u^n/u^\Xi = 1$, may not be satisfied. The total contribution from quark-loops, O_N, contains sea-quark-loop contributions (right hand side of Fig. 1) from u, d and s quarks. By definition

$$O_N = \frac{2}{3}\,{}^\ell G_M^u - \frac{1}{3}\,{}^\ell G_M^d - \frac{1}{3}\,{}^\ell G_M^s , \tag{3}$$

$$= \frac{{}^\ell G_M^s}{3}\left(\frac{1 - {}^\ell R_d^s}{{}^\ell R_d^s}\right), \tag{4}$$

where the ratio of s- and d-quark *loops*, ${}^\ell R_d^s \equiv {}^\ell G_M^s / {}^\ell G_M^d$, is expected to lie in the range (0,1). Note that, in deriving Eq. (4), we have used charge symmetry to set ${}^\ell G_M^u = {}^\ell G_M^d$. Since the chiral coefficients for the d and s loops in the right hand side of Fig. 1 are identical, the main difference comes from the mass of the K compared with that of the π.

With a little algebra, O_N, and hence $G_M^s (\equiv {}^\ell G_M^s)$, may be isolated from Eqs. (2) and (4):

$$G_M^s = \left(\frac{{}^\ell R_d^s}{1 - {}^\ell R_d^s}\right)\left[2p + n - \frac{u^p}{u^\Sigma}\left(\Sigma^+ - \Sigma^-\right)\right], \tag{5}$$

$$G_M^s = \left(\frac{{}^\ell R_d^s}{1 - {}^\ell R_d^s}\right)\left[p + 2n - \frac{u^n}{u^\Xi}\left(\Xi^0 - \Xi^-\right)\right]. \tag{6}$$

After incorporating the experimentally measured baryon moments [18] (in nuclear magnetons, μ_N), Eqs. (5) and (6) become:

$$G_M^s = \left(\frac{{}^\ell R_d^s}{1 - {}^\ell R_d^s}\right)\left[3.673 - \frac{u^p}{u^\Sigma}\left(3.618\right)\right], \tag{7}$$

$$G_M^s = \left(\frac{{}^\ell R_d^s}{1 - {}^\ell R_d^s}\right)\left[-1.033 - \frac{u^n}{u^\Xi}\left(-0.599\right)\right]. \tag{8}$$

We stress that these expressions for G_M^s are exact consequences of QCD, under the assumption of charge symmetry.

Equating (7) and (8) provides a linear relationship between u^p/u^Σ and u^n/u^Ξ which must be satisfied within QCD under the assumption of charge symmetry. Clearly this linear relationship is inconsistent with the assumption of universality of the valence quark moments (*i.e.*, the point (1.0,1.0) does not lie on the empirical line). In addition, it is important from the point of view of reducing systematic errors that the linear relation involves only *ratios* of moments, which can be calculated more accurately than absolute values.

The actual lattice simulations gave the individual valence quark moments as a function of pion mass in quenched QCD. These were then fit using finite range regulated effective field theory on a finite volume, with coefficients for quenched QCD, which were then replaced by full QCD chiral corrections on an infinite volume. From this chiral extrapolation procedure, the ratios of the valence (connected) u-quark contributions, u^p/u^Σ and u^n/u^Ξ were calculated. The final results

$$\frac{u^p}{u^\Sigma} = 1.092 \pm 0.030 \quad \text{and} \quad \frac{u^n}{u^\Xi} = 1.254 \pm 0.124 \tag{9}$$

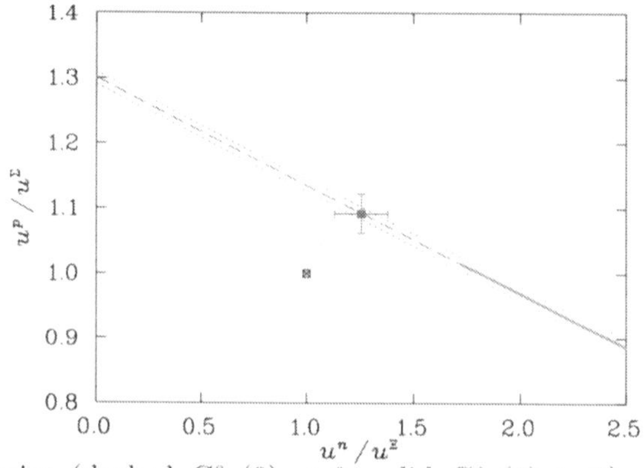

Figure 2. The constraint (dashed $G_M^s(0) < 0$, solid $G_M^s(0) > 0$) on the ratios u^p/u^Σ and u^n/u^Ξ implied by charge symmetry and experimental moments. Experimental uncertainties are indicated by the dotted bounds. The assumption of environment-independent quark moments is indicated by the crossed square. The final result (chiral-corrected extrapolation of lattice results) is illustrated by the filled square, which does indeed lie on the charge symmetry line.

are plotted in Fig. 2. The precision of these results follows from the use of correlated ratios of moments which act to reduce uncertainties associated with the lattice spacing, the regulator mass and statistical fluctuations [19]. This result leaves no doubt that G_M^s is negative. The fact that this point lies exactly on the constraint curve is highly nontrivial, and provides a robust check of the validity of the analysis techniques.

While G_M^s was certainly negative, it remained to set the magnitude. This required an estimate of the strange to light sea-quark loop contributions, ${}^\ell R_d^s$. Earlier estimates of ${}^\ell R_d^s$ had been based on the constituent quark model. A more reliable approach is to estimate the loops using the same successful model invoked to correct the quenched results to full QCD [16, 20]. Allowing the dipole mass parameter to vary between 0.6 and 1.0 GeV provides ${}^\ell R_d^s = G_M^s/G_M^d = 0.139 \pm 0.042$. A complete analysis of the errors associated with the determination of G_M^s using Eqs. (5), (6) and (9) is reported in Ref. [19]. The uncertainty is dominated by the statistical errors included in Eq. (9) and the uncertainty just noted for ${}^\ell R_d^s$. The final result for the strangeness magnetic moment of the nucleon is

$$G_M^s = -0.046 \pm 0.019 \, \mu_N \, . \tag{10}$$

The calculation of the strange quark charge radius involved the same ideas presented in Eqs. (7) and (8) but in this case the octet data are not available, so one could not use ratios of valence properties to reduce errors. Nevertheless the result, $\langle r^2 \rangle_s^p = 0.001 \pm 0.004 \pm 0.004$ fm^2, was remarkably precise. Finally, we note that this technique has also been used to extract the strange magnetic form factor at finite Q^2 [21].

2.2. *Direct method*

The direct calculation of the disconnected strange quark loop has proven extremely difficult in lattice QCD, which is why the indirect techniques described above were applied first. The first direct calculations [22, 23] were unable to extract robust signals for the strange form factors. However, just in the last two years, the Kentucky group has succeeded in beating down the noise to obtain a very convincing signal for the magnetic and electric strange quark form factors of the proton as a function of Q^2 [24]. For the present time, the results have been reported at somewhat heavy light quark masses – corresponding to pion masses around 500 MeV. Nevertheless the strange magnetic moment found by Doi *et al.*, namely $\mu_s = -(0.017 \pm 0.025 \pm 0.007)\mu_N$, is in excellent agreement with the values quoted above. Indeed, a simple estimate of the effect of moving to the physical light quark mass would make the agreement even better. The result for the strange electric charge radius was also in agreement with the value quoted above.

In conclusion, the theoretical status of the calculation of the strange electric and magnetic form factors seems to be very sound, with excellent agreement between the different techniques employed.

3. Experimental overview

As described above, the strange vector form factors are accessible through the precision measurement of the helicity-correlated cross section asymmetry in the elastic scattering of polarized electrons from an unpolarized target, $A_{\rm LR}$. This asymmetry is small, on the order of 1-100 parts per million (ppm) for the kinematics which are typically of interest, $Q^2 < 1$ GeV^2, and must be measured with a precision in the range of 10% or better. Experiments of this nature are optimized to the challenges of precision measurement of very small asymmetries, which require large high count rates and low noise to achieve statistical precision as well as a careful regard for potential systematic errors.

One common feature of all measurements of parity-violation in electron scattering is a rapid flipping of the electron beam helicity, allowing a differential measurement between opposing polarization states on a short timescale. The enabling technology for these measurements lies in the semiconductor photo-emission polarized electron source, which allows rapid reversal of the electron polarization while providing high luminosity, high polarization, and a high degree of uniformity between the two beam helicity states. Developments with the polarized source at Jefferson Lab critical to the success of this program are described elsewhere in this volume [25].

The following sections will describe a series of measurements in parity-violating in electron-nucleus scattering which took place over the first decade of the operation of Jefferson Lab: the HAPPEX measurements in Hall A and the G0 measurements in Hall C.

4. HAPPEX-I

The Hall A Proton Parity Experiment (HAPPEX) ran in 1998 and 1999 and pioneered parity-violation study at Jefferson Lab [26, 27, 28]. HAPPEX measured $A_{\rm LR}$ in elastic electron-proton scattering at the kinematics $\langle\theta_{\rm lab}\rangle = 12.3°$ and $\langle Q^2\rangle = 0.477$ (GeV/c)^2. The Hall A High resolution spectrometers (HRS) [29] were used to focus the scattered electrons onto a total absorption shower counter in the spectrometer focal plane (Fig. 3). The HRS cleanly isolated the elastic scattered electrons, suppressing background from inelastics or low-energy secondaries. The detector in each spectrometer was instrumented with a single photomultiplier (PMT) tube. Rather than counting individual pulses, the anode current from this PMT was simply integrated over the 33 millisecond helicity periods, and the result combined with that from a consecutive window of opposite electron-beam helicity, to form the asymmetry measurement "pair". With a detected rate of 2 MHz, the scattering rate asymmetry was measured at 15 Hz with a precision of 0.38%. The selected kinematics corresponded to the smallest angle and largest energy possible with the Hall A HRS spectrometers, which maximized the figure of merit for a first measurement.

HAPPEX-I stimulated improvements in both polarized source technology and polarimetry at Jefferson Lab. In the 1998 run the experiment used a $I = 100$ μA beam with $P = 38\%$ polarization produced from a bulk GaAs crystal, while in the 1999 run HAPPEX-I became the first experiment to use a strained GaAs photocathode to measure a parity-violating asymmetry in fixed-target electron scattering. This improved the figure-of-merit P^2I with P=70% and I=35 μA but also required several refinements of the techniques used to control systematic errors associated with the laser at the polarized source, in addition to the usual feedback control of the helicity-correlated charge asymmetry. The active layer of the strained photocathode was a thin (~100 nm) layer of GaAs grown on GaAsP. The mismatch between the two lattices produced a strain in the GaS that breaks an energy-level degeneracy, allowing selective photoexcitation of a specific polarization state and a theoretical maximum 100% electron polarization [30, 31]. During HAPPEX-I, a 70% polarization was achieved, and at the present time 85% is available due to further refinements in the technology. The strain, however, introduced an anisotropy in the quantum efficiency of the cathode, making it the dominant source of analyzing power in the system. A rotatable half-wave plate inserted downstream of the Pockels cell provided the ability to rotate the laser beam's polarization ellipse, which reduced the sensitivity to the analyzing power, as determined by scans of the helicity-correlated position differences and charge asymmetries versus the angle of this half-wave plate [28]. A half-wave plate upstream of the Pockels cell was aligned to invert the sign of incident linear polarization, which toggled the sign of the laser circular polarization, and therefore the electron beam helicity, relative to the Pockels cell voltage settings. The state of this waveplate was toggled every 24–48 hours. This method of slow helicity reversal provided a way to cancel out some sources of possible systematic errors, including false asymmetries from electronics pickup and certain helicity-correlated beam asymmetries.

During HAPPEX-I the Hall A Compton polarimeter [32] was commissioned and provided, for the first time, a continuous monitoring of the electron beam polarization with a run-to-run relative error of less than 2%, and a total error on absolute polarization averaged over the run of 3.3%. The dominant systematic uncertainty for this polarimeter lay in the corrections of the theoretical analyzing power for realistic detector performance. A second polarimeter was also used in Hall A, based on Møller scattering from a polarized ferromagnetic foil [29]. The Møller polarimetery results carried a total uncertainty of 3.2%, primarily due to uncertainty in the polarization of the electrons in the target foil. This polarimeter was invasive and could not be used as a continuous monitor, but results from measurements interspersed with data-taking were in good agreement with the results from the Compton polarimeter. These results agree also with measurements by a third polarimeter, based on Mott scattering in the 5 MeV region of the injector, which was quoted with a total error of about 7%.

HAPPEX-I yielded a very clean physics result and was published within a year after the experiment was completed. Due to the high-quality of the Jefferson Lab polarized source and superconducting accelerator, the systematic errors associated with the beam were found to be negligible compared to the statistical error. The accuracy of the result was sufficient to rule out several theoretical estimates of strangeness effects at moderately high Q^2 where it was thought the effects might have been large. The distribution of window-pair asymmetries are shown in Fig. 4; the distribution is Gaussian over 6 orders of magnitude, with a width consistent with counting statistics. Uncorrected asymmetries from the 1999 running period are shown in Fig. 5, averaged over the 24-48 hour periods between half-wave plate insertions. The measured asymmetry flips sign cleanly with the insertable half-wave plate. After all corrections, the HAPPEX-I physics asymmetry was found to be

$$A = -15.05 \pm 0.98 \pm 0.56 \text{ ppm}, \tag{11}$$

where the first error is statistical and the second error is systematic. This latter includes the errors in the beam polarization, background subtraction, helicity-correlated beam properties, and Q^2. Using this result, along with the calculated G_A^p [10] and the known values of the proton and neutron form factors, the experiment determined the linear combination of strange form factors $G_E^s + \eta G_M^s$ where $\eta = \tau G_M^{\gamma p}/\epsilon G_E^{\gamma p} = 0.392$ at the given kinematics. Thus,

$$G_E^s + \eta G_M^s = 0.014 \pm 0.020 \pm 0.010, \tag{12}$$

where the first error is the total experimental error (statistical and systematic errors added in quadrature) and the second error is the error due to the "ordinary" electromagnetic form factors, which was dominated by the uncertainty in G_M^n.

The clean and rapidly published result of HAPPEX-I, encouraged the development of further, more accurate and ambitious parity-violation measurements at Jefferson Lab.

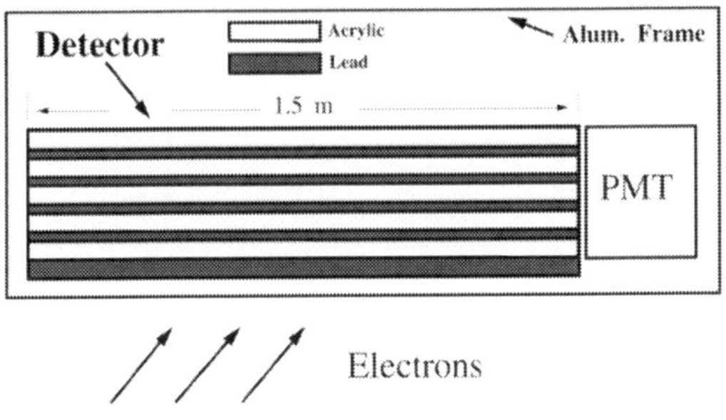

Figure 3. Schematic of the focal plane detector used by HAPPEX-I. The scattered electrons strike a lead-acrylic shower counter whose light is collected by a PMT and integrated over a helicity period.

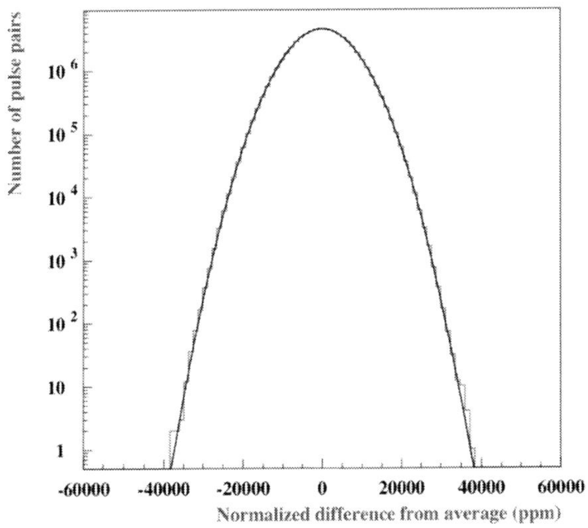

Figure 4. The window pair asymmetries in ppm for HAPPEX-I, normalized by the square root of beam intensity, with the mean value subtracted off. The curve is a Gaussian fit.

5. The second generation HAPPEX experiments

The second generation HAPPEX experiments ran in 2004 [33, 34] and 2005 [35] . Similar to the original HAPPEX experiment in technique, they measured forward-angle elastic scattering from the proton and from the ^{4}He nucleus using the HRS to cleanly separate the elastic events. A smaller scattering angle was achieved, about 6°, by using a superconducting septum dipole magnet to bend the small scattering angles into the 12.5° minimum acceptance for the HRS. The figure of merit was again optimized at the maximum momentum accepted by the HRS, with a $Q^2 \sim 0.1\,\mathrm{GeV}^2$.

Elastic scattering from an isoscaler 0^+ nucleus does not allow contributions from magnetic or axial-vector currents. At tree level, the parity-violating asymmetry for scattering from ^{4}He

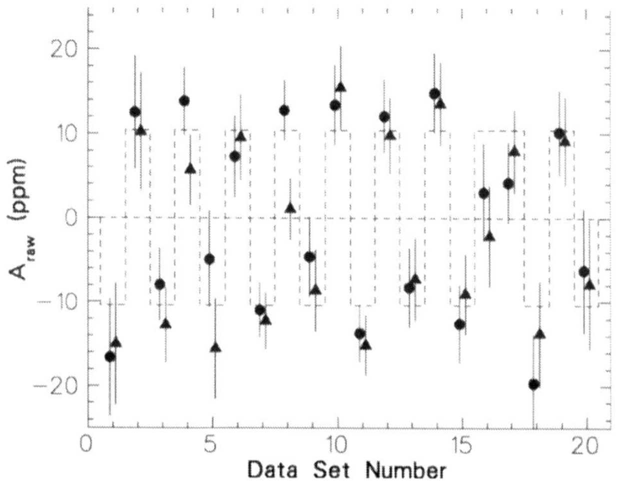

Figure 5. Raw asymmetries (ppm) for HAPPEX-I in the 1999 run. Each data set is ~1 day of running. The circles are for the left spectrometer, triangles for the right spectrometer. The step pattern represents the insertion and removal of the half-wave plate which reverses the sign.

can be written as

$$A_{PV}^{\rm He} = \frac{G_F Q^2}{4\pi\alpha\sqrt{2}}\left(4\sin^2\theta_W + \frac{2G_E^s}{G_E^p + G_E^n}\right). \quad (13)$$

With sufficient precision, measurement of this asymmetry isolates the strange electric form factor G_E^s alone. For ^{4}He, nuclear model corrections to the asymmetry due to isospin mixing [36, 37], D-state admixtures [38], and meson-exchange contributions [39] are all negligible at low Q^2. In addition to the independence from nuclear models, ^{4}He provided a particular advantage of a deeply-bound ground state, which separated the nearest inelastic level by about 20 MeV and allowed the HRS to easily isolate elastic scattering. The strong Q^2 dependence of the Mott cross section supports the figure-of-merit in the region $Q^2 \sim 0.1\,\mathrm{GeV}^2$. Measurements at significantly higher Q^2 would be problematic, both due to figure-of-merit and control of the nuclear corrections.

The measurements used 20 cm long cryogenic targets, with $\sim$ 3 GeV electron beam at currents from 35 to 55 μA. The targets were a novel design, using a transverse flow of cryogen across the beam axis to avoid localized heating in regions of low fluid flow velocity. A superconducting septum magnet captured scattered flux from around 6° into the HRS acceptance, where the elastically scattered electrons in each spectrometer were focused onto a brass/quartz Cerenkov counter. The signal from each counter was integrated over periods of about 33 milliseconds, resulting in an asymmetry measurements made at a rate 15 Hz with a precision of 540 ppm for the ^{1}H target and 1130 ppm for the ^{4}He target.

The high precision of these measurements required careful control of helicity-correlated beam asymmetries. Detailed studies of the laser optics of the polarized source led to improved characterization of the optical elements, as well as an improved algorithm for aligning the electro-optic Pockels cell which is used to create the fast helicity flip. Care was also taken to avoid magnifying any helicity-correlated orbit changes during beam acceleration and transport. As a result, the helicity-correlated position differences, averaged over the course of the run, was held to less than 2 nm for the Hydrogen measurement.

For the Helium measurement, electronics meant to drive additional helicity-correlated feedback systems created an unanticipated problem. An electrical control signal, indicating the

Journal of Physics: Conference Series **299** (2011) 012003 doi:10.1088/1742-6596/299/1/012003

beam polarization state to a hardware driver on the source laser table, interacted with devices on the electron beam line to steer the electron beam. This led to large position differences, as large as 600 nm at the target, which were in fact related only to the electrical signal which indicated helicity, and not to the actual helicity of the electron beam. This was proven by the observation that the differences existed even when the electron beam was left unpolarized by switching off high-voltage to the Pockels cell used to create the laser polarization state. Changing the polarity of this control signal input to the suspect system was also seen to toggle the sign of the steering effect. Careful checks revealed no electrical contamination of the data acquisition in the experimental hall: this effect was confined to the observed steering in the injector. Before the hardware driver was switched off to remove this effect, additional production data was first collected with the control signal flipped to provide a degree of cancellation for effects of this electrical leakage. Because this effect related to the control signal and not to laser or electron beam polarization, it was also seen to cancel well when averaged over the half-wave plate helicity reversal.

In addition to controlling the beam asymmetries, a correction for measured helicity-correlated beam asymmetries was applied. The sensitivities, that is, the changes in detected rate with changing beam parameters, were calibrated using periodic, deliberate beam modulation. Air core magnets in the arc leading to the hall moved the beam in a step-wise, sawtooth pattern, spanning the space of both position and angle at the target. An energy vernier was used to modulate the energy as well. Measuring the response of beam monitors and the detector to these individual modulations allowed for a measurement of the slopes which enforced the orthogonality of the motions, and avoided systematic misinterpretations of the sensitivities due to correlations in the beam motion or electronics noise. The results were consistent with expectations from simulation. The small corrections for helicity-correlated beam asymmetries, and the estimated systematic uncertainty in those corrections, were based on the sensitivities measured using this beam modulation.

The detector, which relied on Cerenkov light generated by high-energy leptons in an electromagnetic shower, preferentially collected light from the direction of primary tracks in the spectrometer. This directional sensitivity and low sensitivity to soft backgrounds, along with the detector location in the heavily shielded detector hut of the HRS, provided a very low background measurement. The largest background in the measurement, due to quasi-elastic scattering from the aluminum windows of the target, was measured to be $0.8 \pm 0.25\%$ of the signal from ^{1}H and $1.8 \pm 0.2\%$ for ^{4}He. This fraction was directly bounded by measurements of the production target with varying cryogenic gas densities. The uncertainty in the background rate and in the asymmetry of the background (due to possible inelastic contributions which would not be detectable at very low fractional rate) combined for a total uncertainty of 1.3% for ^{1}H and 0.6% for ^{4}He. Spectra of the high-resolution kinematic reconstruction in the HRS gave a range of $0.15 \pm 0.15\%$ for the quasielastic and inelastic fractional contributions to the ^{4}He signal. Additional backgrounds due to rescattering of inelastics in the spectrometer were estimated using dedicated calibration runs to measure rescattering probabilities. Rescattering from exposed polarized iron in the spectrometer was estimated to be negligible. The net correction for all backgrounds was $0.6 \pm 1.4\%$ for ^{1}H and $2.8 \pm 0.8\%$ for ^{4}He.

The measured asymmetry was corrected for the electron beam polarization using a Compton polarimeter with a precision of 1%. The dominant systematic uncertainty in the Compton polarimetry arises from the determination of the analyzing power with realistic effects from detector resolution and energy calibration. The key to high precision was the extremely precise beam tune, where the beam halo was so well contained at the Compton electron detector that the silicon strip detector could frequently make measurements with active strips only 5 mm from the primary electron beam. This allowed a precision energy calibration of the electron detector, through measurement of the location of the cross section kinematic limit (the "Compton edge") and the zero-crossing of the asymmetry. The energy calibration of the electron detector could then be transferred to the photon detector through spectra measured with an electron-photon coincidence trigger. This technique reduced the calibration errors relative to previous analyses using the Compton polarimeter [32], with the final uncertainty estimated at the level of 1%. Within the quoted uncertainties, the results were consistent (though somewhat more precise) with the measurements from the Møllerpolarimeter in the hall and the Mott polarimeter in the injector.

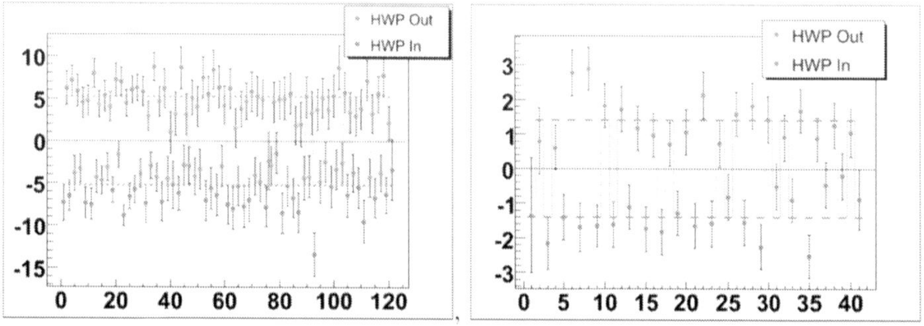

Figure 6. Corrected asymmetries (ppm) for HAPPEX-II in the 2005 run [35]. Each data point represents the measured asymmetry, averaged over the two spectrometers, for a period between half-wave plate reversals, for the ^{4}He (left) and ^{1}H (right) measurements. The step pattern represents the insertion and removal of the half-wave plate which reverses the sign of the asymmetry. Vertical axis labels are in units of parts per million, horizontal axis is the index for periods between half-wave plate changes.

In order to interpret the measured asymmetry, an accurate determination of the accepted kinematics is required. A thin water target with steel beam window foil was used to determine the angular alignment of the spectrometer by observing the momentum difference between hydrogen elastic scattering and elastic or inelastic scattering from heavier nuclei, as shown in Fig. 7. Results from this calibration method agreed well with an independent determination based on survey of a pinhole collimator, which could be inserted between the target and any magnetic elements of the spectrometer. As a result, uncertainties in the kinematics calibration were bounded to less than 1%, leading to a 1.7% systematic error of ^{1}H and 0.9% for ^{4}He. An additional correction of 2.1% $\pm$ 0.2% for the hydrogen measurement, accounting for the finite range of Q^2 accepted, was determined using Monte Carlo simulation. The ^{4}He measurement is exactly proportional to Q^2 at tree level and required no such correction.

The published results for the HAPPEX-II ^{1}H and ^{4}He measurements are given in Table 2. The extracted results for the strange form factors from the hydrogen measurements are necessarily dependent on the values of the electromagnetic form factors, which introduces an uncertainty into the theoretical expectation for the asymmetry in the absence of the strangeness contribution. This extraction is also very weakly dependent on the poorly know "anapole moment" radiative correction, which influences the axial term. The extraction shown here uses the Zhu *et al.* [10] estimate for this correction and corresponding uncertainty. Under the simple assumption that G_E^s is proportional to Q^2 and G_M^s is varying slowly enough in this region to be considered constant, the electric and magnetic strange vector form factors can be extracted from these results to be: $G_E^s = -0.005 \pm 0.019$ and $G_M^s = 0.18 \pm 0.27$, with a correlation coefficient of -0.87.

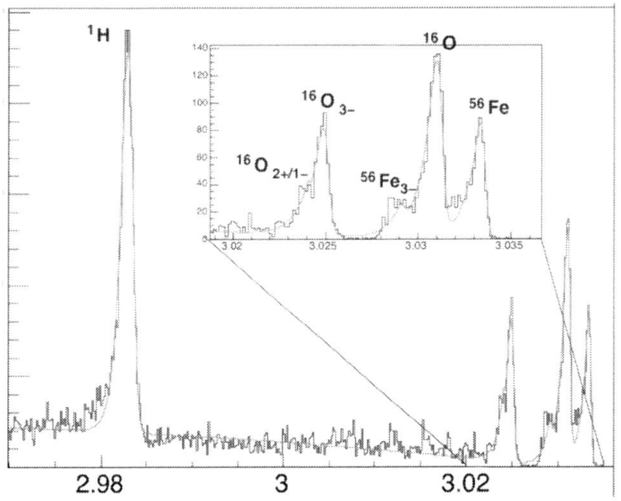

Figure 7. Reconstructed momentum spectrum for the HAPPEX-II watercell target (horizontal axis in units of GeV). Iron and oxygen states can be precisely resolved; measuring the momentum difference between these states and elastic proton scattering provides an accurate measurement of the scattering angle, with greatly reduced sensitivity to uncertainty in the primary beam momentum and spectrometer momentum calibration. The availability of each state from a single target reduces the uncertainties in target energy loss corrections. Using this technique, scattering angles were calibrated with an accuracy of 0.2 mrad in an entirely beam based measurement. Results agreed with survery reports on the experimental apparatus.

Correction (ppb)	Helium	Hydrogen
Beam Asymmetry	183 ± 50	10 ± 17
Target window bkg.	113 ± 32	7 ± 19
Helium QE bkg.	12 ± 20	–
Rescattering bkg.	20 ± 15	2 ± 4
Nonlinearity	0 ± 58	0 ± 15
Scale Factor	Helium	Hydrogen
Acceptance Scale Factor K	1.000 ± 0.001	0.979 ± 0.002
Q^2 Scale	1.000 ± 0.009	1.000 ± 0.017
Polarization P_b	0.844 ± 0.008	0.871 ± 0.009

Table 1. Corrections and systematic error summary for HAPPEX-II and HAPPEX-Helium [35].

6. The G0 experiment

The G0 experiment [40, 41] measured the forward proton asymmetries and backward asymmetries for both the proton and deuteron to provide, within a single apparatus, a complete set of measurements over a broad range of kinematics from which the electric and magnetic strangeness form factors, as well as the axial neutral weak current of the nucleon, could be extracted.

6.1. G0 forward

The forward-angle G0 experiment [40] ran in Hall C in 2003 and 2004, using a novel 8-coil, superconducting toroidal magnet to focus recoil protons from the elastic electron-proton

Journal of Physics: Conference Series **299** (2011) 012003 doi:10.1088/1742-6596/299/1/012003

dataset	Q^2 [GeV^2]	Result	
2004 ^{4}He	0.077	$A^{He}_{NVS} = +7.48$ ppm	$A^{He}_{PV} = +6.72 \pm 0.84 \pm 0.21$ppm
		$G^s_E = -0.038 \pm 0.042 \pm 0.010$	
2005 ^{4}He	0.091	$A^{He}_{NVS} = +6.37$ ppm	$A^{He}_{PV} = +6.40 \pm 0.23 \pm 0.12$ppm
		$G^s_E = 0.002 \pm 0.014 \pm 0.007$	
2004 ^{1}H	0.099	$A^{H}_{NVS} = -1.43 \pm 0.11$ ppm	$A^{H}_{PV} = -1.14 \pm 0.24 \pm 0.06$ppm
		$G^s_E + 0.08G^s_M = 0.030 \pm 0.025 \pm 0.006 \pm 0.012$	
2005 ^{1}H	0.109	$A^{H}_{NVS} = -1.66 \pm 0.05$ ppm	$A^{H}_{PV} = -1.58 \pm 0.12 \pm 0.04$ppm
		$G^s_E + 0.09G^s_M = 0.007 \pm 0.011 \pm 0.004 \pm 0.005$	

Table 2. Published results from the HAPPEX-II measurements on hydrogen and helium [35]. Error bars are listed in order for statistical and systematic uncertainties, and for form factor uncertainties where appropriate.

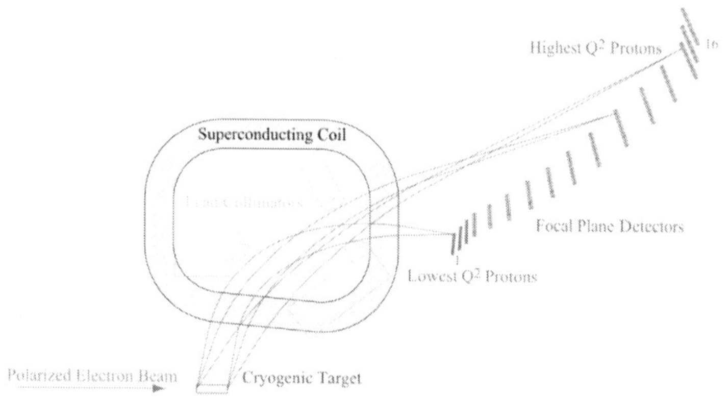

Figure 8. Schematic of the G0 spectrometer as used in the forward angle experiment.

scattering onto one of eight arrays of plastic scintillator detectors located outside the magnet. A 40 μA, 74% polarized beam from a strained GaAs polarized source was used, with 32 ns pulse timing, rather than the standard 2 ns, to allow for precise time-of-flight measurements which discriminated against various backgrounds. The delivery of high average beam current with this pulse timing required very high peak currents, and therefore provided significant challenges in accelerator physics due to the large space-charge effects in each bunch. The acceptance was about 0.9 steradians at recoil angles centered around 70°. With a beam energy of 3 GeV the acceptance corresponded to a Q^2 range $0.12 \leq Q^2 \leq 1.0\,\text{GeV}^2$. Each of the eight detector arrays consisted of 16 different detectors, the Focal Plane Detectors (FPD$_i$). The detectors were each a two-layer sandwich of scintillators, and each layer was read out by two phototubes. Each detector in the range distinguished a unique bin in Q^2, with the exceptions of the range $0.44 \leq Q^2 \leq 0.88\,\text{GeV}^2$, which all lay in FPD 15 and was segmented using time-of-flight, and FPD 16, which was located beyond the kinematic limit for elastic scattering, and was used for monitoring backgrounds. The Q^2 acceptance for each detector was calibrated to 1% precision using time-of-flight measurements between π^+ and elastic protons.

Rather than an integrating method, as adopted by the HAPPEX experiments, the G0

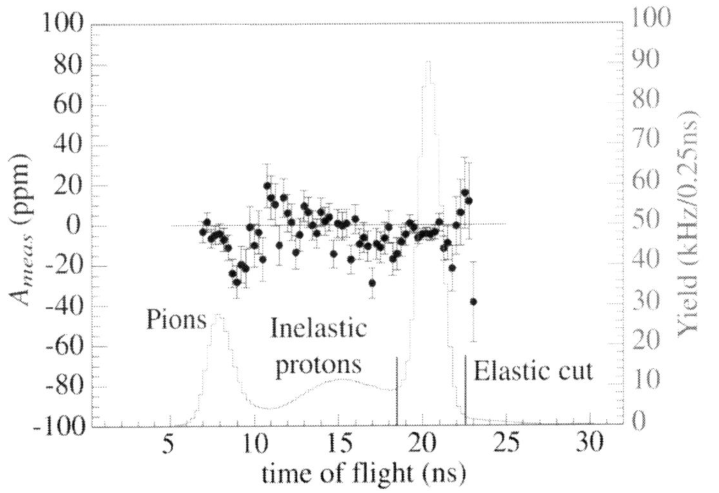

Figure 9. The time-of-flight distribution (histogram) for a typical detector (FPD 8) in the forward angle G0 experiment, along with the raw measured asymmetries (points with statistical errors).

experiment used a counting method to form the experimental asymmetries. Custom electronics allowed time-of-flight spectra to be accumulated for each helicity state. An asymmetry was then formed bin-by-bin in the spectra for each FPD. The pattern of fast helicity reversal was, unlike the HAPPEX experiments, formed not in pairs, but in quartets, where each quartet was either the sequence of helicity states LRRL or RLLR (L and R referring to left and right handed helicity, respectively). The quartet structure was chosen to exactly cancel the effect of any linear drifts in, for example, detector threshold or gain. The measured asymmetry was formed for each helicity quartet, *e.g.* $A_{\rm meas} = \frac{Y_L - Y_R - Y_R + Y_L}{Y_L + Y_R + Y_R + Y_L}$ for LRRL, where Y_L and Y_R are the measured yield in each helicity window. This measured asymmetry can be expressed as $A_{\rm meas} = (1-f)A_{\rm el} + fA_{\rm b}$ where $A_{\rm el}$ is the raw elastic asymmetry, $A_{\rm b}$ the background asymmetry and f the background fraction.

The elastic protons were identified by time-of-flight relative to the electron beam bunch; this allowed the rejection of protons from inelastic scattering, and faster particles such as the π^+, as shown in Fig. 9. Background yields and asymmetries were measured concurrently and used to correct the elastic asymmetries. In the region of the elastic peak, the background is essentially composed only of inelastic protons. Background was subtracted in each detector using a simultaneous fit of the time-of-flight yield and asymmetry spectra near the elastic cut. For this fit, the yield is modeled with a Gaussian elastic peak on a polynomial background, while the elastic asymmetry is fit to a constant and the background to a quadratic function. For higher number detectors the background asymmetry becomes positive, due to a small number of protons from the weak decays of hyperons, that scatter in the magnet, leading to corrections which were 20-110% of the final result. The uncertainties in these corrections were estimated from the variation in results using a variety of different models for the background asymmetries and yields. These corrections lead to a significant correlation between the estimated systematic uncertainties for data points in the range $Q^2 > 0.3\,\mathrm{GeV}^2$.

The beam polarization was measured using a Møller polarimeter [42] to a precision of 1%. Radiative corrections of 1–3% were performed using simulation. Small corrections for electronics deadtime were made on the basis of the observed dependence of the yield on the beam current.

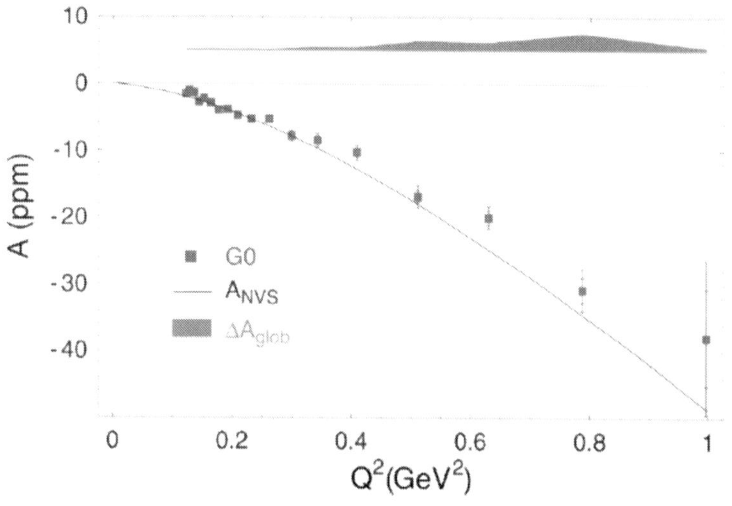

Figure 10. Results from the G0 forward angle measurements of A_{LR} in elastic electron-proton scattering [40]. Data points are shown with statistical and point-to-point systematic error bars combined, while the shaded band represents the magnitude of systematic errors which are highly correlated between the data points. The curve A_{NVS} represents the expected value of A_{LR} in the absence of strange contributions to the vector form factors.

An additional correction was required for an asymmetric "leakage" beam. This was a problem unique to the non-standard time structure: some fraction of the total beam to Hall C was leakage beam, primarily tails of the beam intended for the other two halls, which arrived with the standard 499 MHz time structure and was recorded by the 1497 MHz beam current monitors along with the expected G0 beam. Being off-time, elastic scattered protons from this leakage beam did not contribute to the signal in the elastic cut, leading to a mis-match between the measured beam current asymmetry and the beam current which contributed to the accepted signal. Although a small fraction of the total beam, there was a strong helicity dependence for this leakage beam. The correction for this leakage beam asymmetry was 0.7 ± 0.1 ppm, and nearly uniform over all detectors.

Results for the final measured asymmetries after all corrections are shown as a function of Q^2 in Fig. 10. For each Q^2 point, the combination of strangeness form factors $G_E^s + \eta G_M^s$ as a function of momentum transfer was obtained, as discussed in Section 7 and shown in Fig. 12, where $\eta(Q^2) = \tau G_M^p / \epsilon G_E^p$. The results were in excellent agreement with the earlier HAPPEX-I measurement [28], and displayed a noticeable bias toward a positive strangeness contribution at higher Q^2, which prompted the next round of high-precision experiments at Jefferson Lab and at Mainz.

6.2. G0 backward

The G0 backward-angle phase experiment was conducted in Hall C at Jefferson Lab in 2006 and 2007. The G0 superconducting toroidal spectrometer was turned 180° around with respect to the beam direction compared to the forward-angle configuration, in order to detect the electrons scattered at an angle of about 110° from 20 cm liquid hydrogen and deuterium targets. Polarized electron beams with currents up to 60 μA and energies of 359 and 684 MeV were generated with a strained GaAs polarized source. In this phase, the standard Jefferson Lab time structure was used for the beam, as time-of-flight was not useful for separating the scattered electrons

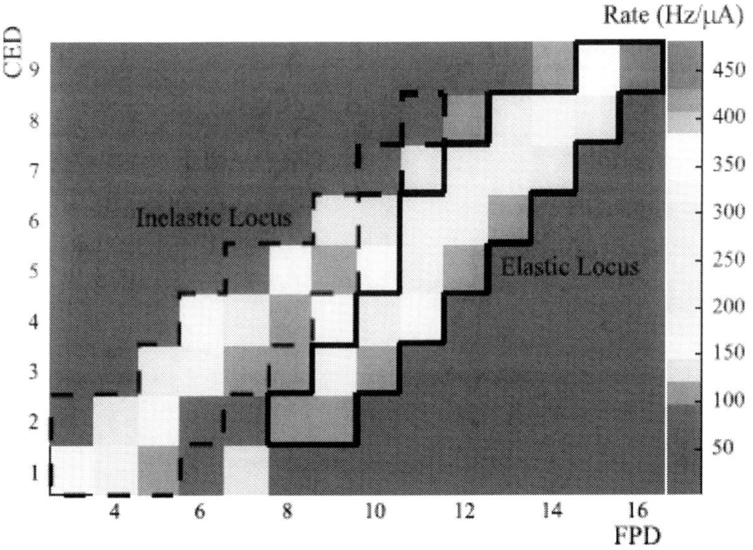

Figure 11. The distribution of measured yields from the backward-angle phase of G0, as a function of FPD and CED, for hydrogen data taken at a beam energy of 687 MeV. (Note: FPDs 1 and 2 are not used,) Electrons from elastic (inelastic) scattering are in the upper right (lower left).

from backgrounds. The quartet helicity reversal pattern was adopted as for the forward-angle experiment. The average beam polarization was $85.8 \pm 2.1(1.4)\%$ at the lower (higher) incident energy. In this phase, the focal-plane detector array (FPD_i) that was used in the forward angle measurement was augmented with a second array of scintillators, near the exit of the magnet (CED_i). Coincidences between these two scintillator arrays allowed electrons from elastic and inelastic scattering to be distinguished, as shown in Figure 11.

Additionally, in each octant an aerogel Cerenkov detector was added in order to separate the electrons from the copious background from pions. The Cerenkov detectors had a pion rejection factor $\geq 85\%$ and an electron efficiency of about 85%. The backgrounds in the region of the elastic locus (see Fig. 11) amounted to 10–15% of the signal. In the elastic locus, the aluminum target windows was the dominant background with misidentified π^- and electrons from π^0 conversion also contributing. The aluminum fraction was measured using runs with cold gaseous hydrogen in the target and the pion contamination was determined from dedicated time-of-flight runs and pulse-shape analysis. Additionally, an acceptance study was performed by sweeping the field of the toroid $\pm 40\%$ of the nominal setting and comparing the measured yield in each coincidence cell with Monte Carlo simulation using GEANT. The aluminum asymmetry was taken to be the same as that of the deuteron (both effectively quasi-elastic scattering only) with an additional uncertainty of 5% for nuclear effects. The asymmetry for the pion production background was small (< 1 ppm) and was measured concurrently with that for the electrons. The background corrections are small because the background asymmetries generally have values close to those of the elastic asymmetry, or, otherwise, the fraction is small.

All asymmetries were corrected for measured rate-dependent effects. For elastic scattering, dead-time corrections generally dominated those from accidentals and amounted to $\sim 15\%$ of the yield based on the measured beam current dependence, and led to an uncertainty of about

Table 3. G0 backward-angle: corrections to the measured asymmetry and the final physics asymmetries. Rate and "Other" corrections are additive; electromagnetic radiative corrections are multiplicative. "Other" corrections include those for helicity-correlated beam parameters, the transverse component of beam polarization, and two-boson exchange. Not listed is the common multiplicative correction for the beam polarization, 1.16 ± 0.02. The uncertainties for the asymmetries are statistical, point-to-point systematic and global systematic, respectively.

Target	Q^2 (GeV^2)	Rate (ppm)	Other (ppm)	EM Radiative	A_{phys} (ppm)
H	0.221	-0.31 ± 0.08	0.22 ± 0.08	1.037 ± 0.002	$-11.25 \pm 0.86 \pm 0.27 \pm 0.43$
D	0.221	-0.58 ± 0.21	0.06 ± 0.10	1.032 ± 0.004	$-16.93 \pm 0.81 \pm 0.41 \pm 0.21$
H	0.628	-1.28 ± 0.18	0.29 ± 0.11	1.037 ± 0.002	$-45.9 \pm 2.4 \pm 0.8 \pm 1.0$
D	0.628	-7.0 ± 1.8	0.34 ± 0.21	1.034 ± 0.004	$-55.5 \pm 3.3 \pm 2.0 \pm 0.7$

0.5 ppm in the asymmetries. In the high-energy deuteron measurement, accidentals from pion signals in the scintillators in coincidence with random signals from the Cerenkov dominated the correction. In this case, the correction to the asymmetry was 7.0 ± 1.8 ppm. Helicity-correlated intensity changes were corrected with active feedback to about 0.3 ppm. Corrections to the measured asymmetry for residual helicity-correlated beam current, position, angle and energy variations of at most 0.2 ± 0.07 ppm were applied *via* linear regression. Electromagnetic radiative corrections of $(3 - -3.5) \pm 0.3\%$ and small two boson exchange effects (1%) were also applied to the asymmetries. Table 3 shows the corrections to the raw elastic asymmetry, A_{el}, as well as the final asymmetries A_{phys} and their statistical and systematic uncertainties. The extraction of strange form factors from these data is discussed in the next section.

7. Summary and outlook

The HAPPEX and G0 experiments at Jefferson Lab have provided precision measurements of the weak form factors of the nucleon over a range of $Q^2 < 1\,\mathrm{GeV}^2$. The global data set on forward-angle scattering from the proton, which is shown in Fig. 12, also includes measurements from the PVA4 collaboration at the Mainz Microtron [43, 44]. For each data point, the combination of strangeness form factors $G_E^s + \eta G_M^s$ as a function of momentum transfer was obtained, where $\eta(Q^2) = \tau G_M^p / \epsilon G_E^p$. This combination represents the net contribution of the strange form factors to A_{LR} in each measurement. Taken as a whole, this data set systematically appears to suggest a small but positive strange form factor combination, however, it does not establish a clear, statistically significant signal. At low Q^2, the highest precision measurements constrain contributions to be very near zero. At higher $Q^2 > 0.5\,\mathrm{GeV}^2$, there remains a suggestion of a possibly measurable contribution.

Figure 13 shows published data obtained at $Q^2 \sim 0.1\,\mathrm{GeV}^2$. In addition to forward-angle data from Jefferson Lab and PVA4 [43, 44] collaborations, this figure displays results from the HAPPEX ^{4}He measurement and backward-angle measurements from SAMPLE at MIT-Bates [45]. Each measurement appears as a band, representing the central value and 1σ error bar for the specific combination of G_E^s and G_M^s. The G0 result plotted here is an average of 3 published data points from the forward-angle data set, covering $Q^2 = 0.12 - 0.14\,\mathrm{GeV}^2$. For each of these measurements the anapole radiative correction to the axial form factor is included with the uncertainty as prescribed by theoretical expectation [10]. These data can be fit to extract $G_E^s = -0.006 \pm 0.016$ and $G_M^s = 0.33 \pm 0.21$, with a correlation coefficient of -0.83 [46]. Also displayed, as the small, filled ellipse, is the result of the Adelaide indirect lattice calculation described in Section 2.1.

In the absence of reliable theoretical guidance on the Q^2 variation of the strange form factors,

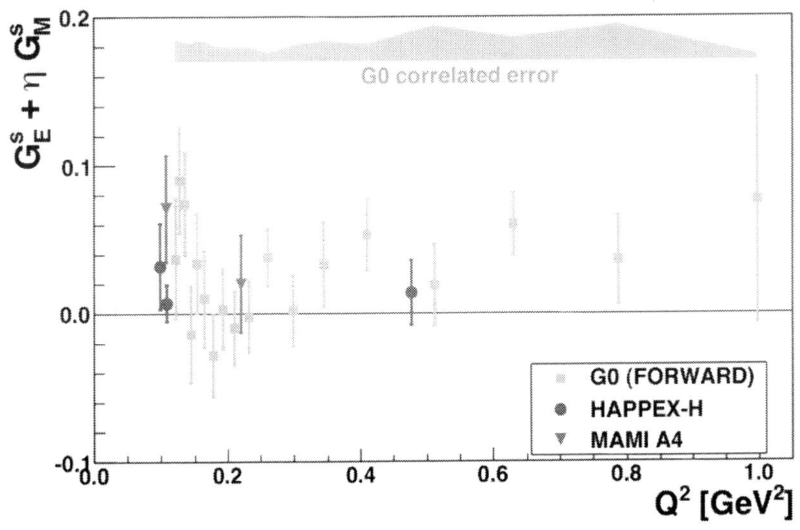

Figure 12. Results from world data measuring A_{PV} from forward-angle scattering from the proton, plotted as the net strange quark contribution $G_E^s + \eta G_M^s$. Here $\eta = \tau G_E^p/\epsilon G_M^p$, and is approximately equal to Q^2 over this plot. In addition to HAPPEX and G0, published results from the PVA4 experiment at MAMI at $Q^2 = 0.1$ and $0.23\,\mathrm{GeV}^2$ are shown [43, 44].

fits to the global data segt have used a first-order power-expansion in Q^2 to fit all data up to about $Q^2 < 0.3\,\mathrm{GeV}^2$. The result of such a fit [47] is shown as the solid elliptical contour (68% confidence level) in Fig. 13. This fit does not include the backward-angle results at $Q^2 \sim 0.22\,\mathrm{GeV}^2$ from G0 and A4, and applies no theoretical constraint on the axial form factor contribution which includes the anapole moment correction. These results are statistically consistent with fits of the data near $Q^2 = 0.1\,\mathrm{GeV}^2$ which incorporate theoretical limits on the anapole correction, as shown above. However, the inclusion of data at higher Q^2 tends to draw the central value of the fit to zero, implying either a strong Q^2 dependence to the strange form factors or that the suggested deviation at $Q^2 \sim 0.15\,\mathrm{GeV}^2$ is a statistical fluctuation.

Figure 14 shows the strange form factors G_E^s, G_M^s, and the isovector axial form factor $G_A^e(T = 1)$, extracted from the combination of the G0 backward angle and forward-angle asymmetries at $Q^2 = 0.221$ and $0.628\,\mathrm{GeV}^2$ [40, 41]. These results utilize a simple interpolation of the G0 forward angle measurements to the exact Q^2 of the backward angle data. The Kelly [48] parameterization of the nucleon electromagnetic form factors, $G_{E,M}^{p,n}$ was adopted for these determinations in order to be consistent with the deuteron model used [49]. Figure 14 also shows the extraction of G_E^s, G_M^s, $Q^2 = 0.1$ GeV^2 described above [46]. Note that the PVA4 points shown [50], in contrast to the G0 results, are not based on measurements on deuterium and therefore assume a value for $G_A^e(T = 1)$ determined by theoretical expectation Zhu *et al.* [10] shown in Fig. 14c), and a dipole form factor with a mass parameter of 1.032 GeV.

Taken as a whole, all these data clearly suggest that the strange vector form factors are a small contribution to the charge and magnetization distributions of the proton, consistent with the recent direct and indirect lattice QCD determinations, discussed above. There remains a suggestion in the experimental data that a significant contribution may be measurable at $Q^2 \sim 0.6\,\mathrm{GeV}^2$. This suggestion will be tested by the HAPPEX-3 experiment [51], which ran in 2009, measuring at a $Q^2 \sim 0.62\,\mathrm{GeV}^2$ and using techniques developed during the preceding HAPPEX experiments. Benefiting from the high luminosity and polarization available at

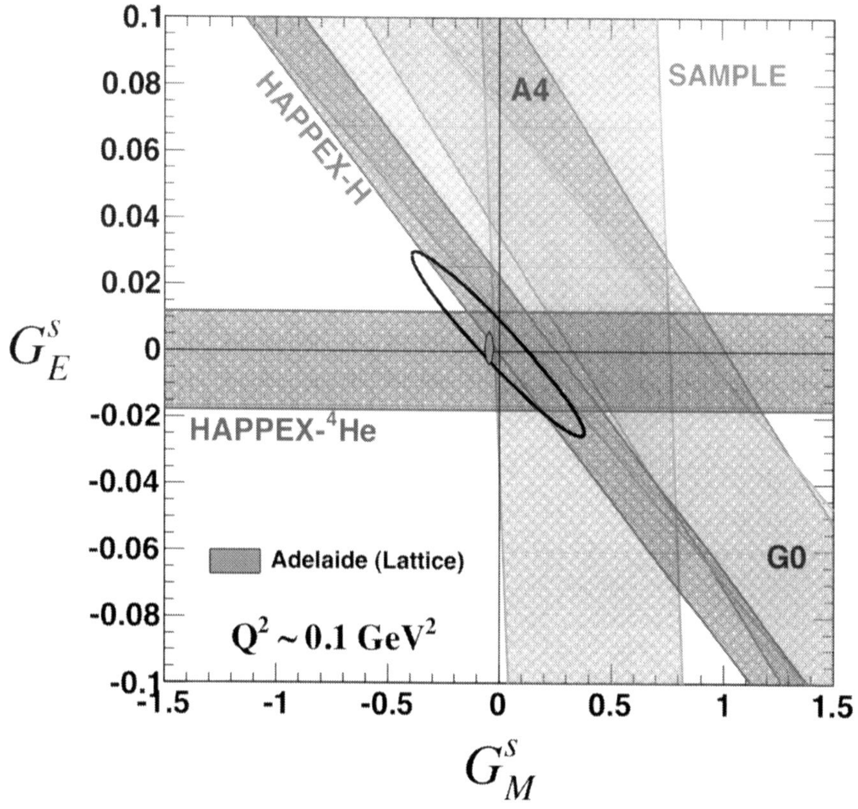

Figure 13. Measurements of G_E^s adn G_M^s in the region of $Q^2 = 0.1\,\mathrm{GeV}^2$. Each band represents the 1σ error bar for a unique combination of the strange form factors. Results of a recent QCD calculation from the Adelaide [13, 14, 21] is shown as a filled ellipse. As described in the text, the 68% confidence level contour of a global fit [47] incorporating data beyond what is shown in this plot is shown as a solid elliptical contour.

Jefferson Lab, along with improved alignment and polarimetry techniques, HAPPEX-3 will measure $G_E^s + 0.52G_M^s$ with an error bar ± 0.010 (stat) $\pm$ 0.005 (syst) $\pm$ 0.010 (ff), where the latter uncertainty arises from imperfect knowledge of the electromagnetic form factors. This result will greatly improve the precision of the G_E^s, G_M^s extraction at this Q^2, providing an opportunity to establish a clear, statistically significant non-zero signal.

This program of measurement was originally motivated by theoretical suggestions of potentially large contributions of strange quarks to the nucleon vector form factors. Over the kinematic range that has been examined, contributions have been found to be small or consistent with zero, with the most precise measurements having now pushed near to the bounds of unique interpretability. One significant limit is the precision in the knowledge of electromagnetic form factors. Presently, uncertainties in the neutron form factors G_E^n and G_M^n contribute an uncertainty to the extraction of the strange form factors of about half of the statistical error bar of the most precise A_{LR} measurements. There are important radiative corrections as well. As discussed above, the anapole moment correction in the axial term is difficult to bound with a precision approaching that of published theoretical estimates [10]. The precision of extraction of

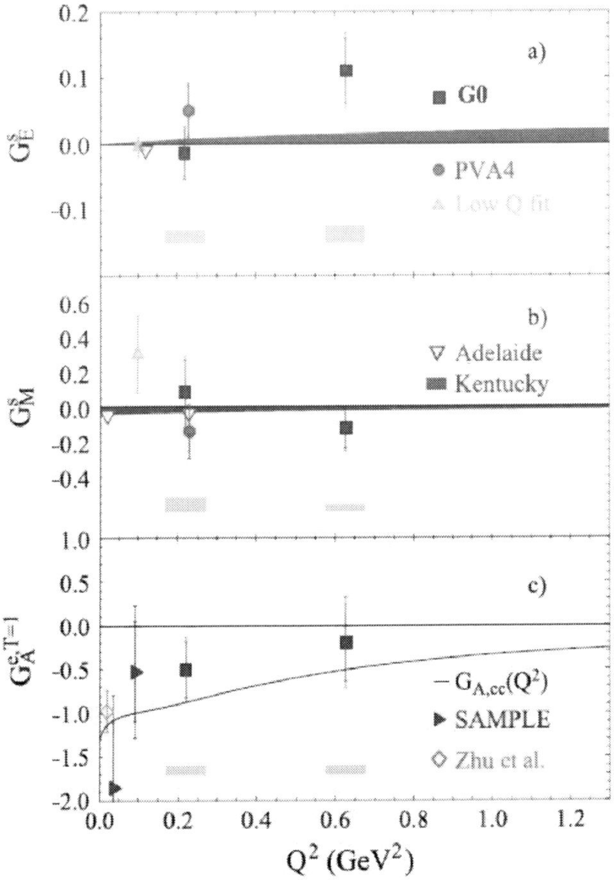

Figure 14. The form factors a) G_E^s, b) G_M^s, and c) $G_A^e(T=1)$ determined by combination of the G0 experiment forward and backward-angle measurements. Error bars show statistical and statistical plus point-to-point systematic uncertainties (added in quadrature); shaded bars below the corresponding points show global systematic uncertainties (for the G0 points). For G_E^s and G_M^s, the extraction from [46] as well as the results of the PVA4 experiment [50] are shown. Recent QCD calculations from groups at Adelaide [13, 14, 21] and Kentucky [24] are also shown; for the former the uncertainties are smaller than the symbols. For $G_A^e(T=1)$, results from the SAMPLE experiment [45] are shown together with the calculation of Zhu *et al.* [10].

G_E^s and G_M^s is dependent on the reliability of calculation of this correction. Similarly, γZ and $\gamma\gamma$ box diagrams present a potentially important correction to the interpretation of measurements of A_{LR}. Recent theoretical results [52, 53, 54, 55, 56] indicate that these uncertainties are sufficiently controlled for the current program, but significant advances beyond the next round of measurements would likely require further investigation into these corrections.

The extraction of the strange vector form factors also depends on the assumption of charge symmetry in the nucleon form factors. While this was once thought to provide a negligible uncertainty [17], recent work in has shown that charge symmetry violation in the nucleon may contribute to the form factor at a level comparable to the statistical error of the most precise current measurements [37, 57]. For this reason, future measurements might best be interpreted more generally, as probing all mechanisms for charge symmetry breaking in the nucleon and not

Journal of Physics: Conference Series **299** (2011) 012003 doi:10.1088/1742-6596/299/1/012003

primarily the strange quark content.

As described later in this volume [25], the studies of low-Q^2 elastic proton electroweak scattering from the strange quark program can be combined to provide a useful constraint on parameters of the Standard Model. The Qweak experiment [58] at Jefferson Lab will take advantage of this in a future measurement of A_{LR} from proton scattering at very low Q^2, in a sensitive new search for physics beyond the Standard Model. Qweak aims for a precision which is more than an order of magnitude beyond any previous Jefferson Lab measurement. This underlines the fact that the HAPPEX and G0 experiments at Jefferson Lab have been successful not only in the measurement of strange form factors, but have forged a path for new measurements of parity-violation in electron scattering. These significant advances in production and control of polarized beam, electron beam polarimetry, high luminosity targets, low noise integration techniques, and electron beam monitoring have enabled a new generation of parity violation experiments. In this way, the legacy of the strange quark program extends to Qweak; to a very challenging measurement of the neutron radius in ^{208}Pb (PREX) [59] which is so crucial in models of neutron-rich nuclei; to a measurement of parity-violation in deep inelastic scattering [60] which will improve precision on the poorly measured electroweak axial quark charges (C_{2q}); and to the measurements (described in [25]) of electroweak couplings and valance parton distributions planned after the upcoming energy upgrade at Jefferson Lab [61, 62].

References

[1] R. L. Jaffe and A. Manohar, Nucl. Phys. **B337**, 509 (1990).
[2] D. B. Kaplan and A. Manohar, Nucl. Phys. **B310**, 527 (1988).
[3] R. D. Young, A. W. Thomas, Phys. Rev. D **81**, 014503 (2010).
[4] S. D. Bass and A. W. Thomas, Phys. Lett. B **684**, 216 (2010).
[5] D. H. Beck and B. R. Holstein, Int. J. Mod. Phys. E **10**, 1 (2001).
[6] K. S. Kumar and P. A. Souder, Prog. Part. Nucl. Phys. **45**, S333 (2000).
[7] R. L. Jaffe, Phys. Lett. B **229**, 275 (1989).
[8] R. D. McKeown, Phys. Lett. B **219**, 140 (1989).
[9] D. H. Beck, Phys. Rev. D **39**, 3248 (1989).
[10] S. L. Zhu, S. J. Puglia, B. R. Holstein and M. J. Ramsey-Musolf, Phys. Rev. D **62**, 033008 (2000).
[11] Ya. B. Zeldovich *et .*, Zh. Eksp. Teor. Fiz. **33**, 1531 (1957) [Sov. Phys. JETP **6**, 1184 (1958)].
[12] D. B. Leinweber, R. M. Woloshyn and T. Draper, Phys. Rev. D **43**, 1659 (1991).
[13] D. B. Leinweber, S. Boinepalli, I. C. Cloet *et al.*, Phys. Rev. Lett. **94**, 212001 (2005).
[14] D. B. Leinweber, S. Boinepalli, A. W. Thomas *et al.*, Phys. Rev. Lett. **97**, 022001 (2006).
[15] D. B. Leinweber and A. W. Thomas, Phys. Rev. D **62**, 074505 (2000).
[16] R. D. Young, D. B. Leinweber, A. W. Thomas *et al.*, Phys. Rev. D **66**, 094507 (2002).
[17] G. A. Miller, B. M. Nefkens and I. Slaus, Phys. Rept. **194**, 1 (1990).
[18] Particle Data Group, Phys. Rev. D **66**, 010001 (2002).
[19] D. B. Leinweber, S. Boinepalli, A. W. Thomas *et al.*, Eur. Phys. J. A **24S2**, 79 (2005).
[20] R. D. Young, D. B. Leinweber and A. W. Thomas, Phys. Rev. D **71**, 014001 (2005).
[21] P. Wang, D. B. Leinweber, A. W. Thomas *et al.*, Phys. Rev. C **79**, 065202 (2009).
[22] S. J. Dong, K. F. Liu and A. G. Williams, Phys. Rev. D **58**, 074504 (1998).
[23] R. Lewis, W. Wilcox, R. M. Woloshyn, Phys. Rev. D **67** 013003 (2003).
[24] T. Doi, M. Deka, S. J. Dong *et al.*, Phys. Rev. D **80**, 094503 (2009).
[25] J. Roche, W. T. H. van Oers and R. D. Young, "Searches for Physics Beyond the Standard Model", this volume.
[26] K. A. Aniol *et al.*, Phys. Rev. Lett. **82**, 1096 (1999).
[27] K. A. Aniol *et al.*, Phys. Lett. B **509**, 211 (2001).
[28] K. A. Aniol *et.al.*, Phys. Rev. C **59**, 065501 (2004).
[29] J. Alcorn *et al.*, Nucl. Instrum. Meth. A **522**, 294 (2004).
[30] T. Nakanishi *et al.*, Phys. Lett. A **158**, 345 (1991).
[31] T. Maruyama *et al.*, Phys. Rev. Lett. **66**, 2376 (1991).
[32] M. Baylac *et al.*, Phys. Lett. B **539**, 8 (2002).
[33] K. A. Aniol *et al.*, Phys. Rev. Lett. **96**, 022003 (2006).
[34] K. A. Aniol *et al.*, Phys. Lett. B **635**, 275 (2006).
[35] A. Acha *et al.*, Phys. Rev. Lett. **98**, 032301 (2007).

Journal of Physics: Conference Series **299** (2011) 012003 doi:10.1088/1742-6596/299/1/012003

[36] S. Ramavataram, E. Hadjimichael and T. W. Donnelly, Phys. Rev. C **50**, 1175 (1994).
[37] M. Viviani *et al.*, Phys. Rev. Lett. **99**, 112002 (2007).
[38] M. J. Musolf and T. W. Donnelly, Phys. Lett. B **318**, 263 (1993).
[39] M. J. Musolf, R. Schiavilla and T. .W. Donnelly, Phys. Rev. C **50**, 2173 (1994).
[40] D. S. Armstrong *et al.*, Phys. Rev. Lett. **95**, 092001 (2005).
[41] D. S. Armstrong *et al.*, Phys. Lett. **104**, 012001 (2010).
[42] M. Hauger *et al.*, Nucl. Instrum. Meth. A **462**, 282 (2001).
[43] F. E. Maas *et al.*, Phys. Rev. Lett. **93**, 022002 (2004).
[44] F. E. Maas *et al.*, Phys. Rev. Lett. **94**, 152001 (2005).
[45] E. J. Beise, M. L. Pitt and D. T. Spayde, Prog. Part. Nucl. Phys. **54**, 289 (2005).
[46] J. Liu, R. D. McKeown and M. J. Ramsey-Musolf, Phys Rev. C **76**, 025202 (2007).
[47] R. D. Young, J. Roche, R. D. Carlini and A. W. Thomas, Phys. Rev. Lett. **97**, 102002 (2006).
[48] J. J. Kelly, Phys. Rev. C **70**, 068202 (2004).
[49] R. Schiavilla, J. Carlson and M. Paris, Phys. Rev. C **70**, 044007 (2004).
[50] S. Baunack *et al.*, Phys. Rev. Lett. **102**, 151803 (2009).
[51] "A Measurement of Nucleon Strange Form Factors at High Q^2", K. A. Aniol *et al*, Jefferson Lab proposal PR-05-009 (2005).
[52] J. A. Tjon and W. Melnitchouk, Phys. Rev. Lett. **100**, 082003 (2008).
[53] M. Gorchtein and C. J. Horowitz, Phys. Rev. Lett. **102**, 091806 (2009).
[54] J. A. Tjon, P. G. Blunden and W. Melnitchouk, Phys. Rev. C **79**, 055201 (2009).
[55] H. Q. Zhou, C. W. Kao, S. N. Yang *et al.*, Phys. Rev. C **81**, 035208 (2010).
[56] A. Sibirtsev, P. G. Blunden, W. Melnitchouk *et al.*, Phys. Rev. D **82**, 013011 (2010).
[57] B. Kubis and R. Lewis, Phys. Rev. C **74**, 015204 (2006).
[58] "The Qweak Experiment: A Search for New Physics at the TeV Scale Via a Measurement of the Proton's Weak Charge", D. Amstrong *et al.*, Jefferson Lab proposal PR-02-020 (2002).
[59] C. J. Horowitz, S. J. Pollock, P. A. Souder and R. Michaels, Phys. Rev. C **63**, 025501 (2001).
[60] "$\vec{e}-{}^2$H Parity Violating Deep Inelastic Scattering at CEBAF 6 GeV", X. Zheng *et al*, Jefferson Lab proposal PR-05-007 (2005).
[61] "An Ultra-precise Weak Mixing Angle Measurement in Møller Scattering", K. Kumar *et al.*, Jefferson Lab proposal PR12-09-005 (2009).
[62] "Precision Measurement of Parity-violation in Deep Inelastic Scattering Over a Broad Kinematic Range", P. Souder *et al.*, Jefferson Lab proposal PR12-09-012 (2009).

Journal of Physics: Conference Series **299** (2011) 012004 doi:10.1088/1742-6596/299/1/012004

Unpolarized Structure Functions at Jefferson Lab

M. E. Christy[1] **and W. Melnitchouk**[2]

[1] Hampton University, Hampton, Virginia 23668, USA
[2] Jefferson Lab, Newport News, Virginia 23606, USA

Abstract. Over the past decade measurements of unpolarized structure functions with unprecedented precision have significantly advanced our knowledge of nucleon structure. These have for the first time allowed quantitative tests of the phenomenon of quark-hadron duality, and provided a deeper understanding of the transition from hadron to quark degrees of freedom in inclusive scattering. Dedicated Rosenbluth-separation experiments have yielded high-precision transverse and longitudinal structure functions in regions previously unexplored, and new techniques have enabled the first glimpses of the structure of the free neutron, without contamination from nuclear effects.

1. Introduction

Throughout the modern history of nuclear and particle physics, measurements of structure functions in high-energy lepton-nucleon scattering have played a pivotal role. The demonstration of structure function scaling in early deep-inelastic scattering (DIS) experiments at SLAC in the late 1960's [1] established the reality of quarks as elementary constituents of protons and neutrons — a feat recognized by the award of the 1990 Nobel Prize in Physics to Friedman, Kendall and Taylor. This paved the way to the development of quantum chromodynamics (QCD) as the theory of strong nuclear interactions, and its subsequent confirmation several years later through the discovery of logarithmic scaling violations in structure functions [2].

The interpretation of structure functions in terms of quark and gluon (or parton) momentum distributions resulted in the emergence of a remarkably simple and intuitive picture of the nucleon [3], allowing a vast amount of scattering data to be described in terms of a few universal functions — the parton distribution functions (PDFs). At leading order (LO) in α_s, the F_2 structure function of the proton, for example, could be simply represented as a charge squared-weighted sum of PDFs,

$$F_2 = \sum_q e_q^2\, x(q + \bar{q}) = \frac{4}{9}x(u + \bar{u}) + \frac{1}{9}x(d + \bar{d}) + \frac{1}{9}x(s + \bar{s}) + \cdots, \tag{1}$$

where q and $\bar{q}$ are the quark and antiquark momentum distribution functions, usually expressed as functions of the momentum fraction x of the nucleon carried by the parton, at a scale given by the momentum transfer squared Q^2.

Over the ensuing decades concerted experimental DIS programs at SLAC, CERN, DESY and Fermilab have provided a detailed mapping of the PDFs over a large range of kinematics, with Q^2 and x spanning several orders of magnitude. To manage the ever increasing number of data sets, from not only inclusive DIS but also other high energy processes such as Drell-Yan, W-boson and jet production in hadronic collisions at Fermilab, sophisticated global fitting efforts were developed [4, 5, 6, 7] that now include perturbative corrections calculated to next-to-leading order (or higher) in the strong coupling constant α_s. With the increasing energies available at facilities such as the DESY ep collider HERA (and in future the Large Hadron Collider at CERN), PDF studies turned their attention primarily to exploring the region of very small x (down to $x \sim 10^{-6}$), where the structure of the nucleon is dominated by its sea quark and gluon distributions. However, while opening the door to exploration of phenomena such as saturation and Q^2 evolution in new kinematic regimes, one can argue whether DIS at low x measures the intrinsic structure of the nucleon or the hadronic structure of the virtual photon, γ^*. Because the virtual photon in the DIS process can fluctuate into $q\bar{q}$ pairs whose coherence length $\lambda \sim 1/Mx$ becomes large at small x, DIS at $x \lesssim 0.1$ really probes the $\gamma^* N$ interaction rather than the structure of the nucleon or the γ^* separately. At high x, in contrast, the virtual photon is point-like and unambiguously probes the structure of the nucleon [8].

Moreover, despite the impressive achievements over the past 4 decades, there are still some regions of kinematics where our knowledge of structure functions and PDFs remains unacceptably poor, with little progress made since the 1970's. A striking example is the region of large x ($x \gtrsim 0.5$), where most of the momentum is carried by valence quarks, and sea quarks and gluons are suppressed. Here the valence quark PDFs can be more directly related to quark models of hadron structure; however, the rapidly falling cross sections have made precision measurements extremely challenging. Another example is the pre-asymptotic region dominated by nucleon resonances, where data on the individual transverse and longitudinal cross sections at intermediate and high values of Q^2 are either nonexistent or have large uncertainties. With the availability of continuous, high luminosity electron beams at the CEBAF accelerator, the first decade of experiments at Jefferson Lab has seen a wealth of high-quality data on unpolarized structure functions of the nucleon, penetrating into the relatively unexplored large-x domain and the transition region between resonances and scaling.

The new data in the resonance region confirmed in spectacular fashion the phenomenon of quark-hadron duality in the proton F_2 structure function, and revealed intriguing details about the workings of duality in a number of other observables. The impact of this has been a re-evaluation of the applicability of perturbative QCD to structure functions at low Q^2, and has allowed a much larger body of data to be used in global PDF analyses [9]. Jefferson Lab has also set a new standard in the determination of Rosenbluth-separated longitudinal and transverse structure functions, which eliminates the need for model-dependent assumptions that have plagued previous extractions of structure functions from cross section data.

On the theoretical front, the region of large x and low Q^2 brings to the fore a number of issues which complicate structure function analysis, such as $1/Q^2$ suppressed target mass and higher twist corrections, and nuclear corrections when scattering from nuclear targets. Controlling these corrections requires more sophisticated theoretical tools to be developed, and has motivated theoretical studies, many of which are still ongoing. It has also paved the way towards the 12 GeV experimental program, in which structure functions will be measured to very high x in the DIS region, addressing some long-standing questions about the behavior of PDFs as $x \to 1$.

In the next section we summarize the kinematics and formalism relevant for inclusive lepton–nucleon scattering, including the key results from the operator product expansion. Measurements of the proton F_2^p structure functions and their moments are reviewed in Sec. 3, together with their role in the verification of quark-hadron duality. Data on the deuteron F_2^d structure function are presented in Sec. 4, and the extraction from these of the neutron F_2^n structure function F_2^n is discussed in Sec. 5. Section 6 reviews new measurements of the longitudinal structure function F_L, while Sec. 7 surveys results from semi-inclusive pion production. Finally, in Sec. 8 we describe the impact that the Jefferson Lab data have had on our understanding of nucleon structure in a global context, and briefly outline prospects for future measurements in the 12 GeV era.

2. Formalism

2.1. Kinematics

Because of the small value of the electromagnetic fine structure constant, $\alpha = e^2/4\pi$, the inclusive scattering of an electron from a nucleon, $e(k) + N(p) \to e'(k') + X$, can usually be approximated by the exchange of a single virtual photon, $\gamma^*(q)$, where $q = k' - k$. In terms of the laboratory frame incident electron energy E, the scattered electron energy E', and the scattering angle θ, the photon virtuality is given by $-q^2 \equiv Q^2 = 4EE'\sin^2\theta/2$, where the electron mass has been neglected. The invariant mass squared of the final hadronic state X is $W^2 = (p+q)^2 = M^2 + 2M\nu - Q^2 = M^2 + Q^2(1-x)/x$, where M is the nucleon mass, $\nu = E - E'$ is the energy transfer, and $x = Q^2/2M\nu$ is the Bjorken scaling variable.

In the one photon exchange approximation the spin-averaged cross section for inclusive electron-nucleon scattering in the laboratory frame can be written as

$$\frac{d^2\sigma}{d\Omega dE'} = \frac{\alpha^2}{Q^4}\frac{E'}{E} L_{\mu\nu} W^{\mu\nu}, \tag{2}$$

where the leptonic tensor $L_{\mu\nu} = 2\left(k_\mu k'_\nu + k'_\mu k_\nu - g_{\mu\nu} k \cdot k'\right)$, and using constraints from Lorentz, gauge and parity invariance the hadronic tensor $W^{\mu\nu}$ can in general be written as

$$MW^{\mu\nu} = \left(-g^{\mu\nu} + \frac{q^\mu q^\nu}{q^2}\right) F_1(x, Q^2) + \left(p^\mu - \frac{p\cdot q}{q^2} q^\mu\right)\left(p^\nu - \frac{p\cdot q}{q^2} q^\nu\right)\frac{F_2(x, Q^2)}{p\cdot q}. \tag{3}$$

The structure functions F_1 and F_2 are generally functions of two variables, but become independent of the scale Q^2 in the Bjorken limit, in which both Q^2 and $\nu \to \infty$ with x fixed. At finite values of Q^2 a modified scaling variable is more appropriate [10, 11],

$$\xi = \frac{2x}{1+\rho}, \quad \text{with } \rho = \frac{|\boldsymbol{q}|}{\nu} = \sqrt{1 + Q^2/\nu^2}, \tag{4}$$

which tends to x in the Bjorken limit.

In terms of cross sections for absorbing helicity ± 1 (transverse) and helicity 0 (longitudinal) photons, σ_T and σ_L, the cross section can be written as

$$\frac{d^2\sigma}{d\Omega dE'} = \Gamma\left(\sigma_T(x, Q^2) + \epsilon\, \sigma_L(x, Q^2)\right), \tag{5}$$

where $\Gamma = (\alpha/2\pi^2 Q^2)(E'/E)K/(1-\epsilon)$ is the flux of transverse virtual photons, with the factor $K = \nu(1-x)$ in the Hand convention [12], and

$$\epsilon = \left[1 + 2\left(1 + \frac{\nu^2}{Q^2}\right)\tan^2\frac{\theta}{2}\right]^{-1} \tag{6}$$

is the relative flux of longitudinal virtual photons. Equating Eqs. (2) and (5), the structure functions can be written in terms of the photoabsorption cross sections as

$$F_1(x, Q^2) = \frac{K}{4\pi^2\alpha} M\sigma_T(x, Q^2), \tag{7}$$

$$F_2(x, Q^2) = \frac{K}{4\pi^2\alpha}\frac{\nu}{(1+\nu^2/Q^2)}\left[\sigma_T(x, Q^2) + \sigma_L(x, Q^2)\right], \tag{8}$$

which reveals that F_1 is related only to the transverse virtual photon coupling, while F_2 is a combination of both transverse and longitudinal couplings. One can also define a purely longitudinal structure function F_L,

$$F_L = \rho^2 F_2 - 2xF_1 = 2xF_1 R, \tag{9}$$

where $R = \sigma_L/\sigma_T$ is the ratio of longitudinal to transverse cross sections.

The separation of the unpolarized structure functions into longitudinal and transverse parts from cross section measurements can be accomplished via the Rosenbluth, or longitudinal-transverse (LT), separation technique [13], by making measurements at two or more values of ϵ for fixed x and Q^2. Fitting the reduced cross section σ/Γ linearly in ϵ yields σ_T (and therefore F_1) as the intercept, while the ratio R is obtained from the slope. Note that F_2 can only be extracted from cross sections either by measuring at $\epsilon = 1$ or by performing LT separations. At typical Jefferson Lab kinematics the contribution of F_L to F_2 can be significant.

The above discussion assumed the dominance of the one-photon exchange amplitude in describing the neutron current electron scattering cross section. In principle there are additional contributions arising from the exchange of a Z boson, and in particular the interference between γ^* and Z exchange. The interference is in fact very relevant for parity-violating electron scattering, discussed elsewhere in this volume in connection with extractions of strange electromagnetic form factors from parity-violating asymmetries.

2.2. Operator product expansion

The theoretical basis for describing the Q^2 dependence of structure functions in QCD is Wilson's operator product expansion (OPE) [14]. The quantities most directly amenable to a QCD analysis are the *moments* of structure functions, the n-th moments of which are defined as

$$M_1^{(n)}(Q^2) = \int_0^1 dx\, x^{n-1} F_1(x,Q^2)\,, \qquad M_{2,L}^{(n)}(Q^2) = \int_0^1 dx\, x^{n-2} F_{2,L}(x,Q^2)\,. \tag{10}$$

As will become relevant in the discussion of duality in Sec. 3 below, at large $Q^2 \gg \Lambda^2_{\rm QCD}$ the moments can be expanded in powers of $1/Q^2$, with the coefficients in the expansion given by matrix elements of local operators corresponding to a certain *twist*, τ, defined as the mass dimension minus the spin, n, of the operator. For the n-th moment of F_2, for instance, one has the expansion

$$M_2^{(n)}(Q^2) = \sum_{\tau=2,4\ldots}^{\infty} \frac{A_\tau^{(n)}(\alpha_s(Q^2))}{Q^{\tau-2}}\,, \qquad n = 2,4,6\ldots \tag{11}$$

where $A_\tau^{(n)}$ are the matrix elements with twist $\le \tau$. As the argument suggests, the Q^2 dependence of the matrix elements can be calculated perturbatively, with $A_\tau^{(n)}$ expressed as a power series in $\alpha_s(Q^2)$. For twist two, the coefficients $A_2^{(n)}$ are given in terms of matrix elements of spin-n operators, $A_2^{(n)} p^{\mu_1} \cdots p^{\mu_n} + \cdots = \langle p|\bar\psi\gamma^{\{\mu_1} iD^{\mu_2} \cdots iD^{\mu_n\}}\psi|p\rangle$, where ψ is the quark field, D^μ is the covariant derivative, and the braces $\{\cdots\}$ denote symmetrization of indices and subtraction of traces.

The leading-twist terms correspond to diagrams such as in Fig. 1 (a), in which the virtual photon scatters incoherently from a single parton. The higher-twist terms in Eq. (11) are proportional to higher powers of $1/Q^2$ whose coefficients are matrix elements of local operators involving multi-quark or quark-gluon fields, such as those depicted in Fig. 1 (b) and (c). The higher twists therefore parametrize long-distance multi-parton correlations, which can provide clues to the dynamics of quark confinement.

The additional terms (referred to as the "trace terms") in the twist-two matrix elements involve structures such as $p^2 g^{\mu_i\mu_j}$ and are thus suppressed by powers of $p^2/Q^2 \sim Q^2/\nu^2$. While negligible in the Bjorken limit, at finite Q^2 these give rise to the so-called *target mass corrections* (TMCs), and are important in the analysis of Jefferson Lab data at large values of x. Because their origin is in the same twist-two operators that give rise to structure function scaling, TMCs

Journal of Physics: Conference Series **299** (2011) 012004 doi:10.1088/1742-6596/299/1/012004

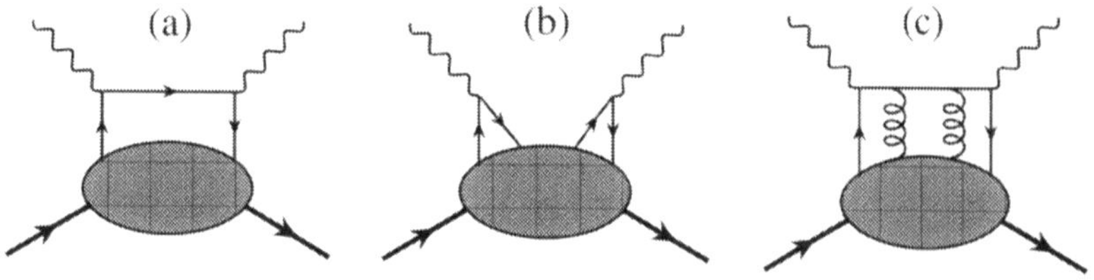

Figure 1. **(a)** Leading-twist ("handbag") contribution to the structure function, **(b)** higher-twist ("cat's ears") four-quark contributions, **(c)** higher-twist quark-gluon interactions.

are formally twist-two effects and are kinematical in origin. Inverting the expressions for the full moments including the trace terms, the resulting target mass corrected F_2 structure function in the OPE is given by [15, 16]

$$F_2^{\mathrm{TMC}}(x,Q^2) = \frac{x^2}{\xi^2\rho^3}F_2^{(0)}(\xi,Q^2) + \frac{6M^2x^3}{Q^2\rho^4}\int_\xi^1 du\left(1+\frac{2M^2x}{Q^2\rho}(u-\xi)\right)\frac{F_2^{(0)}(u,Q^2)}{u^2}, \qquad (12)$$

where $F_2^{(0)}$ is the structure function in the $M^2/Q^2 \to 0$ limit. Similar expressions are found for the F_1 and F_L structure functions [15, 16]. One should note, however, that the OPE result for TMCs to structure functions is not unique; in the collinear factorization approach, for example, in which parton distributions are formulated *a priori* in momentum space, different expressions for TMCs arise [17]. While both formalisms give the same results in the Bjorken limit, the differences between these at finite Q^2 can be seen as representing an inherent prescription dependence and systematic uncertainty in the analysis of structure functions at low Q^2.

3. Proton F_2 structure function

Measurements of the proton F_2^p structure function have been taken at Jefferson Lab over a range of kinematics, from Q^2 as low as 0.1 GeV^2 and below (to study the transition to the photoproduction point, $Q^2 = 0$) and up to $Q^2 = 8$ GeV^2 (to study the large-x behavior and quark-hadron duality). At the larger Q^2 values the high luminosity provided by the CEBAF accelerator has allowed significant improvement in the statistical precision of high-x measurements over all previous experiments. In addition, with the HMS spectrometer in Hall C well understood, LT separated cross sections have been measured with better than 1.6% systematic point-to-point uncertainties and typically less than 1.8% normalization uncertainties.

Precision F_2^p spectra extracted from cross sections measured in Hall C [18, 19, 20] are shown in Fig. 2 (left) as a function of x for several Q^2 values ($Q^2 = 0.5, 1.5, 3$ and 5.5 GeV^2), together with previous SLAC [21] and NMC [22] data at lower x. The data have been bin-centered to the common Q^2 values shown for all measurements within the range of 20% of the central value, utilizing a fit [23] to the DIS data and a global fit [24] to Jefferson Lab resonance region data. In addition to the Hall C measurements, there now also exists a large body of F_2^p data from Hall B covering a significant range of kinematics, which is afforded by the large acceptance of the CLAS spectrometer. An example of the F_2^p spectrum extracted from CLAS is shown in Fig. 2 (right) at $Q^2 = 0.775$ GeV^2. In Table 1 we present a complete list of all unpolarized inclusive and semi-inclusive measurements on the proton and deuteron performed at Jefferson Lab through 2010, including their current status.

3.1. Quark-hadron duality

The proton F_2^p data in Fig. 2 illustrate the intriguing phenomenon of quark-hadron duality, which relates structure functions in the nucleon resonance and DIS regions. First observed by

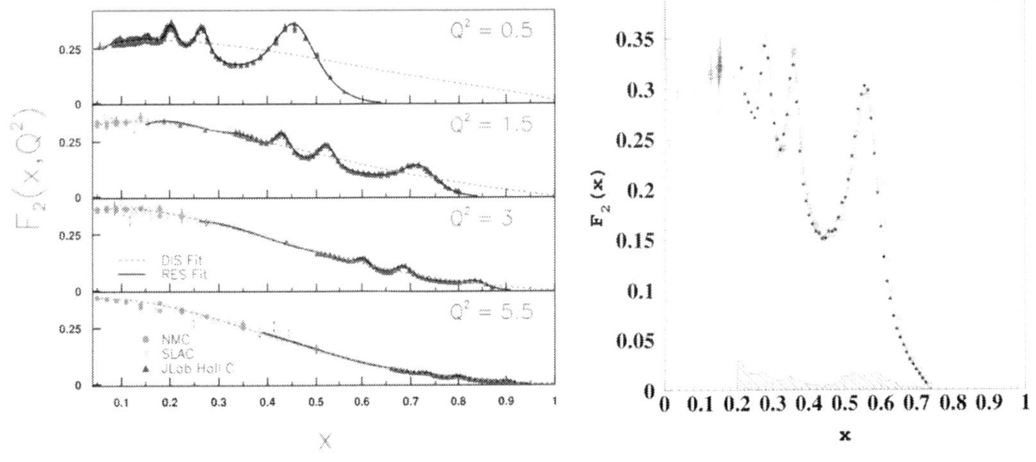

Figure 2. (*Left*) Proton F_2^p structure function data from Hall C [18, 19, 20], SLAC [21], and NMC [22] at $Q^2 = 0.5, 1.5, 3$ and 5.5 GeV^2, compared with a fit [24] to the transverse and longitudinal resonance cross sections from photoproduction to $Q^2 = 9$ GeV^2 (solid), and a global fit [23] to DIS data (dashed). (*Right*) Proton F_2^p from CLAS in Hall B at $Q^2 = 0.775$ GeV^2 (stars) [25], compared to earlier Hall C data (open circles) [26].

Bloom and Gilman [37] in the early inclusive SLAC data (hence also referred to as "Bloom-Gilman duality"), the structure functions in the resonance region are found on average to equal the structure functions measured in the "scaling" region at higher W. The resonance data oscillate around the scaling curve and slide along it with increasing Q^2, as seen in Fig. 2 (left).

The early F_2^p data from SLAC were extracted from cross sections assuming a fixed value for R (= 0.18), and with a scaling curve parametrizing the limited data available in the early 1970's [38]. Since the original measurements, the F_2^p structure function has become one of the best studied quantities in lepton scattering, with data from laboratories around the world contributing to a global data base spanning over five orders of magnitude in both x and Q^2. With the advent of the Jefferson Lab data, precise F_2^p measurements now also exist in the resonance region up to $Q^2 \approx 8$ GeV^2, allowing many new aspects of duality to be quantified for the first time [39].

While the early duality studies considered only the F_2 structure function [37], Jefferson Lab experiments have in addition revealed the presence of duality in other observables. For example, Fig. 3 shows new LT-separated data from Jefferson Lab experiment E94-110 for the proton transverse (F_1^p) and longitudinal (F_L^p) structure functions in the nucleon resonance region [18]. LT-separated data from SLAC, predominantly in the DIS region, are also shown for comparison [42]. Where they refer to the same kinematic values, the Jefferson Lab and SLAC data are in excellent agreement, providing confidence in the achievement of the demanding precision required of this type of experiment. In all cases, the resonance and DIS data merge smoothly with one another in both x and Q^2.

The availability of leading twist PDF-based global fits [4, 5, 6, 7, 9] allows comparison of the resonance region data with leading twist structure functions at the same x and Q^2. The resonance data on the F_1 and F_L structure functions are also found to oscillate around the perturbative QCD (pQCD) curves, down to Q^2 as low as 0.7 GeV^2. Because most of the data lie at large values of x and small Q^2, it is vital for tests of duality to account for the effects of kinematical target mass corrections [15], which give large contributions as $x \to 1$ [16].

Table 1. Listing of Jefferson Lab unpolarized inclusive and semi-inclusive electron-nucleon scattering experiments.

Experiment	Hall	Target	Observable	Reference	Status
E94-110	C	p	R in resonance region	[18, 27, 28, 29]	data taken in 1999, analysis completed
E99-118	C	p, d	nuclear dependence of R at low Q^2	[30]	data taken in 2000, analysis completed
CLAS	B	p, d	inclusive cross sections	[25, 31]	e1/e2 run periods
E00-002	C	p, d	F_2 at low Q^2	[20]	data taken in 2003, analysis completed, pub. in progress
E00-108	C	p, d	semi-inclusive $\pi^{\pm}$ electroproduction	[32, 33]	data taken in 2003, analysis completed
E00-116	C	p, d	inclusive resonance region cross sections at intermediate Q^2	[19]	data taken in 2003, analysis completed
E02-109	C	d	R in resonance region	[34]	data taken in 2005, analysis in progress
E03-012 (BoNuS)	B	$d(n)$	neutron F_2^n *via* spectator tagging	[35]	data taken in 2005, analysis completed, pub. in progress
E06-009	C	d	R in resonance region & beyond: extension of E02-109 to $Q^2 = 4$ GeV^2	[36]	data taken in 2007, analysis in progress

This is clear from Fig. 3, where the data are compared with leading twist structure functions computed from PDFs to next-to-next-to-leading order (NNLO) accuracy from Alekhin [40] and MRST [41, 43]. The latter are shown with (solid) and without (dotted) target mass corrections, and clearly demonstrate the importance of subleading $1/Q^2$ effects at large x. In particular, TMCs give additional strength at large x observed in the data, which would be significantly underestimated by the leading twist functions without TMCs.

The phenomenological results raise the question of how can a scaling structure function be built up entirely from resonances, each of whose contribution falls rapidly with Q^2 [44]? A number of studies using various models have demonstrated how sums over resonances can indeed yield a Q^2 independent function (see Ref. [39] for a review). The key observation is that while

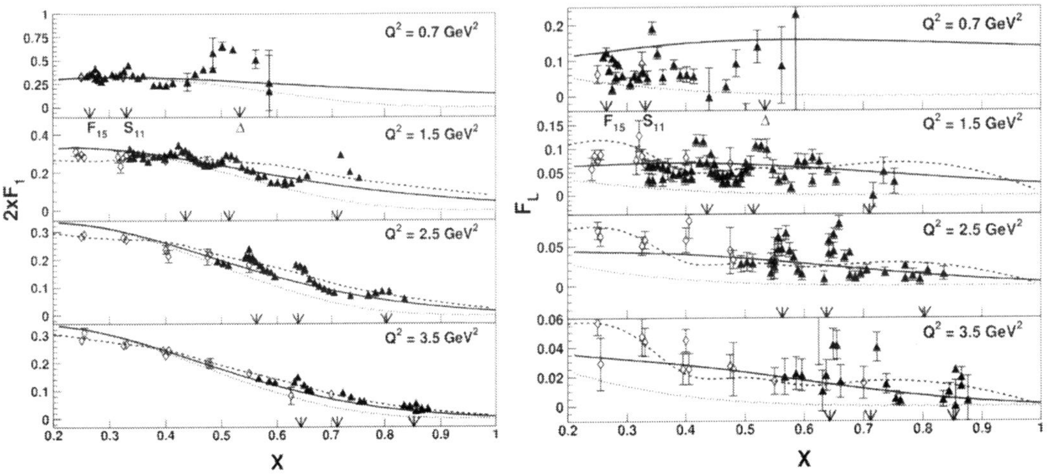

Figure 3. Purely transverse *(left)* and longitudinal *(right)* proton structure functions $2xF_1$ and F_L in the resonance region [18] (triangles), compared with earlier data from SLAC (squares). The curves are leading twist structure functions computed at NNLO from Alekhin (dashed) [40] and MRST [41] with (solid) and without (dotted) target mass corrections. The prominent resonance regions (Δ, S_{11}, F_{15}) are indicated by the arrows along the abscissa.

the contribution from each individual resonance diminishes with Q^2, with increasing energy new states become accessible whose contributions compensate in such a way as to maintain an approximately constant strength overall. At a more microscopic level, the critical aspect of realizing the suppression of the higher twists is that at least one complete set of even and odd parity resonances must be summed over for duality to hold [45]. For an explicit demonstration of how this cancellation takes place in the SU(6) quark model and its extensions, see Refs. [45, 46, 47].

3.2. Structure function moments

The degree to which quark-hadron duality holds can be more precisely quantified by computing integrals of the structure functions over x in the resonance region at fixed Q^2 values, $\int_{x_{\rm th}}^{x_{\rm res}} dx\ F_2(x, Q^2)$, where $x_{\rm th}$ corresponds to the pion production threshold at fixed Q^2, and $x_{\rm res} = Q^2/(W_{\rm res}^2 - M^2 + Q^2)$ indicates the x value at the same Q^2 where the traditional delineation between the resonance and DIS regions at $W = W_{\rm res} \equiv 2$ GeV is made. These integrals can then be compared to the corresponding integrals of the structure functions fitted to the higher-W, deep-inelastic data, at the same Q^2 and over the same interval of x. The early phenomenological findings [26] suggested that the integrated strength of the resonance structure functions above $Q^2 \approx 1$ GeV^2 was indeed very similar to that in the deep-inelastic region, including in each of the individual prominent resonance regions. In this section we explore the duality between the resonance and deep-inelastic structure functions in the context of QCD moments.

According to De Rujula, Georgi and Politzer [48], one can formally relate the appearance of quark-hadron duality to the vanishing suppression of higher twist matrix elements in the QCD moments of the structure functions [14]. Namely, if certain moments of structure functions are observed to be independent of Q^2, as implied by duality, then from Eq. (11) the moments must be dominated by the leading, Q^2 independent term, with the $1/Q^{\tau-2}$ higher twist terms suppressed. Duality is then synonymous with the suppression of higher twists, which in partonic language corresponds to the suppression, or cancellation, of interactions between the scattered quark and the spectator system such as those illustrated in Fig. 1 (b) and (c). Conversely, if the moments display power-law Q^2 dependence, then this implies violation of duality; moreover, if the violation is not overwhelming, the Q^2 dependence of the data can be used to extract information on the higher twist matrix elements [49].

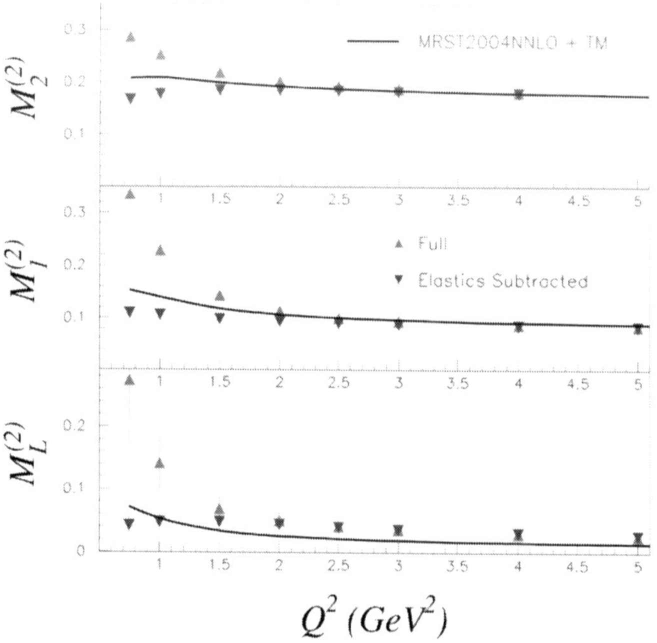

Figure 4. Total $n = 2$ moments of the proton F_2^p *(top)*, F_1^p *(center)* and F_L^p *(bottom)* structure functions determined from global fits to existing DIS data and Jefferson Lab resonance region data [50], compared with moments computed from the leading twist PDFs from MRST at NNLO [41].

The first determination [51] of the F_2^p moments from Jefferson Lab data was made utilizing structure functions measured in Hall C [26], while a later evaluation [25] included the large body of additional data from CLAS. More recently, the extraction of F_2^p has been further enhanced by LT-separated data from Hall C, shown in Fig. 2 along with the fit [24] to the LT separated cross sections. The $n = 2$ moments for the proton F_2^p, F_1^p, and F_L^p structure functions are shown in Fig. 4 versus Q^2, as determined from integrating this fit. For F_2^p they are found to be in very good agreement with the earlier measurements. Also shown is the leading-twist contribution calculated from the MRST parameterization [41], corrected for target mass effects [15].

One of the most striking features of the results in Fig. 4 is that the elastic-subtracted moments exhibit the same Q^2 dependence as the PDF fits down to $Q^2 \approx 1$ GeV^2. Even with the elastic contribution included, which vanishes in the Bjorken limit and is hence pure higher twist, there is excellent agreement between the resonance and DIS data for $Q^2 \gtrsim 2$ GeV^2. Until very recently [6, 9], this fact has not been widely appreciated or utilized in global PDF fitting efforts [4, 5], which typically impose cuts on data of $Q^2 \gtrsim 4$ GeV^2 and $W^2 \gtrsim 12$ GeV^2.

While the OPE provides a systematic approach to identifying and classifying higher twists, it does not reveal *why* these are small or *how* duality is realized globally and locally. To further explore the *local* aspects of duality within a perturbative QCD context, a ground-breaking new

Journal of Physics: Conference Series **299** (2011) 012004 doi:10.1088/1742-6596/299/1/012004

approach using "truncated" moments of structure functions, developed recently by Forte *et al.* [52] and extended by Kotlorz & Kotlorz [53], was applied by Psaker *et al.* [54] to Jefferson Lab data. The virtue of truncated moments is that they obey a similar set of Q^2 evolution equations as those for PDFs themselves, which therefore enables a rigorous connection to be made between local duality and QCD. It allows one to quantify the higher twist content of various resonance regions, and determine the degree to which individual resonances are dominated by the leading twist components.

Defining the n-th truncated moment $\mathcal{M}_n$ of a PDF $q(x, Q^2)$ between $x_{\rm min}$ and $x_{\rm max}$ as

$$\mathcal{M}_n(x_{\rm min}, x_{\rm max}, Q^2) = \int_{x_{\rm min}}^{x_{\rm max}} dx\, x^{n-1}\, q(x, Q^2)\ , \tag{13}$$

the evolution equations for the truncated moments can be written as

$$\frac{d\mathcal{M}_n}{dt} = \frac{\alpha_s(Q^2)}{2\pi} \left(P'_n \otimes \mathcal{M}_n\right)\ , \quad t = \ln\left(Q^2/\Lambda^2_{\rm QCD}\right)\ . \tag{14}$$

The symbol $\otimes$ denotes the Mellin convolution of the truncated moment and the "splitting function" P'_n, which is related to the usual DGLAP evolution splitting function P [55] by $P'_n(z, \alpha_s(Q^2)) = z^n\ P(z, \alpha_s(Q^2))$. The extent to which nucleon structure function data in specific regions in x (or W) are dominated by leading twist can be determined by constructing empirical truncated moments and evolving them to different Q^2. Deviations of the evolved moments from the experimental data at the new Q^2 then reveal any higher twist contributions in the original data.

A next-to-leading order (NLO) analysis [54] of data on the proton F_2^p structure function from Jefferson Lab covering a range in Q^2 from 1 GeV^2 to ≈ 6 GeV^2 revealed intriguing behavior of the higher twists for different nucleon resonance regions. Assuming that F_2^p data beyond a large enough Q^2 (taken to be $Q^2 = Q_0^2 = 25$ GeV^2 in Ref. [54]) are dominated by leading twist, the truncated moments were computed at Q_0^2 and evolved to lower Q^2. Note that the truncated moments are computed over the range $W_{\rm th} \le W \le W_{\rm max}$, where the $W_{\rm th} = M + m_\pi$ is the inelastic threshold. At $Q^2 = 1$ GeV^2 this corresponds to the integration range $x_{\rm min} \le x \le x_{\rm th}$, where $x_{\rm th} = \left[1 + m_\pi(m_\pi + 2M)/Q^2\right]^{-1} \simeq 0.78$.

The ratio of the truncated moments of the data to the leading twist is shown in Fig. 5(a) as a function of $W_{\rm max}$ at $Q^2 = 1$ GeV^2, with and without target mass corrections. Including the effects of TMCs, the leading twist moment differs from the data by $\sim 15\%$ for $W_{\rm max} > 1.5$ GeV. To quantify the higher twist content of the specific resonance regions, and at different values of Q^2, several intervals in W are considered: $W_{\rm th}^2 \le W^2 \le 1.9$ GeV^2 ($\Delta(1232)$ or first resonance region); $1.9 \le W^2 \le 2.5$ GeV^2 ($S_{11}(1535)$ or second resonance region); and $2.5 \le W^2 \le 3.1$ GeV^2 ($F_{15}(1680)$ or third resonance region). The higher twist contributions to $\mathcal{M}_2$ in these regions are shown in Fig. 5(b) as ratios to moments of the data.

The results indicate deviations from leading twist behavior of the entire resonance region data (filled circles in Fig. 5(b)) at the level of $\lesssim 15\%$ for all values of Q^2 considered, with significant Q^2 dependence for $Q^2 \lesssim 4$ GeV^2. The strong Q^2 dependence of the higher twists is evident here in the change of sign around $Q^2 = 2$ GeV^2, with the higher twists going from $\approx -10\%$ at $Q^2 = 1$ GeV^2 to ≈ 10–15% for $Q^2 \sim 3 - 6$ GeV^2. At larger Q^2 the higher twists are naturally expected to decrease, once the leading twist component of the moments begins to dominate.

Interestingly, the magnitude of the higher twist contributions in the Δ region (diamonds) is smallest, decreasing from $\approx -15\%$ of the data at $Q^2 = 1$ GeV^2 to values consistent with zero at larger Q^2. The higher twists are largest, on the other hand, for the S_{11} region (squares), where they vary between $\approx -15\%$ of the data at $Q^2 = 1$ GeV^2 and 20–25% at $Q^2 \sim 5$ GeV^2. Combined, the higher twist contribution from the first two resonance regions (dotted curve) is

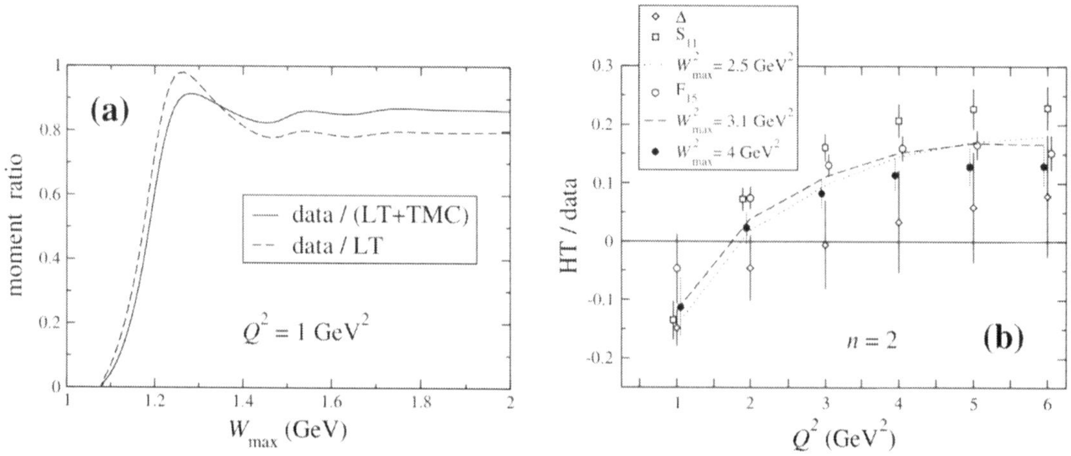

Figure 5. **(a)** Ratio of the $\mathcal{M}_2$ truncated moments of the data to the leading twist + TMC (solid), and data to leading twist without TMC (dashed) at $Q^2 = 1$ GeV^2, as a function of W_{max}. **(b)** Q^2 dependence of the fractional higher twist (HT) contribution to the $n = 2$ truncated moment data, for various intervals in W.

$\lesssim 15\%$ in magnitude for all Q^2. The rather dramatic difference between the Δ and the S_{11}, may, at least in part, be due to the choice of the differentiation point of $W^2 = 1.9$ GeV^2. A lower W^2 choice, for instance, would lower the higher twist content of the S_{11} at large Q^2, while raising that of the Δ. However, this W^2 choice corresponds to the local minimum between these two resonances in the inclusive spectra, and is the one most widely utilized.

The higher twist content of the F_{15} region (open circles) is similar to the S_{11} at low Q^2, but decreases more rapidly for $Q^2 > 3$ GeV^2. The higher twist content of the first three resonance regions combined (dashed curve) is $\lesssim 15$–20% in magnitude for $Q^2 \leq 6$ GeV^2. Integrating up to $W^2_{max} = 4$ GeV^2 (filled circles), the data on the $n = 2$ truncated moment are found to be leading twist dominated at the level of 85–90% over the entire Q^2 range.

The results in Fig. 5 can be compared with quark model expectations [45, 46], which predict systematic deviations of resonance data from local duality. Assuming dominance of magnetic coupling, the proton data are expected to overestimate the DIS function in the second and third resonance regions due to the relative strengths of couplings to odd-parity resonances; the positive higher twists observed in Fig. 5(b) for $Q^2 \gtrsim 2$ GeV^2 indeed support these predictions.

4. Deuteron F_2 measurements

Together with hydrogen, inclusive lepton scattering from deuterium targets has provided an extensive data base of F_2 structure function measurements over a large range of kinematics. While a significant quantity of F_2^d data was collected from experiments at CERN and SLAC, the quasi-elastic and nucleon resonance regions, especially at low and moderate Q^2 ($\approx$ few GeV^2), were only mapped precisely with the advent of Jefferson Lab data [26, 31]. Similarly, Jefferson Lab contributed with precision F_2^d data in the region of $x > 1$ [56], of relevance for constructing higher moments of deuteron structure functions.

Experiments in Hall C [26, 56] have provided F_2^d data in select regions of x and Q^2, while inclusive cross section measurements in CLAS have covered a continuous two-dimensional region over the entire resonance region up to $Q^2 = 6$ GeV^2. This combination is rather useful for determining moments of F_2. In such extractions one usually assumes that the ratio R for the deuteron is similar to that for the proton at scales Q^2 of a few GeV^2 [57]. New measurements which will test this assumption will be reviewed in Sec. 6.

Table 2. Lowest two moments (for $n = 2$ and 4) of the isovector F_2 structure function. Experimental results for $Q^2 \approx 4$ GeV^2 are compared with lattice calculations extrapolated to the chiral limit.

n	Niculescu *et al.* [58] (Hall C)	Osipenko *et al.* [59] (Hall B)	Detmold *et al.* [60] (lattice)
2	0.049(17)	0.050(9)	0.059(8)
4	0.015(3)	0.0094(16)	0.008(3)

The results of the moment analyses are shown in Table 2, expressed as the isovector (proton minus neutron) combination, and compared with isovector moments from lattice QCD [60, 61]. For simplicity the neutron moments here are defined as the difference between the deuteron and proton moments — see, however, Sec. 5.1 below. The $n = 2$ moments from the Hall B [59] and Hall C [58] analyses agree well with each other, and with the lattice extraction, which includes the effects of pion loops and the intermediate $\Delta(1232)$ resonance in the chiral extrapolation. For the $n = 4$ moment the comparison between the Hall B and Halls C results shows a slight discrepancy, which may be reduced once precision high-x data at $Q^2 \sim 4$ GeV^2 are included in the extractions.

Analysis of the Q^2-dependence of the deuteron F_2 moments in Ref. [31] suggests a partial cancellation of different higher twist contributions entering in the OPE with different signs, which is one of the manifestations of quark-hadron duality. The slow variation with Q^2 of the structure function moments, down to $Q^2 \approx 1$ GeV^2, was also found in the analysis of proton data [51], where such cancellations were found to be mainly driven by the elastic contribution. Furthermore, by comparing the proton and (nuclear corrected) neutron structure function moments, the higher twist contributions were found to be essentially isospin independent [62]. This suggests the possible dominance of ud correlations over uu and dd in the nucleon, and implies higher twist corrections that are consistent with zero in the isovector F_2 structure function.

More recently, high precision deuteron cross sections in the resonance region have been measured in Hall C [34, 36] with the aim of providing LT separated deuteron structure functions of comparable precision and kinematic coverage to those performed for the proton. These higher precision LT separated data will allow for a significant further reduction in the uncertainties in the current nonsinglet F_2 extractions.

5. Neutron F_2 structure function

A complete understanding of the valence quark structure of the nucleon requires knowledge of both its u and d quark distributions. While the u distribution is relatively well constrained by measurements of the proton F_2^p structure function, in contrast the d quark distribution is poorly determined due to the lack of comparable data on the neutron structure function F_2^n. The absence of free neutron targets makes it necessary to use light nuclei such as deuterium as effective neutron targets, and one must therefore deal with the problem of extracting neutron information from nuclear data.

5.1. Neutron structure from inclusive F_2 data

In standard global PDF analyses, sensitivity to the d-quark from charged lepton scattering is primarily provided by the neutron in the deuteron. Usually the neutron F_2^n structure function is extracted by subtracting the deuteron and proton structure function data assuming that nuclear corrections are negligible. At large x, however, the ratio of the deuteron to free nucleon structure functions is predicted to deviate significantly from unity [63, 64, 65, 66], which can have significant impact on the behavior of the extracted neutron structure function at large x [9, 67].

Even when nuclear effects are considered, there exist practical difficulties with extracting information on the free neutron from nuclear data, especially in the nucleon resonance region, where resonance structure is largely smeared out by nucleon Fermi motion. A recent analysis [68] used a new method [66] to extract F_2^n from F_2^d and F_2^p data, in which nuclear effects are parameterized via an additive correction to the free nucleon structure functions, in contrast to the more common multiplicative method [69] which fails for functions with zeros or with non-smooth data. In the standard impulse approximation approach to nuclear structure functions, the deuteron structure function can be written as a convolution [65, 66]

$$F_2^d(x, Q^2) = \sum_{N=p,n} \int dy\, f_{N/d}(y, \rho)\, F_2^N(x/y, Q^2), \qquad (15)$$

where $f_{N/d}$ is the light-cone momentum distribution of nucleons in the deuteron (or "smearing function"), and is a function of the momentum fraction y of the deuteron carried by the struck nucleon, and of the virtual photon "velocity" ρ (see Eq. (4)). The smearing function encodes the effects of the deuteron wave function, accounting for nuclear Fermi motion and binding effects, as well as kinematical finite-Q^2 corrections. Although not well constrained, nucleon off-shell effects have also been studied [63, 64, 65]; their influence appears to be small compared with the errors on the existing data, except at very large x.

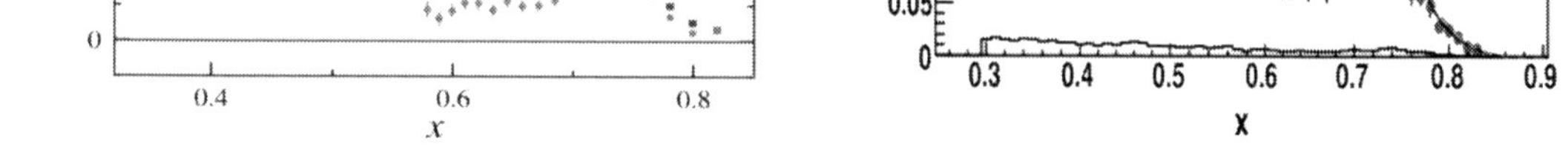

Figure 6. *(Left)* Neutron F_2^n structure function extracted from inclusive deuteron and proton data at $Q^2 = 1.7$ GeV^2 [68], together with the reconstructed F_2^d. *(Right)* Neutron structure function extracted from the BoNuS experiment [35], compared with the Bosted/Christy model of the neutron F_2^n [70] and the corresponding Christy/Bosted parametrization for F_2^p [24].

In Fig. 6 we illustrate the results of a typical extraction of F_2^n from Jefferson Lab F_2^d and F_2^p data at $Q^2 = 1.7$ GeV^2. The proton data show clear resonant structure at large x, which is mostly washed out in the deuteron data. The resulting neutron F_2^n is shown after two iterations of the procedure with an initial guess of $F_2^n = F_2^p$. Clear neutron resonance structure is visible in the first (Δ) and second resonance regions, at $x \sim 0.75$ and 0.55, respectively, with some structure visible also in the third resonance region at $x \sim 0.45$, albeit with larger errors. The deuteron F_2^d reconstructed from the proton and extracted neutron data via Eq. (15) indicates the relative accuracy and self-consistency of the extracted F_2^n.

Of course it is not possible to avoid nuclear model dependence in the inversion procedure, and some differences in the extracted F_2^n will arise using different models for the smearing functions $f_{N/d}$. To remove, or at least minimize, the model dependence in the extracted free neutron structure, several methods have been proposed, such as utilizing inclusive DIS from $A = 3$ mirror nuclei [71, 72, 73], and semi-inclusive DIS from a deuteron with spectator tagging [35, 74]. In addition, experiments involving weak interaction probes [75, 76, 77] can provide the information on the flavor separated valence quark distributions directly. In the next section we discuss in detail one of these methods for determining the free neutron structure, namely the BoNuS experiment at Jefferson Lab [35].

5.2. Tagged neutron structure functions

To overcome the absence of free neutron targets, the Hall B BoNuS (Barely Off-shell NeUtron Structure) experiment has measured inclusive electron scattering on an almost free neutron using the CLAS spectrometer and a recoil detector to tag low momentum protons. The protons are tracked in a novel radial time projection chamber [78] utilizing gas electron multiplier foils to amplify the proton ionization in a cylindrical drift region filled with a mixture of Helium and Di-Methyl Ether, and with the proton momentum determined from the track curvature in a solenoidal magnetic field. Slow backward-moving spectator protons are tagged with momenta as low as 70 MeV in coincidence with the scattered electron in the reaction $D(e, e'p_s)X$. This technique ensures that the electron scattering took place on an almost free neutron, with its initial four-momentum inferred from the observed spectator proton. Cutting on spectator protons with momenta between 70 and 120 MeV and laboratory angles greater than 120 degrees minimizes the contributions from final state interactions and off-shell effects to less than several percent on the extracted neutron cross section.

While inclusive scattering from deuterium results in resonances which are significantly broadened (often to the point of being unobservable), determination of the initial neutron momentum allows for a dramatic reduction in the fermi smearing effects and results in reconstructed resonance widths comparable to the inclusive proton measurements. In addition, the large CLAS acceptance for the scattered electron allowed for the tagged neutron cross section to be measured over a significant kinematic range in both W^2 and Q^2 at beam energies of 4.2 GeV and 5.3 GeV. BoNuS F_2^n data extracted at a beam energy of 5.3 GeV and $Q^2 = 1.7$ GeV^2 are shown in Fig. 6 (right panel) for x values from pion production threshold through the resonance region and into the DIS regime. This will allow for the first time a unambiguous study of the inclusive neutron resonance structure functions.

An extension of BoNuS has been approved [79] to run at a beam energy of 11 GeV after the energy upgrade of the CEBAF accelerator. The kinematic coverage will allow the extraction of the ratio of neutron to proton structure functions F_2^n/F_2^p to x as large as 0.8, and the corresponding d to u large x parton distribution ratio. Other quantities which BoNuS will make possible to measure include the elastic neutron form factor, quark-hadron duality on the neutron, semi-inclusive DIS and resonance production channels, hard exclusive reactions such as deeply-virtual Compton scattering or deeply-virtual meson-production from the neutron, as well as potentially the inclusive structure function of a virtual pion.

6. Longitudinal structure function F_L

The unpolarized inclusive proton cross section contains two independent structure functions. While F_2^p has been measured to high precision over many orders of magnitude in both x and Q^2, measurements of the longitudinal structure function F_L^p (and the ratio R) have been significantly more limited in both precision and kinematic coverage. This is due in part to the challenges inherent in performing LT separations, which typically require point-to-point systematic uncertainties in ϵ to be smaller than 2% to obtain uncertainties on F_L of less than 20%.

While the inclusive cross section is proportional to $2xF_1 + \epsilon F_L$, the F_2 structure function $\sim 2xF_1 + F_L$, so that F_2 is only proportional to the cross section for $\epsilon = 1$. At high Q^2 the scattering of longitudinal photons from spin-1/2 quarks is suppressed, and in the parton model one expects F_L (and R) to vanish as $Q^2 \to \infty$. At low Q^2, however, F_L is no longer suppressed, and could be sizable, especially in the resonance region and at large x. On the other hand, F_L is dominated by the gluon contribution at small x, where new measurements from HERA [80] have shown that it continues to rise. In the kinematic range of Jefferson Lab F_L has been found to be typically 20% of the magnitude of F_2, which is consistent with earlier SLAC measurements where the kinematic regions overlap.

An extensive program of LT separations has been carried out in Hall C, including measurements of the longitudinal strength in the resonance region for $0.3 < Q^2 < 4.5$ GeV^2 for both proton [18] and deuteron [34, 36] targets. The Jefferson Lab experiments listed in Table 1 for which LT separations have been performed for the proton or deuteron are E94-110, E99-118, E00-002, E02-109, and E06-009. The Jefferson Lab data complement well the previous results at smaller x from SLAC and NMC in this Q^2 region, and improve dramatically on the few measurements that existed below $Q^2 = 8$ GeV^2 in this x region, which had typical errors on R and F_L of 100% or more — see Fig. 7 (left).

The recent precision LT separated measurements of proton cross sections [18] have allowed for the first time detailed duality studies in all of the unpolarized structure functions and their moments. The results of the proton separated structure functions in the resonance region were presented in Fig. 3 in Sec.3. Although significant resonant strength is observed in F_L (or R), evidence of duality is nonetheless observed in this structure function, along with F_2 and F_1.

In addition to the proton data, Jefferson Lab experiment E02-109 measured the LT separated F_2 and F_L structure functions of the deuteron, in the same W^2 and Q^2 ranges, and with the same high precision as E94-110 did for the proton. This will allow quantitative studies of duality in both the longitudinal and transverse channels for the deuteron. If duality holds well for both the proton and neutron separately, it will hold to even better accuracy for the deuteron since the Fermi motion effects intrinsically perform some of the averaging over the resonances. However, if duality does not hold for the LT separated neutron structure functions, this should be observable in the deuteron data, and will thus provide a critical test for models of duality.

In addition to the resonance region, measurements of the inclusive longitudinal proton and deuteron structure have also been performed at lower x (higher W^2) and lower Q^2. While the longitudinal strength is significant at Q^2 of several GeV^2, the proton F_L structure function is constrained by current conservation to behave, for fixed W, as $F_L \sim Q^4$ for $Q^2 \to 0$. However, even with the new Jefferson Lab data, which extend down to $Q^2 = 0.15$ GeV^2 [81], the Q^2 at which this behavior sets in has not yet been observed.

Another interesting test provided by the E99-118 data is whether the relative longitudinal contribution to the cross section embodied in R is different in the deuteron and proton at these low Q^2 values. While the higher Q^2 data from SLAC and NMC exhibit no significant difference in the deuteron and proton R, the Jefferson Lab results shown in Fig. 7 (right) suggest a possible suppression of R in the deuteron relative to the proton for $Q^2 < 1$ GeV^2. Although this suppression is consistent with the two lowest Q^2 data points from SLAC, the uncertainties are dominated by systematic errors and the combined significance of the effect is still less than 2 σ. Conclusive experimental evidence for the possible suppression of R in deuterium at low Q^2 will likely be provided when the results from additional data from E00-002 are finalized in very near future.

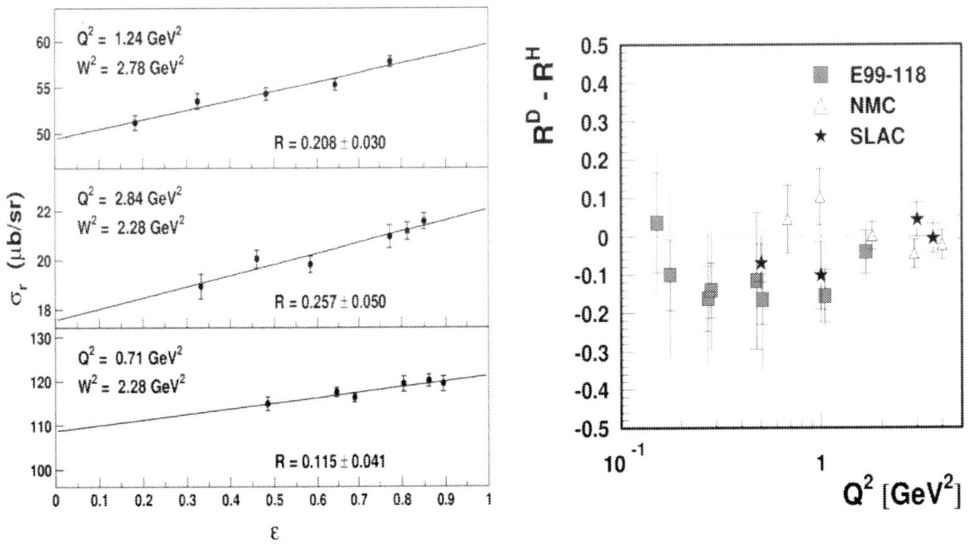

Figure 7. *(Left)* Sample LT separations from Jefferson Lab experiment E94-110 [18]. *(Right)* Difference between the ratios R in deuterium and hydrogen versus Q^2, from E99-118 [81], compared with previous NMC and SLAC measurements.

7. Semi-inclusive deep inelastic scattering

In addition to the traditional F_2 and F_L structure function observables, a number of other processes have been studied at Jefferson Lab over the past decade, with potentially important consequences for our understanding of the workings of QCD at low energy. In this section we focus on semi-inclusive pion electroproduction.

An analysis of the semi-inclusive process $e\,N \to e\,h\,X$, where a hadron h is detected in the final state in coincidence with the scattered electron, has recently been made using data from Hall C in the resonance–scaling transition region [32, 33]. One of the main motivations for studying semi-inclusive meson production is the promise of flavor separation via tagging of specific mesons in the final state. In the valence quark region a produced π^+ (π^-) meson, for example, primarily results from scattering off a u (d) quark in the proton.

The semi-inclusive cross section at LO in α_s is given by a simple product of quark distribution and quark $\to$ hadron fragmentation functions,

$$\frac{d\sigma}{dxdz} \sim \sum_q e_q^2\, q(x)\, D_q^h(z) \equiv \mathcal{N}_N^h(x,z)\,. \tag{16}$$

Here the fragmentation function $D_q^h(z)$ gives the probability for a quark q to fragment to a hadron h with a fraction $z = p_h \cdot p/q \cdot p = E_h/\nu$ of the quark's (or virtual photon's) laboratory frame energy. Although at LO the scattering and particle production mechanisms are independent, higher order pQCD corrections give rise to non-factorizable terms, which involve convolutions of the PDFs and fragmentation functions with hard coefficient functions [82].

For hadrons produced collinearly with the virtual photon, the invariant mass W' of the undetected hadronic system X at large Q^2 can be written [83] $W'^2 \approx M^2 + Q^2(1-z)(1-x)/x$, where the hadron mass is neglected with respect to Q^2. In the elastic limit, $z \to 1$, the hadron carries all of the photon's energy (with $W' \to M$), so z is also referred to as the "elasticity".

While formally the LO factorized expression for the cross section (16) may be valid at large Q^2, at finite Q^2 there are important corrections arising from the finite masses of the target and produced hadron. One can show, however, that the LO factorization holds even at finite Q^2, provided the parton distribution and fragmentation functions are expressed in terms of generalized scaling variables [84], $q(x)D_q^h(z) \to q(\xi_h)D_q^h(\zeta_h)$, where $\zeta_h = (z_h\xi/2x)(1 + \sqrt{1 - 4x^2M^2m_{h\perp}^2/z^2Q^4})$ and $\xi_h = \xi(1 + m_h^2/\zeta_h Q^2)$, with $m_{h\perp}^2 = m_h^2 + p_{h\perp}^2$. Not surprisingly, these effects become large at large x and z when Q^2 is small; however, for heavier produced hadrons such as kaons or protons, significant effects can also arise at small values of z [84].

The validity of the factorized hypothesis in Eq. (16) relies on the existence of a sufficiently large gap in rapidity $\eta = \ln\left[(E_h - p_h^z)/(E_h + p_h^z)\right]/2$ to allow a clean separation of the current fragmentation region (hadrons produced from the struck quark) from the target fragmentation region (hadrons produced from the spectator quark system). At high energies a gap of $\Delta\eta \approx 2$, which is typically required for a clean separation [85], can be achieved over a large range of z; at low energies, however, this can only be reached at larger values of z. On the other hand, at fixed x and Q^2 the large-z region corresponds to resonance dominance of the undetected hadronic system X (corresponding to small W'), so that the factorized description in terms of partonic distributions must eventually break down. It is vital therefore to establish empirically the limits beyond which the simple x and z factorization of Eq. (16) is no longer valid.

It is intriguing in particular to observe whether W' can play a role analogous to W for duality in inclusive scattering, when the undetected hadronic system X is dominated by resonances $W' \lesssim 2$ GeV. In terms of hadronic variables the fragmentation process can be described through the excitation of nucleon resonances, N^*, and their subsequent decays into pions (or other mesons) and lower-lying resonances, N'^*. The hadronic description must be rather elaborate, however, as the production of fast outgoing pions in the current fragmentation region at high energy requires nontrivial cancellations of the angular distributions from various decay channels [44, 45, 86],

$$\mathcal{N}_N^h(x,z) = \sum_{N'^*}\left|\sum_{N^*} F_{\gamma^* N\to N^*}(Q^2, W^2)\, \mathcal{D}_{N^*\to N'^* h}(W^2, W'^2)\right|^2 \qquad (17)$$

where $F_{\gamma^* N\to N^*}$ is the $N \to N^*$ transition form factor, which depends on the masses of the virtual photon and excited nucleon ($W = M_{N^*}$), and $\mathcal{D}_{N^*\to N'^* h}$ is a function representing the decay $N^* \to N'^* h$.

A dedicated experiment (E00-018) to study duality in $\pi^\pm$ electroproduction was performed in Hall C [32, 33], in which a 5.5 GeV electron beam was scattered from proton and deuteron targets at Q^2 between 1.8 and 6.0 GeV^2, for $0.3 \le x \le 0.55$, with z in the range $0.35 - 1$. From the deuterium data the ratio of unfavored to favored fragmentation functions D^-/D^+ was constructed, where D^+ corresponds to a pion containing the struck quark (*e.g.*, π^+ from a struck u or $\bar{d}$ quark), while D^- describes the fragmentation of a quark not contained in the valence structure of the pion (*e.g.*, a d quark for the π^+). Since at moderate x the dependence on PDFs cancels, the fragmentation function ratio is approximately given by $D^-/D^+ = (4 - \mathcal{N}_d^{\pi^+}/\mathcal{N}_d^{\pi^-})/(4\mathcal{N}_d^{\pi^+}/\mathcal{N}_[^{\pi^-} - 1)$, where $\mathcal{N}_d^{\pi}$ is the yield of produced pions in Eq. (17).

The Jefferson Lab data for D^-/D^+ from E00-108 are shown in Fig. 8 as a function of z at fixed $x = 0.32$ and $Q^2 = 2.3$ GeV^2 [32], and compared with earlier HERMES data at higher energies [87]. Despite the different energies, there is good overall agreement between the two measurements, even though the Jefferson Lab data sit slightly higher. Furthermore, the D^-/D^+ ratio extracted from the Jefferson Lab data shows a smooth dependence on z, which is quite remarkable given that the data cover the full resonance region, $0.88 < W'^2 < 4.2$ GeV^2. This

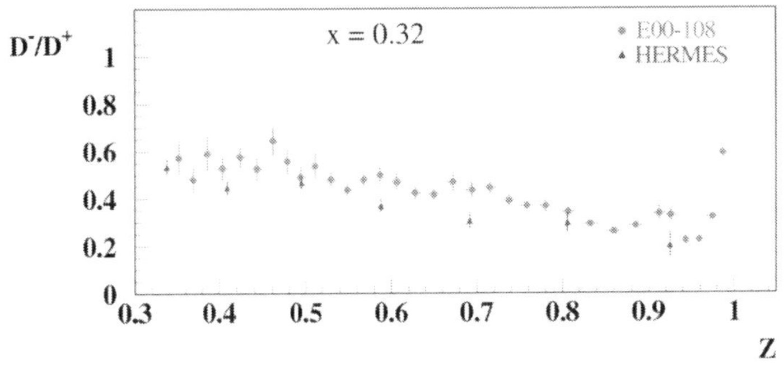

Figure 8. The ratio of unfavored to favored fragmentation functions D^-/D^+ as a function of z extracted from deuterium data, for $x = 0.32$ [32].

strongly suggests a suppression or cancellation of the resonance excitations in the π^+/π^- cross section ratio, and hence in the fragmentation function ratio.

Similar cancellations between resonances naturally arises in quark models, such as those discussed by Close *et al.* [45, 47] for the $\gamma N \to \pi^\pm N'^*$ reaction. The pattern of constructive and destructive interference, which was a crucial feature of the appearance of duality in inclusive structure functions, is also repeated in the semi-inclusive case when one sums over the states N'^*. Moreover, the smooth behavior of the fragmentation function ratio D^-/D^+ in Fig. 8 can be qualitatively understood from the relative weights of the matrix elements, which are always 4 times larger than for π^- production. In this case the resonance contributions to this ratio cancel exactly, leaving behind only the smooth background as would be expected at high energies. This may account for the striking lack of resonance structure in the resonance region fragmentation functions in Fig. 8.

8. Outlook

8.1. Impact of Jefferson Lab data

The first decade of unpolarized structure function measurements at Jefferson Lab has had significant impact on our understanding of nucleon structure, both for leading twist parton distributions and for the resonance–scaling transition and related studies of quark-hadron duality. With most of the data concentrated in the low-W region in the $Q^2 \sim$ few GeV^2 range, the greatest influence on the global data base has naturally been at large x.

Recently a new global PDF analysis was performed [9], exploring the possibility of reducing the uncertainties at large x by relaxing the constraints on the kinematics over which data are included in the fit. The data sets combined proton and deuteron DIS structure functions from Jefferson Lab, SLAC and CERN (NMC) with new *ep* collider data from HERA, as well as new Drell-Yan, W asymmetry and jet cross sections from *pp* and *pd* collisions at Fermilab. The new fit (referred to as "CTEQ6X") allowed for a significant increase in the large-x data set (*e.g.*, a factor of two more DIS data points) by incorporating data for $W^2 > 3$ GeV^2 and $Q^2 > 1.69$ GeV^2, lower than in the standard global fits [5, 88] which typically use $W^2 > 12.25$ GeV^2 and $Q^2 > 4$ GeV^2. The new analysis also systematically studied the effects of target mass and higher twist contributions, and realistic nuclear corrections for deuterium data.

Results from the CTEQ6X fit are shown in Fig. 9 (left) for the u and d quark PDFs, normalized to the earlier CTEQ6.1 fit, which had no nuclear or subleading $1/Q^2$ corrections applied. The biggest change is the $\sim$ 30–40% suppression of the d quark at $x \sim 0.8$, which is found to be stable with respect to variations in the W^2 and Q^2 cuts, provided both TMC and higher twist corrections are included. The effect of the expanded data base is more dramatically

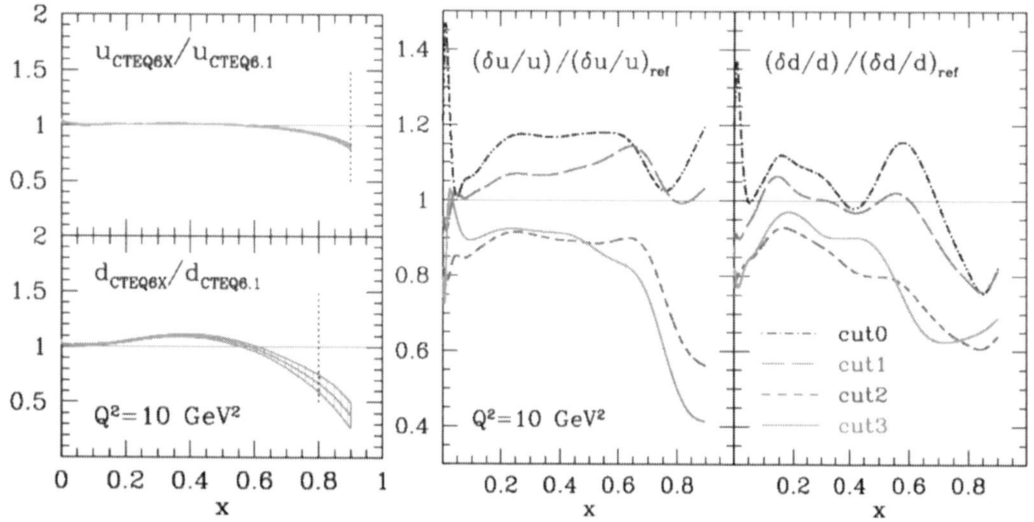

Figure 9. *(Left)* CTEQ6X fit for u and d quark PDFs, normalized to the earlier CTEQ6.1 fit [88]. The vertical lines show the approximate values of x above which PDFs are not directly constrained by data. The error bands correspond to $\Delta\chi = 1$. *(Right)* Relative errors on u and d quark PDFs, normalized to the relative errors in the reference fit.

illustrated in Fig. 9 (right), which shows the relative u and d PDF errors for a range of W and Q^2 cuts ("cut0" being the standard cut, "cut3" the more stringent CTEQ6X cut, and intermediate cuts "cut1" and "cut2"), normalized to those of a reference fit with "cut0" and no nuclear or subleading corrections. The result is a reduction of the errors by up to 40–60% at $x \gtrsim 0.7$, which will have a profound impact on applications of PDFs in high energy processes, such as those as the LHC, as well as in constraining low-energy models of quark distributions.

8.2. Future prospects

Uncertainties in PDFs will be further reduced with the availability of data at even larger x and Q^2 from Jefferson Lab after its 12 GeV energy upgrade, which will determine the d quark distribution up to $x \sim 0.8$ with minimal theoretical uncertainties associated with nuclear corrections. Several of the planned experiments include BoNuS12 [79], which will extend to larger x the earlier measurements of low-momentum, backward protons in semi-inclusive scattering from deuterium (Sec. 5.2); the MARATHON experiment [73], which plans to extract F_2^n/F_2^p from the ratio of ^{3}He to ^{3}H structure functions, in which the nuclear corrections cancel to within $\sim 1\%$ [71, 72]; and the program of parity-violating DIS measurements on hydrogen [76], which will be sensitive to a new combination of d/u in the proton, free of nuclear corrections.

In the resonance region, experiment E12-10-002 [89] will extend proton and deuteron structure function measurements up to $Q^2 \sim 17$ GeV^2 and enable tests of quark-hadron duality over a much larger kinematic range, ultimately providing stronger constraints on large-x PDFs. And finally, a new avenue for exploring nucleon structure at 12 GeV will be opened up with semi-inclusive meson production experiments [90], which will test the factorization of scattering and fragmentation subprocesses needed for a partonic interpretation of semi-inclusive cross sections. A successful program of semi-inclusive measurements tagging specific mesons in the final state would allow unprecedented access to the flavor dependence of PDFs in previously unexplored regions of kinematics.

Acknowledgments

We thank R. Ent and C. E. Keppel for their contributions to this review in the early stages of its development. This work was supported by the U.S. Department of Energy under Contract No. DE-FG02-03ER41231, and DOE contract No. DE-AC05-06OR23177, under which Jefferson Science Associates, LLC operates Jefferson Lab.

References

[1] W. K. H. Panofsky, in Proceedings of the 14th International Conference on High Energy Physics, Vienna, Austria (1968); E. D. Bloom *et al.*, Phys. Rev. Lett. **23**, 930 (1969); M. Breidenbach *et al.*, Phys. Rev. Lett. **23**, 935 (1969).
[2] R. E. Taylor, in International Symposium of Lepton and Photon Interactions, Stanford (1975).
[3] R. P. Feynman, Phys. Rev. Lett. **23**, 1415 (1969); J. D. Bjorken and E. A. Paschos, Phys. Rev. **185**, 1975 (1969).
[4] P. M. Nadolsky *et al.*, Phys. Rev. D **78**, 013004 (2008).
[5] A. D. Martin, W. J. Stirling, R. S. Thorne and G. Watt, Eur. Phys. J. C **63**, 189 (2009).
[6] S. Alekhin, J. Blumlein, S. Klein and S. Moch, Phys. Rev. D **81**, 014032 (2010).
[7] R. D. Ball *et al.*, arXiv:1101.1300 [hep-ph].
[8] A. Levy, arXiv:hep-ph/0002015.
[9] A. Accardi *et al.*, Phys. Rev. D **81**, 034016 (2010); A. Accardi *et al.*, arXiv:1102.3686 [hep-ph].
[10] O. Nachtmann, Nucl. Phys. **B63**, 237 (1973).
[11] O. W. Greenberg and D. Bhaumik, Phys. Rev. D **4**, 2048 (1971).
[12] L. N. Hand, Phys. Rev. **129**, 1834 (1963).
[13] M. N. Rosenbluth, Phys. Rev. **79**, 615 (1950).
[14] K. Wilson, Phys. Rev. **179**, 1499 (1969).
[15] H. Georgi and H. D. Politzer, Phys. Rev. D **14**, 1829 (1976).
[16] I. Schienbein *et al.*, J. Phys. G **35**, 053101 (2008).
[17] R. K. Ellis, W. Furmanski and R. Petronzio, Nucl. Phys. **B212**, 29 (1983); S. Kretzer and M. H. Reno, Phys. Rev. D **66**, 113007 (2002); A. Accardi and J. W. Qiu, JHEP **07**, 090 (2008);
[18] Y. Liang *et al.*, arXiv:nucl-ex/0410027.
[19] S. P. Malace *et al.*, Phys. Rev. C **80**, 035207 (2009).
[20] Jefferson Lab Experiment E00-002, F_2^N *at low* Q^2, C. E. Keppel and M. I. Niculescu spokespersons.
[21] L. W. Whitlow *et al.*, Phys. Lett. B **282**, 475 (1992); L. W. Whitlow, Ph.D. thesis, American University (1990).
[22] M. Arneodo *et al.*, Nucl. Phys. **B483**, 3 (1997).
[23] H. Abramowicz, E. M. Levin, A. Levy and U. Maor, Phys. Lett. B **269**, 465 (1991); H. Abramowicz and A. Levy, hep-ph/9712415.
[24] M. E. Christy and P. E. Bosted, Phys. Rev. C **81**, 055213 (2010).
[25] M. Osipenko *et al.*, Phys. Rev. D **67**, 092001 (2003).
[26] I. Niculescu *et al.*, Phys. Rev. Lett. **85** 1182, 1186 (2000).
[27] M. E. Christy *et al.*, Phys. Rev. C **70**, 015206 (2004).
[28] V. Tvaskis *et al.*, Phys. Rev. C **73**, 025206 (2006).
[29] Y. Liang, M. E. Christy, R. Ent and C. E. Keppel, Phys. Rev. C **73**, 065201 (2006).
[30] V. Tvaskis *et al.*, Phys. Rev. Lett. **98**, 142301 (2007).
[31] M. Osipenko *et al.*, Phys. Rev. C **73**, 045205 (2006).
[32] T. Navasardyan *et al.*, Phys. Rev. Lett. **98**, 022001 (2007).
[33] H. Mkrtchyan *et al.*, Phys. Lett. B **665**, 20 (2008).
[34] V. Tvaskis, J. Steinman and R. Bradford, Nucl. Phys. Proc. Suppl. **159**, 163 (2006).
[35] Jefferson Lab Experiment E03-012, *The structure of the free neutron via spectator tagging*, H. C. Fenker, C. E. Keppel, S. Kuhn, and W. Melnitchouk spokespersons.
[36] Jefferson Lab Experiment E06-009, *Measurement of* $R = \sigma_L/\sigma_T$ *on deuterium in the nucleon resonance region and beyond*, M. E. Christy and C. E. Keppel spokespersons.
[37] E. D. Bloom and F. J. Gilman, Phys. Rev. Lett. **25**, 1140 (1970); Phys. Rev. D **4**, 2901 (1971).
[38] G. Miller *et al.*, Phys. Rev. D **5**, 528 (1972).
[39] W. Melnitchouk, R. Ent and C. E. Keppel, Phys. Rep. **406**, 127 (2005).
[40] S. Alekhin, Phys. Rev. D **68**, 014002 (2003).
[41] A. D. Martin, R. G. Roberts, W. J. Stirling and R. S. Thorne, Phys. Lett. B **604**, 61 (2004).
[42] S. Dasu *et al.*, Phys. Rev. D **49**, 5641 (1994).
[43] A. D. Martin, R. G. Roberts, W. J. Stirling and R. S. Thorne, Eur. Phys. J. C **4**, 463 (1998).

[44] N. Isgur, S. Jeschonnek, W. Melnitchouk and J. W. Van Orden, Phys. Rev. D **64**, 054005 (2001).
[45] F. E. Close and N. Isgur, Phys. Lett. B **509**, 81 (2001).
[46] F. E. Close and W. Melnitchouk, Phys. Rev. C **68**, 035210 (2003).
[47] F. E. Close and W. Melnitchouk, Phys. Rev. C **79**, 055202 (2009).
[48] A. De Rujula, H. Georgi and H. D. Politzer, Annals Phys. **103**, 315 (1977).
[49] X. D. Ji and P. Unrau, Phys. Rev. D **52**, 72 (1995).
[50] M. E. Christy *et al.*, unpublished (2004).
[51] C. S. Armstrong *et al.*, Phys. Rev. D **63**, 094008 (2001).
[52] S. Forte and L. Magnea, Phys. Lett. B **448**, 295 (1999); S. Forte, L. Magnea, A. Piccione and G. Ridolfi, Nucl. Phys. **B594**, 46 (2001); A. Piccione, Phys. Lett. B **518**, 207 (2001).
[53] D. Kotlorz and A. Kotlorz, Phys. Lett. B **644**, 284 (2007).
[54] A. Psaker, W. Melnitchouk, M. E. Christy and C. Keppel, Phys. Rev. C **78**, 025206 (2008).
[55] V. N. Gribov and L. N. Lipatov, Sov. J. Nucl. Phys. **15**, 438, 675 (1972); Y. L. Dokshitzer, Sov. Phys. JETP **46**, 641 (1977); G. Altarelli and G. Parisi, Nucl. Phys. **B126**, 298 (1977).
[56] J. Arrington *et al.*, Phys. Rev. C **64**, 014602 (2001).
[57] V. Tvaskis *et al.*, Phys. Rev. Lett. **98**, 142301 (2007).
[58] I. Niculescu, J. Arrington, R. Ent and C. E. Keppel, Phys. Rev. C **73**, 045206 (2006).
[59] M. Osipenko, W. Melnitchouk, S. Simula, S. A. Kulagin and G. Ricco, Nucl. Phys. **A766**, 142 (2006).
[60] W. Detmold, W. Melnitchouk and A. W. Thomas, Phys. Rev. D **66**, 054501 (2002).
[61] W. Detmold *et al.*, Phys. Rev. Lett. **87**, 172001 (2001).
[62] M. Osipenko *et al.*, Nucl. Phys. Proc. Suppl. **174**, 23 (2007).
[63] W. Melnitchouk, A. W. Schreiber and A. W. Thomas, Phys. Rev. D **49**, 1183 (1994); Phys. Lett. B **335**, 11 (1994).
[64] S. A. Kulagin, G. Piller and W. Weise, Phys. Rev. C **50**, 1154 (1994).
[65] S. A. Kulagin and R. Petti, Nucl. Phys. **A765**, 126 (2006).
[66] Y. Kahn, W. Melnitchouk and S. A. Kulagin, Phys. Rev. C **79**, 035205 (2009).
[67] W. Melnitchouk and A. W. Thomas, Phys. Lett. B **377**, 11 (1996).
[68] S. P. Malace, Y. Kahn, W. Melnitchouk and C. E. Keppel, Phys. Rev. Lett. **104**, 102001 (2010).
[69] A. Bodek *et al.*, Phys. Rev. D **20**, 1471 (1979); A. Bodek and J. L. Ritchie, Phys. Rev. D **23**, 1070 (1981).
[70] P. E. Bosted and M. E. Christy, Phys. Rev. C **77**, 065206 (2008).
[71] I. R. Afnan *et al.*, Phys. Lett. B **493**, 36 (2000); Phys. Rev. C **68**, 035201 (2003).
[72] E. Pace, G. Salme, S. Scopetta and A. Kievsky, Phys. Rev. C **64**, 055203 (2001).
[73] Jefferson Lab Experiment PR12-06-118, *Measurement of the F_2^n/F_2^p, d/u ratios and $A = 3$ EMC effect in DIS off the tritium and helium mirror nuclei*, G. G. Petratos contact person.
[74] L. L. Frankfurt and M. I. Strikman, Phys. Rept. **76**, 215 (1981); S. Simula, Phys. Lett. B **387**, 245 (1996); W. Melnitchouk, M. Sargsian and M. I. Strikman, Z. Phys. A **359**, 99 (1997).
[75] P. A. Souder, AIP Conf. Proc. **747**, 199 (2005).
[76] Jefferson Lab Experiment E12-10-107, *Precision Measurement of parity-violation in DIS over a broad kinematic range*, P. Souder contact person.
[77] T. Hobbs and W. Melnitchouk, Phys. Rev. D **77**, 114023 (2008).
[78] H. C. Fenker *et al.*, Nucl. Instrum. Meth. A **592**, 273 (2008).
[79] Jefferson Lab Experiment PR12-06-113, *The structure of the free neutron at large x-Bjorken*, S. Bültmann contact person.
[80] F. D. Aaron *et al.*, Phys. Lett. B **665**, 139 (2008).
[81] V. Tvaskis *et al.*, Phys. Rev. Lett. **98**, 142301 (2007).
[82] G. Altarelli, R. K. Ellis, G. Martinelli and S. Y. Pi, Nucl. Phys. **B160**, 301 (1979).
[83] A. Afanasev, C. E. Carlson and C. Wahlquist, Phys. Rev. D **62**, 074011 (2000).
[84] A. Accardi, T. Hobbs and W. Melnitchouk, JHEP **0911**, 084 (2009).
[85] E. L. Berger, Nucl. Phys. **B85**, 61 (1975); in Proceedings of the Workshop on Electronuclear Physics with Internal Targets, SLAC (1987).
[86] W. Melnitchouk, AIP Conf. Proc. **588**, 267 (2001).
[87] P. Geiger, Ph.D. dissertation, Heidelberg U. (1998); B. Hommez, Ph.D. dissertation, Gent U. (2003).
[88] J. Pumplin *et al.*, JHEP **0207**, 012 (2002); D. Stump *et al.*, JHEP **0310**, 046 (2003).
[89] Jefferson Lab Experiment E12-10-002, *Precision measurements of the F_2 structure function at large x in the resonance region and beyond*, S. P. Malace contact person.
[90] Jefferson Lab Experiment E12-06-104, *Measurement of the ratio $R = \sigma_L/\sigma_R$ in semi-inclusive DIS*, R. Ent contact person.

Journal of Physics: Conference Series **299** (2011) 012005 doi:10.1088/1742-6596/299/1/012005

Spin Structure Functions

J. P. Chen[1], **A. Deur**[1], **S. Kuhn**[2] **and Z. E. Meziani**[3]

[1] Jefferson Lab, Newport News, Virginia 23606, USA
[2] Old Dominion University, Norfolk, Virginia 23529, USA
[3] Temple University, Philadelphia, Pennsylvania 19122, USA

Abstract. Spin-dependent observables have been a powerful tool to probe the internal structure of the nucleon and to understand the dynamics of the strong interaction. Experiments involving spin degrees of freedom have often brought out surprises and puzzles. The so-called "spin crisis" in the 1980s revealed the limitation of naive quark-parton models and led to intensive worldwide efforts, both experimental and theoretical, to understand the nucleon spin structure. With high intensity and high polarization of both the electron beam and targets, Jefferson Lab has the world's highest polarized luminosity and the best figure-of-merit for precision spin structure measurements. It has made a strong impact in this subfield of research. This chapter will highlight Jefferson Lab's unique contributions in the measurements of valence quark spin distributions, in the moments of spin structure functions at low to intermediate Q^2, and in the transverse spin structure.

1. Introduction

1.1. The spin structure functions g_1 and g_2

For inclusive polarized electron scattering off a polarized nucleon target, the cross section depends on four structure functions, $F_1(x,Q^2)$, $F_2(x,Q^2)$, $g_1(x,Q^2)$ and $g_2(x,Q^2)$, where F_1 and F_2 are the unpolarized structure functions and g_1 and g_2 the polarized ones. In the naive quark parton model, F_1 or F_2 gives the quark momentum distribution and g_1 probes the quark spin distribution [1]:

$$g_1^p = \frac{1}{2}\left(\frac{4}{9}(\Delta u + \Delta\overline{u}) + \frac{1}{9}(\Delta d + \Delta\overline{d}) + \frac{1}{9}(\Delta s + \Delta\overline{s})\right), \tag{1}$$

where Δu ($\Delta\overline{u}$), Δd ($\Delta\overline{d}$), and Δs ($\Delta\overline{s}$), are the polarized $u(\overline{u})$, $d(\overline{d})$, and $s(\overline{s})$ quark (antiquark) distributions (number of quarks with their helicity aligned minus those with their helicity anti-aligned with the nucleon spin), respectively, and heavy quark (c, b and t) contributions are neglected. In naive quark models, it is often assumed that the three flavors of quarks follow the SU(3) flavor symmetry. Then the knowledge from the β decays of the neutron and the hyperons can be combined with the first moment (Γ_1) of g_1 to determine the spin-flavor decomposition. The β-decay constants determine the octet parts of the flavor contributions and, together with Γ_1, can be used to reveal the value of the flavor singlet contribution, the total quark contribution to the proton spin. Consequently, precise measurements of g_1 have been a main goal of the nucleon spin structure study. In the QCD-improved parton model, the simple interpretation of g_1 as the quark spin distribution holds only at leading twist. Gluon radiations come in at higher orders leading to logarithmic scaling violation, which are well described by the QCD evolution equations. Higher twist effects ($1/Q^{t-2}$ power corrections, where t is the twist rank) can become important in the low Q^2 region.

The second spin structure function, g_2, is related to the transverse spin structure, $g_T = g_1 + g_2$. Although g_2 has no simple interpretation in the naive parton model, it provides clean information on the quark-gluon correlations through higher twist effects. The standard operator product expansion (OPE) analysis has shown that at large Q^2, g_2 is related to matrix elements of both twist-2 and twist-3 operators. Neglecting quark masses, g_2 can be separated into a twist-2 (Wandzura-Wilczek) and a twist-3 (and higher) term:

$$g_2(x, Q^2) = g_2^{WW}(x, Q^2) + \overline{g_2}(x, Q^2), \tag{2}$$

where the twist-3 (and higher) term, $\overline{g_2}$, is not $1/Q$ suppressed with respect to the twist-2 term. The twist-2 term can be determined from the twist-2 part of g_1 as [2]

$$g_2^{WW}(x, Q^2) = -g_1(x, Q^2) + \int_x^1 \frac{g_1(y, Q^2)}{y} dy. \tag{3}$$

Therefore, g_2 provides a clean way to study twist-3 effects. In addition, at high Q^2, the x^2-weighted moment, d_2, is a twist-3 matrix element and is related to the color polarizabilities [3]:

$$d_2(Q^2) = \int_0^1 x^2 \left(g_2(x, Q^2) - g_2^{WW}(x, Q^2)\right) dx. \tag{4}$$

Predictions for d_2 exist from lattice QCD [4] and various models [5, 6, 7].

1.2. *Virtual photon asymmetries*

Other physics observables of interest are the virtual photon-nucleon asymmetries A_1 and A_2:

$$A_1 = \frac{g_1 - (Q^2/\nu^2)g_2}{F_1} \approx \frac{g_1}{F_1}, \tag{5}$$

$$A_2 = \frac{Q^2}{\nu^2} \frac{g_1 + g_2}{F_1}. \tag{6}$$

The high-x region is of special interest for A_1 measurements because there the valence quark contributions are expected to dominate. With sea quark and explicit gluon contributions expected to be unimportant, it is a clean region to test our understanding of nucleon structure. Relativistic constituent quark models [8] should be applicable in this region and perturbative QCD (pQCD) [9] can be used to make predictions in the large-x limit. To first approximation, the constituent quarks in the nucleon are described by SU(6) wave functions. SU(6) symmetry leads to the following predictions [10]:

$$A_1^p = 5/9; \quad A_1^n = 0; \quad \Delta u/u = 2/3; \quad \Delta d/d = -1/3. \tag{7}$$

Relativistic constituent quark models (RCQM) with broken SU(6) symmetry, *e.g.*, the hyperfine interaction model [8], predict that as $x \to 1$:

$$A_1^p \to 1; \quad A_1^n \to 1; \quad \Delta u/u \to 1; \quad \text{and} \quad \Delta d/d \to -1/3. \tag{8}$$

In the RCQM, relativistic effects lead to a nonzero orbital angular momentum and reduce the valence quark contributions to the nucleon spin from 1 to 0.6–0.75.

Journal of Physics: Conference Series **299** (2011) 012005 doi:10.1088/1742-6596/299/1/012005

Another approach is leading order pQCD [9], which assumes the quark orbital angular momentum to be negligible and leads to hadron helicity conservation. It yields:

$$A_1^p \to 1; \quad A_1^n \to 1; \quad \Delta u/u \to 1; \quad \text{and} \quad \Delta d/d \to 1. \tag{9}$$

Not only are the limiting values of A_1 and polarized PDFs as $x \to 1$ important, but also their behavior in the high-x region, which is sensitive to the dynamics in the valence quark region. An improved approach [11] in pQCD, including quark orbital angular momentum (and therefore not requiring hadron helicity conservation), shows a different behavior at large x while keeping the same limiting value at $x = 1$. In particular, the approach of $\Delta d/d$ towards unity only sets in at a significantly larger x than predicted by hadron helicity conservation.

1.3. Sum rules

Sum rules involving the spin-structure functions offer an important opportunity to study QCD. Spin sum rules relate the moments of the spin-structure functions to the nucleon's static properties (as in the Bjorken or Gerasimov, Drell and Hearn (GDH) sum rules), or real or virtual Compton amplitudes, which can be calculated theoretically (as in the generalized GDH or the forward spin polarizability sum rules). Refs. [12, 13] provide comprehensive reviews on this subject. The Ellis-Jaffe sum rule [14] was the first to be experimentally studied. By assuming SU(3) flavor symmetry and that the strange sea quarks are unpolarized, this sum rule links, in the Bjorken limit ($Q^2 \to \infty$), the first moment of g_1 to the polarized quark distributions Δu and Δd:

$$\Gamma_1^p \equiv \int_0^1 g_1^p dx \to \frac{1}{2}\left(\frac{4}{9}\Delta u + \frac{1}{9}\Delta d\right), \tag{10}$$

$$\Gamma_1^n \equiv \int_0^1 g_1^n dx \to \frac{1}{2}\left(\frac{4}{9}\Delta d + \frac{1}{9}\Delta u\right), \tag{11}$$

with the values given by the β-decay constants of the neutron and the hyperons. This sum rule has been experimentally found to be violated, resulting in the so-called "spin crisis" of the 1980s and the realization that at least some of the assumptions leading to Eqs. (10)–(11) are incorrect.

A sum rule on the first moment of g_2 also exists: the Burkhardt-Cottingham (BC)sum rule [15] states that the first moment of g_2 is zero at any Q^2:

$$\Gamma_2 \equiv \int_0^1 g_2(x, Q^2) dx = 0. \tag{12}$$

The sum rule assumes that g_2 has good convergence properties and is analytic in the $x \to 0$ limit. In recent years the Bjorken sum rule [16] at large Q^2 and the GDH sum rule [17] at $Q^2 = 0$ have attracted considerable experimental and theoretical efforts.

The Bjorken sum rule links the isovector part of the g_1 moment to g_A, the nucleon axial charge that governs neutron β-decay. It has been extended to finite Q^2 with the OPE technique:

$$\Gamma_1^{p-n}(Q^2) \equiv \int_0^1 (g_1^p - g_1^n) dx = \frac{g_A}{6}\left(1 - \frac{\alpha_s}{\pi} - 3.58\left(\frac{\alpha_s}{\pi}\right)^2 - 20.21\left(\frac{\alpha_s}{\pi}\right)^3 + \cdots\right) + \mathcal{O}\left(\frac{1}{Q^2}\right). \tag{13}$$

It is a rigorous sum rule based on general principles. Studies of Γ_1^N and of the Bjorken sum rule at high Q^2 can be used to extract higher twist information, such as the OPE matrix element f_2. These moments are also well suited for lattice QCD calculations.

The GDH sum rule applies to the opposite end of the Q^2 scale and links the transverse-transverse interference cross section for photoproduction σ_{TT} to the nucleon anomalous moment κ:

$$\int_{\nu_0}^{\infty} \frac{\sigma_{TT}(\nu)}{\nu} d\nu = -\frac{2\pi^2 \alpha \kappa^2}{M^2}. \tag{14}$$

It is, too, a rigorous sum rule. For electroproduction, σ_{TT} can be expressed in term of g_1 and g_2. This allows one to generalize the GDH sum rule [18] to nonzero Q^2 and connect it with the Bjorken sum rule. It provides a clean way to test theories with experimental data over the entire Q^2 range.

The generalized spin polarizabilities γ_0 and δ_{LT} are higher moments of σ_{TT} and σ_{LT} (the longitudinal-transverse interference cross section) and are derived in a similar manner to the generalized GDH sum rule.

$$\gamma_0(Q^2) = \left(\frac{1}{2\pi^2}\right) \int_{\nu_0}^{\infty} \frac{K(\nu, Q^2)}{\nu} \frac{\sigma_{TT}(\nu, Q^2)}{\nu^3} d\nu, \tag{15}$$

$$\delta_{LT}(Q^2) = \left(\frac{1}{2\pi^2}\right) \int_{\nu_0}^{\infty} \frac{K(\nu, Q^2)}{\nu} \frac{\sigma_{LT}(\nu, Q^2)}{Q\nu^2} d\nu, \tag{16}$$

where K is the virtual photon flux. They provide benchmark tests of chiral perturbation theory (χPT) calculations at low Q^2. Since the generalized polarizabilities have an extra $1/\nu^2$ weighting compared to the first moments (GDH sum or Γ_1), these integrals have smaller contributions from the large-ν region and converge much faster, which minimizes the uncertainty due to the experimentally inaccessible region at large ν. At low Q^2, the generalized polarizabilities have been evaluated with next-to-leading (NLO) order χPT calculations [19, 20]. One issue in the χPT calculations is how to properly include the nucleon resonance contributions, especially the Δ resonance. As was pointed out in Refs. [19, 20] , while γ_0 is sensitive to resonances, δ_{LT} is insensitive to the Δ resonance, making it an ideal candidate to test χPT calculations. Measurements of the generalized spin polarizabilities are an important step in understanding the dynamics of QCD in the chiral perturbation region.

1.4. Quark-hadron duality

In 1970, Bloom and Gilman [21] noted that the unpolarized nucleon resonance data averaged to the DIS scaling curve. This phenomenon, called quark-hadron duality, implies that there exists a domain where the fundamental (partonic) and effective (hadronic) descriptions of the strong force are both valid. In an earlier study [22] it was linked to the smallness of higher twists. It is natural to ask whether duality also holds for g_1 which, contrary to F_2 or F_1, can take negative values as well. The investigation of quark-hadron duality will aid in the study of the higher twist effects and the understanding of the high-x behavior in DIS. A detailed review can be found in [23].

1.5. Transversity

The transversity distributions, $\delta q(x, Q^2)$, are fundamental leading twist (twist-2) quark distributions, similar to the unpolarized and polarized parton distributions, $q(x, Q^2)$ and $\Delta q(x, Q^2)$. In quark-parton models, they describe the net transverse polarization of quarks in a transversely polarized nucleon. Several special features of the transversity distributions make them uniquely interesting:

Journal of Physics: Conference Series **299** (2011) 012005 doi:10.1088/1742-6596/299/1/012005

- The difference between the transversity and the longitudinal distributions is purely due to relativistic effects. In the absence of relativistic effects (as in the nonrelativistic quark model, where boosts and rotations commute), the transversity distributions are identical to the longitudinally polarized distributions.
- The quark transversity distributions do not mix with gluonic effects [24] and therefore follow a much simpler evolution and have a valence-like behavior.
- The positivity of helicity amplitudes leads to the Soffer's inequality for the transversity[25]: $|\delta q| \leq \frac{1}{2}(q + \Delta q)$.
- The lowest moment of δq measures a simple local operator analogous to the axial charge, known as the "tensor charge", which can be calculated from lattice QCD.

Due to the chiral-odd nature of the transversity distribution, it cannot be measured in inclusive DIS experiments. In order to measure $\delta q(x, Q^2)$, an additional chiral-odd object is required, such as double-spin asymmetries in Drell-Yan processes, single target-spin azimuthal asymmetries in semi-inclusive DIS (SIDIS) reactions, double-spin asymmetries in Λ production from ep and pp reactions, and single-spin asymmetries in double pion production from ep scattering. The first results, from measurements performed by the HERMES [26] and COMPASS [27] collaborations with SIDIS, offered a first glimpse of possible effects caused by the transversity distributions.

2. Status of spin structure study prior to Jefferson Lab

The doubly polarized inclusive experiments in the DIS domain performed at CERN, SLAC and DESY are the cornerstones of our understanding of the nucleon spin structure. Their precise measurement of g_1 and A_1 (and g_2 and A_2 with less precision) taught us that:

• The strong force, responsible for the nucleon structure, is well described by pQCD at large Q^2, even in the spin sector (where the explicit spin degrees of freedom are important).

• Modeling the nucleon as a sum of quasi-free partons, corrected for QCD gluon radiation, is well suited to describe the longitudinal spin of the nucleon in the DIS domain. Thus, the nucleon spin S_N can be written as [30]:

$$S_N = \frac{1}{2} = \frac{1}{2}\Delta\Sigma + L_q + J_G, \tag{17}$$

where $\Delta\Sigma$ is the contribution of the quark spins to the nucleon spin, L_q is the quark orbital momentum contribution and J_G is the total gluon contribution.

• The quark spin contribution $\Delta\Sigma$ is about 0.3. Consequently, contributions from the gluon and/or the quark orbital momentum are significant [31].

• The Ellis-Jaffe sum rule (10)–(11), is violated for both nucleons. This implies that either the strange quarks are significantly polarized, or that the SU(3) symmetry is broken.

• The overall effects of higher twists are small for Q^2 greater than a few GeV^2.

In addition, the polarized distributions Δu_v, Δd_v, Δs and Δg were extracted through global analysis of inclusive DIS data (usually with certain assumptions), see Fig. 1 for a recent fit. The polarized gluon distribution has large uncertainties since it is accessible only indirectly by inclusive experiments. The Δs is negative, but also with large uncertainties.

From more recent semi-inclusive results we learned that:

• The polarized gluon distribution Δg at $x \simeq 0.1$ is compatible with zero. The various experimental results on $\Delta g/g$ are shown on Fig. 2. Since most of the gluon density lies at small x, this result hints that the integral of Δg, denoted by ΔG, is probably not as large as required by some explanations of the "spin crisis".

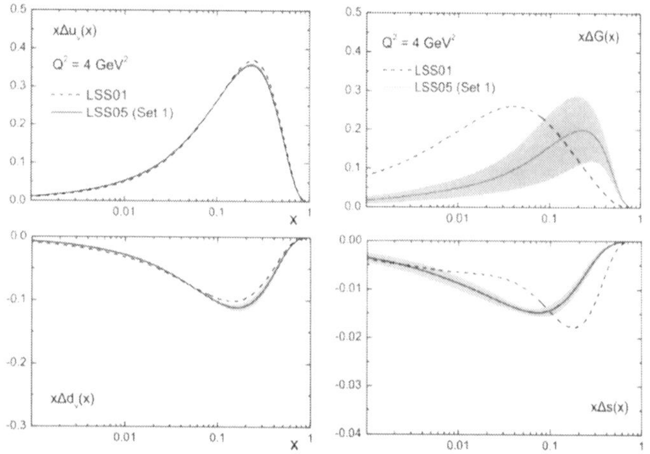

Figure 1. Polarized parton distributions in the proton by Leader, Stamenov and Sidorov [28].

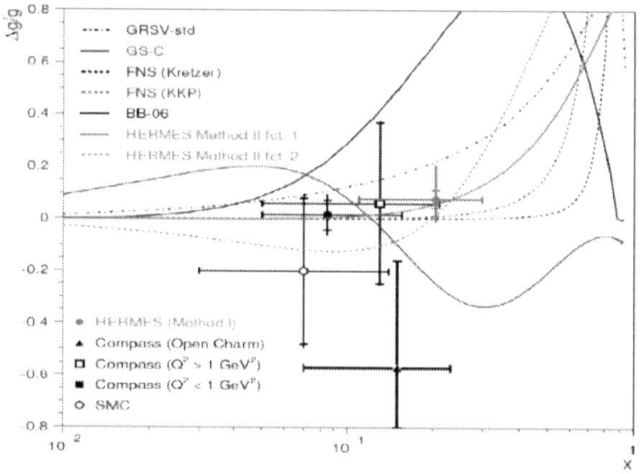

Figure 2. $\Delta g/g$ measurements [29].

• The strange quark polarization over the measured range (and within systematic uncertainties) is consistent with zero. This seems to be in contradiction with the results extracted from global analysis of inclusive data. It is important to independently verify this result.

The results of the high energy experiments left us with some open questions, such as • Is the contribution of L_q significant? • How large is the gluon contribution ΔG? • What is the flavor dependence of the spin structure?

The Jefferson Lab spin program in the last decade has helped answer some of the questions. More importantly, Jefferson Lab experiments have explored the high-x and low- to intermediate-Q^2 regions with high precision. The high-x region is where the valence quarks dominate and pQCD and quark models have predictions. The low-Q^2 region is where predictions from chiral perturbation theory, the leading effective theory of strong interaction at long distance, can be tested. Measurements in the intermediate-Q^2 region explore the transition between the descriptions of the strong force from the perturbative to nonperturbative regime of QCD. In particular, one can learn about the quark-gluon and quark-quark correlations through systematic

Journal of Physics: Conference Series **299** (2011) 012005 doi:10.1088/1742-6596/299/1/012005

study of higher twist effects. It is also where one can test if the quark-hadron duality works for the spin structure functions.

3. Results from Jefferson Lab

With a high current, high polarization electron beam of energy up to 6 GeV and state-of-the-art polarized targets, Jefferson Lab has completed a number of experiments which extended the database on spin structure functions significantly, both in kinematic range (low Q^2 and high x) and in precision. The neutron results are from Hall A using polarized ^{3}He as an effective polarized neutron target and two high resolution spectrometers. The polarized luminosity reached 10^{36} s^{-1}cm^{-2} and in-beam polarization improved from 35% (1998) to over 65% (2008). The proton and deuteron results are from Hall B with the CLAS detector and Hall C with the HMS spectrometer and using polarized NH_3 and ND_3 targets with in-beam polarization of about 80% and 40% respectively.

An example of data on the proton g_1 structure function together with the world data is shown in Fig. 3. The Jefferson Lab measurements are mostly focussing on:

(i) The high-x region and the nucleon resonance region;
(ii) The moments of spin structure functions, especially at low to moderate Q^2;
(iii) The higher twist effects related to parton correlations;
(iv) The contributions to the nucleon spin puzzle.

These results will be discussed in detail in the next sections.

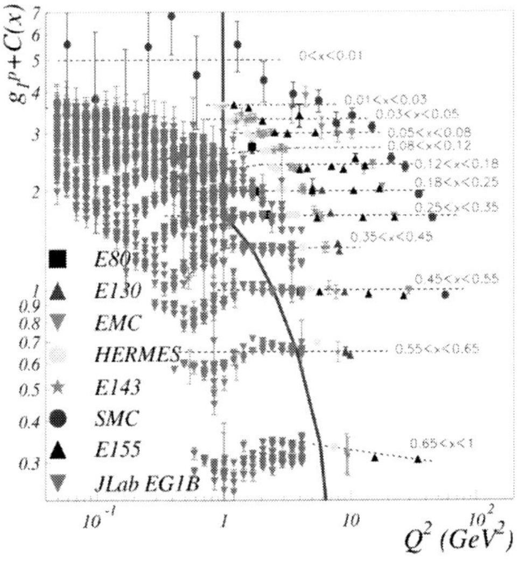

Figure 3. World results on g_1^p. The solid line is the conventional DIS frontier ($W = 2$ GeV, $Q^2 = 1$ GeV^2).

3.1. *Spin structure in the valence quark (high-x) region*

Jefferson Lab Hall A experiment E99-117 [32] measured the neutron asymmetry A_1^n with high precision from a polarized electron beam scattering off a ^{3}He target polarized either longitudinally or transversely. The experiment covered an x region from 0.33 to 0.61 (Q^2 from 2.7 to 4.8 GeV^2). The results on A_1^n are shown in the left panel of Fig. 4. The experiment greatly improved the precision of data in the high-x region, providing the first evidence that A_1^n becomes positive at large x, and clearly inconsistent with SU(6) symmetry. The results are in good agreement with the LSS 2001 pQCD fit to previous world data [28] (solid curve) and the statistical model [33] (long-dashed curve). The trend of the data is consistent with the RCQM [8] predictions (the shaded band). The data disagree with the predictions from the leading order pQCD models [9] (short-dashed and dash-dotted curves).

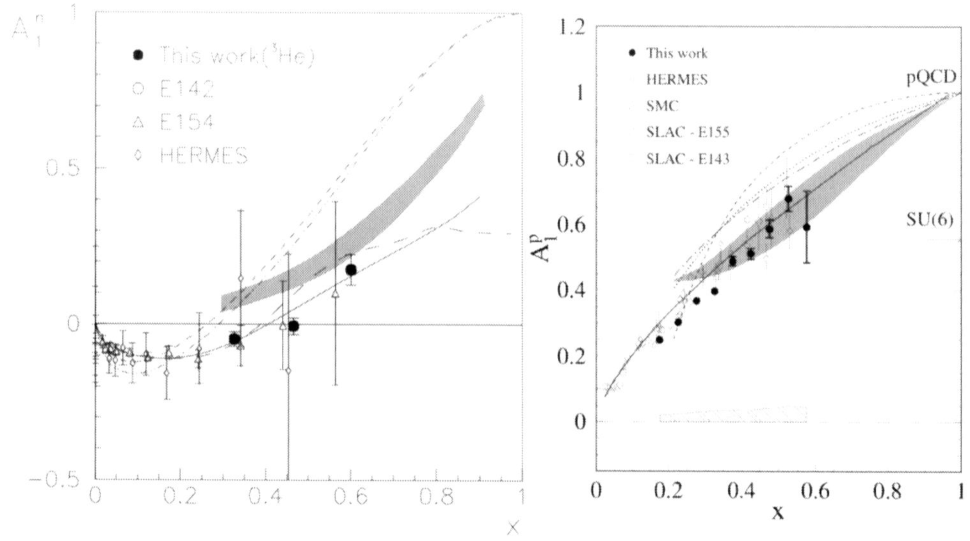

Figure 4. A_1^n (left panel) and A_1^p (right panel) results from Jefferson Lab Hall A E99-117 [32] and CLAS EG1b [34] experiments (filled circles), compared with the world data and theoretical predictions (see text for details).

New results for A_1^p and A_1^d from the Hall B EG1b experiment [34] are also available. The data cover the Q^2 range of 1.4 to 4.5 GeV^2 for x from 0.2 to 0.6 with an invariant mass larger than 2 GeV. The results on A_1^p are shown in the right panel of Fig. 4. The precision of the data improved significantly over that of the existing world data. Similar data also exist on the deuteron and once again exhibit a trend to exceed the asymmetries predicted by SU(6) (7) at large x.

In addition to the prediction by the RCQM [8] (again indicated by the shaded band), several curves based on different scenarios of SU(6) symmetry breaking in the context of quark-hadron duality, as predicted by Close and Melnitchouk [35], are also shown in the right panel of Fig. 4.

These data provide crucial input for the global fits to the world data to extract the polarized parton densities and the extractions of higher twist effects. The polarized quark distribution functions $\Delta u/u$ and $\Delta d/d$ in the high-x region were first extracted, in the leading order approximation, from the Hall A neutron data and the world proton data (left panel of Fig. 5) [32]. A recent leading order extraction used also the CLAS EG1b proton and deuteron data and the world data. The results are shown in the right panel of Fig. 5, along with predictions from leading order pQCD [9] (dashed curves) and a pQCD fit (solid line) including quark orbital angular momentum contributions [11]. The results for $\Delta d/d$ are in significant disagreement with the predictions from the leading order pQCD model assuming hadron helicity conservation. The data agree better with the fit including quark orbital angular momentum, suggesting that

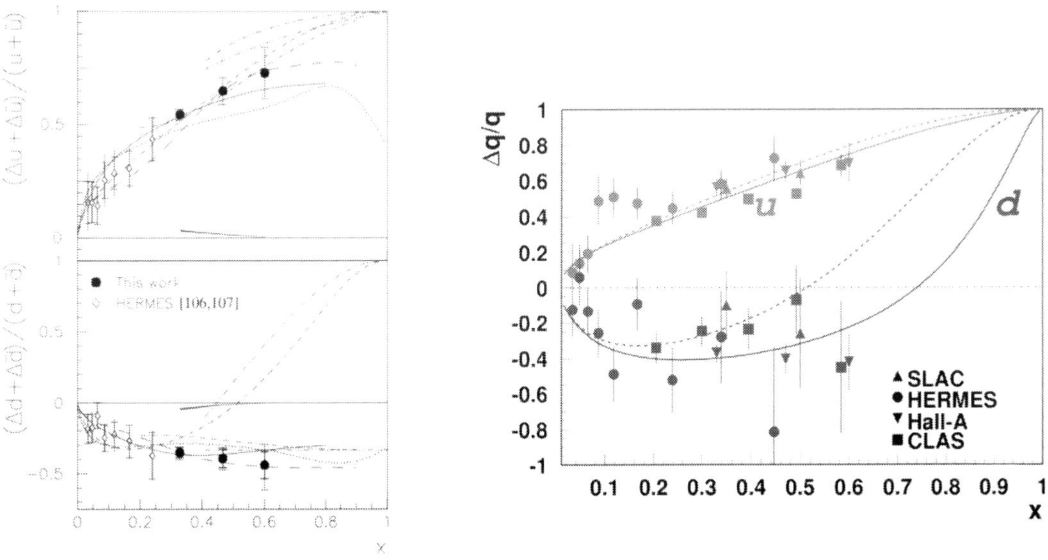

Figure 5. $\Delta u/u$ (upper side of both panels) and $\Delta d/d$ (lower side of both panels) results from Jefferson Lab Hall A E99-117 [32] together with world data (left panel) and CLAS EG1b [34] experiments (right panel), compared with the world data and theoretical predictions (see text for details).

it may play an important role in this kinematic region. However the pQCD fit still predicts that $\Delta d/d \to 1$ when x approaches 1, in contrast to the predictions from the relativistic valence quark models [8] and diquark models [36] that it stays negative.

3.2. Moments of spin structure functions

The spin structure functions g_1 and g_2 (or σ_{TT} and σ_{LT}) were measured in Hall A experiment E94-010 on ^{3}He from breakup threshold to $W = 2$ GeV covering the Q^2 range of 0.10-0.9 GeV^2. The extended GDH integrals $I(Q^2) = \int_{thr.}^{\infty} \sigma_{TT}(\nu, Q^2) d\nu/\nu$ (open symbols) were extracted for ^{3}He [37] (right panel of Fig. 6) and for the neutron [38] (left panel). The solid squares include an estimate of the unmeasured high energy part.

The ^{3}He results rise with decreasing Q^2. Since the GDH sum rule at $Q^2 = 0$ predicts a large negative value, a drastic turn around should happen at Q^2 lower than 0.1 GeV^2. A simple model using MAID plus quasi-elastic contributions estimated from a PWIA model [42] indeed shows the expected turn around. The data at low Q^2 should be a good testing ground for few-body chiral perturbation theory calculations.

The neutron results indicate a drastic yet smooth variation of $I(Q^2)$ to increasingly negative values as Q^2 varies from 0.9 GeV^2 towards zero. The data are more negative than the MAID model calculation [40]. Since the calculation only includes contributions to $I(Q^2)$ for $W \leq 2$ GeV, it should be compared with the open squares. The GDH sum rule prediction, $I(0) = -232.8\,\mu$b, is indicated along with extensions to $Q^2 > 0$ using two NLO χPT calculations, one using the heavy baryon approximation (HBχPT) [39] (dotted line) and the other relativistic baryon χPT (RBχPT) [20] (dot-dashed line). Shown with a band is RBχPT including resonance effects [20], which have an associated large uncertainty due to the resonance parameters used.

The first moment of g_1, Γ_1, was extracted from E94-010 [38] for the neutron and ^{3}He [37]. CLAS experiment EG1 [34] measured g_1 for the proton and the deuteron covering the resonance

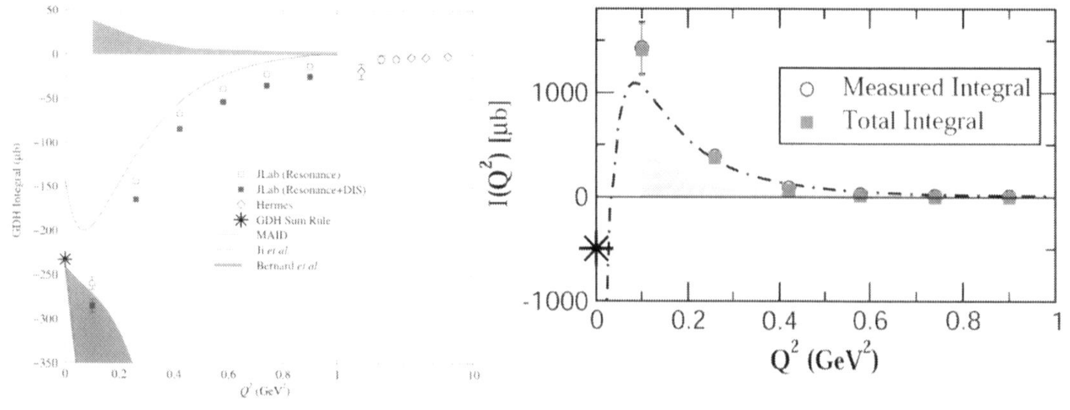

Figure 6. Results of GDH sum $I(Q^2)$ for neutron (left) and ^{3}He (right). The ^{3}He GDH results are compared with the MAID model plus a quasi-elastic contribution. The neutron GDH results are compared with χPT calculations of Ref. [39] (dashed line) and Ref. [20] (shaded band). The MAID model calculation of Ref. [40], is represented by the solid line. Data from HERMES [41] are also shown.

region and beyond over a Q^2 range from 0.05 to 5 GeV^2. The first moments were extracted for the proton, the deuteron and the neutron (from the deuteron with the proton contribution subtracted). The results are plotted in Fig. 7. Also plotted is the first moment for $p-n$ [43], a flavor non-singlet combination, which is the Bjorken sum at large Q^2 (13). These moments show a strong yet smooth variation in the transition region (Q^2 from 1 to 0.1 GeV^2). The Bjorken sum was used to extract the strong coupling constant α_s at high Q^2 (5 GeV^2). An attempt was made to extract an effective coupling α_{s,g_1} at low Q^2 region using the Bjorken sum, see Sec. 3.6.

At large Q^2 the data agree with pQCD models with little higher twist effects. At low Q^2 two calculations using chiral perturbation theory are compared with the Γ_1 data and are in reasonable agreement at the lowest Q^2 value (0.05–0.1 GeV^2).

The x^2 (or equivalently, additional $1/\nu^2$) weighted moments at low Q^2 are related to the generalized spin polarizabilities γ_0 (15) and δ_{LT} (16). They were expected to provide a more reliable test of χPT at low Q^2 because the unmeasured contributions from the high energy (low-x) region are strongly suppressed. In particular, δ_{LT} provides a benchmark test of χPT since the $\Delta(1232)$ contribution, which is difficult to treat in χPT, is suppressed. Figure 8 presents results of the generalized spin polarizabilities γ_0 and δ_{LT} for the neutron [38], and γ_0 for the proton [34] and for the isospin decompositions $p-n$ and $p+n$ [43]. There are strong disagreements between data and χPT calculations. In particular, the disagreement in δ^n_{LT} with both χPT predictions presents a significant challenge to present formulations of chiral perturbation theory. Experimental results at very low Q^2 (down to 0.02 GeV^2) for both the neutron [46] and the proton [47] should be available soon. Further theoretical and experimental efforts will be needed to clarify the situation. In particular, a planned measurement of δ_{LT} for the proton [48] should shed light on this "puzzle".

Lattice QCD results for the transition region are not yet available. An important aspect of the lattice calculation is the extrapolation to the physical mass of the pion. This extrapolation is guided by χPT calculations. Since these calculations have strong disagreements with some of the observables discussed here, one could question the reliability of the chiral extrapolation.

Finally, another important result on the experimental level is the good agreement between the neutron information extracted from a deuteron and a ^{3}He target. It validates the use of

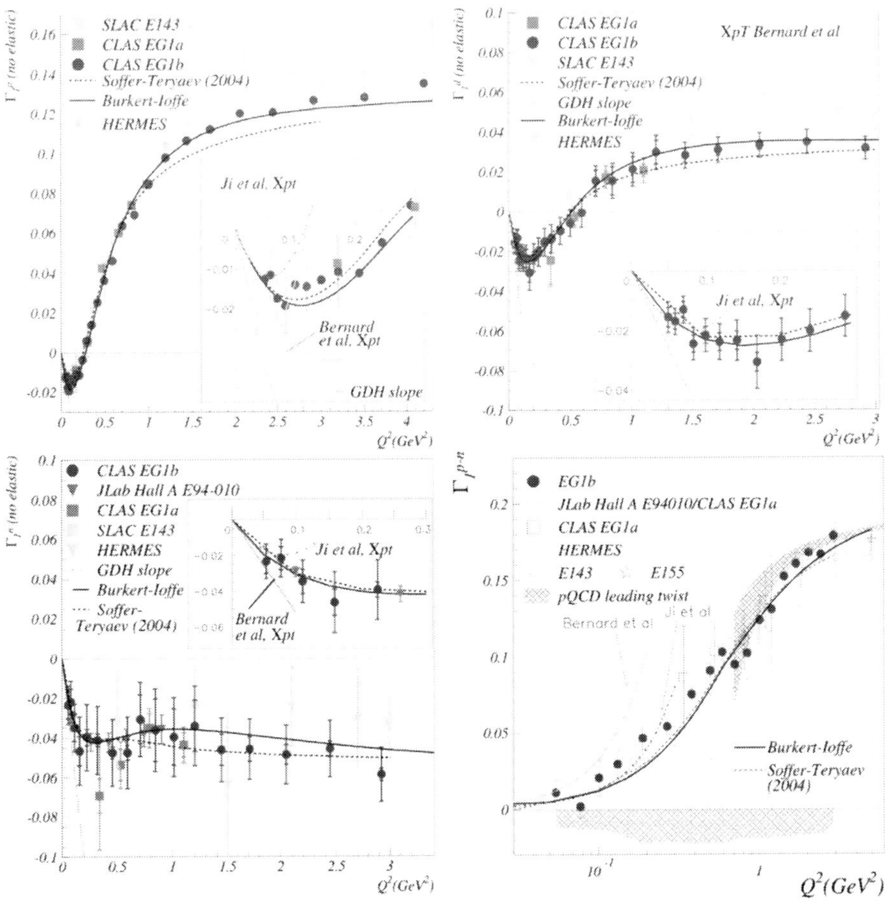

Figure 7. Results of $\Gamma_1(Q^2)$ for p, d, n and $p-n$ from Jefferson Lab Hall A and CLAS EG1. The slopes at $Q^2 = 0$ predicted by the GDH sum rule are given by the dotted lines. The dashed (solid) lines are the predictions from the Soffer-Teryaev [44] (Burkert-Ioffe [45]) model. The leading twist Q^2 evolution of the moments is given by the gray band. The insets show comparisons with χPT calculations by Ji *et al.* [39] and Bernard *et al.* [20].

polarized deuteron and ^{3}He, at least within the precision of the Jefferson Lab data.

3.3. Measurements of g_2, moments of g_2 and higher twist effects

The measurements of g_2 require tranversely polarized targets. SLAC E155x [49] performed the only dedicated g_2 measurement prior to Jefferson Lab. At Jefferson Lab, g_2 and its moments have been extensively measured for the neutron with a polarized ^{3}He target in a wide range of kinematics in several Hall A experiments (E94-010 [38], E97-103 [50], E99-117 [32], E02-012 [51] and E97-110 [46]). The measurement on the proton was performed in the RSS experiment in Hall C at an average Q^2 of 1.3 GeV^2.

Figure 9 shows the world data on g_2^n in the DIS ($W > 2$ GeV) region. The precision data from E97-103 show for the first time a clear deviation from g_2^{WW}, indicating that twist-3 (or higher) effects are important at Q^2 around 1 GeV^2 and lower.

Journal of Physics: Conference Series **299** (2011) 012005 doi:10.1088/1742-6596/299/1/012005

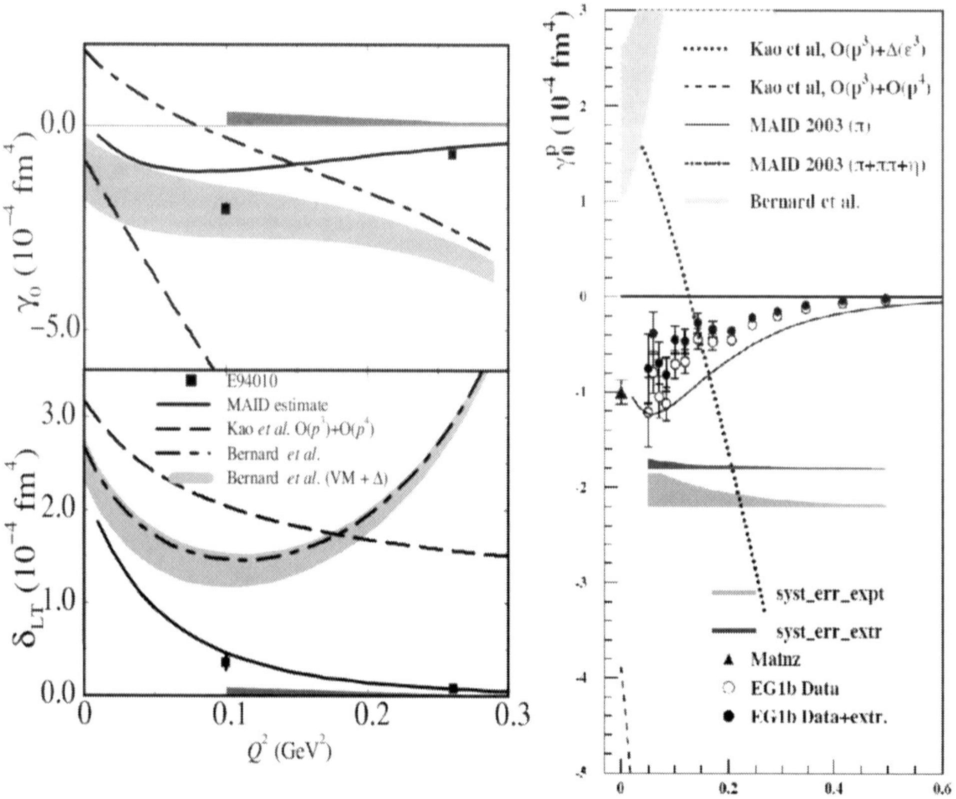

Figure 8. Generalized spin polarizabilities. The neutron generalized forward spin polarizabilities γ_0^n and δ_{LT}^n (left plots) were extracted from the Hall A E94-010 data. γ_0^p (right plot) was extracted from the CLAS EG1 data.

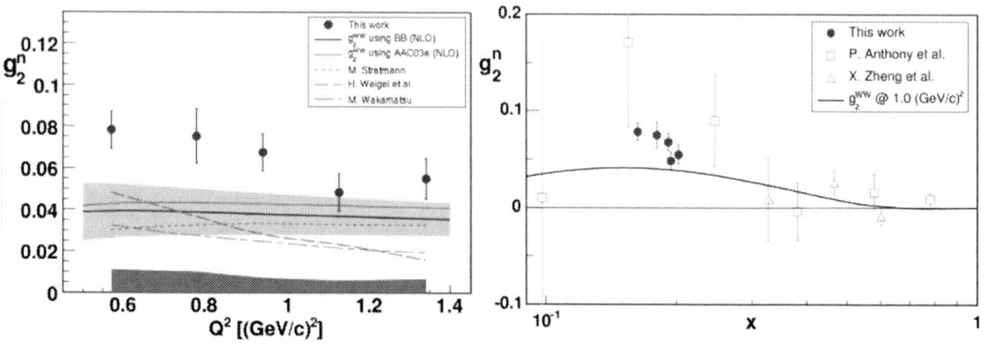

Figure 9. Results from experiment E97-103 on g_2^n. (Left) g_2^n vs Q^2 from E97-103, models and the twist-2 term g_2^{WW}. (Right) g_2^n *vs.* x from experiments E97-103, E99-117 and SLAC E155x. All data are at $W > 2$ GeV.

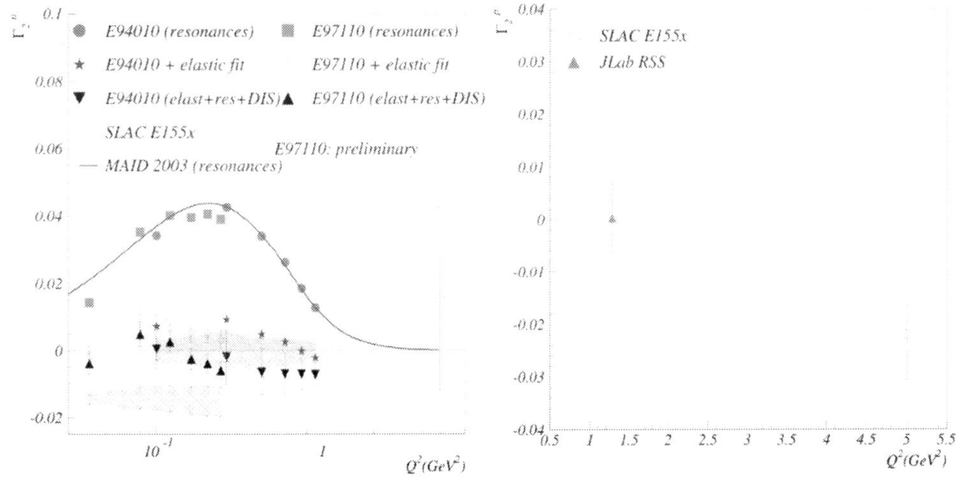

Figure 10. The verification of the BC sum rule from Hall A experiments E94-010 and E97-110 (neutron, left) and Hall C RSS (proton, right), together with SLAC E155x data.

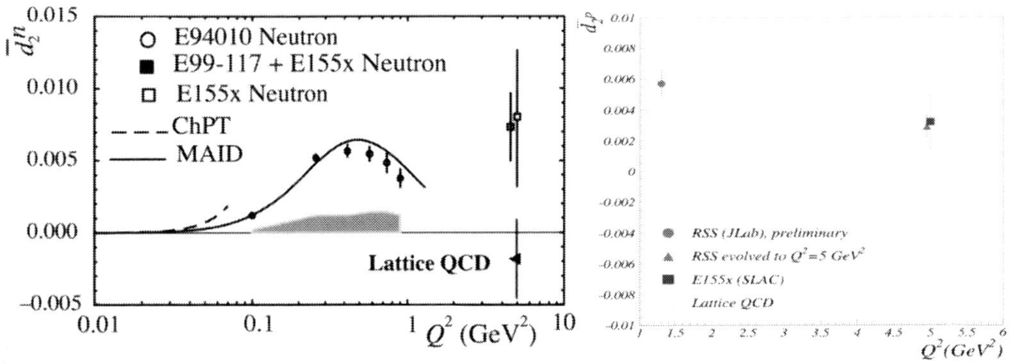

Figure 11. The x^2-weighted moment d_2 of the neutron (left) and proton (right).

The first moment of g_2 is expected to be zero at all Q^2 from the Burkhardt-Cottingham sum rule. The measurements for the proton and the neutron are shown in Fig. 10. The precision data are consistent with the BC sum rule in all cases, indicating that g_2 is a well behaved function (with good convergence at high energy).

The second moment of $g_2 - g_2^{WW}$ has also been formed from E94-010 (neutron) and RSS (proton and neutron), see Fig. 11. The E94-010 results were compared to a chiral perturbation theory calculation [39] and the MAID parametrization [40]. Furthermore, the Hall A experiment E99-117 provided data on g_2^n at high x, see right plot on Fig. 9. Combining these results with the SLAC data, the second moment d_2^n was extracted at an average Q^2 of 5 GeV^2 with much improved uncertainty. While a negative or near-zero value was predicted by lattice QCD and most models, the result for d_2^n at an average Q^2 of 5 GeV^2 is positive.

The high precision data allowed new analyzes [43, 52] to study higher twist contributions from the g_1 moments. In particular, the twist-4 matrix element f_2 was extracted by fitting Γ_1 up to $\mathcal{O}(1/Q^6)$. The same analysis was also performed on the Bjorken sum.

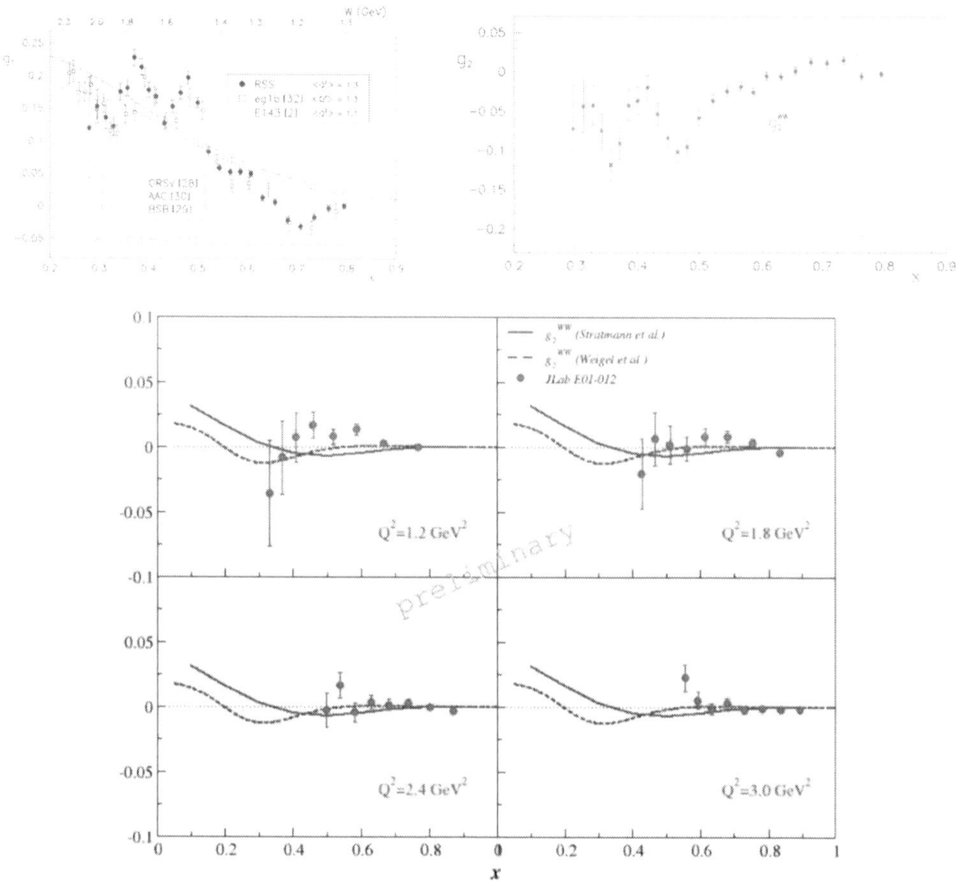

Figure 12. Results from the RSS experiment for g_1 (top-left panel) and g_2 (top-right panel) for the proton in the resonance region measured at $\langle Q^2 \rangle = 1.3$ GeV^2, and results from the Hall A experiment E01-012 (bottom panel) for g_2^n in the resonance region measured at $Q^2 = 1.2 - 3$ GeV^2.

With the high precision data at low Q^2 and high x, the higher twist contributions were included in the new global polarized PDF analysis [28] for the first time. The dynamical twist effects are important at moderate x but become smaller at low and high x. At large x, however, g_1 being small, the relative contributions of the higher twist effects are still significant. At $Q^2 < 1$ GeV^2, the overall effects of higher twists become non-negligible.

3.4. Quark-hadron duality in spin structure functions

A detailed study of duality in the spin structure functions g_1^p and g_1^d has been published by the EG1 collaboration [34]. As shown in Fig. 13, one observes a clear trend of strong, resonant deviations from the scaling curve at lower Q^2, towards a pretty good agreement at intermediate Q^2. The integral of g_1 over the whole resonance region begins to agree with the NLO results above $Q^2 \approx 1.7$ GeV^2. The results on the proton and deuteron from EG1b [34] thus indicate a much slower approach to "global" duality for the polarized structure function g_1 than has been observed for unpolarized structure functions. Local duality seems violated in the Δ resonance region even for Q^2 values as high as 5 GeV^2. The data taken by the RSS collaboration in

Hall C [53] (see Fig. 12) corroborate these observations and add more precise data points for $Q^2 \approx 1.3$ GeV^2.

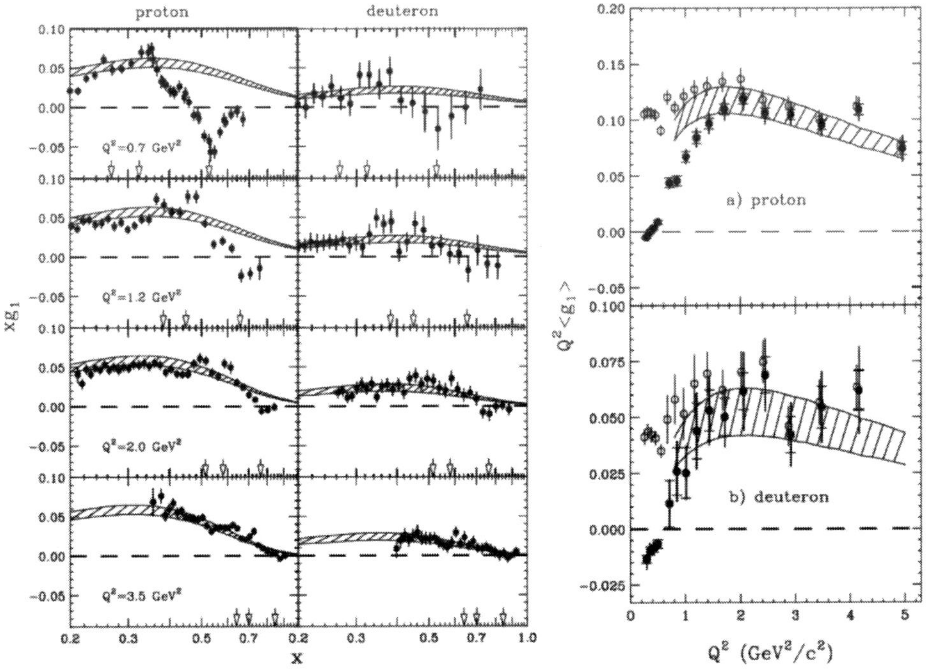

Figure 13. Data on g_1 of the proton and the deuteron from Jefferson Lab Hall B (left panel) and its average over the resonance region (right panel). Prominent resonances are indicated by arrows in the left panel. The hashed curves represent the range of extrapolated DIS results from modern NLO fits (GRSV and AAC), evolved to the Q^2 of the data and corrected for target mass effects. The open circles in the right panel include the elastic contribution, while the filled circles are only integrated over $W > 1.08$ GeV.

The spin structure functions g_1 and A_1 were measured in the resonance region ($W < 2$ GeV) in the Hall A ^{3}He experiments E94010 [38] (in the Q^2 region below 1 GeV^2) and E01012 [51] (from 1 GeV^2 to 4 GeV^2). The results for $A_1^{^3\mathrm{He}}$ above 1 GeV^2 are presented in Fig. 14 (left panel). Also plotted are the world DIS data and a fit to the DIS data. Due to the prominent contributions from the Δ resonance, local duality does not appear to work at low Q^2 (below 2 GeV^2). At high Q^2 (above 2 GeV^2), the Δ resonance contribution starts to diminish. It is interesting to note that the two sets of resonance data at the highest Q^2 (2.9 and 4 GeV^2) agree well, indicating little or no Q^2 dependence, which is a key feature of the DIS data. These data also show the trend of becoming positive at the high-x side, the same trend as observed for DIS data. The resonance data were integrated to study global duality. Figure 14 (right panel) shows the results for both ^{3}He and the neutron in comparison with the DIS fits evolved to the same Q^2. The resonance data agree with the DIS fits at least for Q^2 higher than 1.8 GeV^2, indicating that global duality holds for the neutron and ^{3}He g_1 spin structure function in the high Q^2 region (above 1.8 GeV^2).

The study of quark-hadron duality helps us to study higher twist effects and to extend the kinematic region where one might be able to apply partonic interpretations. From a practical point of view, the good understanding of the higher twist effects and quark-hadron duality allows us to considerably extend the experimental database used to extract the polarized parton

Journal of Physics: Conference Series **299** (2011) 012005 doi:10.1088/1742-6596/299/1/012005

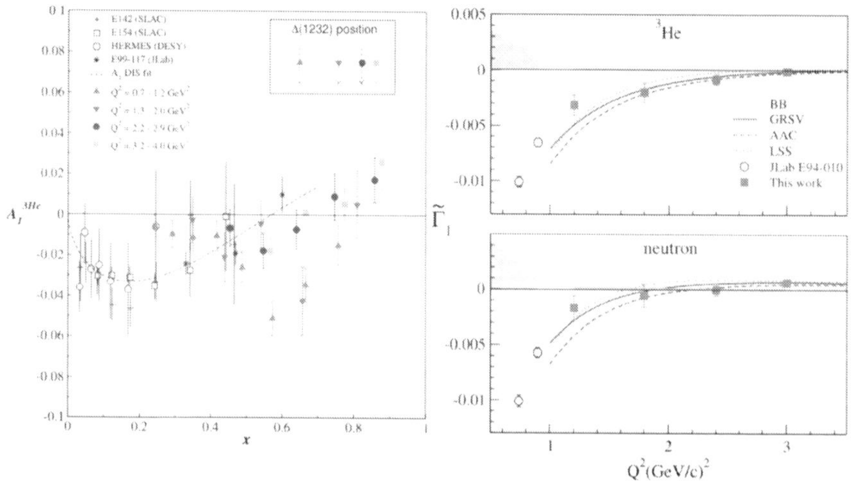

Figure 14. $A_1^{^3\text{He}}$ (left panel) in the resonance region from Jefferson Lab E01-012, compared with the world DIS data and a fit. $\Gamma_1^{^3\text{He}}$ and Γ_1^n (right panel) of the resonance region from Jefferson Lab E01-012, together with lower Q^2 results from Jefferson Lab E94-010, compared with the world DIS fits.

distributions [28], see Fig. 3.

3.5. *The constraint on* ΔG

The doubly polarized inclusive data from Jefferson Lab yield information on the spin of the nucleon carried by gluons, ΔG. The LSS group [28] included the Jefferson Lab data from Hall A (experiment E99-117) and CLAS experiment EG1 in their global analysis. The central values of the polarized parton distributions (Fig. 1) did not change significantly, but their uncertainties were noticeably reduced, see Fig. 15. In particular, the uncertainty on Δg is significantly reduced: by a factor 2 near $x = 0.3$ and 4 near $x = 0.5$ (within the model assumed for Δg by the LSS group).

3.6. *The effective strong coupling at large distance*

In QCD, the magnitude of the strong force is given by the running coupling constant α_s. At large Q^2, in the pQCD domain, α_s is well-defined and can be experimentally extracted, *e.g.* using the Bjorken sum rule (13). The pQCD definition leads to an infinite coupling at large distances, when Q^2 approaches Λ^2_{QCD}. This is not a conceptual problem because here one is outside the validity domain of pQCD. Since the data show no sign of discontinuity when crossing the intermediate Q^2 domain, see *e.g.* Fig. 7, it is natural to look for a definition of an effective coupling α_s^{eff} which works at any Q^2, and matches α_s at large Q^2 but stays finite at small Q^2. The Bjorken sum rule can be used advantageously to define α_s^{eff} at low Q^2 [54]. The data on the Bjorken sum are used to experimentally extract α_s^{eff} following a prescription by Grunberg [55], see Fig. 16. The Bjorken and GDH sum rules also allow us to determine α_s^{eff} at respectively large Q^2 and $Q^2 \simeq 0$. The extracted α_s^{eff} provides for the first time an effective coupling at all Q^2, albeit one that is process dependent. An interesting feature is that α_s^{eff} becomes scale invariant at small Q^2, which was predicted by a number of calculations.

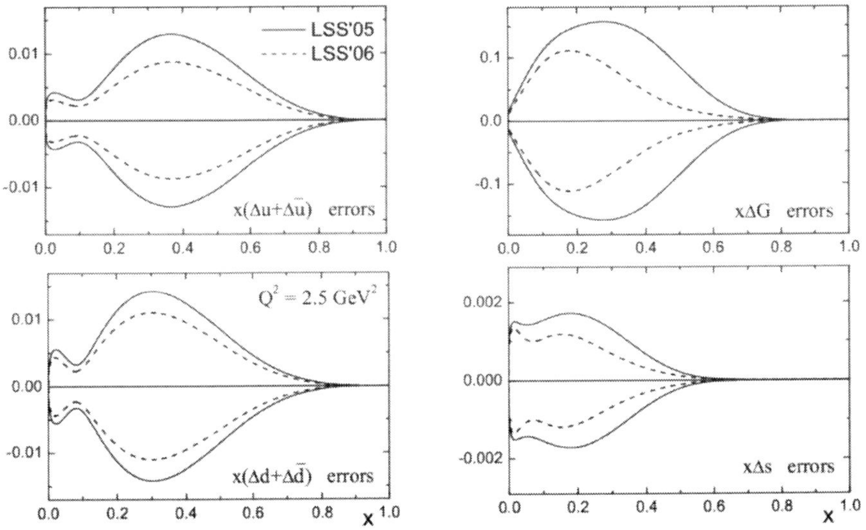

Figure 15. Uncertainty on the polarized parton distributions. Solid curves do not include Jefferson Lab data, and uncertainties are reduced after including the Hall A E99-117 and Hall B EG1 data (dashed curves).

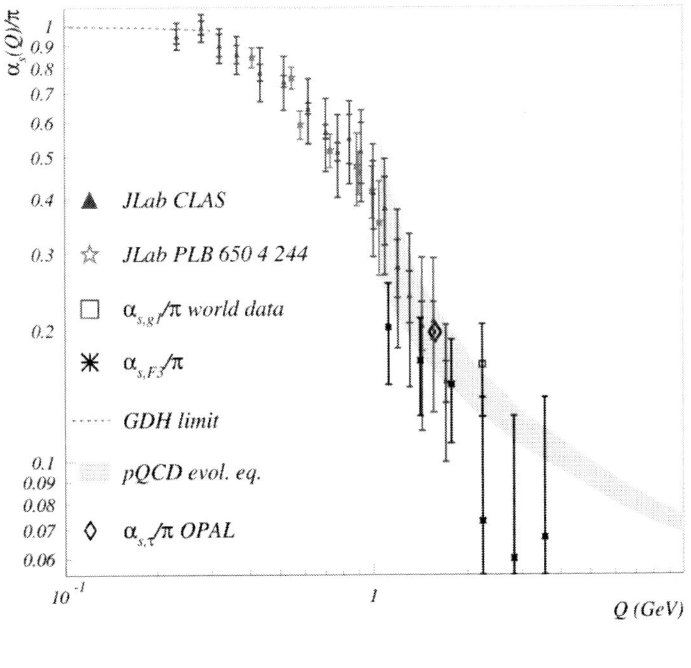

Figure 16. Value of α_s^{eff}/π extracted from the data on the Bjorken sum, on τ decay, and on the Gross-Llewellyn Smith sum rule. The values of α_s^{eff} computed within pQCD and using the Bjorken sum is given by the shaded band. The values of α_s^{eff}/π extracted using the GDH sum rule is given by the dashed line.

4. Transverse spin structure

To fully understand the nucleon structure, the transverse dimension can no longer be neglected. In the last few years, there has been a surge of interest in the nuclear physics community in the transverse spin and transverse momentum dependent distributions [56]. First experimental results of single target spin asymmetries in semi-inclusive DIS from HERMES [26] (for the proton) and COMPASS [27] (for the deuteron and the proton) showed clear nonzero results for both the Collins [57] and Sivers [58] asymmetries. These initial explorations demonstrated the importance of the transverse spin (transversity) and the quark orbital angular momentum.

A recently completed Jefferson Lab experiment [59] is the first one to measure the single-spin asymmetry on the neutron with a transversely polarized ^{3}He target. The goal of this experiment is to provide the first measurement of the neutron transversity and Sivers function, complementary to the HERMES and COMPASS measurements on the proton and deuteron. This experiment focuses on the valence quark region, $x = 0.19 - 0.34$, at $Q^2 = 1.77 - 2.73$ GeV^2. Data from this experiment, when combined with data from HERMES [26], COMPASS [27] and Belle [60], will provide powerful constraints on the transversity distributions of both u-quarks and d-quarks in the valence region. Due to the good particle identification in the HRS, kaon data were collected at the same time, providing a set of data to study the transverse spin asymmetries for semi-inclusive $K^{\pm}$ production.

In Hall B, study of the transverse momentum dependent distributions is one of the main goals of an ongoing experiment [61] with a polarized beam on a longitudinally polarized proton target. Future measurements are planned with tranversely polarized proton and deuteron targets [62].

5. Conclusions and perspective

In summary, the high polarized luminosity available at Jefferson Lab has provided us with high precision data to study the nucleon spin structure in a wide kinematic range. They shed light on the valence quark structure, help to understand quark-gluon correlations, allow us to study the nonperturbative region and to explore the transition between perturbative and nonperturbative regions of QCD. A planned precision study on transverse spin phenomena will open a new window to study the nucleon structure and help us understand the strong interaction.

The 12 GeV energy upgrade [63] opens up a much wider DIS kinematics region to study nucleon spin structure. Planned experiments [64] in the high-x region will definitively establish the contributions of valence quarks to the nucleon structure. A precision measurement of the moment d_2 [65] will provide a benchmark test of lattice QCD predictions. An extensive semi-inclusive DIS program [66] with transversely and longitudinally polarized neutron and proton targets will map out precisely the Collins and Sivers moments and other transverse momentum dependent observables. The tensor charge, a fundamental quantity of the nucleon, will be determined, which will provide another benchmark test of lattice QCD predictions. The 6 and 12 GeV data, together, will complete our picture of nucleon spin structure in the valence region.

A new facility, an electron-ion collider (EIC), is undergoing discussion in the US hadronic-physics community [67], as a long-term future facility. It will provide unique capabilities for the study of QCD well beyond all existing facilities. It will extend the spin structure study over a very wide region.

Acknowledgments

The work is supported by the US Department of Energy (DOE). Jefferson Science Associates operates the Thomas Jefferson National Accelerator Facility for the DOE under contract DE-AC05-84ER-40150.

References

[1] S. E. Kuhn, J. P. Chen and E. Leader, Prog. Part. Nucl. Phys. **63**, 1 (2009).
[2] S. Wandzura and F. Wilczek, Phys. Lett. B **72** (1977).
[3] X. Ji and W. Melnitchouk, Phys. Rev. D **56**, 1 (1997).
[4] M. Göckeler, *et al.*, Phys. Rev. D **63**, 074506 (2001).
[5] M. Stratmann, Z. Phys. C **60**, 763 (1993).
[6] H. Weigel, Pramana **61**, 921 (2003).
[7] M. Wakamatsu, Phys. Lett. B **487**, 118 (2000).
[8] N. Isgur, Phys. Rev. D **59**, 034013 (1999); F. E. Close and A. Thomas, Phys. Lett. B **212**, 227 (1988).
[9] G. R. Farrar and D. R. Jackson, Phys. Rev. Lett. **35**, 1416 (1975); S. Brodsky, M. Burkardt and I. Schmidt, Nucl. Phys. **B441**, 197 (1995).
[10] F. Close, Nucl. Phys. **B80**, 269 (1974).
[11] H. Avakian, S. Brodsky, A. Deur and Y. Feng, Phys. Rev. Lett. **99**, 082001 (2007).
[12] J. P. Chen, A. Deur and Z. E. Meziani, Mod. Phys. Lett. A **20**, 2745 (2005).
[13] D. Drechsel, B. Pasquini and M. Vanderhaeghen, Phys. Rep. **378**, 99 (2003); D. Drechsel and L. Tiator, Ann. Rev. Nucl. Part. Sci. **54**, 69 (2004).
[14] J. Ellis and R. L. Jaffe, Phys. Rev. D **9**, 1444 (1974); *ibid.*, D **10**, 1669 (1974).
[15] H. Burkhardt and W. N. Cottingham, Ann. Phys. (N.Y.) **56**, 453 (1970).
[16] J. D. Bjorken, Phys. Rev. **148**, 1467 (1966).
[17] S. B. Gerasimov, Sov. J. Nucl. Phys. **2**, 598 (1965); S. D. Drell and A. C. Hearn, Phys. Rev. Lett. **162**, 1520 (1966).
[18] X. Ji and J. Osborne, J. Phys. G **27**, 127 (2001).
[19] C. W. Kao, T. Spitzenberg and M. Vanderhaeghen, Phys. Rev. D **67**, 016001 (2003).
[20] V. Bernard, T. Hemmert and U.-G. Meissner, Phys. Rev. D **67**, 076008 (2003).
[21] E. D. Bloom and F. J. Gilman, Phys. Rev. Lett. **25**, 1140 (1970).
[22] A. De Rujula, H. Georgi and H. D. Politzer, Phys. Lett. B **64**, 428 (1976).
[23] W. Melnitchouk, R. Ent and C. Keppel, Phys. Rep. **406**, 127 (2005).
[24] C. Bourrely, J. Soffer and O. V. Teryaev, Phys. Lett. B **420**, 375 (1998).
[25] J. Soffer, Phys. Rev. Lett. **74**, 1292 (1995).
[26] A. Airapetian, *et al.*, Phys. Rev. Lett. **94**, 012002 (2005).
[27] V. Y. Alexakhin *et al.*, Phys. Rev. Lett. **94**, 202002 (2005); E. S. Ageev *et al.*, Nucl. Phys. **B765**, 31 (2007).
[28] E. Leader, A. V. Sidorov and D. B. Stamenov, Eur. Phys. J. C **23**, 479 (2002); Phys. Rev. D **73**, 034023 (2006); Phys. Rev. D **75**, 074027 (2007).
[29] See *e.g.* N. Bianchi, in talk presented at Pacific-SPIN07, www.triumf.info/hosted/pacspin07/talks/Bianchi.ppt.
[30] B. W. Filippone and X. Ji, Adv. Nucl. Phys. **26**, 1 (2001)
[31] F. Myhrer and A. W. Thomas, Phys. Lett. B **663**, 302 (2008)
[32] X. Zheng *et al.*, Phys. Rev. Lett. **92**, 012004 (2004); Phys. Rev. C **70**, 065207 (2004).
[33] C. Bourrely, J. Soffer and F. Buccella, Eur. Phys. J. C **23**, 487 (2002).
[34] R. Fatemi *et al.*, Phys. Rev. Lett. **91**, 222002 (2003); J. Yun *et al.*, Phys. Rev. C **67**, 055204 (2003); K. V. Dharmawardane *et al.*, Phys. Lett. B **641** 11 (2006); Y. Prok *et al.*, Phys. Lett. B **672**, 12 (2009); P. Bosted *et al.*, Phys. Rev. C **75**, 035203 (2007).
[35] F. E. Close and W. Melnitchouk, Phys. Rev. C **68**, 035210 (2003).
[36] I. C. Cloet, W. Bentz and A. W. Thomas, Phys. Lett. B **621**, 246 (2005).
[37] K. J. Slifer *et al.*, Phys. Rev. Lett **101**, 022303 (2008).
[38] M. Amarian *et al.*, Phys. Rev. Lett. **89**, 242301 (2002); Phys. Rev. Lett. **92**, 022301 (2004).
[39] X. Ji, C. Kao and J. Osborne, Phys. Lett. B **472**, 1 (2000).
[40] D. Drechsel, S. Kamalov and L. Tiator, Phys. Rev. D **63**, 114010 (2001)
[41] A. Airapetian *et al.*, Phys. Rev. Lett. **90**, 092002 (2003); Eur. Phys. J. C **26**, 527 (2003); Phys. Rev. D **75**, 012007 (2007).
[42] C. Ciofi degli Atti, E. Pace and G. Salme, Phys. Rev. C **51**, 1108 (1995).
[43] A. Deur *et al.*, Phys. Rev. Lett. **93**, 212002 (2004); Phys. Rev. D **78**, 032001 (2008).
[44] J. Soffer and O. V. Teryaev, Phys. Rev. D **70**, 116004 (2004).
[45] V. D. Burkert and B. L. Ioffe, Phys. Lett. B **296**, 223 (1992).
[46] Jefferson Lab Experiment E97-110, J. P. Chen, A. Deur and F. Garibaldi, spokespersons.
[47] Jefferson Lab Experiment E03-006, M. Battaglieri, A. Deur, R. De Vita and M. Ripani, spokespersons.
[48] Jefferson Lab Experiment E08-027, A. Camsonne, J. P. Chen and K. Slifer, spokespersons.
[49] P. L. Anthony *et al.*, Phys. Lett. B **553**, 18 (2003).
[50] K. Kramer *et al.*, Phys. Rev. Lett. **95**, 142002 (2005).
[51] P. Solvignon *et al.*, Phys. Rev. Lett. **101**, 182502 (2008).

[52] Z. E. Meziani *et al.*, Phys. Lett. B **613**, 148 (2005); M. Osipenko *et al.*, Phys. Lett. B **609**, 259-264 (2005).
[53] F. R. Wesselmann *et al.*, Phys. Rev. Lett. **98**, 132003 (2007).
[54] A. Deur, V. Burkert, J. P. Chen and W. Korsch, Phys. Lett. B **650**, 244 (2007); Phys. Lett. B **665**, 349 (2008).
[55] G. Grunberg, Phys. Lett. B **95**, 70 (1980); Phys. Rev. D **29**, 2315 (1984); Phys. Rev. D **40**, 680 (1989).
[56] P. Mulders and R. D. Tangerman, Nucl. Phys. **B461**, 197 (1996).
[57] J. Collins, Nucl. Phys. **B396**, 161 (1993).
[58] D. W. Sivers, Phys. Rev. D **41**, 83 (1990).
[59] Jefferson Lab Experiment E06-010, J. P. Chen, E. Cisbani, H. Gao, X. Jiang and J.-C. Peng, spokespersons.
[60] A. Ogawa *et al.*, AIP Conf. Proc. **792**, 949 (2005).
[61] Jefferson Lab Experiment E05-113, H. Avakian, P. Bosted, D. Crabb and K. Griffioen, spokespersons.
[62] Jefferson Lab Experiment E08-015, H. Avakian, P. Bosted, K. Hafidi, N. Makins and P. Rossi, spokespersons.
[63] Conceptual Design Report for the Science and Experimental Equipment for the 12 GeV Upgrade of CEBAF (2005).
[64] Jefferson Lab Experiment E12-06-110, G. Cates, J. P. Chen, Z. E. Meziani and X. Zheng, spokespersons; Jefferson Lab E12-06-109, S. Kuhn *et al.*, spokespersons.
[65] Jefferson Lab Experiment E12-06-121, T. Averett, W. Korsch, Z. E. Meziani and B. Sawatzky, spokespersons.
[66] A. Afanasev *et al.*, hep-ph/0703288; Jefferson Lab Experiment PR09-014, J. P. Chen, H. Gao, X. Jiang, J. C. Peng and X. Qian, spokespersons; Jefferson Lab Experiment E12-07-107, H. Avakian *et al.*, spokespersons; Jefferson Lab Experiment E12-09-009, H. Avakian *et al.*, spokespersons.
[67] The Nuclear Science Advisory Committee Report: "The Frontiers of Nuclear Science - A Long Range Plan" (2007).

Journal of Physics: Conference Series **299** (2011) 012006 doi:10.1088/1742-6596/299/1/012006

Deeply Virtual Exclusive Processes and Generalized Parton Distributions

Charles E. Hyde[1,2], **Michel Guidal**[3], **and Anatoly V. Radyushkin**[2,4]

[1] Laboratoire de Physique Corpusculaire, Université Blaise Pascal, 63177 Aubière, France
[2] Old Dominion University, Norfolk, Virginia 23529, USA
[3] Institut de Physique Nucléaire ORSAY, 91406 Orsay, France
[4] Jefferson Lab, Newport News, Virginia 23606, USA

Abstract. The goal of the comprehensive program in deeply virtual exclusive scattering at Jefferson Lab is to create transverse spatial images of quarks and gluons as a function of their longitudinal momentum fraction in the proton, the neutron, and in nuclei. These functions are the generalized parton distributions (GPDs) of the target nucleus. Cross section measurements of the deeply virtual Compton scattering (DVCS) reaction $ep \to ep\gamma$ in Hall A support the QCD factorization of the scattering amplitude for $Q^2 \geq 2$ GeV^2. Quasi-free neutron-DVCS measurements on the deuteron indicate sensitivity to the quark angular momentum sum rule. Fully exclusive H$(e, e'p\gamma)$ measurements have been made in a wide kinematic range in CLAS with polarized beam, and with both unpolarized and longitudinally polarized targets. Existing models are qualitatively consistent with the Jefferson Lab data, but there is a clear need for less constrained models. Deeply virtual vector meson production is studied in CLAS. The 12 GeV upgrade will be essential for for these channels. The ρ and ω channels reactions offer the prospect of flavor sensitivity to the quark GPDs, while the ϕ-production channel is dominated by the gluon distribution.

1. Deeply virtual exclusive scattering

In the past decade, deep exclusive scattering (DES) has emerged as a powerful new probe of the partonic structure of the nucleon, hadrons and nuclei. These are reactions of the type:

$$\begin{aligned} e + \text{Nucleon} &\to e + \text{Nucleon} + \gamma \\ &\to e + \text{Nucleon} + \text{meson}. \end{aligned} \tag{1}$$

The $eN \to eN\gamma$ reaction is the coherent sum of the Bethe-Heitler and virtual Compton amplitudes, as illustrated in Fig. 1. The electron scattering kinematics of deeply virtual processes corresponds to the deep inelastic scattering (DIS), or Bjorken scaling region of inclusive electron scattering, with Q^2 large and W^2 above the resonance region. In addition, the forward exclusive limit is defined kinematically by $-t \lesssim \Lambda^2_{\rm QCD} \lesssim 1$ GeV^2. Thus the deeply virtual Compton scattering (DVCS) amplitude, $\gamma^* N \to \gamma N$ is an "off-forward" generalization of the forward Compton amplitude which defines the DIS cross section *via* the optical theorem.

The intense interest in DVCS started after the article by Ji [1], linking DVCS to the total contribution of quarks to the proton spin. It was found [1, 2, 3, 4] that, in analogy with DIS, in the limit of large Q^2 and small t, amplitudes of DVCS and deeply virtual meson production (DVMP) can be expressed in a power series of $1/Q^2$, with the power determined by the twist (dimension minus spin) of each operator in the expansion. Detailed proofs of factorization for DVCS and DVMP were given in Refs. [5, 6, 7, 8]. This factorization is depicted in Fig. 2 for the DVCS and DVMP amplitudes. In these processes, the leading power term of the amplitude is the convolution of the perturbative kernel with a new class of nonlocal bilinear twist-2 quark (or gluon) operators, called generalized parton distributions (GPDs) [1, 2]. These were first described by Müller *et al.* [9]. In the case of deeply virtual meson production, the hard kernel is also convoluted with the meson distribution amplitude (DA).

1.1. Generalized parton distributions

The GPDs parameterize Fourier transforms of nucleon matrix elements of bilinear quark (and gluon) operators separated by a light-like interval $z^2 = 0$ [9]. The kinematics are commonly defined in terms of symmetric variables (see Figs. 1 and 2):

$$P = (p + p')/2, \qquad \overline{q} = (q + q')/2$$
$$\xi = \frac{-\overline{q}^2}{2\overline{q} \cdot P} \longrightarrow \frac{x_{\mathrm{Bj}}}{2 - x_{\mathrm{Bj}}} \qquad \text{as} \quad t/Q^2 \to 0. \tag{2}$$

The generalized Bjorken variable ξ has the same form with respect to the symmetrized variables P and $\overline{q}$ as does x_{Bj} with respect to the DIS variables p and q.

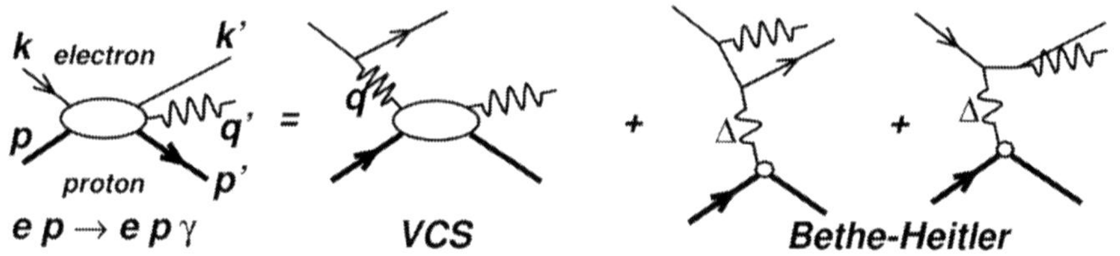

Figure 1. Lowest order QED amplitude for the $ep \to ep\gamma$ reaction. The momentum four-vectors of all external particles are labeled at left. The net four-momentum transfer to the proton is $\Delta_\mu = (q - q')_\mu = (p' - p)_\mu$. In the virtual Compton scattering (VCS) amplitude, the (space-like) virtuality of the incident photon is $Q^2 = -q^2 = -(k - k')^2$. In the Bethe-Heitler (BH) amplitude, the virtuality of the incident photon is $-\Delta^2 = -t$. Standard (e, e') invariants are $s_e = (k + p)^2$, $x_{\mathrm{Bj}} = Q^2/(2q \cdot p)$ and $W^2 = (q + p)^2$.

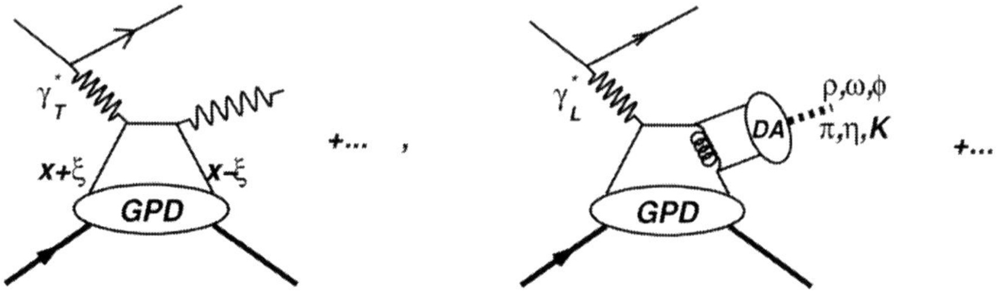

Figure 2. Factorization of the $\gamma^* p \to \gamma p$ DVCS amplitude and the $\gamma^* p \to MN$ deep virtual meson production amplitude in the Bjorken limit of large Q^2 and $-t \ll Q^2$. All permutations of the photon and gluon vertices should be included in the amplitudes. The labels on the quark lines are the light-cone momentum fractions, relative to $P^+ = (p^+ + p'^+)/2$.

It is convenient to use a reference frame in which P^μ has only time- and z-components, both positive. We define light-cone vectors

$$n^\mu = [1, 0, 0, -1]/(\sqrt{2}P^+), \qquad \tilde{p}^\mu = [1, 0, 0, 1]\, P^+/\sqrt{2}. \tag{3}$$

Then in the forward limit of either DVCS or deeply virtual production of a light meson, -2ξ is the "+" fraction of both the momentum transfer to the target and the virtual photon:

$$\Delta^+ \;=\; \Delta\cdot n \;\approx\; -2\xi P^+ \;\approx\; q\cdot n \;=\; q^+. \tag{4}$$

The quark GPDs H and E are the nucleon helicity-conserving and helicity-flip matrix elements of the vector operator containing $\gamma\cdot n = \gamma^+$. Suppressing the QCD scale dependence and the Wilson-line gauge link one can write the flavor-f dependent GPDs as (see, *e.g.*, Ref. [10]):

$$\begin{aligned}\int \frac{dz^- P^+}{2\pi} e^{ixP^+z^-} \langle p', s'|\, \overline{\Psi}_f\left(-z^-/2\right)\gamma\cdot n\Psi_f\left(z^-/2\right)|p, s\rangle \\ = \overline{U}(p', s')\left[H_f(x,\xi,t)\,\gamma\cdot n + E_f(x,\xi,t)\,\frac{i}{2M}n_\alpha\sigma^{\alpha\beta}\Delta_\beta\right] U(p, s)\,,\end{aligned} \tag{5}$$

where the $U(p, s)$ are the nucleon spinors. The factorization proofs demonstrate that the initial and final + momenta of the active parton are $(x \pm \xi)P^+$. The GPDs $\widetilde{H}$, $\widetilde{E}$ are defined similarly as the matrix elements of the axial operator containing $n\cdot\gamma\gamma_5 = \gamma^+\gamma_5$:

$$\begin{aligned}\int \frac{dz^- P^+}{2\pi} e^{ixP^+z^-} \langle p', s'|\, \overline{\Psi}_f\left(-z^-\right)\gamma\cdot n\gamma_5\Psi_f\left(z^-\right)|p, s\rangle \\ = \overline{U}(p', s')\left[\widetilde{H}_f(x,\xi,t)\,n\cdot\gamma\gamma_5 + \widetilde{E}_f(x,\xi,t)\,\frac{n\cdot\Delta}{2M}\gamma_5\right] U(p, s)\,.\end{aligned} \tag{6}$$

With the convention that positive and negative momentum fractions refer to quarks and antiquarks, respectively, we observe the following kinematic regions of the GPDs (Fig. 2): $x > \xi > 0$: the initial and final partons are quarks; $x < -\xi < 0$: the initial and final partons are antiquarks; $|x| < \xi$: a $q\bar{q}$ pair is exchanged in the t-channel. This identification is reflected in the QCD evolution equations of the GPDs. For $|x| > \xi$, the evolution of GPDs is similar to the DGLAP evolution of the forward parton distributions, whereas for $|x| < \xi$, the GPDs evolve according to the ERBL equations of a meson DA [2].

The GPDs combine the momentum fraction information of the forward parton distribution functions (PDFs) of DIS with the transverse spatial information of the elastic electroweak form factors. The forward limits of the helicity conserving GPDs are

$$\left.\begin{matrix} H_f(x,0,0) \\ \widetilde{H}_f(x,0,0)\end{matrix}\right\} = \begin{cases} q_f(x)\theta(x) - \overline{q}_f(-x)\theta(-x) \\ \Delta q_f(x)\theta(x) + \Delta\overline{q}_f(-x)\theta(-x), \end{cases} \tag{7}$$

where $q_f(x)$ and $\overline{q}_f(x)$ are the flavor-f dependent quark and antiquark momentum-fraction distributions; and $\Delta q_f(x)$ and $\Delta\overline{q}_f(x)$ are the quark and antiquark helicity distributions. There are no specific analogous constraints on the forward limits of E and $\widetilde{E}$.

The first moments of the GPDs are equal to the corresponding elastic form factors,

$$\int dx \begin{Bmatrix} H_f(x,\xi,t) \\ E_f(x,\xi,t)\end{Bmatrix} = \begin{cases} F_{1,f}(-t) \\ F_{2,f}(-t), \end{cases} \tag{8}$$

where $F_{1,f}(-t)$, $F_{2,f}(-t)$ are flavor-f components of the Dirac and Pauli form factors of the proton, defined with positive arguments in the space-like regime. Similarly,

$$\int dx \left\{ \begin{matrix} \widetilde{H}_f(x,\xi,t) \\ \widetilde{E}_f(x,\xi,t) \end{matrix} \right\} = \left\{ \begin{matrix} g_{A,f}(-t) \\ g_{P,f}(-t), \end{matrix} \right. \tag{9}$$

where $g_{A,f}$ and $g_{P,f}$ are the flavor components of the axial and pseudoscalar form factors of the proton.

The ξ-independent integrals of Eqs. (8) and (9) are examples of the polynomiality condition required by Lorentz invariance. Specifically, the x^N moment of a GPD is polynomial in even powers of ξ, with maximal power $\leq (N+1)$ for H, E, and maximal power $\leq N$ for $\widetilde{H}$, $\widetilde{E}$. The second moment of the GPD sum $H+E$ leads to the important angular momentum sum rule of Ji [1]:

$$\lim_{t\to 0} \int_{-1}^{1} dx\, x\, [H_f(x,\xi,t) + E_f(x,\xi,t)] = 2J_f\,, \tag{10}$$

where $2J_f$ is the fraction of the spin of the proton carried by quarks of flavor f, including both spin and orbital angular momentum. More generally, at nonzero t, the individual second moments are form factors (*i.e.*, Fourier transforms of spatial distributions) of the nucleon's energy-momentum tensor [1].

$$\begin{aligned} \int dx\, x H_f(x,\xi,t) &= M_{2,f}(t) + \frac{4}{5}\xi^2 d_{1,f}(t)\,, \\ \int dx\, x E_f(x,\xi,t) &= [2J_f(t) - M_{2,f}(t)] - \frac{4}{5}\xi^2 d_{1,f}(t)\,. \end{aligned} \tag{11}$$

The forward limit $M_{2,f}(0)$ is the ordinary momentum sum rule

$$M_{2,f}(0) = \int_0^1 dx x \left[q_f(x) + \overline{q}_f(x)\right]\,. \tag{12}$$

The term $d_{1,f}(t)$ is Fourier conjugate to the spatial distribution of the time-averaged shear stress on quarks in the nucleon [11, 12].

The GPDs themselves, and not just the moments, provide unique spatial information about partons in the hadronic target. The parton impact parameter $\mathbf{b}$ is Fourier conjugate to $\Delta_\perp$. The Fourier transform of $H_f(x,0,\Delta^2)$ determines a positive-definite probability distribution of quarks of flavor f as a function of longitudinal momentum fraction x and spatial coordinate $\mathbf{b}$ in the transverse plane [13, 14]. As a consequence, the Dirac form factors $F_1^{p,n}$ are the 2D Fourier transforms of the charge densities of the proton and neutron in impact parameter space. A recent analysis reveals the presence of a negative charge density at the heart of the neutron [15]. In the context of the Double Distribution models of the GPDs (see Section 1.3), this can be understood by the excess of down quarks over up quarks in the neutron at large x. The second (or x-weighted) moment of E_f describes by how much the transverse center of momentum of quarks of flavor f is shifted away from the origin in a transversely polarized proton [16]. These distributions strongly break azimuthal symmetry about the longitudinal axis. In particular, the centroid of the up and down distributions are displaced in opposite directions, as required by the fact that the Lorentz boost of a magnetic dipole produces an electric dipole field. For $\xi \neq 0$, the Fourier transform of the GPDs determines overlap matrix elements for partons of initial and final impact parameter $\mathbf{b}/(1\pm\xi)$ with respect to the center-of-momentum of the initial and final proton [17]. In the particular case of $x=\xi$, the variable $\mathbf{r}$, Fourier conjugate to $\Delta_\perp$, is the transverse separation of the active parton from the center-of-momentum of the spectator partons [18].

The experimental program to establish the domain of factorization in DES and to extract GPDs promises new insight into the quark-gluon structure of hadrons. The GPDs offer for the first time a probe of the rich correlations between spatial and momentum degrees of freedom of quarks and gluons in hadronic systems.

1.2. Scattering amplitude and observables

The $ep \to ep\gamma$ cross section has the form (Fig. 1)

$$\frac{d^5\sigma}{dQ^2 dx_{\rm Bj} d\phi_e d\Delta^2 d\phi_{\gamma\gamma}} = \frac{\alpha_{QED}^3}{4(2\pi)^2}\frac{x_{\rm Bj}y^2}{Q^4}\frac{1}{\sqrt{1+\epsilon_{DVCS}^2}}\left[|\mathcal{T}_{\rm BH}|^2 + \mathcal{I} + |\mathcal{T}_{\rm VCS}|^2\right] , \tag{13}$$

where $\epsilon^2_{DVCS} = 4x^2_{\rm Bj}M^2/Q^2$ and $\mathcal{I}$ is the BH·VCS interference term. The pure Bethe-Heitler term $|\mathcal{T}_{\rm BH}|^2$ is exactly calculable in terms of the nucleon form factors [19, 20, 21]. The full VCS amplitude is expressed as

$$\mathcal{T}_{\rm VCS}(e^{\pm}) = \overline{u}(k', \lambda)\gamma_\mu u(k,\lambda)\frac{(\pm e)}{q^2}H^{\mu\nu}\epsilon^\dagger_\nu . \tag{14}$$

The general VCS hadronic tensor H has 12 independent terms. In the leading order twist-2 approximation, H reduces to just four terms [1]

$$\begin{aligned} H^{\mu\nu}_{\rm LO, twist\,2} = \; & \tfrac{1}{2}(-g^{\mu\nu})_\perp \overline{U}(p')\left[(n\cdot\gamma)\mathcal{H}(\xi,t) + \tfrac{i}{2M}n_\kappa\sigma^{\kappa\lambda}\Delta_\lambda\mathcal{E}(\xi,t)\right]U(p) \\ & -(\epsilon^{\mu\nu})_\perp \overline{U}(p')\left[(n\cdot\gamma\gamma_5)\widetilde{\mathcal{H}}(\xi,t) + (\gamma_5 n\cdot\Delta)\widetilde{\mathcal{E}}(\xi,t)\right]U(p) , \end{aligned} \tag{15}$$

where the transverse tensors are defined as

$$(-g^{\mu\nu})_\perp = -g^{\mu\nu} + n^\mu\tilde{p}^\nu + \tilde{p}^\mu n^\nu , \qquad (\epsilon^{\mu\nu})_\perp = \epsilon^{\mu\nu\alpha\beta}n_\alpha\tilde{p}_\beta . \tag{16}$$

The Compton form factors (CFF) $\mathcal{H}\ldots$ in (15) are defined by the integration over the quark loop in Fig. 2:

$$\begin{aligned} \left[\mathcal{H},\mathcal{E}\right](\xi,t) &= \int_{-1}^{+1} dx \left[\frac{1}{x-\xi+i\epsilon} + \frac{1}{x+\xi-i\epsilon}\right]\left[H,E\right](x,\xi,t) , \\ \left[\widetilde{\mathcal{H}},\widetilde{\mathcal{E}}\right](\xi,t) &= \int_{-1}^{+1} dx \left[\frac{1}{x-\xi+i\epsilon} - \frac{1}{x+\xi-i\epsilon}\right]\left[\widetilde{H},\widetilde{E}\right](x,\xi,t) . \end{aligned} \tag{17}$$

Thus the imaginary parts of the CFFs are proportional to the GPDs at the point $x = \pm\xi$. Complete expressions for the VCS hadronic tensor to twist 3 accuracy are given in *e.g.*[21, 12].

The importance of the azimuthal distributions of the DVCS and interference terms, both for testing factorization and extracting constraints on the GPDs, was first pointed out by Diehl *et al.*[22]. The azimuthal distribution of the DVCS and interference terms has the general form [20]

$$|\mathcal{T}_{\rm DVCS}|^2 = \frac{e^6(s_e-M^2)^2}{x^2_{\rm Bj}Q^6}\left\{\sum_{n=0}^{2} c_n^{\rm DVCS}\cos(n\phi_{\gamma\gamma}) + \sum_{n=1}^{2} s_n^{\rm DVCS}\sin(n\phi_{\gamma\gamma})\right\} , \tag{18}$$

$$\mathcal{I}(e^{\pm}) = \frac{\pm e^6 x^2_{\rm Bj}(s_e-M^2)^3}{\Delta^2 Q^2(k-q')^2(k'+q')^2}\left\{\sum_{n=0}^{3} c_n^{\mathcal{I}}\cos(n\phi_{\gamma\gamma}) + \sum_{n=1}^{3} s_n^{\mathcal{I}}\sin(n\phi_{\gamma\gamma})\right\} . \tag{19}$$

The azimuthal angle $\phi_{\gamma\gamma}$ of the hadronic $\mathbf{q}' \otimes \mathbf{p}'$ plane relative to the electron scattering plane is defined such that $\sin(\phi_{\gamma\gamma}) > 0$ when $(\mathbf{k} \wedge \mathbf{k}') \cdot \mathbf{q}' > 0$ and $\phi_{\gamma\gamma} = 0$ when the final photon is on the beam side of $\mathbf{q}$.

The leading order twist-2 DVCS amplitude couples only to transverse photons. Consequently, there is a specific twist-hierarchy to the terms in Eqs. (18) and (19) [20]:

(i) $c_0^{\rm DVCS}$, $c_0^{\mathcal{I}}$ are twist-2;
(ii) $(c,s)_1^{\mathcal{I}}$ are twist-2 (transverse VCS in interference with longitudinal BH amplitudes);
(iii) $(c,s)_1^{\rm DVCS}$ are twist-3 (LT electroproduction interference terms);
(iv) $(c,s)_2^{\rm DVCS}$ are bilinear combinations of ordinary twist-2 and gluon transversity terms;
(v) $(c,s)_2^{\mathcal{I}}$ are twist-3;
(vi) $(c,s)_3^{\mathcal{I}}$ are linear in the twist-2 gluon transversity terms.

For example, the $c, s_1^{\mathcal{I}}$ terms for unpolarized and longitudinally polarized targets are, respectively:

$$\left.\begin{array}{c} c^{\mathcal{I}}_{1,\rm unp} \\ s^{\mathcal{I}}_{1,\rm unp}\end{array}\right\} \propto \left\{\begin{array}{c} \Re{\rm e} \\ \lambda\Im{\rm m}\end{array}\right\} \mathcal{C}^{\mathcal{I}}_{\rm unp}\,,$$
$$\mathcal{C}^{\mathcal{I}}_{\rm unp} = \left[F_1\mathcal{H} + \xi G_M\widetilde{\mathcal{H}} + \tau_C F_2\mathcal{E}\right]\,, \tag{20}$$
$$\left.\begin{array}{c} c^{\mathcal{I}}_{1,\rm LP} \\ s^{\mathcal{I}}_{1,\rm LP}\end{array}\right\} \propto \Lambda\left\{\begin{array}{c} \lambda\Re{\rm e} \\ \Im{\rm m}\end{array}\right\} \mathcal{C}^{\mathcal{I}}_{\rm LP}\,,$$
$$\mathcal{C}^{\mathcal{I}}_{\rm LP} = \left[\xi G_M\left(\mathcal{H} + \frac{x_{\rm Bj}}{2}\mathcal{E}\right) + F_1\widetilde{\mathcal{H}} - \xi\left(\frac{x_{\rm Bj}}{2}F_1 + \tau_C F_2\right)\widetilde{\mathcal{E}}\right]\,, \tag{21}$$

where Λ is the target polarization and $\tau_C = -\Delta^2/(4M^2)$ [20]. The electromagnetic form factors F_1, F_2, G_M are evaluated at $-t$ and the Compton form factors $\mathcal{H}$, $\mathcal{E}\ldots$ are evaluated at (ξ, t). In particular, we expect that on the proton the $F_1\mathcal{H}$ and $F_1\widetilde{\mathcal{H}}$ terms will dominate the unpolarized and longitudinal target polarization observables of Eqs. (20) and (21), respectively. On the other hand, on the neutron both F_1 and $g_1(x) = \widetilde{H}(x,0,0)$ are small. We anticipate a greater sensitivity to E from the interference term on an unpolarized neutron and to the combination of $\mathcal{H}$ and $\widetilde{\mathcal{E}}$ on a longitudinally polarized neutron. Transverse target observables depend on different combinations of CFFs [20]. Dynamic and kinematic twist-3 terms in the scattering amplitude cause kinematically suppressed twist-3 terms to mix into the "twist-2" observables listed above [23, 24]. In addition, any "twist-2" observable will naturally contain contributions of all higher even-twist in a power series in $1/Q^2$. The precise twist content of DES observables must be determined *via* a Q^2-dependent analysis at fixed (ξ, t). A complete analysis must also include the logarithmic Q^2 dependence from QCD evolution.

1.3. Models

Several models of the GPDs exist. To varying degrees, they incorporate the theoretical and empirical constraints on the GPDs. In the valence region, the most widely used models are based on the double distribution (DD) *ansatz* proposed by Radyushkin [25]. Detailed versions of this model are presented by Vanderhaeghen, Guichon and Guidal (VGG) [26] and Goeke, Polyakov and Vanderhaeghen [21]. The DDs re-parameterize the (x,ξ) dependence of the GPDs in terms of the momentum fractions β and α of P^+ and Δ^+, respectively. Thus the initial and final parton + components of momentum are $\beta P^+ \mp (1 \pm \alpha)\Delta^+/2$. The H, E, and $\widetilde{H}$ DDs are

Journal of Physics: Conference Series **299** (2011) 012006 doi:10.1088/1742-6596/299/1/012006

parameterized as:

$$GPD_{f,DD}(x,\xi,t) = \int_{-1}^{+1} d\beta \int_{-1+|\beta|}^{1-|\beta|} d\alpha\, \delta(x-\beta-\alpha\xi) F_f(\beta,\alpha,t),$$

$$F_f(\beta,\alpha,0) = h(\beta,\alpha) \begin{cases} q_f(\beta); \\ \kappa_f q_f(\beta)(1-\beta)^{\eta_f}/A_f; \\ \Delta q_f(\beta), \end{cases} \tag{22}$$

for H, E, and $\widetilde{H}$, respectively. In Eq. (22) q_f and Δq_f are the helicity-averaged and helicity-dependent PDFs of flavor f, and κ_f is the flavor anomalous magnetic moment of the proton. The normalization of E is such that $A_f = \int d\beta (1-\beta)^{\eta_f} q_f(\beta)$. The profile function h is commonly parameterized as

$$h(\beta,\alpha) = \frac{\Gamma(2b+2)}{2^{2b+1}\Gamma(b+1)} \frac{[(1-|\beta|)^2-\alpha^2)]^b}{(1-|\beta|)^{2b+1}}. \tag{23}$$

For $b=1$, this form reduces at $\beta=0$ to an asymptotic meson DA, $\Phi(\alpha)=3(1-\alpha^2)/4$, with support $-1<\alpha<1$. This connection is suggested by the ERBL evolution equations for $|x|<\xi$. In general, the exponent b is a free parameter and for $b\to\infty$, the GPD is ξ-independent.

The DD form of Eq. (22) ensures that the polynomiality conditions are automatically satisfied. However, it was pointed out by Polyakov and Weiss that for H and E, an additional "D-term" must be included [27] to produce the highest ξ^{N+1} power for x^N moment. This term, which only has support in the ERBL region, is an isosinglet and enters with opposite sign to H and E:

$$H_f(x,\xi,t) = H_{f,DD}(x,\xi,t) + \theta(\xi-|x|)\frac{1}{N_f}D(x/\xi,t)\ ,$$

$$E_f(x,\xi,t) = E_{f,DD}(x,\xi,t) - \theta(\xi-|x|)\frac{1}{N_f}D(x/\xi,t)\ . \tag{24}$$

In practice, the D-term has been taken as an expansion in odd Gegenbauer polynomials, with the first few terms fitted to a Chiral Soliton model calculation [21, 28]. The t dependence is introduced into the model *via* a Regge inspired *ansatz* [21]:

$$F_f(\beta,\alpha,t) \quad = h(\beta,\alpha)q_f(\beta)|\beta|^{-\alpha' t}\ . \tag{25}$$

As found in [29], the high-t form factor is dominated by the contribution in the DD at large β, for which the simple Regge forms must be modified to describe the data. A fit to all of the nucleon form factors data was obtained with the *ansatz*

$$F_f(\beta,\alpha,t) \quad = h(\beta,\alpha)q_f(\beta)|\beta|^{-(1-\beta)\alpha' t}. \tag{26}$$

Aside from the choice of b-parameter for each flavor and GPD, variants of the model exist with different choices of including the valence or valence plus sea contributions to E and H. The GPD $\widetilde{E}$ is generally parameterized separately, as the pion-pole in the t-channel. In this framework, $\widetilde{E}$ is directly related to the pion form factor [21, 30].

The family of models sketched above, and generically labeled "VGG", is qualitatively successful in describing the DVCS data. However, the model is highly constrained and does not have the full degrees of freedom of the GPDs. More general parameterizations will be needed as the data improves in precision and covers both a broader kinematic range and a more complete set of spin and flavor observables.

Another approach [31, 32] is to construct valence generalized parton distributions as an overlap of light-cone wave functions. However, a model involving the lowest Fock state component only produces GPDs vanishing at the border points $x = \pm\xi$ and in the whole central region $-\xi < x < \xi$ [31, 33]. It was shown [34] that inclusion of the higher Fock state components gives GPDs that are nonzero in the central region and at the border points. In particular, one may assume that overlap of the lowest Fock state components gives model GPDs at a low normalization point $Q_0 \sim 300$ MeV, and then evolve them to hard scales $Q \gtrsim 1$ GeV: the evolution will induce nonzero values for GPDs in the central region. Originally [35], the evolution approach was used to build a model for the gluon GPD, assuming that for a low normalization point Q_0 it coincides with the usual ("forward") gluon density, $H^g(x,\xi;Q_0) = xG(x,Q_0)$. In Ref. [36] this *ansatz* was also applied for quark distributions in an attempt to describe HERA DVCS data at low $x_{\rm Bj}$, for which predictions based on the double distribution *ansatz* are too large in magnitude.

More recently, the "dual parameterization" (DP) framework developed by Polyakov and collaborators [37, 38] was used to address this issue. In this approach, GPDs are expanded in terms of the partial waves exchanged in the t-channel. It was expected that for low $x_{\rm Bj}$ of the HERA DVCS data, the expansion may be truncated to the first "forward-like" functions [38]. However, a detailed analysis [39] demonstrated that the minimal model of the dual parameterization significantly (by a factor of 4) overestimates the HERA data. The relation between the DP approach and the DD *ansatz* was investigated in Ref. [40]. It was shown that GPDs built from DD-based models with $b = 1$ in Eq. (23) and small ξ may be reproduced just by the first term of the dual parameterization expansion, *i.e.*, the minimal DP- and DD-based models give similar results for DVCS at small $x_{\rm Bj}$, and both give a rather large value ~ 1.8 for the ratio $R^\Sigma(\xi) \equiv H^\Sigma(\xi,\xi)/\Sigma(\xi)$ of singlet quark distributions for small ξ, while experimental data favor the value close to 1. In a model developed by Müller and collaborators [41] it is possible to keep the value of $R^\Sigma(\xi)$ "flexible", *i.e.* to adjust it to describe the data. The "flexibility" may be achieved also in the dual parametrization approach, if one adds the second "forward-like" function.

2. Initial DVCS experiments

The richness of physical information in the GPDs has sparked an intense experimental effort. The H1, HERMES, and CLAS Collaborations published the first evidence for the DVCS reaction in 2001.

2.1. DVCS at HERA

The H1 [42, 43, 44] and ZEUS [45, 46] Collaborations measured the $p(e,e'\gamma)X$ cross section, integrated over $\phi_{\gamma\gamma}$. The exclusive $p(e,e'\gamma)p$ channel is enhanced over $p(e,e'\gamma)N^*$ channels by vetoing on forward detectors [44, 46]. In ZEUS, a subset of $p(e,e'\gamma p)$ events was tagged in a forward tracker [46]. The HERA data cover a wide kinematic range at low $x_{\rm Bj}$, with central values of Q^2 and W from 8 to 85 GeV^2 and 45 to 130 GeV, respectively.

The HERMES Collaboration measured the $\phi_{\gamma\gamma}$ distribution of the relative beam-helicity asymmetry in the $\mathrm{H}(\vec{e},e'\gamma)X$ reaction at average kinematics $\langle Q^2, x_{\rm Bj}, t\rangle = (2.6\,\mathrm{GeV}^2, 0.11, -0.27\,\mathrm{GeV}^2)$ [47]. The FWHM of the M_X^2 distribution was $\approx 1\,\mathrm{GeV}^2$, therefore covering the majority of the resonance region. However, at low $x_{\rm Bj}$ and $-t$, model estimates indicate that the exclusive $\mathrm{H}(\vec{e},e'\gamma)p$ channel is dominant [47]. The HERMES Collaboration has also recently measured the beam-charge asymmetry [48], transversely polarized target asymmetries [49], longitudinally polarized target asymmetries [50], and a more extensive set of beam spin asymmetries (BSAs) [51]. The final 2006-2007 HERMES run utilized a new recoil detector [52], to establish exclusivity *via* $\mathrm{H}(\vec{e},e'\gamma p)$ triple coincidence [53].

2.2. *Initial CLAS DVCS data*

The Jefferson Lab CLAS Collaboration first measured the relative beam-helicity asymmetry in the $H(\vec{e}, e'p)x$ reaction with 4.25 GeV incident electrons [54]. The exclusive photon was detected in only a fraction of the acceptance, due to the limited small angle acceptance of the standard CLAS calorimeter. The M_x^2 distribution is shown in Fig. 3 (left). The position and width of the exclusive $H(\vec{e}, e'p)\gamma$ event distribution was constrained to fit a subsample of $H(\vec{e}, e'p\gamma)$ data at small $(e'\gamma)$ opening angle, such that the events are dominated by the BH process. Similarly, exclusive π^0 event distributions were constrained to a subset of $H(\vec{e}, e'p\gamma\gamma)$ events from π^0 decay. Thus the $H(\vec{e}, e'p)x$ events in the exclusive region were fitted with two gaussians, for the $x = \gamma$ and $x = \pi^0$ channels. The widths and positions of these two gaussians are *a priori* constrained. In this way the exclusive $H(\vec{e}, e'p)\gamma$ channel was isolated.

The resulting DVCS beam-helicity asymmetry is shown in Fig. 3 (right). The shaded band is a one-sigma fit of the form

$$\text{BSA} = \alpha \sin(\phi_{\gamma\gamma}) + \beta \sin(2\phi_{\gamma\gamma}). \tag{27}$$

The α coefficient contains the twist-2 physics. The β coefficient contains the twist-3 physics, as well as contributions from $\cos(\phi_{\gamma\gamma})$ terms in the unpolarized cross section in the denominator of the beam spin asymmetry. The dashed and dotted curves in Fig. 3 are leading twist calculations of the VGG model, in the ξ-independent (at fixed x) and ξ-dependent versions, respectively [26]. The solid curve includes an estimate of twist-3 effects [55, 56]. The models, though constrained by fundamental principles, are still very preliminary. It is remarkable that the data and models are in as good agreement as indicated by Fig. 3. Within the VGG model, the largest contribution to the beam-helicity asymmetry on the proton comes from the $H(\pm\xi, \xi, t)$ GPD.

A second CLAS experiment, still with the standard CLAS configuration [57], measured the longitudinal target spin asymmetry in the $\vec{p}(e, e'p\gamma)$ reaction on a polarized NH_3 target [58]. In order to isolate the exclusive $ep \to ep\gamma$ events from the nuclear continuum, the statistics were limited to the triple coincidence $(e, e'p\gamma)$ events, with the photons detected in the standard CLAS calorimeter. The resulting exclusivity spectrum in Fig. 4 shows a 10:1 signal to background ratio. The longitudinal target spin asymmetry, averaged over the acceptance, is displayed in Fig. 5, for $\langle Q^2 \rangle = 1.82$ GeV^2, $\langle \xi \rangle = 0.16$, and $\langle t \rangle = -0.31$ GeV^2. The solid curve is a fit of the same form as (27). The resulting $\sin(\phi_{\gamma\gamma})$ moments are plotted in Fig. 6. The error bars in Figs. 5 and 6 are statistical, with the systematic errors displayed as a band at the bottom. The dashed and dotted curves in Figs. 5 and 6 indicate the sensitivity of the longitudinal target spin asymmetry to $\widetilde{H}$.

The initial success of the CLAS DVCS analysis, and the limited small angle acceptance of the CLAS calorimeter, led to the construction of a small angle "Inner Calorimeter". The ongoing dedicated DVCS program in CLAS will be described in Section 4.

3. The Hall A DVCS program at 6 GeV

The Hall A DVCS program started with experiments E00-110 [59] and E03-106 [60]. These experiments measured, respectively, the cross sections of the $H(\vec{e}, e'\gamma)p$ and $D(\vec{e}, e'\gamma)pn$ reactions at $x_{Bj} = 0.36$ with an incident beam of 5.75 GeV. In both experiments the scattered electron was detected in the standard High Resolution Spectrometer (HRS) [61], and the photon was detected in a new 132 element PbF_2 calorimeter, subtending ~ 0.1 sr. PbF_2 is a pure Cerenkov medium, thereby minimizing the hadronic background and delivering the fastest timing pulses. All PbF_2 channels were readout by a custom 1 GHz digitizer [62], based on the ANTARES ARS0 chip [63]. The luminosity of 1–4 $\cdot 10^{37}$ Hz/cm^2 per nucleon was unprecedented for open detectors in a non-magnetic environment. The halo-free CW beam of CEBAF was essential to this success.

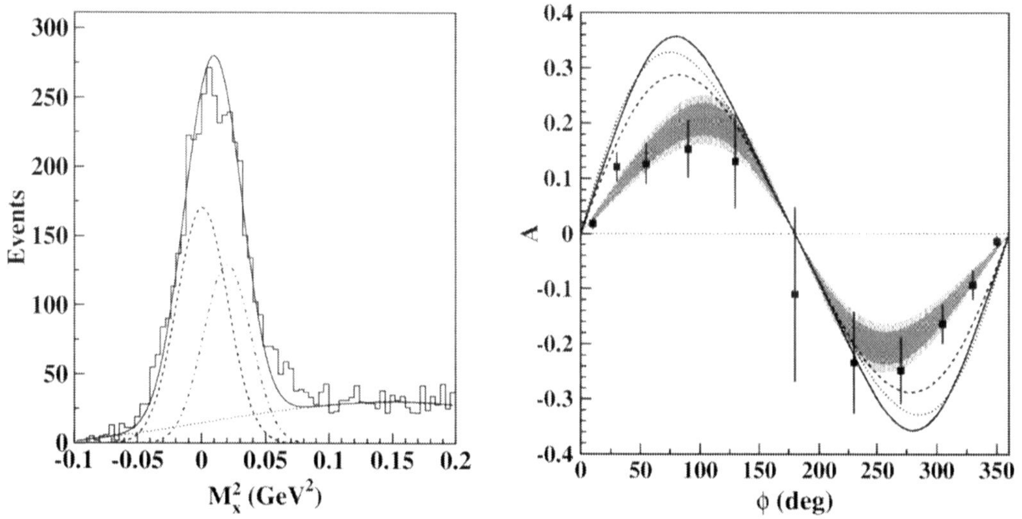

Figure 3. $H(\vec{e}, e'p)x$ analysis from CLAS at 4.25 GeV incident electron energy [54]. **Left:** Missing mass squared M_x^2 distribution. The two Gaussian distributions representing the $H(\vec{e}, e'p)\gamma$ and $H(\vec{e}, e'p)\pi^0$ events are described in the text. The smooth polynomial background represents the contribution of processes such as $ep \to eN^*\gamma$. **Right:** DVCS beam-helicity asymmetry. The kinematics are integrated over $Q^2 \in [1.0, 1.75]\,\mathrm{GeV}^2$ and $-t \in [0.1, 0.3]\,\mathrm{GeV}^2$. The shaded region is the fit described in the text, in Section 2.2. The curves, described in the text, are evaluated at the fixed values $Q^2 = 1.25\,\mathrm{GeV}^2$, $x_{\mathrm{Bj}} = 0.19$, and $t = -0.19\,\mathrm{GeV}^2$.

3.1. Proton DVCS

Hall A experiment E00-110 measured DVCS on the proton at $Q^2 = 1.5$, 1.9, and 2.3 GeV^2. The isolation of the exclusive $H(\vec{e}, e'\gamma)p$ signal is illustrated in Fig. 7. The helicity dependent cross sections as a function of $\phi_{\gamma\gamma}$ in four bins in Δ^2 are displayed in Figs. 8 and 9.

The latter figure also displays the helicity independent cross sections for $Q^2 = 2.3$ GeV^2. The helicity dependent cross sections demonstrate the dominance of the effective twist-2 term $s_1^{\mathcal{I}}$ of Eq. (19). The helicity independent cross sections (Fig. 9) show significant contributions from the sum of the interference and DVCS terms, in addition to the pure BH cross section. Thus the analysis of relative asymmetries of the form $\Delta\sigma/\sigma$ requires the inclusion of the full DVCS terms in both the numerator and denominator. The effective "twist-2" interference term $\Im\mathrm{m}\mathcal{C}^{I}$ of Eq. (20) is presented in Fig. 10. The VGG model calculation, described in Section 1.3 agrees in slope with the data, but lies roughly 30% above the data. Within statistics, the results in Fig. 10 are close to Q^2-independent in all bins in Δ^2. This provides support to the conjecture that DVCS factorization results in leading twist dominance at the same scale of $Q^2 \geq 2\,\mathrm{GeV}^2$ as in DIS.

3.2. Neutron DVCS

The Jefferson Lab Hall A experiment E03-106 measured the helicity dependent DVCS cross section on deuterium, $D(\vec{e}, e'\gamma)X$ at $Q^2 = 1.9\,\mathrm{GeV}^2$ and $x_{\mathrm{Bj}} = 0.36$. Within the impulse approximation, the cross section is described as the incoherent sum of coherent deuteron and quasi-free proton and neutron channels:

$$D(\vec{e}, e'\gamma) = d(\vec{e}, e'\gamma)d + n(\vec{e}, e'\gamma)n + p(\vec{e}, e'\gamma)p + \cdots \tag{28}$$

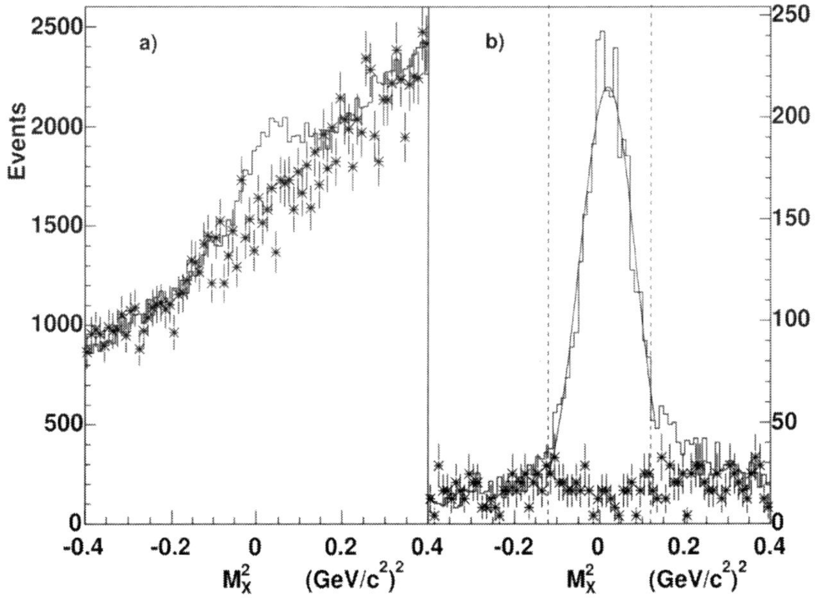

Figure 4. Missing mass squared M_x^2 distributions of the $(e, e'p)x$ reaction for $(e, e'p\gamma)$ events on a longitudinally polarized $N\vec{H}_3$ target. **Left:** Raw distribution; **Right:** Distribution after requiring detected photon within 2° of the predicted direction of an exclusive photon from the $(e, e'p)\gamma$ kinematics. In both figures the stars are the corresponding distributions from a C target, normalized to the negative M_x^2 region.

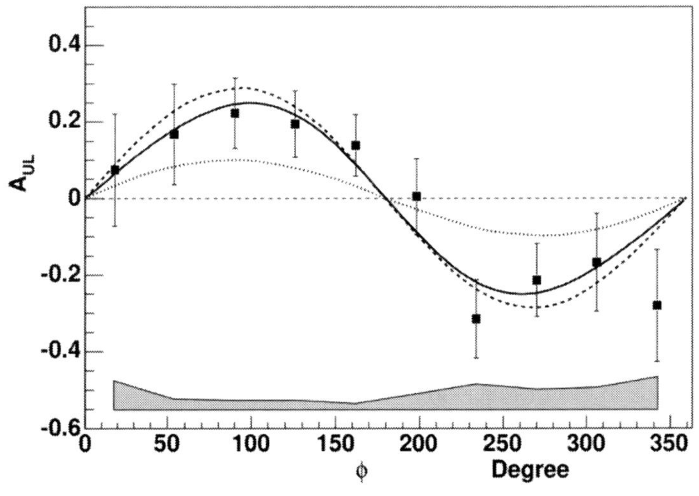

Figure 5. CLAS Longitudinal target spin asymmetry, A_{UL} [58], for $\langle Q^2 \rangle = 1.82\,\text{GeV}^2$, $\langle \xi \rangle = 0.16$, and $\langle t \rangle = -0.31\,\text{GeV}^2$. The solid curve is a fit of the form of (27). The dashed and dotted curves are from the ξ-dependent VGG model with $E = \widetilde{E} = 0$. The dotted curve includes only H. The dashed curve includes both H and $\widetilde{H}$.

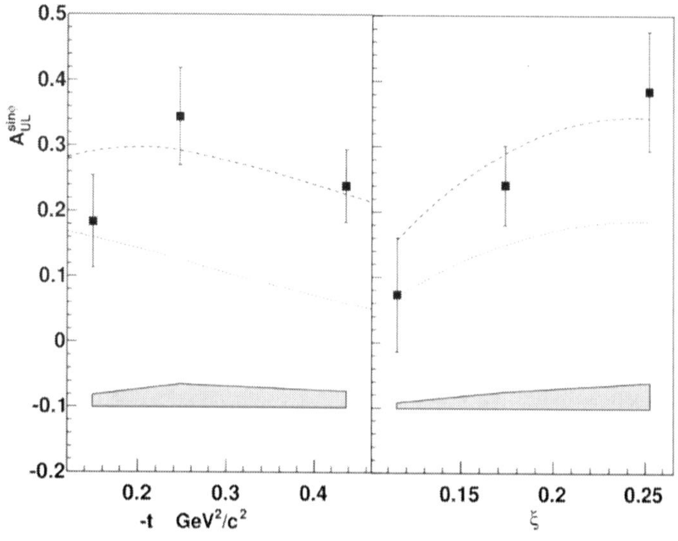

Figure 6. The $\sin(\phi_{\gamma\gamma})$ moments of A_{UL} [58]. **Left:** Three bins in $-t$, integrated over ξ; **Right:** Three bins in ξ, integrated over t. The curves are described in Fig. 5.

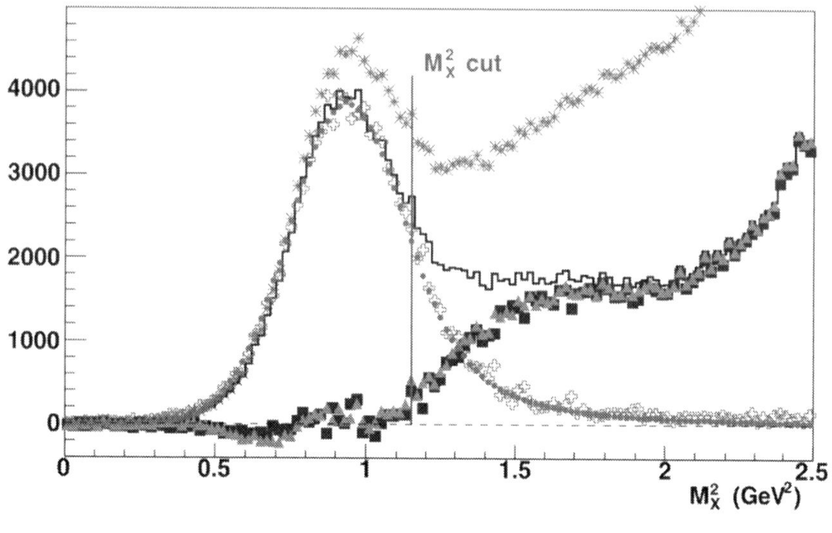

Figure 7. Missing mass squared distribution of the $H(e, e'\gamma)X$ reaction in Jefferson Lab Hall A experiment E00-110 [64]. The [green] stars are the raw data after accidental subtraction. The continuous [black] histogram is the data after subtracting the statistical sample of $H(e; e'\gamma)\gamma X'$ events inferred from the measured $H(e, e'\pi^0)X'$ sample. The open [red] cross histogram is a normalized sample of $H(e, e'\gamma p)$ events. The [magenta] dots are the exclusive simulation. The [blue] triangles and squares are obtained by subtracting the last two histograms from the solid [black] histogram.

Journal of Physics: Conference Series **299** (2011) 012006 doi:10.1088/1742-6596/299/1/012006

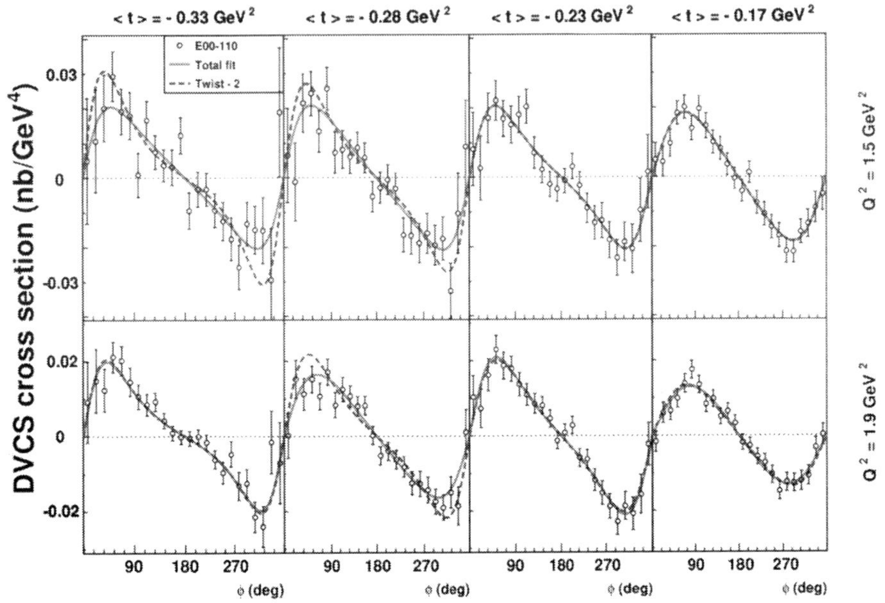

Figure 8. Helicity dependent DVCS cross sections $\Delta^4\sigma/[dQ^2 dx_{\rm Bj} d\Delta^2 d\phi_{\gamma\gamma}]$ from JLab Hall A E00-110 [64] at $Q^2 = 1.5$, and 1.9 GeV^2. The mean values of $-\Delta^2$ are, from right to left 0.17, 0.23, 0.28, and 0.33 GeV^2. Each distribution is fitted with the form of (19), with the (effective) twist-2 term in solid (red) and the complete fit in dashed (blue).

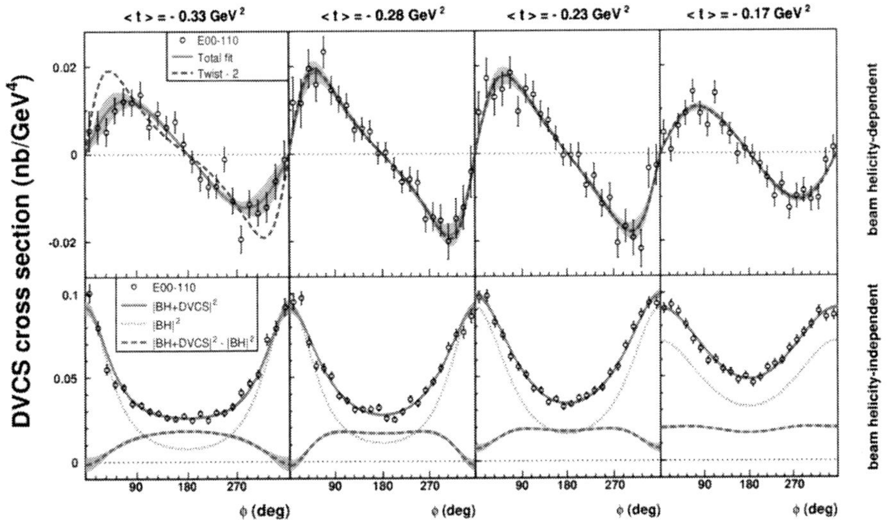

Figure 9. Helicity dependent (top) and independent (bottom) DVCS cross sections from Jefferson Lab Hall A at $Q^2 = 2.3$ GeV^2 [64]. The bins are the same as Fig. 8, as are the curves in the top plot. In the bottom plot, the green curve (mostly concave up) is the pure $|\mathrm{BH}|^2$ cross section. The magenta curve (mostly concave down) is a fit including the twist-2 and twist-3 terms of Eqs. (18) and (19).

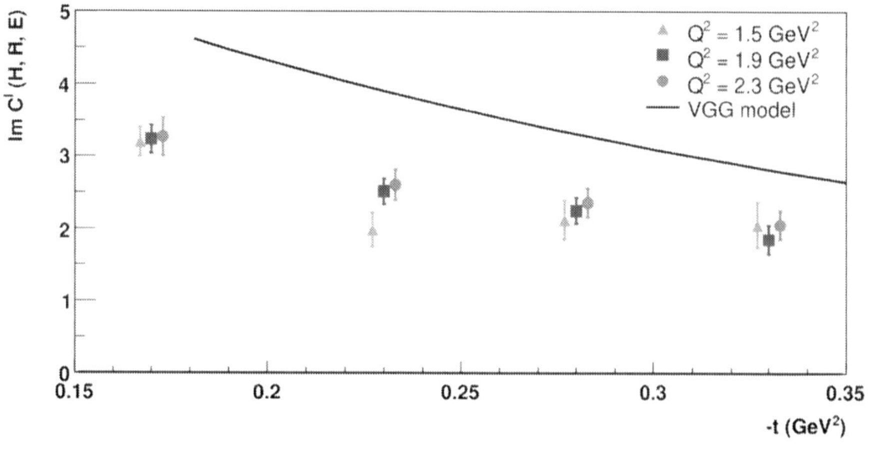

Figure 10. Imaginary part of the effective interference term $C^{\mathcal{I}}_{\rm unp}$ extracted from the helicity dependent data of Figs. 8 and 9.

Meson production channels contribute as background. The proton-DVCS contribution is calculated by smearing the $H(e,e'\gamma)X$ data by the nucleon momentum distribution in the deuteron. This statistical estimate of the proton contribution is subtracted from the data. The coherent deuteron and quasi-free neutron channels were separated, within statistics, by fitting the missing mass distribution with a Monte Carlo simulation of these two channels. This separation exploits the fact that for M_X^2 calculated relative to a nucleon target, the quasi-free neutron spectrum peaks at $M_X^2 \approx M^2$ whereas the coherent deuteron peak lies at $M^2 + t/2$. This analysis produced constraints on the neutron and deuteron DVCS†·BH interference terms $\Im m[C^{\mathcal{I}}_{\rm unp}]$ [65].

Mazouz *et al.* [65] fitted the neutron interference signal by varying the parameters of the E GPD within the VGG model of [21]. This results in a model dependent constraint on the Ji sum rule values of (J_d, J_u), illustrated in Fig. 11. A similar constraint obtained by the HERMES Collaboration in DVCS on a transversely polarized proton target is also illustrated in the figure. Both of these experimental determinations are essentially constraints on the model at one value of $x_{\rm Bj}$, and then the model is integrated over x at fixed ξ to obtain the sum rule estimate. Measurements over a more extensive range in ξ with a more complete set of spin observables, and models with more degrees of freedom, are necessary in order to more fully constrain the sum rule with realistic error bars. Lattice QCD calculations, and other phenomenological estimates, are also illustrated in Fig. 11.

3.3. Future Hall A program at 6 GeV

The unpolarized cross sections in Fig. 9 are not fully dominated by the pure BH process. The harmonic $\phi_{\gamma\gamma}$ structure of the cross section does not allow the full separation of the $\Re{\rm e}[{\rm DVCS}^\dagger{\rm BH}]$ and $|{\rm DVCS}|^2$ contributions. These terms can be separated either by the beam charge dependence (*e.g.* Ref. [48]) or by measuring the incident energy dependence of the cross sections. At fixed Q^2 and $x_{\rm Bj}$, the DVCS, interference, and BH terms in the cross section scale roughly as $s_e^2 : s_e : 1$ (see Eqs. (18) and (19)). Jefferson Lab Experiment E07-007 [71] will measure the DVCS helicity independent cross sections in the three kinematics of Figs. 8 and 9 at two separate beam energies in each kinematics. This will measure the Q^2 dependence of the separated leading twist-2 and twist-3 observables of the $\Re{\rm e}[{\rm DVCS}^\dagger{\rm BH}]$ and $|{\rm DVCS}|^2$ terms.

Experiment E08-025 [72] will measure the DVCS cross sections on the deuteron at the same

Journal of Physics: Conference Series **299** (2011) 012006 doi:10.1088/1742-6596/299/1/012006

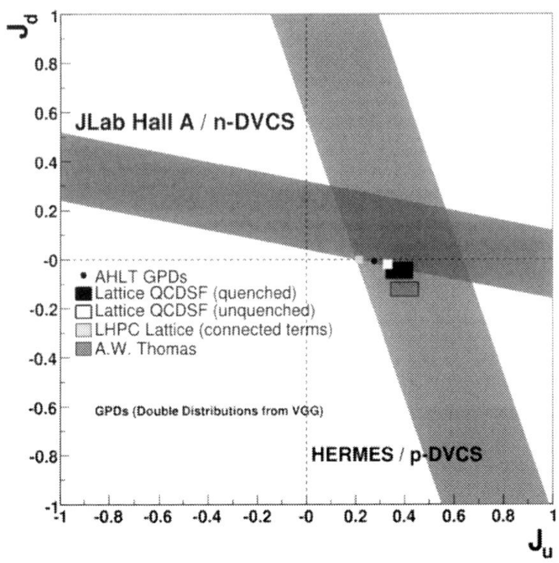

Figure 11. Experimental constraints on the total up and down quark contributions to the proton spin from Jefferson Lab Hall A neutron [65] and HERMES transversely polarized proton [49] results. The theory/model values are from AHLT [66], QCDSF quenched [67] and unquenched [68] LHPC [69], and Thomas [70].

$(Q^2, x_{\rm Bj})$ value as in E03-106, but at two incident beam energies. Together with an expanded calorimeter to improve the neutral pion subtraction, the two beam energies will allow a more complete separation of the $\rm DVCS^2$ and real and imaginary parts of the DVCS·BH interference on a quasi-free neutron. This will be an important step towards a full flavor separation of DVCS. Both experiments E07-007 and E08-025 ran in Autumn of 2010.

4. The CLAS DVCS program at 6 GeV

4.1. Unpolarized proton targets

A new calorimeter of 424 tapered $\rm PbWO_4$ crystals was constructed to provide complete 2π photon coverage for polar angles from 4.5° to 15°, relative to the beam line. An approximately 5 Tesla superconducting solenoid was added at the target, to confine Møller electrons. The new calorimeter is located 60 cm from the target where the solenoid fringe field is still a few Tesla. Therefore, the individual crystals were read-out by Avalanche Photo-Diodes. Having strongly benefited from the CERN CMS pioneering research and development effort on this recent technology, the present CLAS experiment is the first one to use such photodetectors in a physics production mode.

All particles of the $ep \to ep\gamma$ reaction final state were detected in CLAS. To ensure exclusivity, several cuts were made, a couple of them being illustrated in Fig. 12. In spite of these very constraining cuts, some contamination from the $ep \to ep\pi^0$ reaction remained. Indeed, when one of the two γ's originating from the decay $\pi^0 \hookrightarrow \gamma\gamma$ escapes detection and/or has little energy (below the 150 MeV threshold of the calorimeter), an event $ep \to ep\gamma(\gamma)$ may pass all DVCS cuts and become a perfect candidate to be selected as an $ep \to ep\gamma$ event. Such "1-γ" π^0 backgrounds can be estimated from Monte Carlo combined with the actual number of detected "2-γ" π^0's, resulting, depending on the kinematics, in contaminations ranging from 1 to 25%, being 5% in average.

The extensive CLAS dataset of DVCS beam spin asymmetries is illustrated in Fig. 13. The

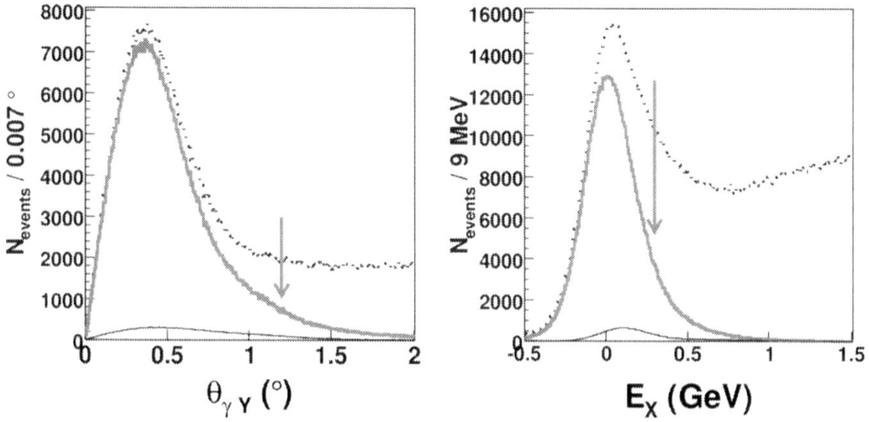

Figure 12. Example of exclusivity cuts (given by the location of the arrow) for the CLAS DVCS experiment[73]. Distribution in cone angle $\theta_{\gamma Y}$ for the $ep \to epY$ reaction (left) and in missing energy E_X for the $ep \to ep\gamma X$ reaction (right) before (black-dotted curve) and after (red solid) kinematic cuts (including others not displayed here). The thin solid black line represents the background from the $ep \to ep\pi^0$ events. These distributions are integrated over all kinematics variables.

blue solid curves are the result of the twist-2 handbag GPD calculation (VGG) including just the H GPD [26, 29]. The blue dashed curves include the associated twist-3 calculation. Although the general trends of the data are reproduced, the model tends to overestimate the BSAs. These too large BSAs by the VGG model can come from either an overestimation of $\mathcal{H}(\xi, t)$ (the dominant factor in the numerator of the BSA) or an underestimation of the CFFs associated to the real part of the DVCS amplitude, which contribute predominantly to the denominator of the BSA [74]. The dashed black curve (third curve in some panels of Fig. 13) is the result of a Regge model [75] for the DVCS process. As Q^2 increases, the Regge model drops significantly below both the data and the VGG calculations. This experiment was continued in 2008–2009, which will significantly improve the statistical precision relative to Fig. 13 [76].

4.2. Polarized targets

In 2009, a new DVCS experiment completed data taking with the longitudinally polarized NH_3 target [77]. Relative to the previous experiment (Ref. [58] and Figs. 4–6), this new experiment will improve both the statistics and acceptance by the addition of the new electromagnetic calorimeter mentioned in the previous section. We recall that the target spin asymmetry is mostly sensitive to $\widetilde{\mathcal{H}}(\xi, t)$ and that, therefore, strong constraints on this CFF should arise from this experiment, as discussed *e.g.* in Ref. [78].

After decades of development, the HD-ice target ran successfully at the BNL-LEGS facility in 2005 and 2006. This target has now been transferred to Jefferson Lab and is being prepared for a photoproduction run in 2011 [79]. Initial studies of local depolarization by microwaves suggest that the spin relaxation times of this target are sufficiently long for the target to operate with electron beams in CLAS. An electron beam test is projected for the end of the 2011 photoproduction run. If successful, a full suite of transverse polarization observables for the DVCS process will be feasible in CLAS in 2011. When combined with the cross section and longitudinal polarization data, along with the resulting double polarization observables, a full separation of the real and imaginary parts of all four Compton form factors $\mathcal{H}(\xi, t)$, $\mathcal{E}(\xi, t)$, $\widetilde{\mathcal{H}}(\xi, t)$, and $\widetilde{\mathcal{E}}(\xi, t)$ is in principle possible [20, 74].

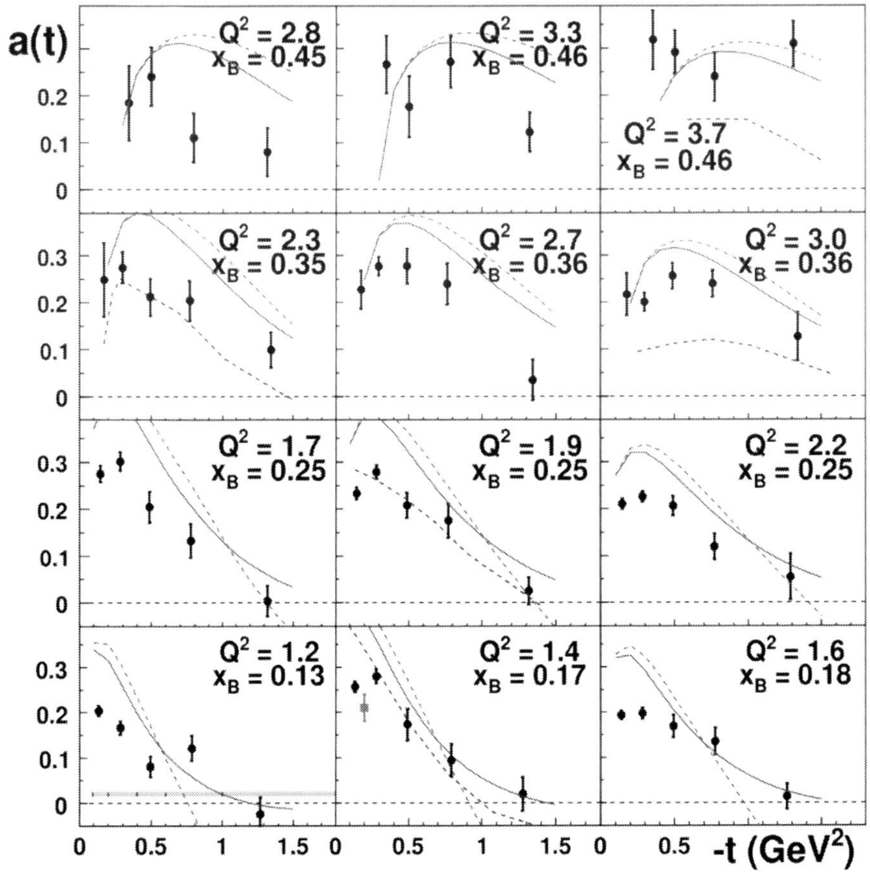

Figure 13. Beam spin asymmetry $\sin(\phi_{\gamma\gamma})$ moments from the CLAS DVCS experiment [73]. Curves are described in the text.

4.3. Nuclear targets

GPDs are also defined for nuclei [80, 81]. One can study effects similar to the nuclear EMC effect observed for standard inclusive parton distributions, where the PDF of a nucleus is not simply the sum of the individual nucleon PDFs. A pioneering experiment [82, 83] of coherent DVCS on a ^{4}He target ran with the CLAS detector in 2010. ^{4}He is a very good starting case study as it is dense enough to generate nuclear medium effects, many microscopic calculations for its nuclear structure and dynamics exist and, as a global spin-0 object, at leading twist there is only one GPD. This ^{4}He-DVCS experiment detected the scattered electron in CLAS, the final state photon with the $PbWO_4$ and standard calorimeters mentioned in the previous sections, and the recoil nucleus with a radial time-projection chamber [84]. The ϕ-distribution of the coherent DVCS BSA, up to twist-3 corrections, can yield the real and imaginary parts of the Compton form factor of the coherent GPD.

5. Deeply virtual meson production

GPDs are in principle also accessible through exclusive meson electroproduction (see Fig. 2). With respect to the DVCS process, a few features are proper to the meson channels:

- The factorization holds only for the longitudinal part of the amplitude which implies to separate, experimentally, the transverse and longitudinal parts of the cross section. For

Journal of Physics: Conference Series **299** (2011) 012006 doi:10.1088/1742-6596/299/1/012006

pseudoscalar mesons, this can be done through a Rosenbluth separation. For vector mesons, this separation can be carried out, relying on the s-channel helicity conservation (SCHC) concept, by measuring the angular distribution of the vector meson decay products.

- In comparison to the DVCS handbag diagram, there is now a perturbative gluon exchange. This suggests that factorization will be obtained at a higher scale in exclusive meson production than in DVCS.
- Besides the GPDs, there is another nonperturbative object entering the meson handbag diagram, the meson distribution amplitude. It is usually taken as the asymptotic DA, but it potentially adds a further unknown in the process.
- As a positive point, the meson channels have the advantage of filtering certain GPDs: the vector meson channels are sensitive, at leading twist, only to the H and E GPDs, while the pseudoscalar mesons are sensitive only to the $\tilde{H}$ and $\tilde{E}$ GPDs. Deeply virtual meson production also offers a flavor filter of the GPDs. For example, ρ^0 and ω electroproduction are sensitive to different combinations of the up- and down-quark GPDs.

Exclusive π^0 electroproduction results have been published from CLAS [85] and Hall A [86]. The Hall C results on exclusive π^+ production are discussed elsewhere in this volume [87]. In this section, we focus on exclusive vector meson production.

5.1. The ρ^0 channel

Deeply virtual electroproduction of the ρ^0 was studied by the CLAS Collaboration at incident energies of 4.2 GeV [88] and 5.75 GeV [89]. To select the $e^-p \to e^-p\rho^0 \hookrightarrow \pi^+\pi^-$ channel, the scattered electron, the recoil proton and the π^+ were detected. A cut on the missing mass $ep \to ep\pi^+X$ was then used to identify the $e^-p \to e^-p\pi^+\pi^-$ final state. The main challenge in this analysis was to subtract under the broad ($\Gamma_{\rho^0} \approx 150$ MeV) ρ^0 peak, the non-resonant $e^-p \to e^-p\pi^+\pi^-$ (physical) background, arising for instance from processes such as $e^-p \to e^-\pi^-\Delta^{++} \hookrightarrow p\pi^+$. These "background" channels led to uncertainties of the order of 20% to 25% on the extracted cross sections.

The separation of the longitudinal and transverse parts of the cross section was carried out, as mentioned earlier, by studying the angular distribution of the decay pions in the center-of-mass of the $\pi^+\pi^-$ system. At the same time, by the analysis of various azimuthal angular distributions, SCHC was verified experimentally at the $\approx 20\%$ level. The longitudinal part of the $\gamma^*p \to p\rho^0$ cross section, which can in principle lend itself to a GPD interpretation, is displayed in Fig. 14 along with the world data.

The cross sections clearly exhibit two different behaviors as a function of W. At low W, the cross sections decrease as W increases (x_{Bj} decreases) and then begin to rise slowly for $W > 10$ GeV. These two kinematic regimes can be identified, simply speaking, with regimes of t-channel exchange of Reggeon or $q\bar{q}$ exchange in the former case and of Pomeron or 2-gluon exchange in the latter case. The results of the calculations of the JML model [93], based on Reggeon exchange and hadronic degrees of freedom, and of the VGG [26] and GK [90, 91] models based on GPDs and on the handbag diagram of Fig. 2, are shown in Fig. 14. At lower W values, where the new CLAS data lie, it is striking that both the GK and VGG models fail to reproduce the data even though they are very successful at large W, even at $Q^2 \approx 2.3\,\mathrm{GeV}^2$. In the high-$W$ (low-x_{Bj}) region, the gluon GPD calculations already contain large higher twist effects in the form of intrinsic $k_\perp$ effects. The question then arises whether the higher twist effects have a different nature in the region dominated by quark GPDs (low W), or whether the double distribution based GPD models are missing an essential contribution. Ideas for such "missing" contribution in the D-term of the GPDs are speculated in [89, 92], leading to the thick solid curve in Fig. 14.

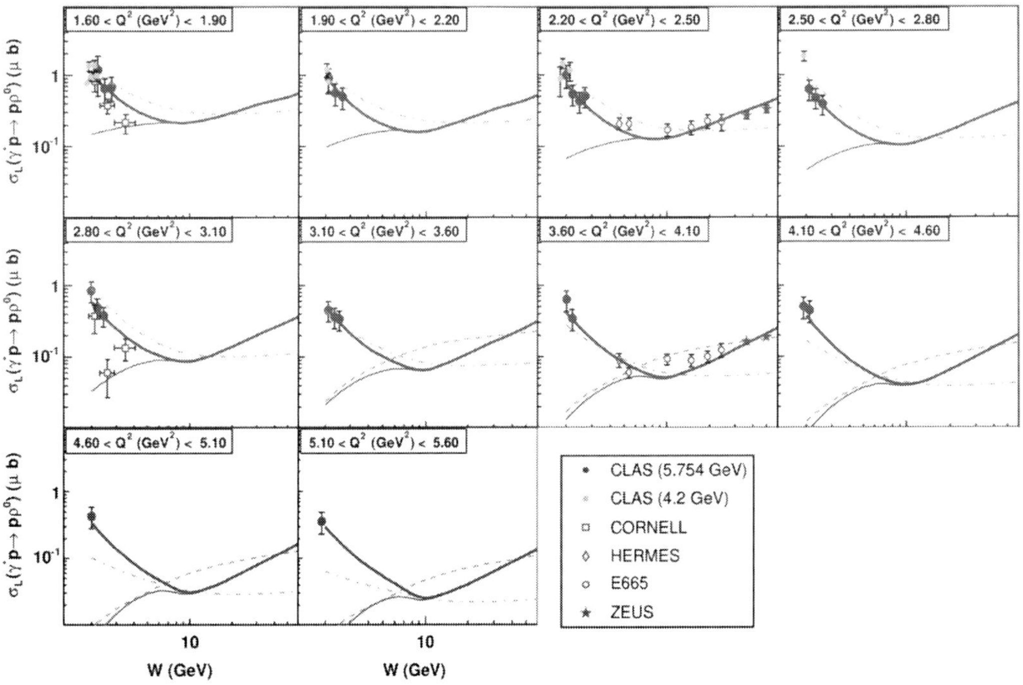

Figure 14. World data (for $W < 50$ GeV) for the reduced cross sections $\gamma_L^* p \to p\rho_L^0$ as a function of W for constant Q^2 bins (μb). The dashed curve (GK) [90, 91] and the thin solid curve (VGG) [26] are GPD calculations. The thick solid curve is the VGG calculation with the addition of the D-term inspired contribution [89, 92]. The dot-dashed curve is the Regge JML [93] calculation. The 5.75 GeV CLAS, 4.2 GeV CLAS, Cornell, HERMES, Fermilab and ZEUS data are, respectively, from Refs. [89], [88], [94], [95], [96] and [97].

5.2. The ω channel

The ω channel was studied in CLAS by detecting the $e^-p \to e^-p\pi^+X$ and $e^-p \to e^-p\pi^+\pi^-X$ topologies [98]. The former is advantageous to determine total cross sections with high statistics and the latter is necessary to measure the distribution of the decay products of the ω to separate the longitudinal and transverse parts of the cross section if SCHC is verified. However, one important result of this experiment was that many SCHC-violating spin density matrix elements were measured to be significantly different from 0 in ω electroproduction. Therefore, the longitudinal and transverse parts of the cross sections were not separated. Also, the angular analysis revealed the importance of unnatural parity exchange in the t-channel, such as π^0 exchange.

In terms of quantum numbers, the π^0 exchange contribution can be identified with the $\tilde{E}$ GPD. In the framework of the JML model [104], t-channel π^0 exchange is a major contributor to the cross section. The suggested importance of π^0 exchange is in apparent contradiction with the theoretical prediction that, at sufficiently large Q^2, exclusive vector meson production should be mostly longitudinal and sensitive only to H and E. Therefore, in order to study the GPD formalism in ω production, it is essential to experimentally isolate the purely longitudinal cross section, *via* a Rosenbluth separation. The VGG calculation of σ_L, shown in Fig.15, lies well below the unseparated data. This suggests that a precision extraction of σ_L *via* a Rosenbluth separation will be a difficult experimental challenge.

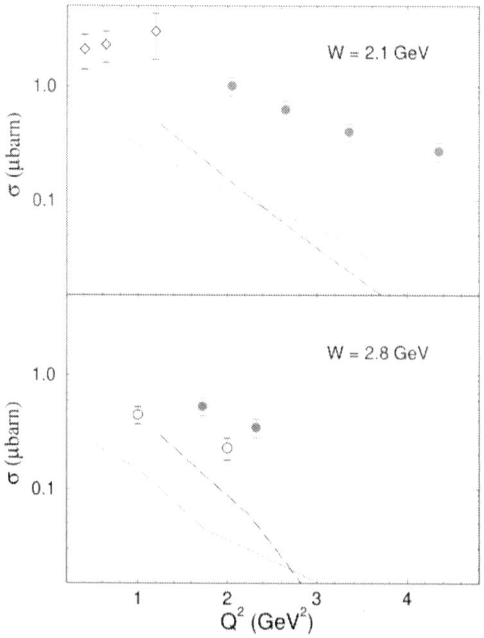

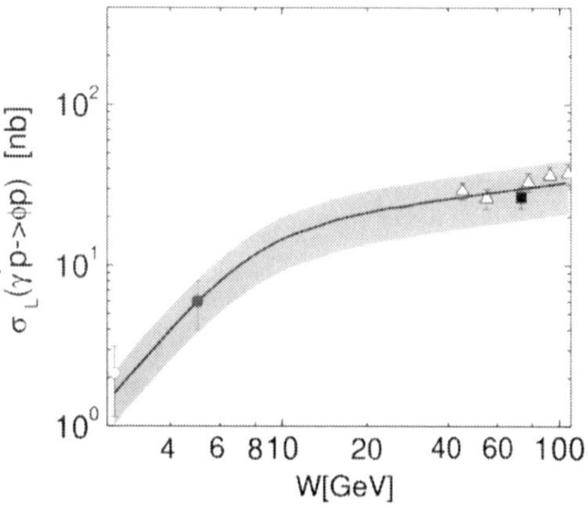

Figure 15. **Left:** Total (unseparated) cross section for the reaction $\gamma^* p \to p\omega$, for $\langle W\rangle$=2.1 GeV (top) and $\langle W\rangle$=2.8 GeV (bottom), as functions of Q^2. The dotted curve is the JML model for the total cross section $\sigma_T + \epsilon\sigma_L$ and the solid (dashed) curves shows the result of the JML (VGG) calculation for $\epsilon\sigma_L$. Data are from CLAS (full circles) [98], DESY (open diamonds) [99] and Cornell (open circles) [94]. **Right:** Longitudinal $\gamma^* p \to p\phi$ cross section as a function of W at Q^2=3.8 GeV^2. Data are from CLAS (open circle) [100], HERMES (filled circle) [101], ZEUS (open triangles) [102] and H1 (filled square) [103]. (Courtesy of P. Kroll and S. Goloskokov.)

5.3. The ϕ channel

The $e^- p \to e^- p\phi \hookrightarrow K^+K^-$ reaction was identified in CLAS by detecting the scattered electron, the recoil proton and the positive kaon and cutting around the missing mass of a kaon [100]. Relying on the SCHC concept, which was experimentally verified to hold in this channel, the longitudinal/transverse separation of the cross section was carried out. Figure 15 shows the resulting longitudinal total cross section at $Q^2 = 3.8$ GeV^2, along with higher energy HERMES and HERA data at comparable Q^2.

Exclusive ϕ electroproduction on the proton can be interpreted in terms of the handbag diagram with *gluonic* GPDs. Figure 15 shows the result of such calculation in the framework of the GK model [90, 91]. The very good agreement between this GPD calculation and the data gives confidence in the way higher twists corrections are handled, *i.e.* by taking into account the intrinsic transverse momentum dependence of the partons in the handbag calculation.

This set of three experiments delivered the largest ever dataset on vector meson production in the large Q^2 valence region. Although conclusions for the meson channels are more challenging than for DVCS, there may be the possibility to interpret the ρ^0 and ϕ channels in terms of the handbag diagram, though with large higher twist corrections and possibly modifications of the double distribution based GPD parametrisations. The higher-Q^2 data from Jefferson Lab at 12 GeV, as well as a global analysis including the larger DVCS dataset anticipated in the coming years should greatly clarify the role of factorization in deep virtual vector meson production.

Journal of Physics: Conference Series **299** (2011) 012006 doi:10.1088/1742-6596/299/1/012006

6. Outlook

6.1. Jefferson Lab at 12 GeV

The Jefferson Lab 12 GeV project offers an unprecedented frontier of intensity and precision for the study of deep exclusive scattering. The design luminosity of the upgraded CLAS12 detector is $10^{35}s^{-1}cm^{-2}$, with a large phase space acceptance for simultaneous detection of DVCS and deeply virtual meson production channels. At this luminosity, the Hall B dynamic nuclear polarization NH_3 target will achieve a longitudinal proton polarization of 80%. The Hall A and Hall C spectrometers will allow dedicated studies at luminosities $\geq 10^{37}s^{-1}cm^{-2}$ for neutral channels γ, π^0 at low t and up to $4 \cdot 10^{38}s^{-1}cm^{-2}$ for charged channels $\pi^{\pm}$, $K^{\pm}$. Specific 12 GeV experiments on hydrogen are approved in Hall A for DVCS (E12-06-114), in Hall B for DVCS (E12-06-119), and deep virtual π^0, η production (E12-06-108), and in Hall C for deep virtual π^+ production (E12-06-101, E12-07-105). Detailed descriptions of these experiments are available on the Hall A, B, and C web pages at `www.jlab.org`. The projected kinematic range of the DVCS programs in Hall A and B is illustrated in Fig. 16. With CLAS12, additional studies are in progress for measurements of deep virtual vector meson production, neutron DVCS *via* $D(e, e'\gamma n)p$ (LOI-09-001), coherent deuteron DVCS (PR-06-015) and DVCS on transversely polarized protons.

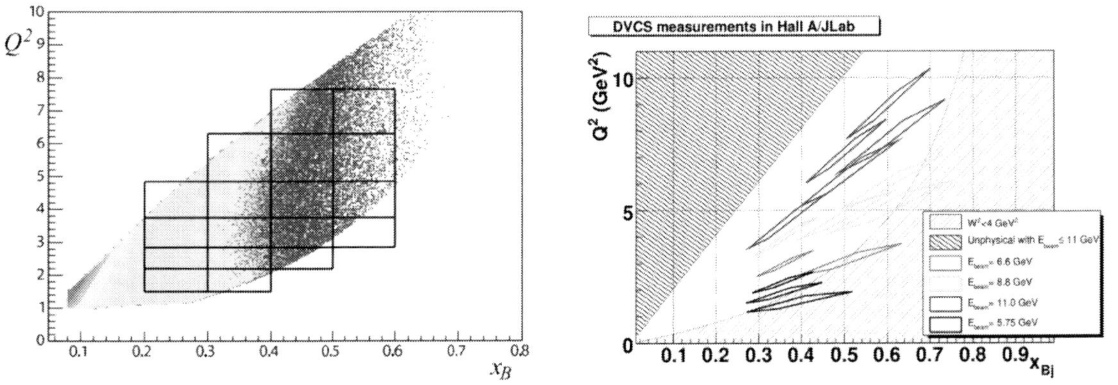

Figure 16. Projected kinematic bins at Jefferson Lab 12 GeV; **Left:** CLAS12 kinematics for DVCS and DVMP on unpolarized H_2 and longitudinally polarized NH_3 targets. The colors and density are proportional to the relative count rates. **Right:** Hall A kinematics for DVCS and π^0 electroproduction. Beam time is adjusted for roughly equal counts in all bins.

6.2. Beyond 12 GeV

The COMPASS experiment at CERN proposes to measure DVCS in high energy muon scattering at low x_{Bj} *via* triple coincidence $H(\vec{\mu}^{\pm}, \mu^{\pm}, \gamma p)$ detection [105]. The muon beams have the particularity that the muon spin and charge are correlated, enabling measurements of the DVCS†·BH interference *via* correlated beam charge spin asymmetries. In addition to the COMPASS spectrometer, exclusivity will be determined by detecting the recoil protons in a scintillation array surrounding the target. The expected (correlated) range for DVCS and exclusive vector meson production is $x_{Bj} \in (0.03, 0.25)$ and $Q^2 \in (1.5, 7.5)\,\text{GeV}^2$. A future electron-ion collider, with luminosity several orders of magnitude higher than HERA, would greatly expand the reach of GPD studies. Maximizing the luminosity is essential to measure fully differential cross sections in all kinematic variables. A collider can deliver both longitudinally and transversely polarized beams without the accompanying background of unpolarized nuclei of polarized targets. A collider would also offer enhanced opportunities for spectator tagging to measure neutron GPDs, and recoil tagging for nuclear GPDs.

6.3. Conclusions

Deep virtual exclusive scattering offers the tantalizing prospect of forming spatial images of quarks and gluons in the nucleon. The GPD formalism has already given us new insight into nucleon structure, with evidence for quark angular momentum emerging from GPD models and lattice calculations, and global analysis of forward parton distributions and electromagnetic form factors. A very important study of DVCS and DES in the valence region has started with Jefferson Lab at 6 GeV and will expand with the 12 GeV upgrade. Several systematic analysis demonstrate the important constraints on individual GPDs of the proton and neutron obtained from the data [41, 78, 106]. The unprecedented quality of the CEBAF continuous wave beam is essential to achieving full exclusivity at high luminosity. The revolution in polarized beams and targets over the past two decades allows us a full study of the spin degrees of freedom of DES. Over the next decade, Jefferson Lab and COMPASS will obtain new precision DVCS data spanning a factor of 20 in $x_{\rm Bj}$, and at each value of $x_{\rm Bj}$, a factor of two in Q^2, with maximal Q^2 from 4 to 10 GeV^2. The present Jefferson Lab data are fully differential in Q^2, $x_{\rm Bj}$ and t, allowing a systematic study of the approach to scaling in both cross section and asymmetry observables.

Acknowledgments

This work was supported by US DOE and French CNRS/IN2P3 and ANR. The authors thank our colleagues whose spirited conversations have sharpened and deepened our understanding of this subject. This paper is authored by Jefferson Science Associates, LLC under U.S. DOE Contract No. DE-AC05-06OR23177.

References

[1] X.-D. Ji, Phys. Rev. Lett. **78**, 610 (1997).
[2] A. V. Radyushkin, Phys. Lett. B **380**, 417 (1996).
[3] A. V. Radyushkin, Phys. Lett. B **385**, 333 (1996).
[4] X.-D. Ji, Phys. Rev. D **55**, 7114 (1997).
[5] J. C. Collins, L. Frankfurt and M. Strikman, Phys. Rev. D **56**, 2982 (1997).
[6] A. V. Radyushkin, Phys. Rev. D **56**, 5524 (1997).
[7] X.-D. Ji and J. Osborne, Phys. Rev. D **58**, 094018 (1998).
[8] J. C. Collins and A. Freund, Phys. Rev. D **59**, 074009 (1999).
[9] D. Müeller, D. Robaschik, B. Geyer, F. M. Dittes and J. Horejsi, Fortschr. Phys. **42**, 101 (1994).
[10] M. Diehl, Phys. Rep. **388**, 41 (2003).
[11] M. V. Polyakov, Phys. Lett. B **555**, 57 (2003).
[12] A. V. Belitsky and A. V. Radyushkin, Phys. Rep. **418**, 1 (2005).
[13] M. Burkardt, Phys. Rev. D **62**, 071503 (2000).
[14] J. P. Ralston and B. Pire, Phys. Rev. D **66**, 111501 (2002).
[15] G. A. Miller, Phys. Rev. Lett. **99**, 112001 (2007).
[16] M. Burkardt, Int. J. Mod. Phys. A **18**, 173 (2003).
[17] M. Diehl, Eur. Phys. J. C **25**, 223 (2002).
[18] M. Burkardt, hep-ph:0711.1881.
[19] P. A. M. Guichon and M. Vanderhaeghen, Prog. Part. Nucl. Phys. **41**, 125 (1998).
[20] A. V. Belitsky, D. Müller and A. Kirchner, Nucl. Phys. **B629**, 323 (2002).
[21] K. Goeke, M. V. Polyakov and M. Vanderhaeghen, Prog. Part. Nucl. Phys. **47**, 401 (2001).
[22] M. Diehl, T. Gousset, B. Pire and J. P. Ralston, Phys. Lett. B **411**, 193 (1997).
[23] A. V. Belitsky and D. Müller, Phys. Rev. D **79**, 014017 (2009).
[24] P. Guichon and M. Vanderhaeghen, in preparation.
[25] A. V. Radyushkin, Phys. Rev. D **59**, 014030 (1999).
[26] M. Vanderhaeghen, P. A. M. Guichon and M. Guidal, Phys. Rev. D **60**, 094017 (1999).
[27] M. V. Polyakov and C. Weiss, Phys. Rev. D **60**, 114017 (1999).
[28] V. Y. Petrov *et al.*, Phys. Rev. D **57**, 4325 (1998).
[29] M. Guidal, M. V. Polyakov, A. V. Radyushkin and M. Vanderhaeghen, Phys. Rev. D **72**, 054013 (2005).
[30] M. Penttinen, M. V. Polyakov and K. Goeke, Phys. Rev. D **62**, 014024 (2000).
[31] M. Diehl, T. Feldmann, R. Jakob and P. Kroll, Eur. Phys. J. C **8**, 409 (1999).

[32] S. J. Brodsky, M. Diehl and D. S. Hwang, Nucl. Phys. **B596**, 99 (2001).
[33] S. Boffi, B. Pasquini and M. Traini, Nucl. Phys. **B649**, 243 (2003).
[34] C.-R. Ji, Y. Mishchenko and A. Radyushkin, Phys. Rev. D **73**, 114013 (2006).
[35] L. Frankfurt, A. Freund, V. Guzey and M. Strikman, Phys. Lett. B **418**, 345 (1998).
[36] A. Freund, M. McDermott and M. Strikman, Phys. Rev. D **67**, 036001 (2003).
[37] M. V. Polyakov and A. G. Shuvaev, hep-ph/0207153.
[38] V. Guzey and M. V. Polyakov, Eur. Phys. J. C **46**, 151 (2006).
[39] V. Guzey and T. Teckentrup, Phys. Rev. D **79**, 017501 (2009).
[40] M. V. Polyakov and K. M. Semenov-Tian-Shansky, Eur. Phys. J. A **40**, 181 (2009).
[41] K. Kumericki and D. Müller, Nucl. Phys. **B841**, 1 (2010).
[42] C. Adloff *et al.*, Phys. Lett. B **517**, 47 (2001).
[43] A. Aktas *et al.*, Eur. Phys. J. C **44**, 1 (2005).
[44] F. D. Aaron *et al.*, Phys. Lett. B **659**, 796 (2008).
[45] S. Chekanov *et al.*, Phys. Lett. B **573**, 46 (2003).
[46] S. Chekanov *et al.*, arXiv:0812.2517.
[47] A. Airapetian *et al.*, Phys. Rev. Lett. **87**, 182001 (2001).
[48] A. Airapetian *et al.*, Phys. Rev. D **75**, 011103 (2007).
[49] A. Airapetian *et al.*, JHEP **06**, 066 (2008).
[50] A. Airapetian *et al.*, JHEP **06**, 019 (2010).
[51] A. Airapetian *et al.*, JHEP **11**, 083 (2009).
[52] B. Seitz, Nucl. Instrum. Meth. A **535**, 538 (2004).
[53] I. Lehmann, *Proceedings of 11th International Conference on Meson-Nucleon Physics and the Structure of the Nucleon (MENU 2007)*, Jülich, Germany, Sep. 10-14, 2007.
[54] S. Stepanyan *et al.*, Phys. Rev. Lett. **87**, 182002 (2001).
[55] N. Kivel, M. V. Polyakov and M. Vanderhaeghen, Phys. Rev. D **63**, 114014 (2001).
[56] A. V. Belitsky, A. Kirchner, D. Müller and A. Schafer, Phys. Lett. B **510**, 117 (2001).
[57] B. A. Mecking *et al.*, Nucl. Instrum. Meth. A **503**, 513 (2003).
[58] S. Chen *et al.*, Phys. Rev. Lett. **97**, 072002 (2006).
[59] Jefferson Lab Experiment E00-110, "Deeply virtual Compton scattering at 6 GeV," Y. Roblin *et al.*
[60] Jefferson Lab Experiment E03-106, "Deeply virtual Compton scattering on the neutron," E. Voutier *et al.*.
[61] J. Alcorn *et al.*, Nucl. Instrum. Meth. A **522**, 294 (2004).
[62] A. Camsonne, Ph.D. thesis, Université Blaise Pascal, Clermont-Ferrand, France (2005).
[63] F. Feinstein, Nucl. Instrum. Meth. A **504**, 258 (2003).
[64] C. Munoz Camacho *et al.*, Phys. Rev. Lett. **97**, 262002 (2006).
[65] M. Mazouz *et al.*, Phys. Rev. Lett. **99**, 242501 (2007).
[66] S. Ahmad, H. Honkanen, S. Liuti and S. K. Taneja, Phys. Rev. D **75**, 094003 (2007).
[67] M. Gockeler *et al.*, Phys. Rev. Lett. **92**, 042002 (2004).
[68] D. Brommel *et al.*, PoS **LAT2007**, 158 (2007).
[69] P. Hagler *et al.*, Phys. Rev. D **77**, 094502 (2008).
[70] A. W. Thomas, Phys. Rev. Lett. **101**, 102003 (2008).
[71] Jefferson Lab Experiment E07-007, "Complete separation of deeply virtual photon and neutral pion electroproduction observables off unpolarized protons," C. M. Camacho *et al.*
[72] Jefferson Lab Experiment E08-025, "Measurement of the deeply virtual Compton scattering cross section off the neutron," M. Mazouz *et al.*
[73] F. X. Girod *et al.*, Phys. Rev. Lett. **100**, 162002 (2008).
[74] M. Guidal, Eur. Phys. J. A **37**, 319 (2008).
[75] J. M. Laget, Phys. Rev. C **76**, 052201 (2007).
[76] Jefferson Lab Experiment E06-003, "Deeply virtual Compton scattering with CLAS at 6 GeV," V. Burkert, L. Elouadrhiri, M. Garcon, R. Niyazov, S. Stepanyan *et al.*
[77] Jefferson Lab Experiment E05-114, "Deeply virtual Compton scattering at 6 GeV with polarized target and polarized beam using the CLAS detector," A. Biselli, L. Elouadrhiri, K. Joo, S. Niccolai *et al.*
[78] M. Guidal, Phys. Lett. B **689**, 156 (2010).
[79] Jefferson Lab Experiment E06-101, "N^* resonances in pseudoscalar meson photoproduction from polarized neutrons in $\vec{H} \cdot \vec{D}$ and a complete determination of the $\gamma n \to K^0\Lambda$ amplitude," F. Klein, A. Sandorfi *et al.*
[80] E. R. Berger, F. Cano, M. Diehl and B. Pire, Phys. Rev. Lett. **87**, 142302 (2001).
[81] F. Cano and B. Pire, Eur. Phys. J. A **19**, 423 (2004).
[82] Jefferson Lab Experiment E08-024, "Deeply virtual Compton scattering off ^{4}He," H. Egiyan, F.-X. Girod, K. Hafidi, S. Liuti, E. Voutier *et al.*

[83] E. Voutier, in *Proceedings of the International Workshop on Nuclear Theory*, Rila Mountains, Bulgaria, Jun. 23-28, 2008.
[84] H. C. Fenker *et al.*, Nucl. Instrum. Meth. A **592**, 273 (2008).
[85] R. De Masi *et al.*, Phys. Rev. C **77**, 042201 (2008).
[86] E. Fuchey *et al.*, arXiv:1003.2938.
[87] R. Gilman, R. Holt and P. Stoler, "Transition to Perturbative QCD," this volume.
[88] C. Hadjidakis *et al.*, Phys. Lett. B **605**, 256 (2005).
[89] S. A. Morrow *et al.*, Eur. Phys. J. A **39**, 5 (2009).
[90] S. V. Goloskokov and P. Kroll, Eur. Phys. J. C **50**, 829 (2007).
[91] S. V. Goloskokov and P. Kroll, Eur. Phys. J. C **42**, 281 (2005).
[92] M. Guidal and S. Morrow, arXiv:0711.3743.
[93] J. M. Laget, Phys. Lett. B **489**, 313 (2000).
[94] D. G. Cassel *et al.*, Phys. Rev. D **24**, 2787 (1981).
[95] A. Airapetian *et al.*, Eur. Phys. J. C **17**, 389 (2000).
[96] M. R. Adams *et al.*, Z. Phys. C **74**, 237 (1997).
[97] S. Chekanov *et al.*, PMC Phys. A **1**, 6 (2007).
[98] L. Morand *et al.*, Eur. Phys. J. A **24**, 445 (2005).
[99] P. Joos *et al.*, Nucl. Phys. **B122**, 365 (1977).
[100] J. P. Santoro *et al.*, Phys. Rev. C **78**, 025210 (2008).
[101] A. B. Borissov, Nucl. Phys. Proc. Suppl. **99A**, 156 (2001).
[102] S. Chekanov *et al.*, Nucl. Phys. **B718**, 3 (2005).
[103] C. Adloff *et al.*, Phys. Lett. B **483**, 360 (2000).
[104] J. M. Laget, Phys. Rev. D **70**, 054023 (2004).
[105] N. d'Hose, E. Burtin, P. A. M. Guichon and J. Marroncle, Eur. Phys. J. A **19**, s01, 47 (2004).
[106] H. Moutarde, Phys. Rev. D **79**, 094021 (2009).

Journal of Physics: Conference Series **299** (2011) 012007 doi:10.1088/1742-6596/299/1/012007

Lattice QCD and the Jefferson Lab Program

Jozef J. Dudek[1,2], **Robert G. Edwards**[1], **Kostas Orginos**[1,3], **and David G. Richards**[1]

[1]Jefferson Lab, Newport News, Virginia 23606, USA
[2]Old Dominion University, Norfolk, Virginia 23529, USA
[3]College of William and Mary, Williamsburg, Virginia 23187, USA

Abstract. Lattice gauge theory provides our only means of performing *ab initio* calculations in the nonperturbative regime. It has thus become an increasingly important component of the Jefferson Lab physics program. In this paper, we describe the contributions of lattice QCD to our understanding of hadronic and nuclear physics, focusing on the structure of hadrons, the calculation of the spectrum and properties of resonances, and finally on deriving an understanding of the QCD origin of nuclear forces.

1. Introduction

Lattice QCD currently provides our only means of *solving* QCD in the low-energy regime, and thus has a natural place in the program of Jefferson Lab. Nathan Isgur recognised the potential of lattice QCD to greatly enhance our understanding of hadronic and nuclear physics when he established, in collaboration with John Negele of MIT, a lattice initiative at the laboratory. That initiative spans the three major components of a successful lattice effort. Firstly, the development of the computational and software infrastructure conducted by both the Theory Center and the High-Performance Computing Group under the auspices of the Department of Energy's SciDAC (*Scientific Discovery through Advanced Computing*) Initiative. Secondly, the development and operation of hardware, based on commodity clusters and latterly heterogeneous architectures, optimized for lattice QCD, and now serving as a national facility. Finally, a rigorous research effort dedicated to addressing the key scientific questions posed by the Jefferson Lab program.

Lattice gauge calculations solve QCD on a four-dimensional lattice, or grid, of points in Euclidean space. The quarks reside on the points of the grid, whilst the gluons are associated with the links joining those points. Lattice calculations proceed through a Monte Carlo method, in which ensembles of gauge configurations are generated with a probability distribution prescribed by the Euclidean QCD action. At the start of the lattice program at Jefferson Lab, most lattice calculations were performed in the so-called quenched approximation to QCD, in which quarks would propagate through a gluonic sea, but themselves did not affect the distribution of the gluons. Such an approximation reduced the computational cost by several orders of magnitude and had some theoretical justification, notably from large-N_c QCD. The claim to be *solving*, rather than modelling, QCD has only been truly justified with the elimination of the quenched approximation.

The remainder of this paper is laid out as follows. We conclude the introduction with an outline of some of the algorithmic and computational advances that have taken place at the laboratory. We then describe the lattice studies of hadron structure, including those of generalized parton distributions (GPDs) and of the pion form factor. The next section focus on the resonance spectrum of QCD, and the relation of the program both to the current search for baryon resonances, and the future quest to photoproduce exotics at GlueX. We then describe how lattice QCD can be used to delve into the origin of the nuclear force. Finally, we conclude with the prospects for lattice QCD in the 12 GeV era.

2. Computation challenges

Lattice QCD calculations have historically been at the leading edge of exploiting the most powerful supercomputing resources available. In part, this is due to the natural enthusiasm of the proponents of the field, but is also conditioned by the highly regular nature of the problem. The lattice can be decomposed across a grid of processing nodes, with each node responsible for a given sub-lattice and executing exactly the same code, and with communication of the surfaces of the sub-lattices to the neighboring nodes; the problem is inherently data-parallel. The cost of most lattice calculations is dominated by the solution of a large, sparse system of equations, corresponding to the discretized Dirac equation, and therefore an important task for lattice practitioners is the development of efficient solvers, both for the generation of the gauge configurations, and for the subsequent measurement of physics observables. An important advance for the latter has been the development of eigenvalue deflation methods, enabling physics measurements to be made with greatly increased efficiency [1].

Our ability to exploit the latest and most powerful architectures benefits strongly by the availability of portable software that can be optimized for the target platform. The development of such software has been the aim of the *National Computational Infrastructure for Lattice Gauge Theory* funded under the SciDAC initiative. A particularly notable achievement has been the development, principally at Jefferson Lab, of the *chroma* software suite [2], which has been adopted by many lattice gauge groups both within the US and worldwide.

The development of high-performance, cost-optimized facilities for lattice QCD calculations has been an important activity of the High-Performance Computing Group. The first Intel-based cluster was installed in 2002, and Jefferson Lab now provides a national facility for lattice QCD, funded by the Department of Energy (DOE); the cluster installed in 2007, comprising nearly 400 nodes of quad-core AMD Opteron CPUs connected *via* Infiniband, is shown in Figure 1.

Recently, General Purpose Graphical Processing Units (GPGPUs) have become available that provide significant floating point performance at a relatively low cost. As part of the 2009 and 2010 ARRA project in lattice QCD, Jefferson Lab deployed a cluster with NVIDIA GPUs. A single GPU achieved 178 Gflops in single precision for one of the key numerical kernels, the inverter, used in lattice QCD. By comparison, a cluster node with dual quad-core CPUs could achieve only 18 Gflops. Thus GPUs admitted a ten-fold increase in computational performance for the inverter compared with that obtained on the host node at a relatively small increment in cost.

In order to perform calculations on the spatial volumes needed to confront experiment, it is necessary to spread the calculation across more than a single CPU. Researchers from Boston University, Harvard University and Jefferson Lab developed a parallel lattice QCD solver running on multiple GPUs across multiple cluster nodes. With this development, GPUs are now central to the lattice effort at the laboratory, enabling lattice calculations to address a range of questions previously inaccessible.

3. Hadron structure

The ability for lattice QCD to make increasingly precise, *ab initio* calculations of hadron structure has been driven by advances in three areas: algorithmic advances, enabling both chiral symmetry to be more faithfully represented on the lattice and gauge configurations to

Figure 1. The left-hand panel shows the cluster installed at Jefferson Lab in 2007, consisting of 396 quad-core AMD Opteron-based nodes connected *via* a fast Infiniband network. The right-hand panel shows an NVIDIA GTX-285 General Purpose Graphical Processing Unit used in the 2009-2010 ARRA cluster at Jefferson Lab.

be generated more efficiently, computational advances represented by the availability of special-purpose and leadership-class computers and by commodity clusters, and theoretical advances enabling extrapolations to be performed from the quark masses employed in the computation to the physical up- and down-quark masses. Members of the Theory Center, in collaboration with the Lattice Hadron Physics Collaboration (LHPC), have been at the forefront of applying these advances to aid our understanding of hadron structure and to the benefit of the Jefferson Lab program. The layout of the rest of this section is as follows. We begin by describing how hadron-structure calculations proceed in lattice QCD. We will then review benchmark calculations, such as that of the nucleon's electromagnetic form factors. Subsection 3.3 will describe calculations of moments of generalized parton distributions and the impact such calculations are having on our understanding of hadron structure. We will conclude with prospects for future calculations.

3.1. Anatomy of a nucleon-structure calculation

Hadron structure is expressed through quantities such as the electromagnetic form factors, describing the distribution of charge and currents within a hadron, and the polarized and unpolarized parton structure functions, describing how the longitudinal momentum fraction and spin is apportioned amongst the constituents. Generalized parton distributions [3, 4, 5], encompass both of these concepts, and allow a three-dimensional picture of the nucleon to be constructed.

GPDs are expressed as the matrix elements of light-cone correlation functions $\mathcal{O}_\Gamma(x)$:

$$\mathcal{O}_\Gamma(x) = \int \frac{d\lambda}{4\pi} e^{i\lambda x} \bar{q}\left(\frac{-\lambda n}{2}\right) \Gamma \mathcal{P} e^{-ig\int_{-\lambda/2}^{\lambda/2} d\alpha n\cdot A(\alpha n)} q\left(\frac{\lambda}{2}\right) \tag{1}$$

where n is a light-cone vector, the parallel-transporter is necessary to ensure gauge invariance, and the flavor indices on the quarks are suppressed. The familiar quark polarized and unpolarized parton distributions are then forward matrix elements of this operator, with $\Gamma = \gamma_\mu$ or $\gamma_\mu\gamma_5$ for the unpolarized and polarized distribution functions, respectively; the GPDs correspond to matrix elements in the off-forward direction, with different momenta for the incoming and outgoing partons.

The use of a Euclidean lattice precludes the measurement of these matrix elements directly. Instead we appeal to the operator product expansion to expand the operator $\mathcal{O}(x)$ about the light cone, yielding a set of local operators which can be measured on a Euclidean lattice, and furthermore analytically continued to Minkowski space. In particular, the moments, with respect to Bjorken-x, of the quark distributions are obtained in terms of the forward matrix elements

$$\mathcal{O}_\Gamma^{\{\mu_1 \dots \mu_n\}} = \bar{q}(\gamma_5)\gamma^{\{\mu_1} iD^{\mu_2} \dots D^{\mu_n\}} q. \tag{2}$$

Specifically, for the unpolarized distribution for a nucleon carrying momentum $\vec{p}$, we have

$$\langle \vec{p} | \mathcal{O}_q^{\mu_1 \dots \mu_n} \mid \vec{p} \rangle \longrightarrow \int_0^1 dx\, x^{n-1} q(x), \tag{3}$$

where the Lorentz structure on the right-hand-side is suppressed.

The spectrum is obtained in lattice QCD through the measurement of so-called two-point functions representing the Euclidean correlators of two operators that interpolate between a state and the vacuum. Nucleon structure is investigated through the measurement of *three-point* functions, illustrated in Figure 2, representing the Euclidean correlation functions of three operators:

$$C_{3\rm pt}(t_{\rm snk}, \tau; \vec{p}, \vec{q}) = \sum_{\vec{x},\vec{y}} e^{-i\vec{p}\cdot\vec{x} - i\vec{q}\cdot\vec{y}} \langle 0 | J(\vec{x}, t_{\rm snk}) \mathcal{O}(\vec{y}, \tau) J^\dagger(0) | 0 \rangle, \tag{4}$$

where J is an operator that interpolates between the state and the vacuum. Note that the disconnected contribution in the right-hand panel of Figure 2 is far more computationally demanding that the connected piece; for most of the following, we will emphasise the calculation of flavor non-singlet, or isovector, quantities, for which the disconnected piece vanishes. We insert complete sets of states between the operators in Eq. (4); the time-slice sum projects onto states of definite momentum, and we obtain:

$$\begin{aligned} C_{3\rm pt}(t_{\rm snk}, \tau; \vec{p}, \vec{q}) = & \sum_{n=1,\dots} e^{-E_n(\vec{p})(t_{\rm snk}-\tau)} e^{-E_n(\vec{p}+\vec{q})\tau} \\ & \times \langle 0 | J(0) | n, \vec{p} \rangle \langle n, \vec{p} | \mathcal{O}(0) | n, \vec{p}+\vec{q} \rangle \langle n, \vec{p}+\vec{q} | J^\dagger(0) | 0 \rangle. \end{aligned} \tag{5}$$

For sufficiently large τ and $t_{\rm snk} - \tau$, $C_{3\rm pt}$ is dominated by the lowest-lying state, and we can obtain $\langle n = 1, \vec{p} | \mathcal{O} | n = 1, \vec{p} \rangle$ after elimination of the vacuum-to-state matrix elements from the corresponding two-point functions.

Computations of hadron structure are being performed by several groups, using a variety of fermion and gauge-field discretizations.

The common theme is calculations at decreasingly small values of the pion mass, enabling lattice calculations, together with chiral effective theory, to provide insight into physics at the physical quark masses; recent reviews are contained in Ref. [6, 7]. In the remainder of this section, we will in large part quote results from the LHP Collaboration. We employ a hybrid approach, using lattices generated by the MILC Collaboration [8] with 2 + 1 flavors of Asqtad quarks, corresponding to degenerate u/d and strange, but with domain-wall fermions (DWF) for the valence quarks, computed on HYP-smeared lattices [9]. The generation of lattices using

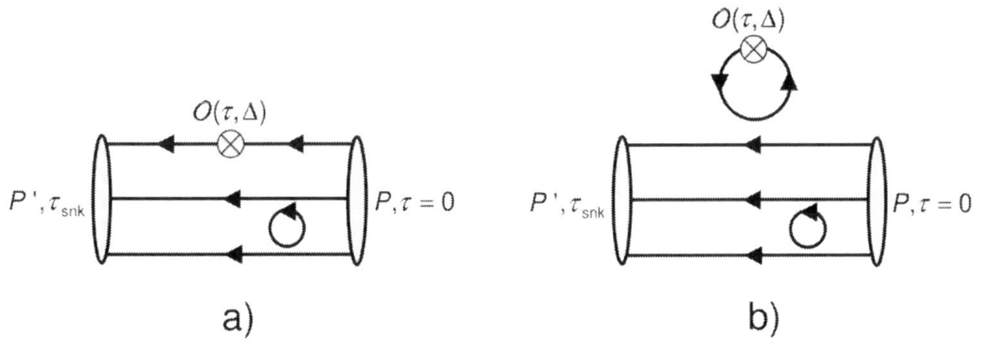

Figure 2. **a)** Connected and **b)** disconnected contributions to the three-point function of Eq. (4).

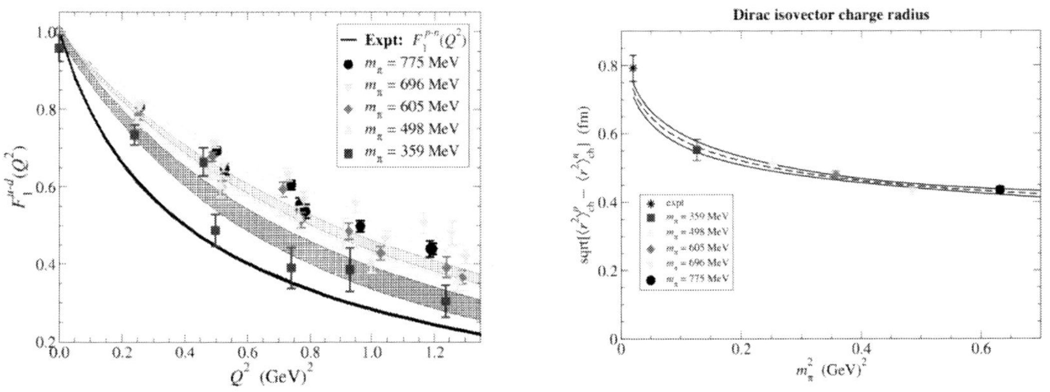

Figure 3. The left-hand panel shows measurements of the isovector F_1 form factor at five values of the pion mass; the bands are uncertainties from a pole-dominance fit to the data. The solid line is a parametrization of the experimental data [12]. The right-hand panel shows the isovector charge radius using a mixed Asqtad/DWF calculation [11], together with the chiral behaviors using a finite-range regulator [13].

Asqtad fermions is very computationally efficient, whilst the DWF valence quarks have the desired chiral-symmetry properties, simplifying operator mixing and hence the matching of our results to the continuum. Whilst this approach violates unitarity at finite lattice spacing, we expect this approach to yield the correct theory in the continuum limit [10].

3.2. Electromagnetic form factors

These describe the distribution of charge and current within the nucleon, and remain the subject of intense experimental and theoretical interest. They are the most straightforward quantity to measure in a lattice calculation, corresponding to the matrix element of the electromagnetic current $V_\mu = \bar{q}\gamma_\mu q$. Figure 3 shows the form factor $F_1(Q^2)$ determined by the LHP Collaboration [11], together with a parametrization of the isovector experimental data [12]. The data show the approach to the experimental parametrization at decreasing values of the pion mass used in the lattice calculation.

The slope of the form factor at $Q^2 = 0$ is related to the isovector charge radius $\langle r^2\rangle^{1/2}$. The right-hand panel in Figure 3 shows the charge radius obtained from the calculation above. A naive linear fit in m_π^2 would clearly be far below the experimental points, illustrating the need to correctly describe the non-analytic behaviour in the approach to the physical light-quark masses. The curve shows the chiral extrapolation of the charge radius, using the finite-range regulator [13].

An important corollary to this work has been the investigation of the electromagnetic properties of baryons containing the strange valence quark [14]. The computation of the contribution of the strange quarks, and more generally sea quarks, to hadron structure is very challenging. However, the combination of lattice QCD, the constraints of charge symmetry and chiral-extrapolation techniques has enabled the strange-quark contributions to the electric and magnetic form factors to be determined with high precision [15, 16], providing an important theoretical complement to the experimental measurements.

3.3. Generalized parton distributions

The electromagnetic form factors described above are particular cases of generalized parton distributions. There are two GPDs corresponding to the choice $\Gamma = \gamma \cdot n$ in Eq. (1), $H(x, \xi, t)$ and $E(x, \xi, t)$, and a further two corresponding to the $\Gamma = \gamma \cdot n\gamma_5$, $\tilde{H}(x, \xi, t)$ and $\tilde{E}(x, \xi, t)$. The invariants are:

$$
\begin{aligned}
t &= -\Delta^2 = -(P - P')^2, \\
\xi &= -n \cdot \Delta/2, \qquad (6)
\end{aligned}
$$

where n_μ is a light-cone vector, and P, P' are the four-momenta of the incoming and outgoing states. Once again, we appeal to the operator product expansion to obtain the matrix elements of local operators, and both the form factors described above and the familiar parton distributions emerge as special cases:

$$H(x, 0, 0) = q(x), \qquad (7)$$

$$\tilde{H}(x, 0, 0) = \Delta q(x), \qquad (8)$$

and

$$\int_{-1}^{1} dx\, H(x, \xi, t) = F_1(t). \qquad (9)$$

Note here that the x runs from -1 to 1, corresponding to antiquark and quark momentum fraction.

Before proceeding to discuss the new insights enabled by the study of GPDs, we begin by discussing the polarized and unpolarized distributions as a means of benchmarking our calculation. A fundamental measure of QCD is the nucleon's axial-vector charge g_A, corresponding to the choice $\Gamma = \gamma_\mu \gamma_5$; it has additional importance in its rôle as a fundamental low-energy constant of the theory that appears in the chiral expansion of other quantities. Figure 4 shows a calculation of the axial-vector charge by the LHP Collaboration, together with the chiral extrapolation to the physical quark masses [17]. The consistency between the extrapolated lattice data and the experimental value is very encouraging, though the chiral behaviour for this quantity is very mild.

The non-analytic behaviour with decreasing quark mass is more strongly exhibited in the unpolarized distributions, and in particular in the calculation of the momentum fraction carried by the valence quarks in the nucleon. An analysis of the forward matrix elements for both the unpolarized and polarized distributions, as well as for the transversity, has been performed using a renormalized chiral expansion [21]; data for the non-singlet momentum fraction, and a summary of the results for a range of benchmark quantities, is shown in Figure 5.

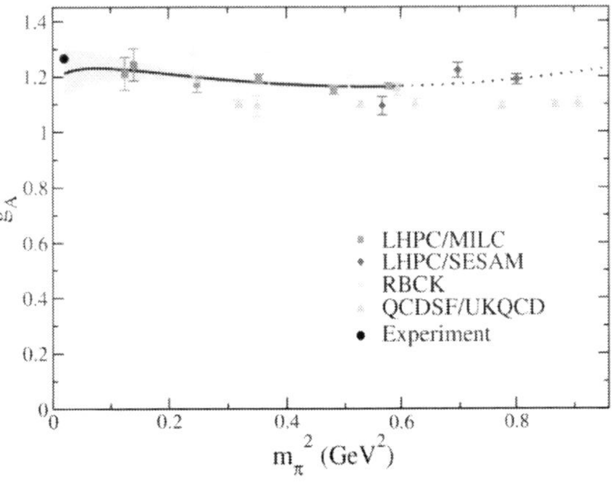

Figure 4. The squares show LHPC/MILC data for the axial-vector charge g_A; the two data points at the lowest pion mass correspond to to different volumes. The remaining lattice points are taken from QCDSF/UKQCD [18], RBCK [19] and SESAM [20].

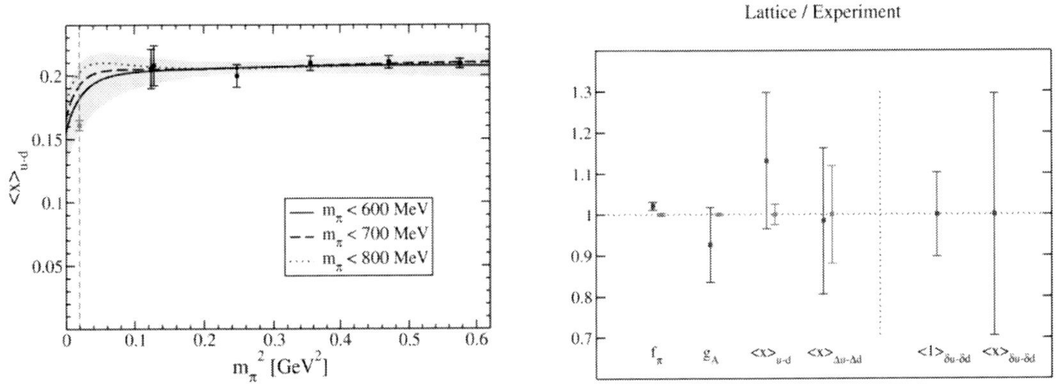

Figure 5. The left-hand figure shows the momentum fraction carried by the valence quarks in a nucleon, together with the chiral extrapolation to the physical quark masses. The right-hand figure is a summary which shows the lattice values at the physical pion masses for a range of benchmark quantities, normalized to the experimental value where known (left-hand points) [21]; the errors on the corresponding experimental measurements are shown to the right of the experimental points.

3.4. New insights: origin of nucleon spin

The excitement and significance of GPDs arises from the new insights they can provide into hadron structure, spurring new experimental initiatives such as the 12 GeV upgrade at Jefferson Lab. There is thus a rigorous effort both by the LHP Collaboration in a hybrid DWF/Asqtad approach [22, 21], and by the QCDSF/UKQCD Collaboration using two flavors of dynamical, improved Wilson quarks [23], to measure moments of GPDs.

A salient example of the new insights facilitated through the study of GPDs is how the spin of the nucleon is distributed amongst its constituents, and in particular the rôle of orbital angular momentum of the quarks. The spin carried by the quarks in a nucleon has long been the pursuit of lattice calculations, but a means to calculate the orbital angular momentum of the quarks eluded theorists. However, it was realized that the total angular momentum carried by the

Journal of Physics: Conference Series 299 (2011) 012007 doi:10.1088/1742-6596/299/1/012007

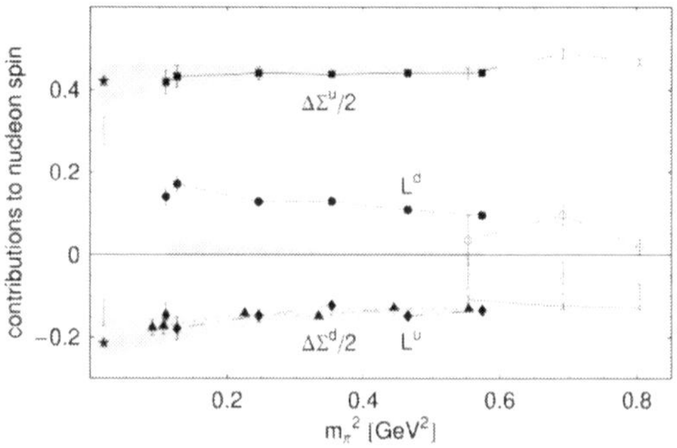

Figure 6. The figure shows the decomposition of the spin between the quark flavors in a hybrid Asqtad/DWF approach [22], together with an experimental determination of the spin from HERMES [26, 27].

quarks within a nucleon could be related to moments of GPDs through [24]

$$J_q = \frac{1}{2}\int dx\, x\,(H(x,\xi,0) + E(x,\xi,0))\,. \tag{10}$$

A pioneering effort to measure the quark orbital angular momentum was performed in the quenched approximation to QCD [25]; Figure 6 shows a more recent calculation by the LHP Collaboration. Note that the calculation includes only the connected pieces. None-the-less, it suggests that the total orbital angular momentum carried by the quarks within a nucleon is small, but that the orbital angular momentum carried by the individual quark flavors may be substantial.

3.5. Pion form factor

The measurement of the form factor of the pion is an important part of the Jefferson Lab program both at 6 GeV and after the future 12 GeV upgrade. The pion is the simplest, and lightest, hadron and provides an important theatre in which to test many of our ideas about hadron structure, notably the range of validity of vector-meson dominance with increasing Q^2, and the approach to asymptotic freedom in form factors at high Q^2. The experimental study of the pion form factor is obscured by the lack of a free pion target, except at very low momenta, requiring an extrapolation to the pion pole. In contrast, the study of pion structure in lattice QCD is subject to no such ambiguities.

The pion form factor in the hybrid calculation using DWF for the valence quarks and the asqtad formulation for the sea quark is shown in Figure 7 [28]. We see that the lattice calculation is able to explore a momentum region beyond that explored experimentally, and the lattice data are well described by vector-meson dominance over the region of the calculation. Also shown is the expectation for the asymptotic behavior of the form factor, it is clear that the lattice calculation is far from the perturbative regime at the current momentum transfers.

4. Spectroscopy

In order to really understand QCD and hence test whether it is the complete theory of the strong interaction, we must know the spectrum of mesons and baryons that it implies and test those spectra against high quality data. The complete combined analysis of available experimental data on the photoproduction of nucleon resonances was the 2009 milestone in Hadronic Physics (HP), and the measurement of the electromagnetic properties of the low-lying baryons is an HP 2012 milestone. The observed spectrum of QCD provides little direct evidence of the presence

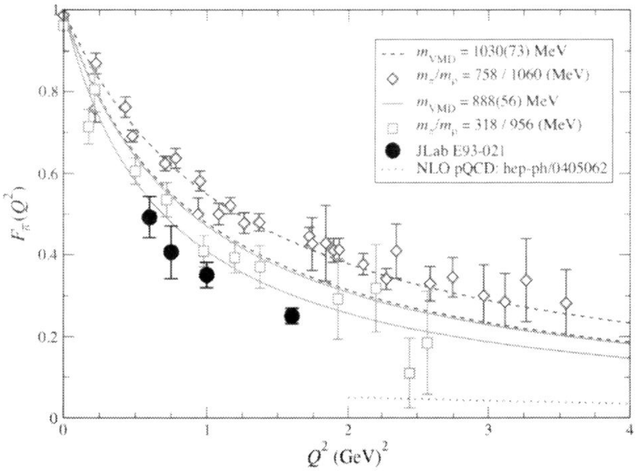

Figure 7. The open symbols show the pion form factor computed in a hybrid DWF/Asqtad calculation at two different values of light quark mass; the shaded bands are VMD fits to the calculation. The closed symbols show the experimental data, whilst the dashed line is the perturbative QCD expectation [29].

of the gluons. However, QCD admits the possibility of exotic mesonic states of matter in which the gluonic degrees of freedom are explicitly exhibited. The search for such states will be an important component of the Jefferson Lab 12 GeV program.

Given the intense experimental efforts in hadron spectroscopy, the need to predict and understand the hadron spectrum from first principles calculations in QCD is clear. Hence, an important goal of the lattice effort since its inception has been the study of the resonance spectrum of QCD, and more recently the calculation of the electromagnetic properties of low-lying resonances.

A comprehensive picture of resonances requires that we go beyond a knowledge of the ground state mass in each channel, and obtain the masses of the lowest few states of a given quantum number. This we can accomplish through the use of the variational method [30, 31]. Rather than measuring a single correlator $C(t)$, we determine a matrix of correlators

$$C_{ij}(t) = \sum_{\vec{x}} \langle O_i(\vec{x}, t) O_j^\dagger(\vec{0}, 0) \rangle,$$

where $\{O_i; i = 1, \ldots, N\}$ is a basis of interpolating operators with given quantum numbers. We then solve the generalized eigenvalue equation

$$C(t)u = \lambda(t, t_0)C(t_0)u$$

to obtain a set of real (ordered) eigenvalues $\lambda_n(t, t_0)$, where $\lambda_0 \geq \lambda_1 \geq \cdots \geq \lambda_{N-1}$. At large Euclidean times, these eigenvalues then delineate between the different masses

$$\lambda_n(t, t_0) \longrightarrow e^{-M_n(t-t_0)} + \mathcal{O}\left(e^{-\Delta M_n(t-t_0)}\right),$$

where $\Delta M_n = \min\{| M_n - M_i |: i \neq n\}$. The eigenvectors u are orthogonal with metric $C(t_0)$, and a knowledge of the eigenvectors can yield information about the partonic structure of the states.

The (hyper-) cubic lattice does not possess the full rotational symmetry of the continuum. Thus in a lattice calculation, states at rest are classified not according to the spin (J, J_z), but rather according to the irreducible representations (irreps) of a cube; for states of higher spin, the different continuum degrees of freedom are distributed across several lattice irreps.

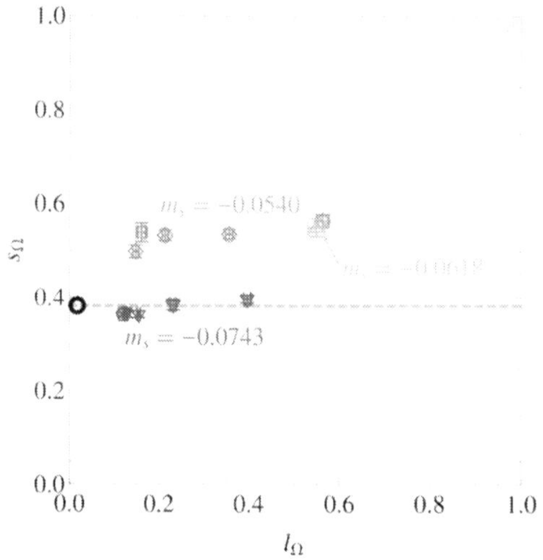

Figure 8. The figure shows the dynamical ensembles generated in [34] plotted in terms of the dimensionless coordinates l_Ω and s_Ω, as described in the text. The solid line corresponds to the theory with three degenerate quark flavors, whilst the dashed horizontal line shows the approach to the physical quark masses; the open circle is the physical value of the coordinates.

Thus a crucial ingredient in our spectroscopy program has been the use of a large basis of interpolating operators, decomposed into their lattice irreps, enabling the many spins to be reliably determined. Our ability to calculate correlation functions efficiently for such a large basis of operators has been greatly advanced through the development of a new method, "distillation" [32], for the construction of the interpolating operators.

The final ingredient in the spectroscopy effort has been the generation of so-called anisotropic lattices, with finer temporal than spatial discretization, enabling the fall-off of the principle correlators introduced above to be discerned at short temporal separations.

4.1. Anisotropic clover lattice generation

Precise calculations that can truly confront the experimental program require that we generate lattice with the correct number of light-quark flavors. The use of anisotropic lattices has proved essential in the reliable determination of the energy levels necessary to extract the resonance spectrum. Thus a crucial activity has been the generation of lattices with two flavors of fully dynamical light and a dynamical strange quark, using the clover fermion action, designed both for studies of the spectrum, and for the calculation of the scattering lengths important for understanding the nucleon-nucleon interaction. An important milestone was the tuning of the parameters of the action, beginning with the three-flavor theory [33].

A major challenge in calculations in lattice QCD is a procedure for specifying the values of the quark masses, and in particular those of the light u/d and s quarks, in a way that enables lattice calculations to be extrapolated to the physical values of these masses. This was accomplished through the introduction of a novel pair of dimensionless coordinates l_Ω and s_ω that are primarily sensitive to the light and strange quark masses, respectively [34]:

$$\begin{aligned} l_\Omega &= \frac{9m_\pi^2}{4m_\Omega^2}, \\ s_\Omega &= \frac{9(2m_K^2 - m_\pi^2)}{4m_\Omega^2}. \end{aligned} \tag{11}$$

The use of these coordinates is illustrated in Figure 8, showing the approach at fixed strange-quark mass to the physical values of the light (u/d) quark masses.

Whilst the anisotropic lattices are designed to enable the investigation of the resonance spectrum of QCD, the low-lying spectrum, corresponding to the lightest states for each

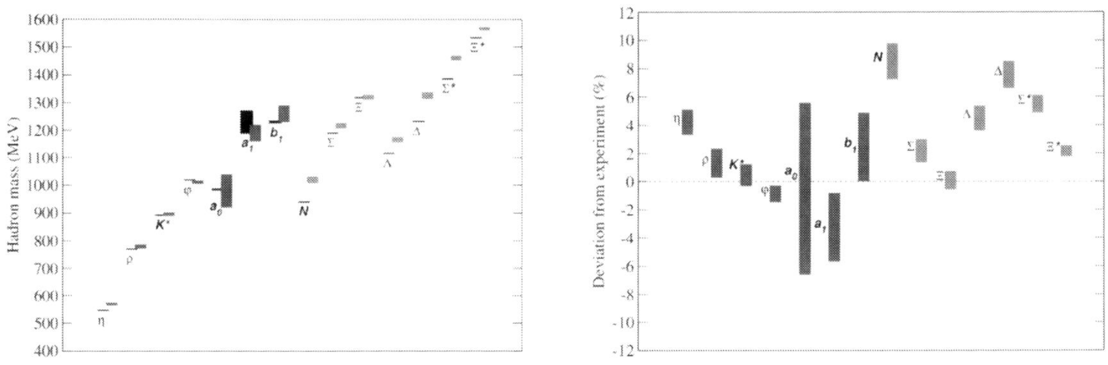

Figure 9. The left-hand panel shows a comparison between the calculated values of the low-lying hadrons and their physical values; the right-hand panel shows the percentage deviation from experiment.

quantum number, provides an important benchmark for future calculations. The left-hand panel of Figure 9 shows a summary of the low-lying light-hadron masses compared with their experimental values; the right-hand panel shows the relative errors on the results compared to experiment. It is this lattice-generation program that is at the core of our efforts to understand QCD spectroscopy, and has been a key element of the national USQCD program under the DOE INCITE awards since 2007.

4.2. Meson resonances

The new Hall D of the Jefferson Lab 12 GeV upgrade centers on the study of meson states produced in photoproduction reactions in the GlueX. Photoproduction has been proposed, within QCD-motivated models, as a favorable method for the production of exotic hybrid mesons, those mesons having J^{PC} outside the set allowed to a fermion-antifermion pair. The hybrid hypothesis is that an excited gluonic field in addition to a quark-antiquark pair can give rise to these quantum numbers.

Over the past few years the Jefferson Lab lattice group has developed a methodology for the extraction of the spectrum and photocouplings including those involving excited states and exotics. The initial calculations were performed using charm mass quarks [35]. The charmonium system is particularly attractive, in that it is both computationally far less demanding than systems composed of the light (u, d, s) quarks, and because there is a wealth of high-precision experimental data. These calculations were performed in the quenched approximation to QCD, using anisotropic lattices; here the use of an anisotropic lattice, with temporal lattice spacing $a_t < a_s$, performs a further rôle in controlling the discretization uncertainties, $a_t m_c \ll 1$.

A formidable barrier to the interpretation of lattice calculations has been the reduced cubic symmetry imposed by the lattice which makes the assignment of the continuum spins to the energy eigenstates problematic. We discovered that, by judiciously constructing operators so as to have a known continuum behavior, the spins of the excited states could be determined, and the barrier imposed by the reduced cubic symmetry of the lattice overcome, and this discovery has been core both to investigation of the meson spectrum in both the light- and charm-quark sectors.

The availability of the large ensembles of $N_f = 2 + 1$ anisotropic clover lattices, together with the efficient operator-construction admitted by "distillation", has enabled us to compute the isovector meson spectrum for states composed of the light and strange quarks [36, 37], at pion masses down to around 400 MeV. Masses of highly excited states were calculated, with the

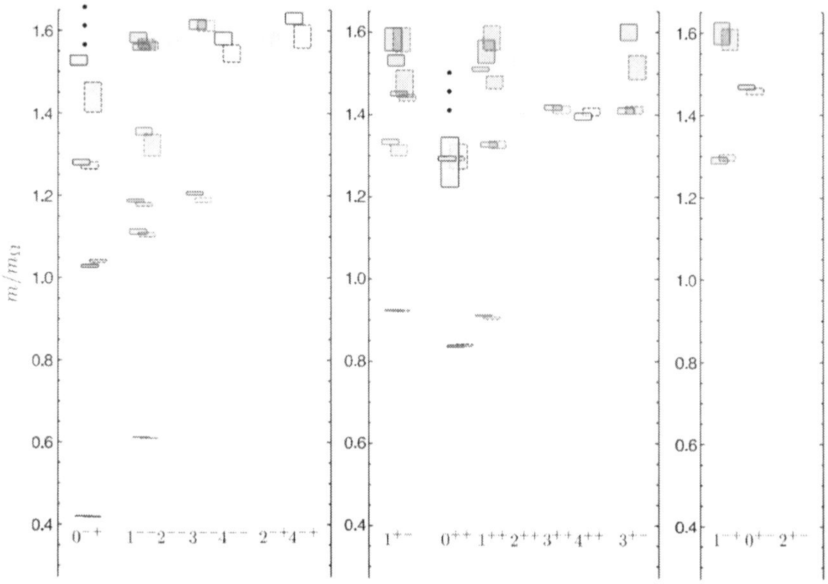

Figure 10. Spin-identified spectrum of isovector (octet) mesons in a calculation with three degenerate flavors of quarks of mass around that of strange quark; 16^3(solid) and 20^3(dashed) spectra agree well. Ellipses indicate that there are heavier states with a given J^{PC} but that they are not well determined in this calculation. Note that the right-most panel corresponds to states of exotic quantum numbers.

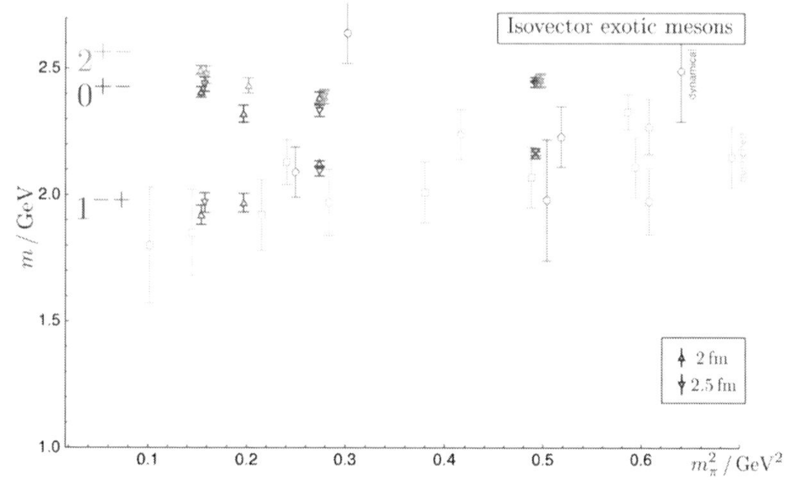

Figure 11. Summary of extracted isovector exotic states. For comparison 1^{-+} results from Refs. [38, 39, 40, 41, 42, 43] are also plotted.

quantum numbers reliably delineated, including states of up to spin 4, illustrated in Figure 10. Figure 11 summarizes the results on exotic states and compares those results with previous lattice QCD results from Refs. [38, 39, 40, 41, 42, 43]. The experimental implications of these results are important: there should be mesons of exotic quantum numbers in a mass region accessible to the GlueX experiment.

Lattice QCD can go beyond the calculation of the spectrum to compute the transition

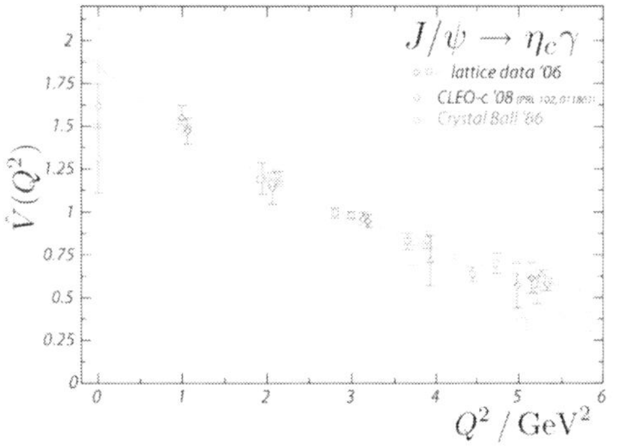

Figure 12. The lattice calculation of the radiative transition form factor; the band is a quark-model fit to the lattice data, yielding the value for the photocoupling at $Q^2 = 0$. Also shown are the experimental extractions of the photocouplings.

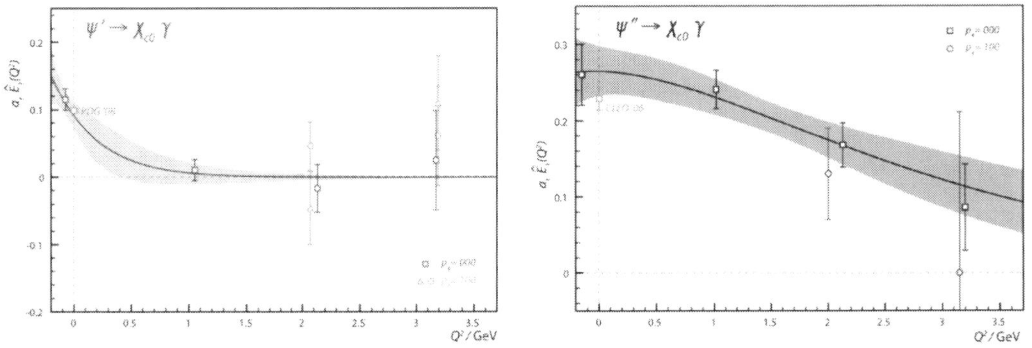

Figure 13. First and second excited vector state electric dipole transition form factors to the ground state scalar in charmonium. Experimental results for real photons from PDG and CLEO-c.

form factors between states, and the photocouplings, that can guide experiment as to expected photoproduction rates. The calculation of the transition form factors follows the same procedure as that used to study hadron structure, and once again the charmonium system provides an important theatre both to develop the tools and to compare with experiment. Figure 12 shows the radiative transition form factor $J/\psi \to \eta_c\gamma$, obtained using domain-wall fermions on an anisotropic lattice [44]. Also shown is value for the photocoupling obtained from the experimental measurement of the partial width; the more recent 2008 analysis by CLEO-c was inspired by this lattice calculation. The optimum excited state operators determined from the study of the resonance spectrum can be used to extract photocouplings involving excited and exotic mesons [45], some examples of which are presented in Figure 13. Most importantly, this paper showed, for the first time, that radiative transitions to states with exotic quantum numbers were calculable, and furthermore extracted a large partial width for the exotic decay $\Gamma(\eta_{c1} \to J/\psi\gamma)$; should these results be continued to the light-quark sector, they would suggest copious photoproducion rates for exotic mesons.

4.3. N resonances in lattice QCD*

Baryons, containing three quarks, are emblematic of the non-Abelian nature of QCD, and of the three colors of the theory. An important goal in exploring baryons is attempting to discern the effective degrees of freedom that describe the spectrum; the search for so-called "missing resonances" focuses on whether the spectrum can be well described by a quark model, or whether an effective theory with fewer degrees of freedom, such as a quark-diquark picture, provides a more faithful description of the baryon spectrum.

The procedure for constructing interpolating operators for baryons was described in two papers, the first of which employed a Clebsch-Gordan approach mirroring that of the continuum [46], and the second an automated method allowing a very general basis of operators to be constructed [47]. There are three double-valued irreducible representations of the cubic group, denoted $G_{1u/g}(2)$, $H_{u/g}(4)$ and $G_{2u/g}(2)$, where g and u refer to positive and negative parity, respectively, and the brackets contain the dimension of the irrep. G_1 contains continuum spins $1/2, 7/2, \ldots$, H_g spins $3/2, 5/2, \ldots$ and G_2 spins $5/2, 7/2, \ldots$. Thus, at any fixed lattice spacing a, a state corresponding to spin 5/2 has four degrees of freedom in H, and two in G_2, with degeneracies between the energies in the two irreps, emerging in the continuum limit.

Our ability to extract the baryon resonances using the variational method was first shown in a calculation in the quenched approximation to QCD, using an anisotropic clover fermion action. This method has been extended to a calculation with 2 mass-degenerate flavors of light quarks, with gauge configurations generated on an anisotropic lattice using the Wilson fermion action. Results were obtained for two pion masses, 416(36) MeV and 578(20) MeV [48]. The lowest four energies were reported in each of the six irreducible representations of the octahedral group, at each pion mass, as illustrated in Figure 14. Most notably, evidence was found for a $5/2^-$ state in the pattern of negative-parity excited states, agreeing with the pattern seen in experiment, and the first time a spin-5/2 state has been realized in a lattice QCD calculation. The "distillation" method has now been used to extend this to a calculation on the anisotropic clover lattices, for each of the nucleon, Δ and Ω channels; as in the case of the mesons, there is little evidence for the multi-hadron states that must be present in the spectrum.

A major goal of both the 6 GeV and future 12 GeV programs is the extraction of electromagnetic transition form factors between nucleon resonances. Once the spectrum of energies has been established, these properties can be computed in lattice QCD. Figure 15 shows the $N - P_{11}$ transition form factors measured by CLAS, together with a lattice calculation by the Jefferson Lab group [49]. Whilst the calculation is in the quenched approximation to QCD, and at too large a pion mass to provide a reliable comparison, it demonstrates the ability of lattice QCD to go beyond the calculation of the spectrum.

5. Hadronic interactions from lattice QCD

A long-standing quest of nuclear physics has been gleaning an *ab initio* understanding of how the nuclear force arises from QCD, and to rigorously compute the properties and interactions of nuclei. With the presently available computational resources, this would seem an intractable problem: the scale of the typical nuclear binding energies of a few MeV is far from the typical scale of QCD of the order of 1 GeV. There had been various attempts to gain an understanding of the nuclear force from simpler systems, beginning with an investigation of the emergence of π and ρ exchange in the $B^{(*)}B^{(*)}$ system [50, 51]. However, the foundation of recent successes at studying the nature of the hadron-hadron interaction was the realization that lattice computations in the foreseeable future could be used to derive rigorous results for nuclear physics, opening up a new avenue for lattice computations that are at the core of this proposal [52].

The scattering lengths between mesons provides an important avenue in which to demonstrate our ability to understanding the origin of nuclear forces, since the signal-to-noise ratio is subject to far less degradation with decreasing pion mass than baryon systems. Meson-meson scattering lengths were calculated with domain-wall fermions on $N_f = 2 + 1$ dynamical MILC sea configurations with pion masses down to $m_\pi \sim 290$ MeV [53, 54], obtaining the results shown in Figure 16. The first prediction of the $K\pi$ scattering lengths in both isospin channels was made possible by combining the lattice QCD calculation in the $I = 3/2$ channel with chiral

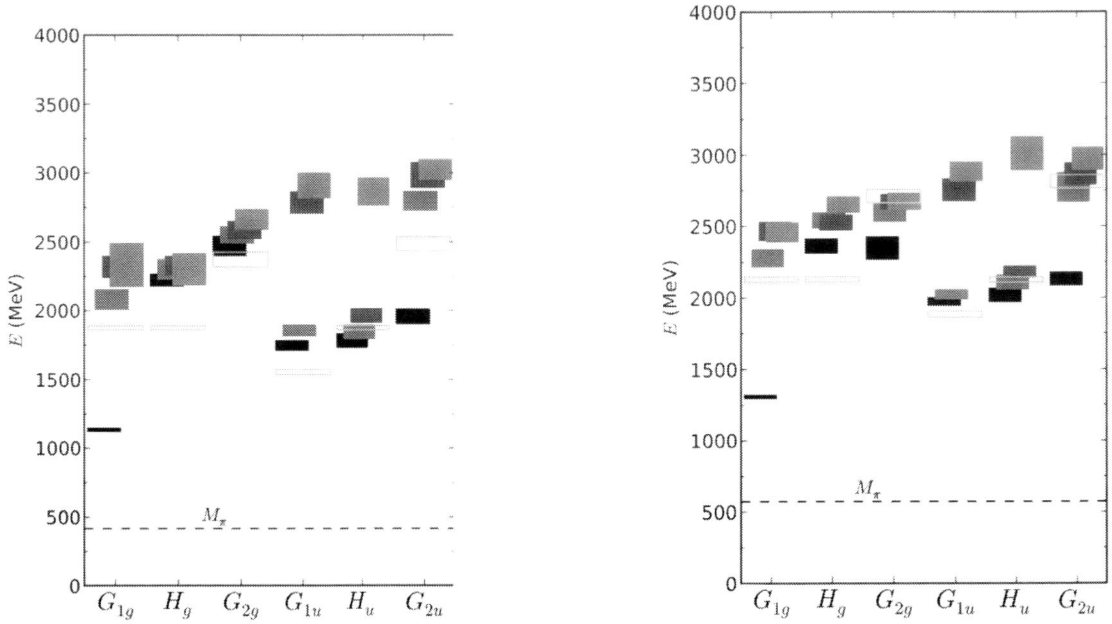

Figure 14. The left- and right-hand panels show the spectrum of $I = 1/2$ baryon resonance, indicated by the solid boxes, obtained on $N_f = 2$ Wilson fermion lattices at $m_\pi = 416$ and 578 MeV respectively [48]; the errors are indicated by the vertical width of the box. The open boxes show the expected thresholds for multi-particle states.

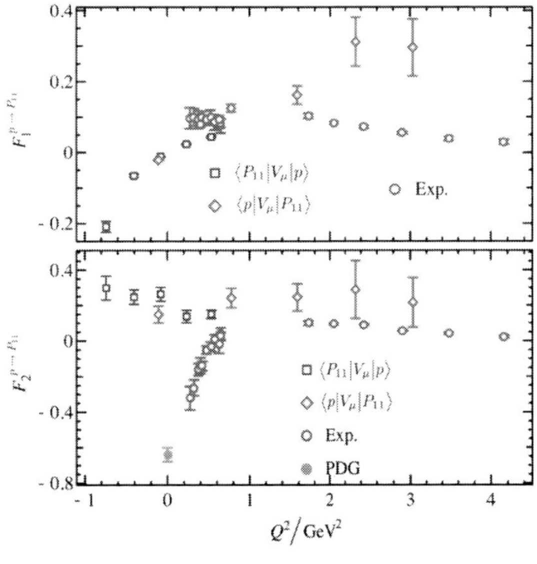

Figure 15. Proton-Roper form factors $F_{1,2}^*$ obtained from CLAS experiments and PDG (circles) and the lattice calculation (squares, diamonds).

perturbation theory, the result of which is also shown in Figure 16. An important goal is to establish the presence of a three-hadron interaction; recently, the formalism employed above was extended to include many pions in a finite volume [56, 57]. An important outcome of this work was indeed the identification of a three-body $\pi^+\pi^+\pi^+$ interaction.

The investigation of the interaction between baryons is considerably more computationally

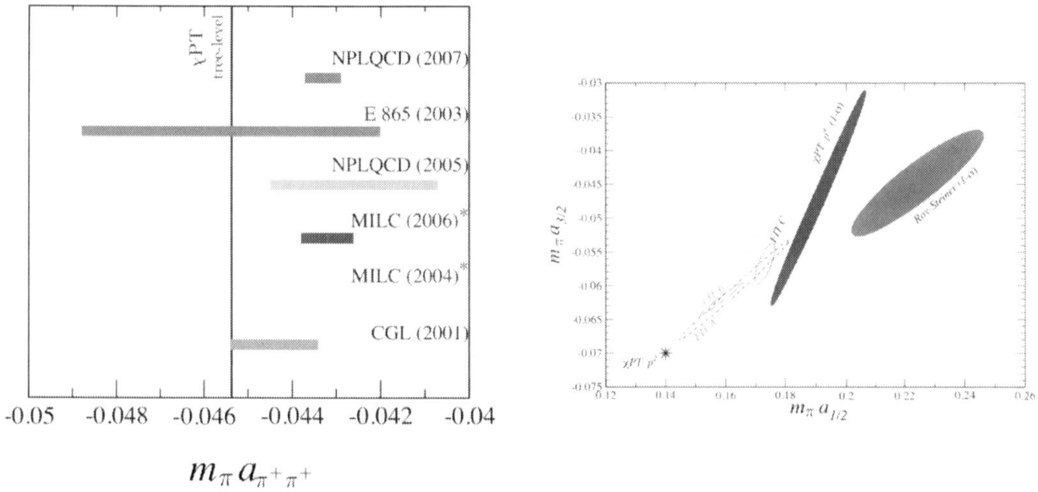

Figure 16. Meson-meson scattering lengths. The left-hand panel shows various determinations of the $I = 2\pi\pi$ scattering lengths; the bar denotes the lattice computation in Ref. [54]. The right panel shows the prediction for the $K\pi$ scattering lengths at the physical pion mass [55].

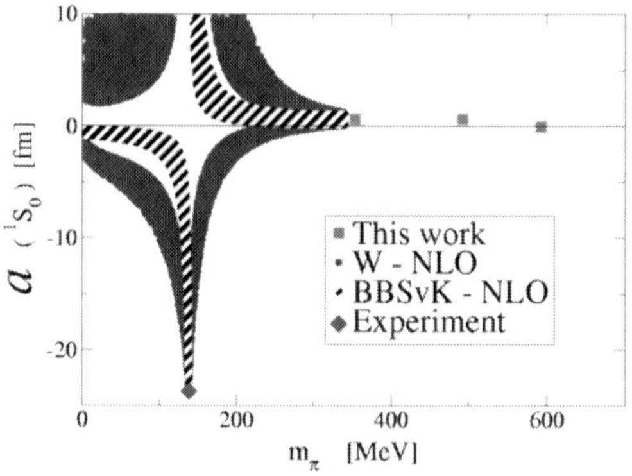

Figure 17. Allowed regions for the scattering length in the $^1S_0$ channel as a function of the pion mass. The experimental value of the scattering length and NDA have been used to constrain the extrapolation in both BBSvK [59, 60, 61] and W [62, 63, 64] power-countings at NLO.

demanding. The signal-to-noise ratio is inherently worse for even a single baryon than is typical for a meson, and this situation is only compounded for multi-hadron states, thus requiring very high statistics. Figure 17 shows allowed regions for the scattering length in the $^1S_0$ channel as a function of the pion mass, in a hybrid DWF/Asqtad calculation analogous to that used for our studies of hadron structure [58]; the pion masses used in this study are insufficient to constrain the scattering length at the physical quark masses, but demonstrate the future promise of the method. The need for very high statistical accuracy encouraged the NPLQCD Collaboration to use the anisotropic clover lattices to measure, for the first time, the binding energy of a three-baryon $\Xi^0\Xi^0 n$ system [65]; most strikingly, a window of timeslices was identified where the required statistics were lower than might naively be expected.

6. Conclusions

The last ten years have seen lattice studies incorporated as essential elements in the physics program of the laboratory, spanning the areas of hadron structure, spectroscopy and the physics of nuclei. The emergence of ever more powerful computational resources, reaching perhaps the exascale regime in the 12 GeV era, can only strengthen the role of lattice QCD in describing the physics of Jefferson Lab, and providing the *ab initio* calculations of a quality that can confront the experimental program.

For our understanding of hadron structure, the amalgam of experimental studies of transverity and generalized parton distributions, together with lattice calculations, will provide far greater insight into the distribution of mass, charge, spin and currents within a hadron than either experiment or calculation can provide alone. An important challenge will be the development of improved methods to incorporate the contributions both of the sea quarks and of the gluons to hadron structure. In the study of spectroscopy, lattice studies of the radiative transitions between light mesons will provide vital input for the GlueX experiment, and calculation of the spectrum and properties of meson and baryon resonances confronting experiment will enable us to truly understand QCD. Here an important challenge will be the developing the tools to investigate unstable resonances, and the extension of the method for understanding elastic decays to the inelastic regime. Finally, the combination of lattice calculation and effective field theory will enable the origin of the nucleon force, and of the light nuclei, to be understood.

Acknowledgements

This work was done using the Chroma software suite [2] on clusters at Jefferson Lab using time awarded under the USQCD Initiative. This research used resources of the National Center for Computational Sciences at Oak Ridge National Laboratory, which is supported by the Office of Science of the Department of Energy under Contract DE-AC05-00OR22725. This research was supported in part by the National Science Foundation through TeraGrid resources provided by Pittsburgh Supercomputing Center (PSC), San Diego Supercomputing Center (SDSC) and the Texas Advanced Computing Center (TACC). We are grateful to the MILC Collaboration for the use of the lattices used in the study of hadron structure. Authored by Jefferson Science Associates, LLC under U.S. DOE Contract No. DE-AC05-06OR23177. The U.S. Government retains a non-exclusive, paid-up, irrevocable, world-wide license to publish or reproduce this manuscript for U.S. Government purposes.

We are grateful for a very productive collaboration with colleagues both at Jefferson Lab and elsewhere; in particular, we are grateful to current and past members of the High-Performance Computing Group who have been so essential to the success of the Jefferson Lab Lattice Program.

References

[1] A. Stathopoulos and K. Orginos, arXiv:0707.0131.
[2] R. G. Edwards and B. Joo, Nucl. Phys. Proc. Suppl. **140**, 832 (2005).
[3] D. Müller, D. Robaschik, B. Geyer, F. M. Dittes and J. Horejsi, Fortschr. Phys. **42**, 101 (1994).
[4] X.-D. Ji, Phys. Rev. D **55**, 7114 (1997).
[5] A. V. Radyushkin, Phys. Rev. D **56**, 5524 (1997).
[6] P. Hägler, PoS **LAT2007**, 13 (2007).
[7] K. Orginos, PoS **LAT2006**, 018 (2006).
[8] C. W. Bernard *et al.*, Phys. Rev. D **64**, 054506 (2001).
[9] A. Hasenfratz and F. Knechtli, Phys. Rev. D **64**, 034504 (2001).
[10] S. R. Sharpe, PoS **LAT2006**, 22 (2006).
[11] R. G. Edwards *et al.*, PoS **LAT2006**, 121 (2006).
[12] J. J. Kelly, Phys. Rev. C **70**, 068202 (2004).
[13] D. B. Leinweber, A. W. Thomas and R. D. Young, Phys. Rev. Lett. **86**, 5011 (2001).
[14] H.-W. Lin and K. Orginos, Phys. Rev. D **79**, 074507 (2009).
[15] D. B. Leinweber *et al.*, Phys. Rev. Lett. **94**, 212001 (2005).
[16] D. B. Leinweber *et al.*, Phys. Rev. Lett. **97**, 022001 (2006).
[17] R. G. Edwards *et al.*, Phys. Rev. Lett. **96**, 052001 (2006).
[18] A. A. Khan *et al.*, Nucl. Phys. Proc. Suppl. **140**, 408 (2005).
[19] S. Ohta and K. Orginos, Nucl. Phys. Proc. Suppl. **140**, 396 (2005).
[20] D. Dolgov *et al.*, Phys. Rev. D **66**, 034506 (2002).
[21] D. B. Renner *et al.*, PoS **LAT2007**, 160 (2007).

[22] P. Hagler *et al.*, Phys. Rev. D **77**, 094502 (2008).
[23] D. Brömmel *et al.*, PoS **LAT2007**, 158 (2007).
[24] X.-D. Ji, Phys. Rev. Lett. **78**, 610 (1997).
[25] N. Mathur, S. J. Dong, K. F. Liu, L. Mankiewicz and N. C. Mukhopadhyay, Phys. Rev. D **62**, 114504 (2000).
[26] A. Airapetian *et al.*, Phys. Rev. D **75**, 012007 (2007).
[27] K. Ackerstaff *et al.*, Phys. Lett. B **464**, 123 (1999).
[28] F. D. R. Bonnet, R. G. Edwards, G. T. Fleming, R. Lewis and D. G. Richards, Phys. Rev. D **72**, 054506 (2005).
[29] A. P. Bakulev, K. Passek-Kumericki, W. Schroers and N. G. Stefanis, Phys. Rev. D **70**, 033014 (2004).
[30] C. Michael, Nucl. Phys. **B259**, 58 (1985).
[31] M. Luscher and U. Wolff, Nucl. Phys. **B339**, 222 (1990).
[32] M. Peardon *et al.*, Phys. Rev. D **80**, 054506 (2009).
[33] R. G. Edwards, B. Joo and H.-W. Lin, Phys. Rev. D **78**, 054501 (2008).
[34] H.-W. Lin *et al.*, Phys. Rev. D **79**, 034502 (2009).
[35] J. J. Dudek, R. G. Edwards, N. Mathur and D. G. Richards, Phys. Rev. D **77**, 034501 (2008).
[36] J. J. Dudek, R. G. Edwards, M. J. Peardon, D. G. Richards and C. E. Thomas, Phys. Rev. Lett. **103**, 262001 (2009).
[37] J. J. Dudek, R. G. Edwards, M. J. Peardon, D. G. Richards and C. E. Thomas, Phys. Rev. D **82**, 034508 (2010).
[38] C. Bernard *et al.*, Phys. Rev. D **68**, 074505 (2003).
[39] C. W. Bernard *et al.*, Phys. Rev. D **56**, 7039 (1997).
[40] P. Lacock and K. Schilling, Nucl. Phys. Proc. Suppl. **73**, 261 (1999).
[41] P. Lacock, C. Michael, P. Boyle and P. Rowland, Phys. Lett. B **401**, 308 (1997).
[42] C. McNeile and C. Michael, Phys. Rev. D **73**, 074506 (2006).
[43] J. N. Hedditch *et al.*, Phys. Rev. D **72**, 114507 (2005).
[44] J. J. Dudek, R. G. Edwards and D. G. Richards, Phys. Rev. D **73**, 074507 (2006).
[45] J. J. Dudek, R. Edwards and C. E. Thomas, Phys. Rev. D **79**, 094504 (2009).
[46] S. Basak *et al.*, Phys. Rev. D **72**, 074501 (2005).
[47] S. Basak *et al.*, Phys. Rev. D **72**, 094506 (2005).
[48] J. M. Bulava *et al.*, Phys. Rev. D **79**, 034505 (2009).
[49] H.-W. Lin, S. D. Cohen, R. G. Edwards and D. G. Richards, Phys. Rev. D **78**, 114508 (2008).
[50] D. G. Richards, D. K. Sinclair and D. W. Sivers, Phys. Rev. D **42**, 3191 (1990).
[51] W. Detmold, K. Orginos and M. J. Savage, Phys. Rev. D **76**, 114503 (2007).
[52] S. R. Beane, P. F. Bedaque, A. Parreno and M. J. Savage, Phys. Lett. B **585**, 106 (2004).
[53] S. R. Beane, P. F. Bedaque, K. Orginos and M. J. Savage, Phys. Rev. D **73**, 054503 (2006).
[54] S. R. Beane *et al.*, Phys. Rev. D **77**, 014505 (2008).
[55] S. R. Beane *et al.*, Phys. Rev. D **74**, 114503 (2006).
[56] S. R. Beane *et al.*, Phys. Rev. Lett. **100**, 082004 (2008).
[57] W. Detmold *et al.*, Phys. Rev. D **78**, 014507 (2008).
[58] S. R. Beane, P. F. Bedaque, K. Orginos and M. J. Savage, Phys. Rev. Lett. **97**, 012001 (2006).
[59] D. B. Kaplan, M. J. Savage and M. B. Wise, Nucl. Phys. **B534**, 329 (1998).
[60] D. B. Kaplan, M. J. Savage and M. B. Wise, Phys. Lett. B **424**, 390 (1998).
[61] S. R. Beane, P. F. Bedaque, M. J. Savage and U. van Kolck, Nucl. Phys. **A700**, 377 (2002).
[62] S. Weinberg, Phys. Lett. B **251**, 288 (1990).
[63] S. Weinberg, Nucl. Phys. **B363**, 3 (1991).
[64] C. Ordonez, L. Ray and U. van Kolck, Phys. Rev. C **53**, 2086 (1996).
[65] S. R. Beane *et al.*, Phys. Rev. D **80**, 074501 (2009).

Journal of Physics: Conference Series **299** (2011) 012008 doi:10.1088/1742-6596/299/1/012008

Results from the N^* Program at Jefferson Lab

I.G. Aznauryan[1,2], **V.D. Burkert**[1], **T.-S. H. Lee**[1,3], **V.I. Mokeev**[1,4]
[1]Jefferson Lab, Newport News, Virginia 23606, USA
[2]Yerevan Physics Institute, Yerevan 375036, Armenia
[3]Argonne National Laboratory, Argonne, Illinois 60439, USA
[4]Moscow State University, 119899, Moscow, Russia

Abstract. We discuss the results on the fundamental degrees of freedom underlying the nucleon excitation spectrum and how they evolve as the resonance transitions are investigated with increasingly better space-time resolution of the electromagnetic probe. Improved photocouplings for a number of resonant states, those for the N(1720)P_{13} being significantly changed, have been determined and entered into the 2008 edition of the RPP. Strong sensitivity to the N(1900)P_{13} state, listed now as a 2-star state in the same edition of RPP, has been observed in $K\Lambda$ and $K\Sigma$ photoproduction. None of the earlier observations of a $\Theta_5^+(1540)$ was confirmed in a series of three Jefferson Lab high statistics dedicated measurements, and stringent upper limits on production cross sections were placed in several channels. For the four lowest excited states, the $\Delta(1232)P_{33}$, N(1440)P_{11}, N(1520)D_{13}, and N(1535)S_{11}, the transition amplitudes have been measured in a wide range in photon virtuality Q^2. The amplitudes for the $\Delta(1232)$ show the importance of the pion-cloud contribution and do not show any sign of approaching the pQCD regime for $Q^2 < 7$ GeV^2. For the Roper resonance, N(1440)P_{11}, the data provide strong evidence for this state as a predominantly radial excitation of the nucleon as a 3-quark ground state. For the N(1535)S_{11}, comparison of the results extracted from π and η photo- and electroproduction data allowed one to specify the branching ratios of this state to the πN and ηN channels; they entered into the 2010 edition of the RPP. Measured for the first time, the longitudinal transition amplitude for the N(1535)S_{11} became a challenge for quark models and can be indicative of large meson-cloud contributions or alternative representations of this state. The N(1520)D_{13} clearly shows the rapid changeover from helicity-3/2 dominance at the real photon point to helicity-1/2 dominance at $Q^2 > 0.5$ GeV^2, confirming a long-standing prediction of the constituent quark model. The search for undiscovered but predicted states continues to be pursued with a vigorous experimental program. While recent data from Jefferson Lab and elsewhere provide intriguing hints of new states, final conclusions will have to wait for the results of the broad experimental effort currently underway with CLAS, and subsequent analyses involving the EBAC at Jefferson Lab.

1. Introduction

Over 98% of the mass of the visible universe is made of nucleons. They are at the core of atoms and nuclei in stars that provide the energy that heat the planets and allows for life to exist on at least one of them, our earth. Understanding their internal structure has been at the center of nuclear and particle physics for decades. Similar to atomic nuclei, nucleons are complex systems of confined quarks and gluons and exhibit characteristic spectra of excited states that encode information on symmetry properties of baryonic matter. Highly excited nucleon states (N^*) are sensitive to details of quark confinement, which is poorly understood within Quantum Chromodynamics (QCD), the fundamental theory of strong interactions. Measurements of excited nucleon states are needed to come to a better understanding of how confinement works in nucleons. The clustering of quarks in the nucleon can lead to an excitation spectrum with fewer states than if quarks do not cluster. The N^* states couple strongly to the meson-baryon continuum to form nucleon resonances of characteristic masses and decay widths, and can be most effectively investigated by using meson production reactions on the nucleon. With the high precision electron beam and the CEBAF Large Acceptance Spectrometer (CLAS) in Hall B, Jefferson Lab has developed an excited baryon program and has made important contributions in this direction.

The excited baryon program at Jefferson Lab has two main objectives. The first one is to provide information from electromagnetic probes to improve our understanding of the systematics of the N^* spectrum and the nature of the effective degrees of freedom in low energy QCD. The second objective is to measure the resonance transition form factors from the nucleon ground state to the excited baryon states. These measurements probe the internal structure of excited states and provide information about the confining forces of the 3-quark system and the spatial distribution of quarks in the transverse plane.

The experimental program makes wide use of the CLAS detector [1], which is depicted in Fig. 1. CLAS provides particle detection and identification, as well as momentum analysis in a polar angle range from 8° to 140°. For operation of energy-tagged photon beams, the photon energy tagging facility provides photons with an energy resolution of $\sigma(E)/E = 10^{-3}$, and covers an energy range from 20% to 95% of the incident electron beam energy. Other equipment essential for the N^* program includes the coherent bremsstrahlung facility that has been used to produce linearly polarized photons with polarizations up to 90%. There are also two frozen-spin polarized targets, one using butanol as target material for polarized protons (FROST), and the other using HD as an efficient neutron target (HD-Ice). FROST has been operated successfully in longitudinal polarization mode, in conjunction with linearly and circularly polarized photons, and is planned to be used in transverse polarization mode in 2010. HD-Ice is planned to be used as a polarized neutron target in 2010/2011. In addition, the highly polarized electron beam, $P_e = 0.85$, generates circularly polarized photons when scattered off an amorphous radiator. The polarization transfer is maximum at the highest photon energies.

CLAS is also being employed for the measurement of electroproduction of single and double pions, and other pseudoscalar mesons, *e.g.* K^+, K^-, K_s^0. Some of these experiments make use of a dynamically polarized nucleon target [2]. Some experiments, focusing on specific kinematics, make use of the magnetic spectrometer setups in Hall A, using proton recoil polarimeters for the measurement of single and double polarization observables, and in Hall C to study high Q^2 pion and eta production. The list of experiments approved to study N^* excitations or search for new states is given in Table 1.

In addition to analyzing the electromagnetic meson production data, it is important to interpret the extracted N^* parameters in terms of QCD. The Excited Baryon Analysis Center (EBAC) was established at Jefferson Lab in 2006 to address this problem. This effort is making rapid progress by making a dynamical coupled-channels analysis of the world data of both the hadronic and electromagnetic meson production data. In addition, effort has been made to make contact with hadron structure calculations, in particular the lattice QCD effort at Jefferson Lab.

In Sections 2 and 3, the results from analyzing the Jeferson Lab photoproduction and electroproduction data will be described. The progress and status of EBAC will be given in Section 4. Section 5 is devoted to the discussion of the future prospects of the N^* program at 6 GeV and of the possibilities with the 12 GeV upgrade.

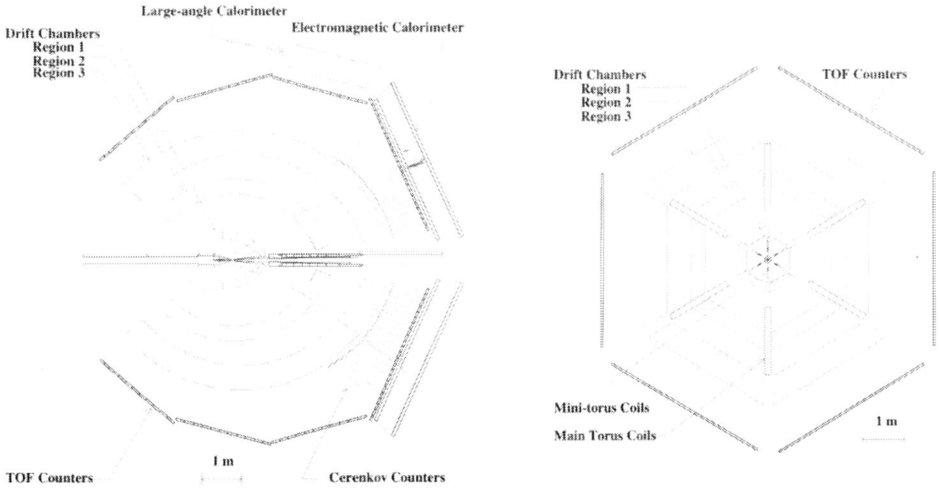

Figure 1. The CEBAF Large Acceptance Spectrometer (CLAS). A cut along the beamline (with the beam incident from the left) is shown on the l.h.s., while the r.h.s. shows a cut perpendicular to the beam. A superconducting toroidal magnet provides the magnetic analysis in six independent sectors. Each of the six sectors is independently instrumented with tracking chambers and with detectors for particle identification.

Table 1. Experiments at Jefferson Lab that are part of the N^* program.

$p(\gamma, K)X$	[3]	$p(e, e'K^+K^-)p, p(e, e'K^+\pi^-)X$	[17]
$p(\gamma, \eta, \eta')p$	[4]	$p(e, e'\pi)N$	[18]
$D(\gamma, \eta),\ D(\gamma, \eta')$	[5]	$\vec{p}(\vec{e}, e'p)\pi^0$	[19]
$p(\gamma, \pi)N$	[6]	$p(e, e'\omega)p$	[20]
$p(\vec{\gamma}, \omega)p$	[7]	$p(e, e'\pi^+\pi^-)p$	[21]
$\vec{\gamma}\vec{p} \to K^+\Lambda, K^+\Sigma, K^0\Sigma^+$	[8]	$\vec{p}(\vec{e}, e'\pi)n$	[22]
$\vec{p}(\vec{\gamma}, \pi^+)n, \vec{p}(\vec{\gamma}, p)\pi^0$	[9]	$p(\vec{e}, e'p)\pi^0,\ p(\vec{e}, e'\pi^+)n$	[23]
$\vec{p}(\vec{\gamma}, \eta)p$	[10]	$p(e, e'p)\pi^0$	[24]
$\vec{p}(\vec{\gamma}, \pi^+\pi^-p)$	[11]	$p(e, e'K^+)\Lambda, \Sigma$	[25]
$n(\vec{\gamma}, K\Lambda)$	[12]	$p(e, e'\pi^0)p, p(e, e'\pi^+)n$	[26]
$p(e, e'\pi)N$	[13]	$p(e, e'\pi^+\pi^-p)$	[27]
$p(e, e'\pi^+)n,\ p(e, e'p)\pi^0,\ n(e, e'\pi^-)p$	[14]	$p(\vec{e}, e'K, \vec{\Lambda}, \vec{\Sigma})$	[28]
$p(e, e'p)\eta$	[15]	$p(e, e'p)\pi^0, \eta$	[29]
$p(\vec{e}, e'p)\pi^0,\ p(\vec{e}, e'\pi^+)n$	[16]		

2. Results from photoproduction experiments

2.1. Photoproduction of pseudoscalar mesons

Precise differential cross sections on proton targets for final states $p\pi^0$ [30], $n\pi^+$ [31], $p\eta$ [32], $p\eta'$ [33], $K^+\Lambda$ [34, 35], and $K^+\Sigma^0$ [34, 35] have been measured. As an example, Fig. 2 shows differential cross sections for $\gamma p \to n\pi^+$ for different hadronic mass W. These data, together with the $\pi N \to \pi N$ elastic data and the cross sections from the $\gamma p \to p\pi^0$ channel also measured with CLAS, have been used to determine improved helicity amplitudes for a number of resonant states with significant couplings to the single pion channel. They have been also entered into the 2008 edition of the Review of Particle Physics (RPP) [36]. For example, the photocoupling of the N(1720)P$_{13}$ has changed significantly with the new precise data. More complex processes such as $\gamma p \to p\pi^+\pi^-$ are also being studied. The latter are particularly sensitive to higher-mass nucleon resonances that couple dominantly to the $\Delta\pi$ or $N\rho$ isobars with two pions in the final state.

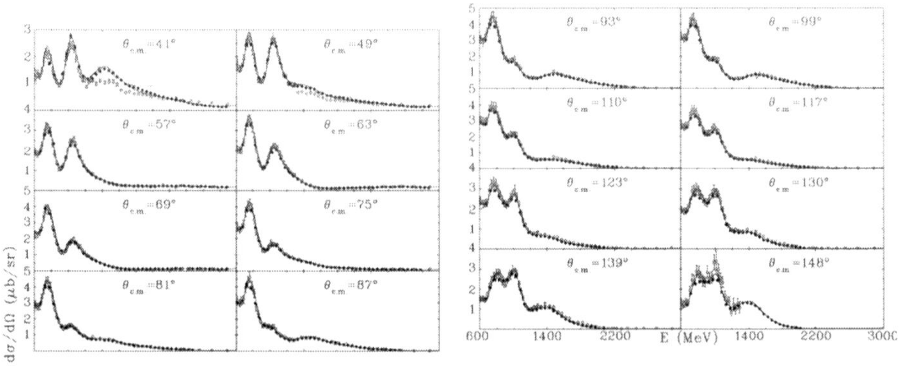

Figure 2. Differential cross sections from CLAS for $\gamma p \to n\pi^+$ [31] *vs.* photon energy E ($W^2 = m_p^2 + 2m_pE$). Solid and open circles show, respectively, the CLAS and previous data. New measurements [31] have improved the world data base significantly in energy range and polar angle coverage. Curves show the results of the SAID FA08 solution.

In photoproduction of single pions or etas, partial wave analyses of differential cross sections alone result in ambiguous solutions for the contributing resonant partial waves. The CLAS N^* program is therefore aimed at complete or nearly complete measurements for all of these processes. Complete information may be obtained using a combination of linearly and circularly polarized photon beams, measurement of the hyperon recoil polarization, and using longitudinally and transversely polarized targets. Single pseudoscalar meson production is fully described by 4 complex, parity conserving amplitudes, requiring eight [37] well-chosen combinations of beam, target, and recoil polarization measurements for an unambiguous extraction of the production amplitude. Table 2 shows the observables that can be determined from these measurements. If all possible combinations are measured, 16 observables can be extracted. In measurements that involve nucleons in the final state with no recoil polarization being measured, 7 independent observables can be measured directly. In addition, the recoil polarization asymmetry P can be inferred from the double polarization asymmetry with a linearly polarized beam and transverse target polarization.

In addition to precise $K\Lambda$ and $K\Sigma$ cross section data, recoil polarization and polarization transfer data have been measured [38]. The recoil polarization data in the $K^+\Lambda$ sector showed a highly unexpected behavior: the spin transfer from the circularly polarized photon to the Λ hyperon is complete, creating the Λ hyperon 100% polarized, as can be seen in Fig. 3. The sum of all polarization components $R^2 \equiv P^2 + C_x^2 + C_z^2$, which has an upper bound of 1, is consistent with $R^2 = 1$ throughout the region covered by the measurement. The combined analysis of the CLAS cross section and polarization transfer data by the Bonn-Gatchina group [39] shows strong sensitivity to a N(1900)P$_{13}$ candidate state. The decisive ingredient in this analysis was the CLAS spin transfer data. An N(1900)P$_{13}$ state is listed as a 2-star candidate state in the 2008 edition of the RPP [36]. If this assignment is further corroborated, the existence of an N(1900)P$_{13}$ state will provide strong evidence against the quark-diquark model that has no place for such a state in this mass range [40].

New precise high statistics data with a linearly polarized photon beam on a proton target are becoming also available. Preliminary results for the process $\vec{\gamma}p \to K^+\Lambda$ in just one energy bin are shown in Fig. 4. For the partial wave analysis, complete angle coverage in all observables is especially important.

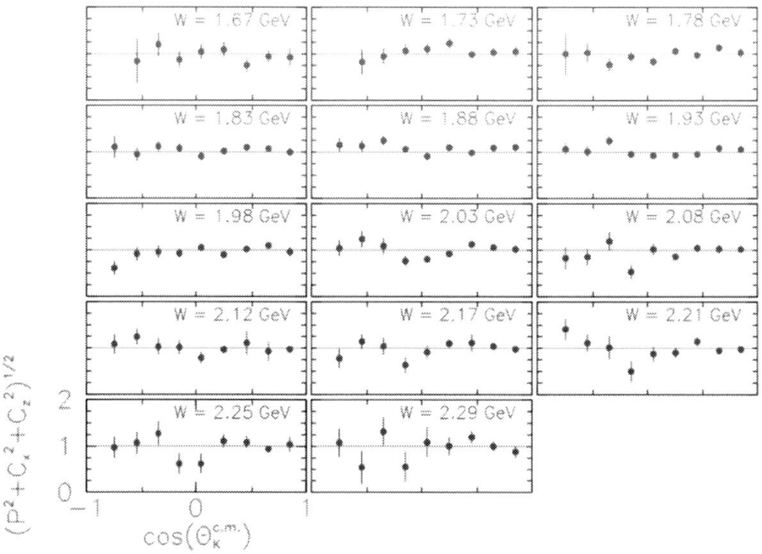

Figure 3. CLAS data [38] on the polarization observable vector combining recoil polarization P and transferred polarization components C_x and C_z for the Λ hyperons produced in $\vec{\gamma}p \to K^+\vec{\Lambda}$. Lower-left axis scales apply to all plots. The data show the Λ is 100% polarized. These data were crucial when fitted simultaneously with the CLAS $K^+\Lambda$ and $K^+\Sigma$ differential cross section data. They provided evidence for an N(1900)P$_{13}$ candidate state, which is required for a good simultaneous fit to the cross section and spin transfer data.

2.2. Search for new cascade (Ξ) baryons

The production of Ξ baryons, *i.e.* strangeness $S = -2$ excited states, provides another promising way of searching for new baryon states due to the expected narrower widths of these states compared to $S = 0$ and $S = -1$ resonances. However, the disadvantage of using a photon beam for detection of the $S = -2$ states is that they require production of at least two kaons in the final state. Possible production mechanisms include t-channel K or K^* exchanges on proton targets with an excited hyperon Y^* (Λ^* or Σ^*) as an intermediate state and with subsequent decays $Y^* \to K^+ + \Xi^*$ and $\Xi^* \to \Xi\pi$ or $\Xi \to \Lambda(\Sigma)\bar{K}$. The missing mass technique may work for $\gamma p \to K^+K^+X$ if the state is sufficiently narrow to be observed as a peak in the missing mass spectrum. Fig. 5 shows that one can identify the lowest two cascade states in this way. To identify the higher-mass states, higher energies are needed.

Another possibility to isolate excited cascade baryons is to measure additional particles in the final state resulting from the decays of the excited Ξ, *e.g.* $K^+K^+\Xi^0\pi^-$. The invariant mass of the $\Xi\pi^-$ system shows the first exited state $\Xi(1530)$ and indications of additional structure

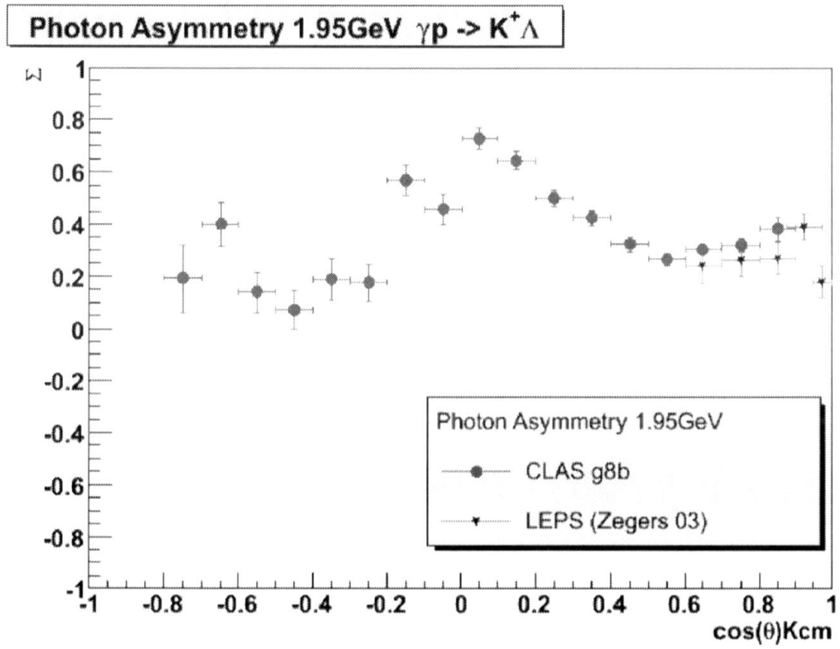

Figure 4. Beam asymmetry for $\vec{\gamma} p \to K^+\Lambda$ with linearly polarized photons. Preliminary CLAS data are shown for $W = 1.95$ GeV. The previous data are from GRAAL [41] and LEPS [42].

near 1620 MeV. A state near that mass is predicted as a dynamically generated $\Xi\pi$ system in the model of Ref. [43]. The data have insufficient statistics and were taken at too low of an energy to allow further investigations. New data taken at 5.5 GeV electron energy and with higher statistics are currently being analyzed, and should allow more definite conclusions.

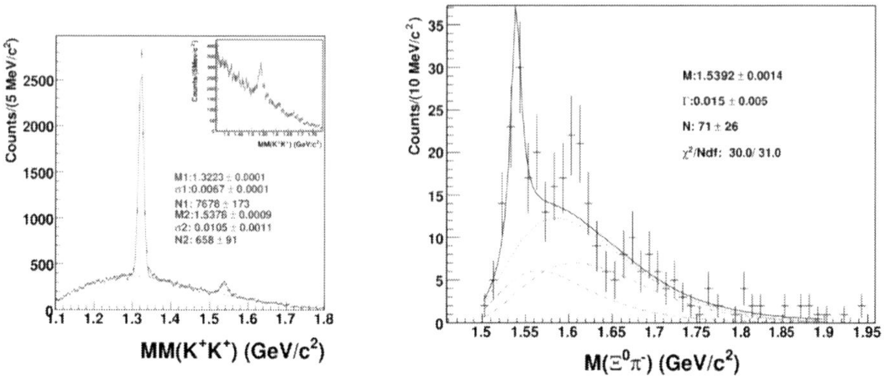

Figure 5. Missing mass spectrum M_X for $\gamma p \to K^+K^+X$. The Ξ ground state and first excited state are clearly identified. The right panel shows the invariant mass of $\Xi^0(1320)\pi^-$. The low mass peak is the $\Xi(1530)$. The other structure near 1620 MeV is statistically not significant.

Journal of Physics: Conference Series **299** (2011) 012008 doi:10.1088/1742-6596/299/1/012008

Table 2. Polarization observables in $J^\pi = 0^-$ meson photoproduction that can be extracted from complete experiments using all combinations of beam, target, and recoil polarizations. I, II and III correspond, respectively, to unpolarized, linearly and circularly polarized photon beams. The observables given without hats and tildes are directly extractable. Other observables provide additional redundant information: letters with hats show the single-polarization observables measured from double-polarization asymmetries, letters with tildes show the double-polarization observables measured from triple-polarization asymmetries. For example, the target polarization asymmetry T can be extracted in a single-polarization measurement with unpolarized beam and a transversely polarized target, and also in a double-polarization measurement with a linearly polarized beam and measurement of the recoil polarization $P_{y'}$. This allows multiple cross checks of measurements with different systematics.

Photon	T			R			T	-	R						
				x'	y'	z'	x	y	z	x	y	z	x	y	z
	x	y	z				x	y	z	x	y	z	x	y	z
I		T			P		$T_{x'}$		$L_{x'}$		$\hat{\Sigma}$		$T_{z'}$		$L_{z'}$
II	H	$\hat{P}$	G	$O_{x'}$	$\hat{T}$	$O_{z'}$	$\tilde{L}_{z'}$	$\tilde{C}_{z'}$	$\tilde{T}_{z'}$	$\tilde{E}$		$\tilde{F}$	$\tilde{L}_{x'}$	$\tilde{C}_{x'}$	$\tilde{T}_{x'}$
III	F		E	$C_{x'}$		$C_{z'}$		$\tilde{O}_{z'}$		$\tilde{G}$		$\tilde{H}$		$\tilde{O}_{x'}$	

2.3. Demise of the pentaquarks?

In 2003 and 2004, about a dozen experiments in nuclear physics and high energy physics laboratories claimed the observation of several states with flavor exotic quantum numbers, $\Theta_5^+(1540)$, $\Phi_5(1860)$, and $\Theta_c^0(3100)$. For a summary see Ref. [44]. These claims were exclusively based on non-dedicated experimental observations. At Jefferson Lab a series of three high statistics dedicated measurements was conducted in 2004 and 2005 with CLAS to verify some of these claims. None of the earlier observations of a $\Theta_5^+(1540)$ could be confirmed and stringent upper limits on production cross sections were placed in several channels, in particular, $\gamma p \to \bar{K}^0 K^0$, $\bar{K}^0 K^+ n$ [45, 46], $\gamma n \to K^- K^+ n$ [47], $\gamma D \to \Lambda n K^+$ [48], $\gamma p \to K^+ K^- p$ [49]. Other channels are still being evaluated.

3. Results from electroproduction experiments

3.1. Jefferson Lab/CLAS data on single pion electroproduction

The CLAS detector at Jefferson Lab is the first full acceptance instrument designed for the comprehensive investigation of exclusive electroproduction of mesons with the goal to study the electroexcitation of nucleon resonances in detail. In recent years, a variety of measurements of single pion electroproduction on protons have been performed at CLAS in a wide range of Q^2 from 0.16 to 6 GeV^2 [50]-[57]. The obtained experimental data include about 119,000 data points of differential cross sections, longitudinally polarized beam asymmetries, and longitudinal target and beam-target asymmetries. A comprehensive analysis of these data was performed using two approaches: one, based on fixed-t dispersion relations (DR), and another, based on the effective Lagrangian approach — Unitary Isobar Model (UIM).

The DR approach was developed for pion photo- and electroproduction on nucleons in the 1950s [58, 59] and played an extremely fruitful role in the analyses of the data for these processes.

In recent publications [60, 61, 62, 63], this approach was further developed by extending it into the wide kinematical region in W and Q^2 covered by the new data. DR allowed also to obtain strict constraints on the multipole amplitudes $M_{1+}^{3/2}$, $E_{1+}^{3/2}$, $S_{1+}^{3/2}$ that correspond to the contribution of the prominent $\Delta(1232)P_{33}$ resonance [61]. The constraint on the large $M_{1+}^{3/2}$ amplitude plays an important role in the reliable extraction of the amplitudes for the $\gamma^* N \to \Delta(1232)P_{33}$ transition. It also impacts the analysis of the second resonance region, because resonances from this region are overlapping with the $\Delta(1232)P_{33}$.

Starting in the late 1990s, the UIM approach [64] (also known as MAID), became widely used for the analyses of single-pion photo- and electroproduction data. This approach was modified in Ref. [61] by the incorporation of Regge poles, which enabled a good description of all photoproduction multipole amplitudes up to an invariant mass $W = 2$ GeV using a unified Breit-Wigner parametrization of the resonance contributions.

Within DR and UIM, the analyses of the data [50]-[57] were performed in Refs. [62, 63, 65, 66]. As a result, the helicity amplitudes for the electroexcitation of the low mass resonances $\Delta(1232)P_{33}$, $N(1440)P_{11}$, $N(1520)D_{13}$, and $N(1535)S_{11}$ were extracted from the experimental data in a range of Q^2 up to 6 GeV^2. The non-resonant contributions are built in DR and UIM in conceptually different ways; this allowed to draw conclusions on the model sensitivity of the resulting electroexcitation amplitudes. Model uncertainties caused by the higher-mass resonances and by the non-resonant amplitudes due to the uncertainties in the Born terms (the s- and u-channel nucleon exchanges and t-channel pion contribution) and in the ρ and ω t-channel contributions were also evaluated. All uncertainties were added in quadrature and presented as model uncertainties of the extracted electroexcitation amplitudes.

In Figs. 6-10, some examples that show the description of the data are presented. Shown is the W dependence of the differential cross sections for different reactions at several Q^2, polar and azimuthal angles. The W dependence of σ_{tot}, $\sigma_{tot}^{LT'}$, and of the target (A_t) and beam-target (A_{et}) asymmetries for $\vec{e}\vec{p} \to ep\pi^0$ integrated over cosθ, ϕ and Q^2 is also shown.

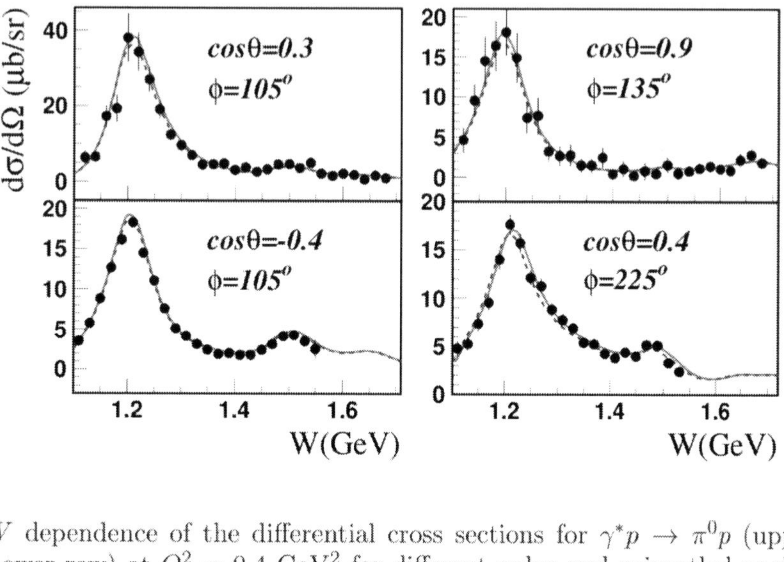

Figure 6. W dependence of the differential cross sections for $\gamma^* p \to \pi^0 p$ (upper row) and $\gamma^* p \to \pi^+ n$ (lower row) at $Q^2 = 0.4$ GeV^2 for different polar and azimuthal angles. Data are from Refs. [50] (π^0) and [53] (π^+). The solid (dashed) curves correspond to the results obtained using the DR (UIM) approach [63].

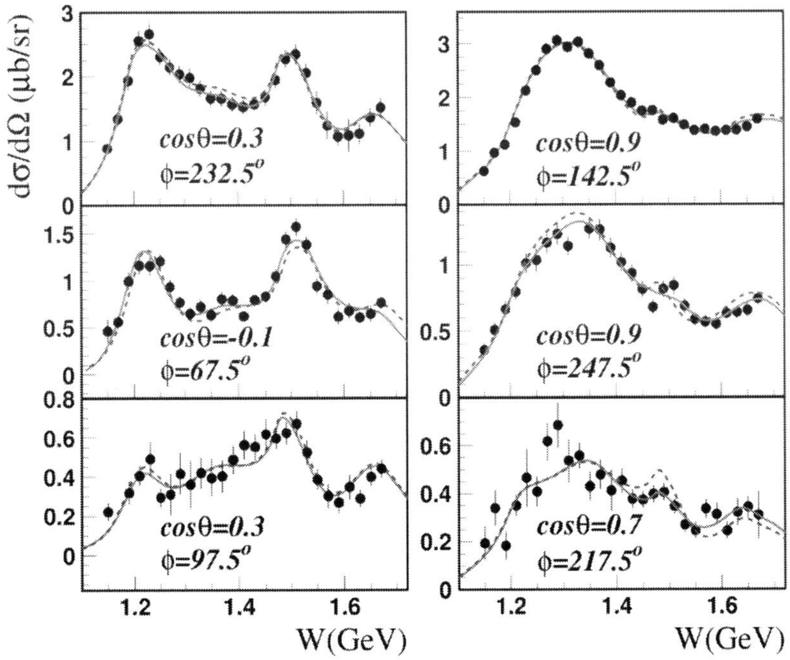

Figure 7. W dependence of the differential cross section for $\gamma^* p \to \pi^+ n$ for different polar and azimuthal angles. Upper, middle, and lower rows are for $Q^2 = 1.72$, 2.44 and 3.48 GeV^2, respectively. Data are from Ref. [56]. Other notations as in Fig. 6.

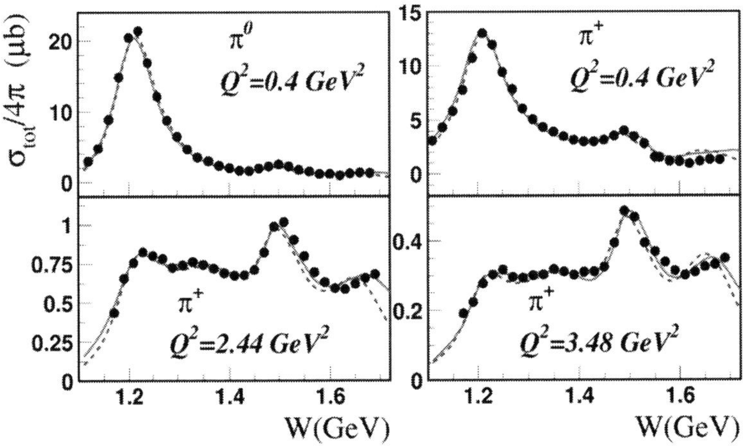

Figure 8. W dependence of the $\gamma^* p \to \pi^0 p, \pi^+ n$ total cross sections. Data are from Refs. [50, 53, 56]. Other notations as in Fig. 6.

3.2. Jefferson Lab/CLAS data on two pion electroproduction

Two pion electroproduction is one of the biggest contributors to N^* electroexcitation. This exclusive channel is sensitive to almost all well established excited states and in particular to high lying resonances with masses above 1.6 GeV. According to the quark model expectations [67, 68], two pion electroproduction has a great potential for the discovery of the so-called

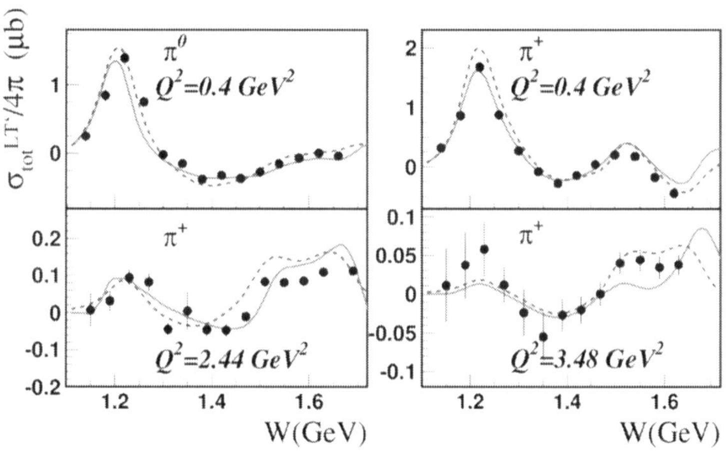

Figure 9. W dependence of the structure function $\sigma_{tot}^{LT'}$ corresponding to a longitudinally polarized electron beam integrated over polar angle. Data are from Refs. [51, 52, 56]. Other notations as in Fig. 6.

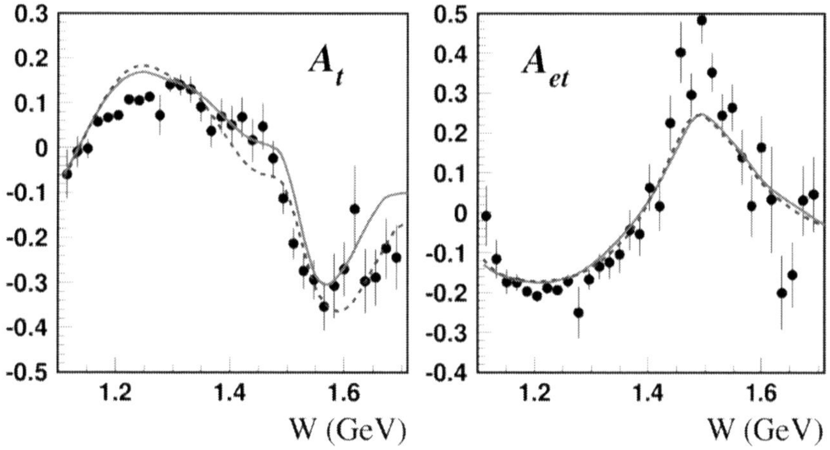

Figure 10. W dependence of the target (A_t) and beam-target (A_{et}) asymmetries for $\vec{e}\vec{p} \to ep\pi^0$ integrated over $\cos\theta$, ϕ and Q^2 ($0.252 < Q^2 < 0.611$ GeV^2). Data are from Ref. [57]. Other notations as in Fig. 6.

"missing" baryon states.

A combination of the CEBAF continuous electron beam and CLAS detector makes it possible for the first time to measure nine independent one-fold differential $\gamma^* p \to \pi^+\pi^- p$ cross sections, as well as fully integrated cross sections, in the kinematical areas presented in Table 3. The high statistics and good momentum resolution of the measurements allowed W and Q^2 bin sizes to be used that are several times smaller than the ones used in previous measurements.

All essential mechanisms contributing to $\pi^+\pi^- p$ electroproduction were established from the analysis of these data within the framework of the reaction model presented in Ref. [71]. This model provides a good description of data, allowing resonant contributions to observables to be isolated, which are needed for the evaluation of $\gamma^* NN^*$ electrocouplings.

The CLAS data [70] for the first time revealed the structure in the fully integrated $\pi^+\pi^- p$

Table 3. Kinematical areas covered by the CLAS measurements of the $\pi^+\pi^-p$ photo- and electroproduction cross sections.

Q^2 coverage , GeV^2	W coverage, GeV	Bin size over W/Q^2 , GeV/GeV^2	Data status
0.20-0.60	1.30-1.57	0.025/0.050	Completed [69]
0.50-1.50	1.40-2.10	0.025/0.3-0.4	Completed [70]
2.0-5.0	1.40-2.00	0.025/0.5	In progress
0.	1.60-2.80	0.025	In progress

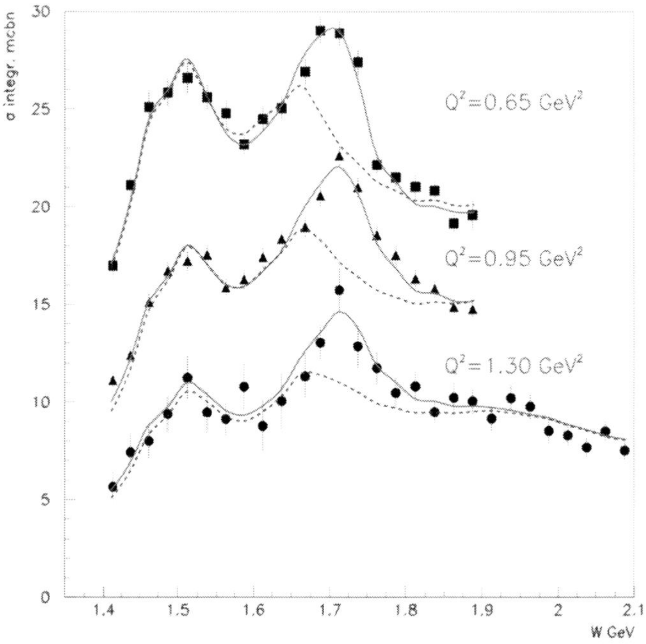

Figure 11. Description of the CLAS data [70] on fully integrated $\pi^+\pi^-p$ electroproduction cross sections within the framework of the reaction model [71] with parameters fitted to one-fold differential cross sections. The calculations, taking into account the contributions from the conventional N^*s only, are shown by the dashed lines, while the solid lines correspond to the fit after implementation of the $3/2^+(1720)$ candidate state. Difference between solid and dashed lines represents a signal from the possible new state.

electroproduction cross sections at $W \approx 1.7$ GeV, shown in Fig. 11. Analysis of these data within the framework of the reaction model [71] offers an evidence for the candidate state $3/2^+(1720)$. Signals from this possible new state can be seen in Fig. 11. If these signals will be confirmed in the future combined analysis of the CLAS $\pi^+\pi^-p$ photo- and electroproduction data (Table 3), we would obtain an unambiguous evidence for long awaited "missing" baryon state.

Evaluation of γ^*NN^* electrocouplings from CLAS $\pi^+\pi^-p$ electroproduction data is in progress. The γ^*NN^* electrocouplings for almost all N^* states will be determined at photon virtualities $Q^2 < 5.0$ GeV^2. Preliminary results may be found in Refs. [72, 73]. Single and charge double pion electroproduction channels have completely different non-resonant contributions and provide independent information on the extracted resonance electrocouplings. Consistent results obtained from analyses of the dominant πN and $\pi^+\pi^-p$ exclusive channels will offer evidence for reliable extraction of these fundamental quantities from meson electroproduction data.

Finally, a combined analysis of single and charge double pion electroproduction data is planned within the framework of a coupled-channel approach, as described in Sec. 4.

3.3. Other Jefferson Lab data on exclusive meson electroproduction

There are Jefferson Lab Hall A and Hall C measurements of π^0 electroproduction on the proton in the $\Delta(1232)P_{33}$ resonance region that include Hall A data on recoil polarization response functions at $Q^2 = 1$ GeV^2 [74, 75] and Hall C data on differential cross sections at $Q^2 = 2.8$, 4.2 GeV^2 [76] and 6.4, 7.7 GeV^2 [77]. Measurements of η electroproduction on the proton were performed in Hall B with CLAS and in Hall C; they include CLAS data at $Q^2 = 0.165 - 3.1$ GeV^2 [78, 79] and Hall C data at $Q^2 = 2.4$, 3.6 GeV^2 [80], and 5.7, 7 GeV^2 [81].

3.4. The $\Delta(1232)P_{33}$ resonance

The excitation of the $\Delta(1232)P_{33}$ has been studied for more than 50 years with various probes. But only in the past decade have the experimental tools become available in electron scattering to enable precise measurements in exclusive π production from protons with photon virtualities up to $Q^2 = 8$ GeV^2. Benchmark results from CLAS and other laboratories are shown in Fig. 12. New exclusive measurements confirm a rapid falloff of the $\gamma^*p \to \Delta(1232)P_{33}$ magnetic-dipole form factor relative to that for the proton seen previously in inclusive experiments. There has been a long-standing underestimation of the data for this form factor by the constituent quark model. Within dynamical reaction models [86, 89, 90], the pion-cloud contribution was identified as the source of this discrepancy. The importance of the pion-cloud contribution for the $\gamma^*p \to \Delta(1232)P_{33}$ transition is confirmed also by the lattice QCD calculations [91]. In Fig. 12, the results of the dynamical model [86] are given that show that the pion-cloud contribution makes up more than 30% of $G^*_M(Q^2)$ at the photon point, and remains sizeable at the highest Q^2.

The Q^2 dependence of R_{EM} and R_{SM} is of great interest as a measure of the Q^2 scale where the approach to the asymptotic domain of QCD may set in. In the pQCD asymptotics $R_{EM} \to 1$ and $R_{SM} \to$ constant. The data on R_{EM} and R_{SM} show that in the range $Q^2 < 7$ GeV^2 there is no sign of an approach to the asymptotic pQCD regime in either of these ratios.

3.5. The Roper resonance $N(1440)P_{11}$ — a puzzle resolved

By quantum numbers, the simplest and most natural classification of the Roper resonance in the constituent quark model is as a first radial excitation of the $3q$ ground state. However, difficulties of quark models to describe the low mass and large width of the $N(1440)P_{11}$, and also its electroexcitation amplitudes on the proton and neutron at the real photon point, gave rise to numerous speculations around the Roper. Alternative descriptions of this state as a gluonic baryon excitation [92, 93], or as a hadronic $N\sigma$ molecule [94], were suggested. The CLAS measurements allowed, for the first time, the determination of the electroexcitation amplitudes of the Roper resonance on the proton for $Q^2 < 4.5$ GeV^2 (Fig. 13).

These results are crucial for the understanding of the nature of this state. There are several specific features in the extracted $\gamma^*p \to N(1440)P_{11}$ amplitudes that are very important for testing models. First, the specific behavior of the transverse amplitude $A_{1/2}$, which, being large and negative at $Q^2 = 0$, becomes large and positive at $Q^2 \simeq 2$ GeV^2, and then drops smoothly with Q^2. Second, the relative sign between the longitudinal $S_{1/2}$ and transverse $A_{1/2}$

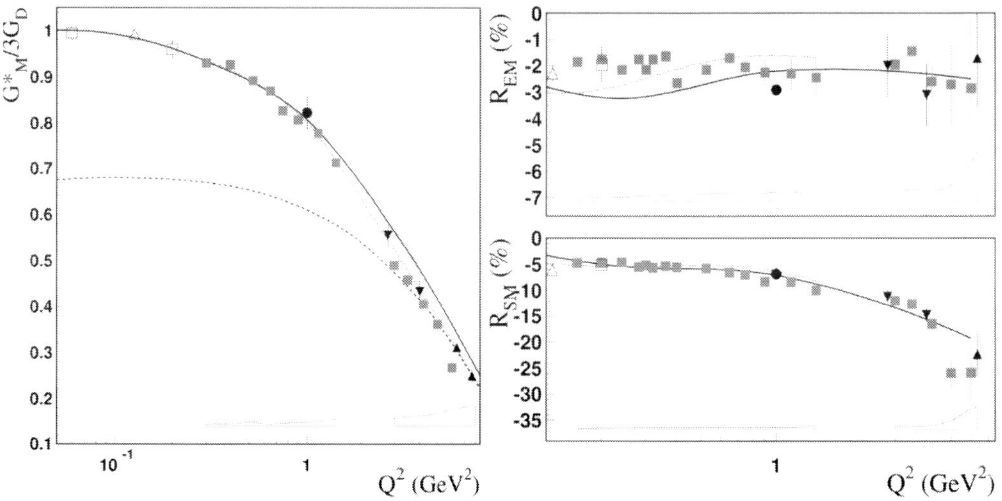

Figure 12. Left panel: the form factor G^*_M [82] for the $\gamma^* p \to \Delta(1232)\mathrm{P}_{33}$ transition relative to $3G_D(Q^2)$: $G_D(Q^2) = 1/(1 + Q^2/0.71\ \mathrm{GeV}^2)$. Right panel: the ratios $R_{EM} \equiv E^{3/2}_{1+}/M^{3/2}_{1+}$ and $R_{SM} \equiv S^{3/2}_{1+}/M^{3/2}_{1+}$. The filled sqaures are the results extracted from the CLAS data [50, 51, 54, 55] in Ref. [63]. The bands show the model uncertainties of the results. Also shown are the results from Jefferson Lab/Hall A [74, 75] (filled circles) and Jefferson Lab/Hall C [76, 77] (filled triangles). The low Q^2 results are from MAMI [83, 84] (open squares) and MIT/Bates [85] (open triangles). The solid and dashed curves correspond to the 'dressed' and 'bare' contributions from Ref. [86]; for R_{EM}, R_{SM}, only the 'dressed' contributions are shown; the 'bare' contributions are close to zero. The dotted curves are the predictions obtained in the large-N_c limit of QCD [87, 88].

amplitudes. And third, the common sign of the amplitudes $A_{1/2}$ and $S_{1/2}$ extracted from the data on $\gamma^* p \to \pi N$ includes signs from the $\gamma^* p \to$N(1440)P$_{11}$ and N(1440)P$_{11} \to \pi$N vertices; both signs should be taken into account while comparing with model predictions. All these characteristics are described by the light-front relativistic quark models [95, 96] assuming that the N(1440)P$_{11}$ is the first radial excitation of the $3q$ ground state. The presentation of the Roper resonance as a gluonic baryon excitation is definitely ruled out. No predictions exist for the Nσ molecule model.

3.6. The N(1535)S$_{11}$ state

For the first time the $\gamma^* N \to$ N(1535)S$_{11}$ transverse helicity amplitude has been extracted from the π electroproduction data in a wide range of Q^2 (Fig. 14), and the results confirm the Q^2 dependence of this amplitude observed in η electroproduction. Comparison of the results extracted from π and η photo- and electroproduction data have allowed one to specify the relation between the branching ratios of N(1535)S$_{11}$ to the πN and ηN channels: from the fit to the amplitudes at $0 \leq Q^2 < 4.5$ GeV^2, we found

$$\frac{\beta_{\eta N}}{\beta_{\pi N}} = 0.95 \pm 0.03. \tag{1}$$

Furthermore, taking into account the branching ratio to the $\pi\pi N$ channel $\beta_{\pi\pi N} = 0.01 - 0.1$ [36], which accounts practically for all channels different from πN and ηN, we found

$$\beta_{\pi N} = 0.485 \pm 0.008 \pm 0.023, \tag{2}$$

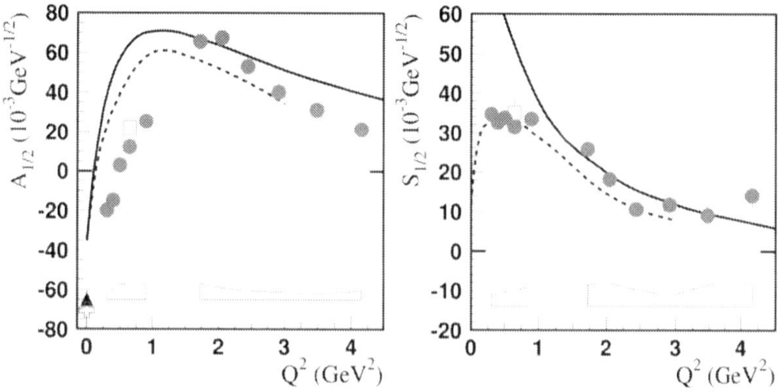

Figure 13. Helicity amplitudes for the $\gamma^* p \rightarrow$ N(1440)P$_{11}$ transition. The solid circles are the results extracted from the CLAS data [50, 51, 52, 53, 56, 57] in Ref. [63]. The bands show the model uncertainties. The cross at $Q^2 = 0$ is from SAID analysis that includes CLAS data on π^0 and π^+ photoproduction cross sections [30, 31]. The open squares are obtained in the combined analysis of CLAS single π and 2π electroproduction data [66]. The filled triangle at $Q^2 = 0$ is the RPP estimate [36]. All amplitudes, except RPP values, correspond to $M = 1440$ MeV, $\Gamma_{tot} = 350$ MeV, and $\beta_{\pi N} = 0.6$. The predictions of the light-front (LF) relativistic quark models [95, 96], assuming the N(1440)P$_{11}$ is a first radial excitation of the $3q$ ground state, are shown by solid and dashed curves. The dotted curves are obtained assuming that the N(1440)P$_{11}$ is a gluonic baryon excitation (q^{3}G hybrid state) [92, 93].

$$\beta_{\eta N} = 0.460 \pm 0.008 \pm 0.022. \tag{3}$$

The first error corresponds to the fit error in Eq. (1) and the second error is related to the uncertainty of $\beta_{\pi\pi N}$. The ratio (1) entered into the 2010 edition of the RPP.

Due to the CLAS measurements of π electroproduction, for the first time the $\gamma^* N \rightarrow$ N(1535)S$_{11}$ longitudinal helicity amplitude was extracted from experimental data. These results are crucial for testing theoretical models. It turned out that at $Q^2 < 3$ GeV^2, the sign of $S_{1/2}$ is not described by the quark models. Combined with the difficulties of quark models to describe the substantial coupling of N(1535)S$_{11}$ to the ηN channel [36] and to strange particles [99, 100], this can be indicative of a large meson-cloud contribution and/or additional $q\bar{q}$ components in this state [101]. Alternative representations of the N(1535)S$_{11}$ as a meson-baryon molecule have been also discussed [102, 103, 104, 105].

3.7. The N(1520)D$_{13}$ resonance

The CLAS data allowed the determination of the longitudinal amplitude of the $\gamma^* p \rightarrow$ N(1520)D$_{13}$ transition, and much more accurate results are obtained for the transverse amplitudes (Fig. 15). There is a longstanding prediction of constituent quark models for the $\gamma^* p \rightarrow$ N(1520)D$_{13}$ transition that follows from the structure of the nucleon and the N(1520)D$_{13}$ wave functions. It consists in the rapid helicity switch from the dominance of the $A_{3/2}$ amplitude at the photon point to the dominance of $A_{1/2}$ at $Q^2 > 1$ GeV^2. This prediction is definitely confirmed by the data, although quark models fail to describe the details of the Q^2 dependence of the amplitudes.

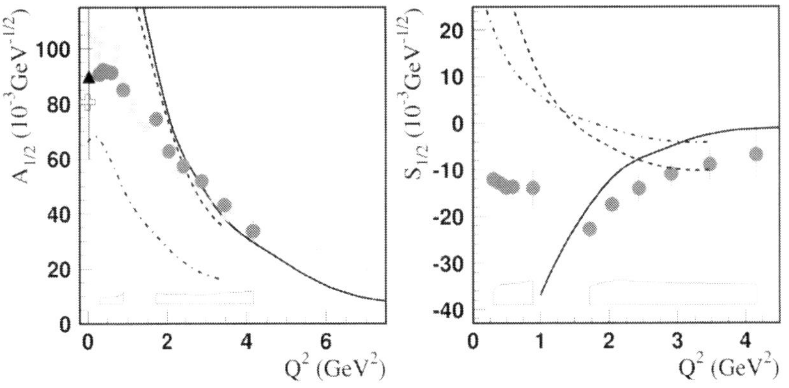

Figure 14. Helicity amplitudes for the $\gamma^* p \to$ N(1535)S$_{11}$ transition. The legend is partly as for Fig. 13. The filled squares and stars are the CLAS [78, 79] and Hall C [80, 81] results, respectively, obtained in η electroproduction. All amplitudes, except RPP values, correspond to M = 1535 MeV, Γ_{tot} = 150 MeV, and to the branching ratios $\beta_{\pi N} = 0.485$, $\beta_{\eta N} = 0.46$ (2,3). The results of the LF relativistic quark models are given by the dashed [95] and dashed-dotted [97] curves. The solid curves show the amplitudes found within light-cone sum rules using lattice results for the light-cone distribution amplitudes of the N(1535)S$_{11}$ resonance [98].

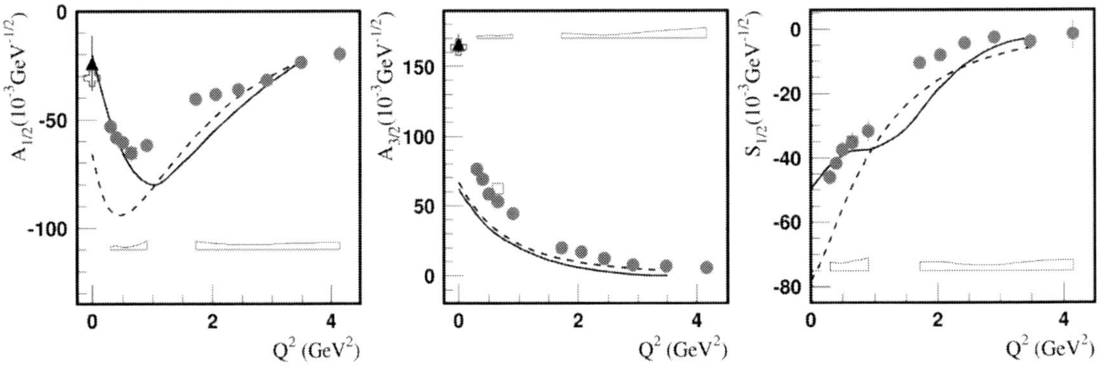

Figure 15. Helicity amplitudes for the $\gamma^* p \to$ N(1520)D$_{13}$ transition. The legend is partly as for Fig. 13. All amplitudes, except RPP values, correspond to M = 1520 MeV, Γ_{tot} = 112 MeV, and $\beta_{\pi N} = 0.6$. The curves correspond to the predictions of quark models: [106] (solid), [107] (dashed).

3.8. Transverse charge densities in the nucleon — resonance transitions

The accurate information now available for all amplitudes of the $\gamma^* p \to \Delta(1232)P_{33}$, N(1440)P$_{11}$, N(1520)D$_{13}$, and N(1535)S$_{11}$ transitions allowed, for the first time, to obtain spatial images of the transitions by mapping out the transition charge densities in impact parameter space. The corresponding result for the Roper resonance is presented in Fig. 16.

4. Excited Baryon Analysis Center

The Excited Baryon Analysis Center (EBAC) was established at Jefferson Lab in January, 2006 to provide theoretical support to the excited baryon program. EBAC's program has two objectives. First, to establish the spectrum of excited nucleon states and to extract the N^*

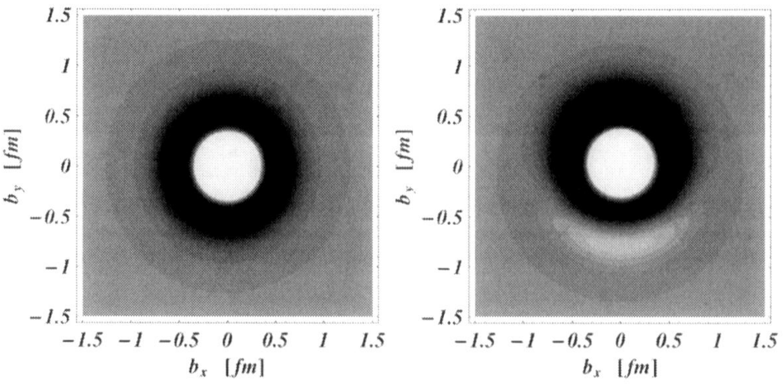

Figure 16. Quark transverse charge density in the $\gamma^* p \to$ N(1440)P$_{11}$ transition [108]. Left panel: unpolarized p and N(1440)P$_{11}$. Right panel: p and N(1440)P$_{11}$ polarized along the positive x-axis. The light (dark) areas are dominated by up (down) quarks and correspond to dominantly positive (negative) charges.

parameters from the πN, γN and $N(e, e'\pi)$ data. The second objective is to develop theoretical interpretations of the extracted N^* parameters. To achieve these two goals, a dynamical coupled-channel (DCC) reaction model has been developed and is being used to perform analyses of the meson production data from Jefferson Lab and other facilities. The essential feature of the DCC model is to account for the coupled-channel effects resulting from the unitarity conditions and the reaction mechanisms in the short-range (off-shell) region where we want to map out the structure of baryons.

The DCC model is based on a Hamiltonian formulation [90] of multi-channel and multi-resonance reactions. The first work at EBAC was to determine the hadronic parameters of the model by fitting the world data on the $\pi N \to \pi N$, $\pi N \to \eta N$, and $\pi N \to \pi\pi N$ reactions. These were completed [109, 110, 111] with the results on the total πN cross sections shown in Fig. 17. The model was then applied to analyze pion photoproduction [112] and electroproduction [113] data. An important feature of the DCC model is the possibility to distinguish the N^* excitation due to the quark-gluon core from that due to the meson-cloud (meson-baryon dressing) effects. This is illustrated in Fig. 12 for the G^*_M form factor of the $\gamma N \to \Delta(1232)P_{33}$ transition. It is further found [86, 112] that the contribution due to the quark-gluon core is close to the prediction of constituent quark models.

In parallel, an exact analytic continuation method for extracting nucleon resonances from multi-channel reactions has been developed [114] and applied [115] to extract the nucleon resonances from this model. As an example, Fig. 18 shows the positions of the extracted P_{11} resonances in the complex energy plane. Within the DCC model, it is found that these two P_{11} resonances evolve from the same bare state with a mass of about 1700 MeV, which could be identified with the hadron structure calculations in the absence of meson-baryon degrees of freedom, such as the constituent quark model.

The plan at EBAC is to complete the analysis of single π and η production up to $W = 2$ GeV and $Q^2 \leq 6$ GeV^2, to analyze the 2π production data, and to extract the N^* form factors at the resonance poles. With these, EBAC will complete its first-stage analysis.

EBAC's second-stage work is to analyze the data of $K\Lambda$ and $K\Sigma$ production. EBAC is collaborating with the CLAS Collaboration to develop methods for extracting $\gamma N \to K\Lambda$ multipole amplitudes as model independently as possible from the forthcoming Jefferson Lab data from the over-complete measurements of $K\Lambda$ and $K\Sigma$ production.

Efforts are also being made to develop interpretations of the extracted N^* parameters in terms of hadron structure calculations. In addition to considering constituent quark models, we are also investigating how lattice QCD results can be related to the dynamical coupled-channel analysis [116].

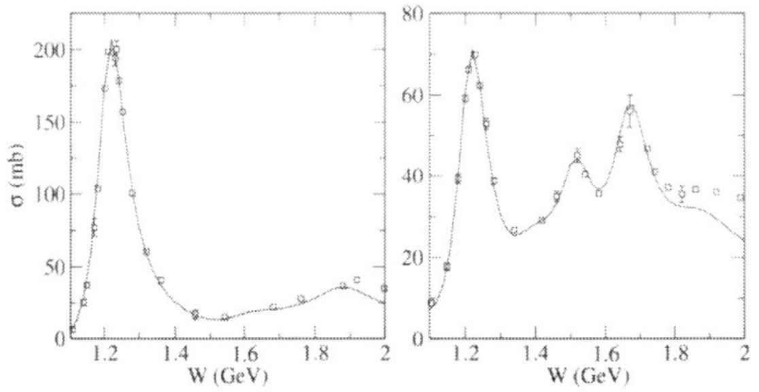

Figure 17. Total cross sections of π^+p (left) and π^-p (right) reactions. The curves are from the dynamical coupled-channel model [109, 111].

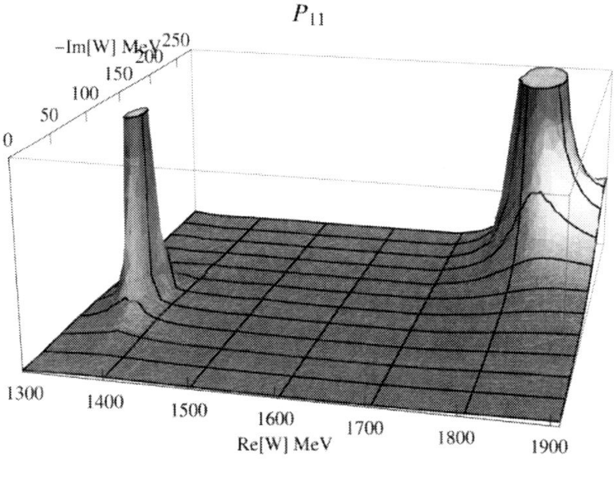

Figure 18. The P_{11} resonance poles extracted [114] from the dynamical coupled-channel model of Refs. [109, 111].

5. Future prospects of the N^* program at 6 GeV and with the 12 GeV upgrade.

The main part of the experimental program to search for predicted new baryon states still lies in the future. The polarized frozen spin target facility FROST is being installed in CLAS to complete the extensive program to measure double and triple polarization observables with protons that are polarized transversely to the production plane and to use linearly and circularly polarized photons. For the kaon-hyperon final states, the recoil polarization will be measured as well. Data taking with FROST will be followed with the equivalent program on polarized neutrons. This part will use the polarized HD target and provide complementary information on the isospin content of the reaction and of the excited state. This information is required to separate isospin 1/2 from isospin 3/2 excitations. Moreover, selection rules allow the photo-excitation of some states, *e.g.* $N(1675)D_{15}$, from neutrons, but not from protons.

With the 12 GeV upgrade, the N^* program at Jefferson Lab is being extended as an essential part of the comprehensive program of exclusive electroproduction measurements with the CLAS12 detector. A proposal for measuring π^0, η, and charged multi-pion final states in the unexplored domain of Q^2, from 5.0 to 12 GeV^2, has recently been approved by Jefferson Lab's Program Advisory Committee in 2009.

The proposed experiments have three objectives [117]. The first is to map out the quark structure of N^*s from the data on exclusive meson electroproduction reactions. In the considered $Q^2 > 5.0$ GeV^2 domain, the meson-baryon dressing is weak, as illustrated in Fig. 12 for the $\Delta(1232)P_{33}$, and hence the extracted $\gamma^* N \to N^*$ form factors can be used more directly to probe the quark substructure of N^*s.

The second objective is to investigate the dynamics of dressed quark interactions inside the nucleon core and to understand how these interactions emerge from QCD. We are motivated by the recent advance in developing hadron models based on the Dyson-Schwinger equations (DSE) of QCD [118, 119], as well by exploratory attempts to evaluate transition $\gamma^* N N^*$ from factors starting from the QCD Lagrangian within the framework of lattice QCD [91, 98, 120, 121].

The third objective is to study the Q^2 dependence of the nonperturbative dynamics of QCD. This is based on the recent investigation of the momentum dependence of the dressed quark mass of the quark propagator within lattice QCD [122] and DSE [123]. Our focus is on the important question of how baryon structure emerges from confinement and dynamical chiral symmetry breaking of QCD.

The proposed experiments are closely related to the GPD program at Jefferson Lab. The specification of exclusive reactions at high momentum transfer in terms of GPDs is a major goal of the CLAS12 upgrade. Experiments that are already approved as part of this program include deeply virtual Compton scattering and deeply virtual meson production.

References

[1] B. A. Mecking *et al.*, Nucl. Inst. Meth. A **503**, 513 (2003).
[2] C. A. Keith *et al.*, Nucl. Inst. Meth. A **501**, 327 (2003).
[3] Jefferson Lab Experiment E89-004, *Electromagnetic production of hyperons*, R. Schumacher *et al.*
[4] Jefferson Lab Experiment E91-008 *Photoproduction of η, η' mesons*, B. Ritchie *et al.*
[5] Jefferson Lab Experiment E94-008, *Photoproduction of η, η' mesons from deuterium*, B. Ritchie *et al.*
[6] Jefferson Lab Experiment E94-103, *The photoproduction of pions*, J. Ficenec *et al.*
[7] Jefferson Lab Experiment E99-013, *Photoproduction of omega mesons off protons with linearly polarized photons*, F. Klein *et al.*
[8] Jefferson Lab Experiment E02-112, *Search for missing resonances in the photoproduction of hyperons using a polarized photon beam and a polarized target*, F. Klein *et al.*
[9] Jefferson Lab Experiment E03-105, *Pion photoproduction from a polarized target*, S. Strauch *et al.*
[10] Jefferson Lab Experiment E05-012, *Measurements of polarization observables in eta-photoproduction with CLAS*, E. Pasyuk *et al.*
[11] Jefferson Lab Experiment E06-013, *Measurement of π^+, π^- photoproduction in double- polarization experiments using CLAS*, V. Crede *et al.*
[12] Jefferson Lab Experiment E06-103, *Kaon production in the deuteron using polarized photons*, P. Nadel-Turonski *et al.*
[13] Jefferson Lab Experiment E89-037, *Electroproduction of the $P_{33}(1232)$ resonance*, V. Burkert *et al.*
[14] Jefferson Lab Experiment E89-038, *Measurements of $p(e,e'\pi^+)n$, $p(e,e'p)\pi^0$, and $n(e,e'\pi^-)p$ in the second resonance region*, R. Minehart *et al.*
[15] Jefferson Lab Experiment E89-039, *Amplitudes for the $S_{11}(1535)$ and $P_{11}(1710)$ resonances from an $ep \to e'p\eta$ experiment*, S. Dytman *et al.*
[16] Jefferson Lab Experiment E89-042, *Measurement of the electron asymmetry in $p(e,e'\pi^+)n$ and $p(e,e'p)\pi^0$ in the mass region of the $P_{33}(1232)$*, V. Burkert *et al.*
[17] Jefferson Lab Experiment E89-043, *Measurements of the electroproduction of the $\gamma^*(1520)$ and $f_0(975)$ via the K^+k^-p and $K^+\pi^-p$ final states*, H. Funsten *et al.*

[18] Jefferson Lab Experiment E91-002, *The study of excited baryons at high momentum transfer with the CLAS spectrometer*, P. Stoler *et al.*
[19] Jefferson Lab Experiment E91-011, *Recoil polarization measurements for neutral pion electroproduction at $Q^2 = 1\ GeV^2$ near the Delta resonance*, R. Lourie *etQal.*
[20] Jefferson Lab Experiment E91-024, *Search for "missing" resonances in the electroproduction of ω mesons*, H. Funsten *et al.*
[21] Jefferson Lab Experiment E93-006, *Two pion decay of electroproduced baryon resonances*, M. Ripani *et al.*
[22] Jefferson Lab Experiment E93-036, *Measurement of of single pion electroproduction from the proton with polarized beam and polarized target using CLAS*, H. Weller *et al.*
[23] Jefferson Lab Experiment E94-003, *Study of the $\Delta(1232)$ using double polarization asymmetries*, P. Stoler *et al.*
[24] Jefferson Lab Experiment E94-014, *The $\Delta(1232)$ form factor at high momentum transfer*, P. Stoler *et al.*
[25] Jefferson Lab Experiment E99-006, *Polarization observables in the $p(e, e'K^{\pm})\Lambda, \Sigma$ reaction*, D. Carman *et al.*
[26] Jefferson Lab Experiment E99-107, *N^* excitations at high Q^2 in the $p\pi^0$, $n\pi^+$ channels*, V. Burkert, P. Stoler *et al.*
[27] Jefferson Lab Experiment E99-108, *N^* excitations at high Q^2 in the two-pion channel*, V. Burkert *et al.*
[28] Jefferson Lab Experiment E00-112, *Exclusive kaon electroproduction in Hall B at 6 GeV*, D. Carman *et al.*
[29] Jefferson Lab Experiment E01-002, *Baryon resonance electroproduction at high momentum transfer*, P. Stoler *et al.*
[30] M. Dugger *et al.*, Phys. Rev. C **76**, 025211 (2007).
[31] M. Dugger *et al.*, Phys. Rev. C **79**, 065206 (2009).
[32] M. Dugger *et al.*, Phys. Rev. Lett. **89**, 222002 (2002), *Erratum-ibid.* **89**, 249904 (2002).
[33] M. Dugger *et al.*, Phys. Rev. Lett. **96**, 062001 (2006); *Erratum-ibid.* **96**, 169905 (2006).
[34] J. W. C. McNabb *et al.*, Phys. Rev. C **69**, 042201 (2004).
[35] R. Bradford *et al.*, Phys. Rev. C **73**, 035202 (2006).
[36] C. Amsler *et al.*, Phys. Lett. B **667**, 1 (2008).
[37] G. Keaton and R. Workman, Phys. Rev. C **53**, 1434 (1996).
[38] R. Bradford *et al.*, Phys. Rev. C **75**, 035205 (2007).
[39] V.A. Nikonov *et al.*, Phys. Lett. B **662**, 245 (2008).
[40] E. Santopinto, Phys. Rev. C **72**, 022201 (2005).
[41] A. Lleres *et al.*, Eur. Phys. J. A **31**, 79 (2007).
[42] R. G. T. Zegers *et al.*, Phys. Rev. Lett. **91**, 092001 (2003).
[43] A. Ramos *et al.*, arXiv:nucl-th/0502053.
[44] V. Burkert, Int. J. Mod. Phys. A **21**, 1764 (2006).
[45] R. De Vita *et al.*, Phys. Rev. D **74**, 032001 (2006).
[46] M. Battaglieri *et al.*, Phys. Rev. Lett. **96**, 042001 (2006).
[47] B. McKinnon *et al.*, Phys. Rev. Lett. **96**, 212001 (2006).
[48] S. Niccolai *et al.*, Phys. Rev. Lett. **97**, 032001 (2006).
[49] V. Kubarovski *et al.*, Phys. Rev. Lett. **97**, 102001 (2006).
[50] K. Joo *et al.*, Phys. Rev. Lett. **88**, 122001 (2002).
[51] K. Joo *et al.*, Phys. Rev. C **68**, 032201 (2003).
[52] K. Joo *et al.*, Phys. Rev. C **70**, 042201 (2004).
[53] H. Egiyan *et al.*, Phys. Rev. C **73**, 025204 (2006).
[54] M. Ungaro *et al.*, Phys. Rev. Lett. **97**, 112003 (2006).
[55] L. C. Smith *et al.*, Proceedings of the Workshop *Shape of Hadrons*, p. 222, Athens (2006).
[56] K. Park *et al.*, Phys. Rev. C **77**, 015208 (2008).
[57] A. Biselli *et al.*, Phys. Rev. C **78**, 045204 (2008).
[58] G. F. Chew, M. L. Goldberger, F. E. Low, and Y. Nambu, Phys. Rev. **106**, 1345 (1957).
[59] S. Fubini, Y. Nambu, and V. Watagin, Phys. Rev. **111**, 329 (1958).
[60] I. G. Aznauryan, Phys. Rev. C **57**, 2727 (1998).
[61] I. G. Aznauryan, Phys. Rev. C **67**, 015209 (2003).
[62] I. G. Aznauryan *et al.*, Phys. Rev. C **78**, 045209 (2008).
[63] I. G. Aznauryan *et al.*, Phys. Rev. C **80**, 055203 (2009).
[64] D. Drechsel, O. Hanstein, S. Kamalov, and L. Tiator, Nucl. Phys. **A645**, 145 (1999).
[65] I. G. Aznauryan *et al.*, Phys. Rev. C **71**, 015201 (2005).
[66] I. G. Aznauryan *et al.*, Phys. Rev. C **72**, 045201 (2005).
[67] S. Capstick and W. Roberts, Prog. Part. Nucl. Phys. **45**, S241 (2000).
[68] S. Capstick and W. Roberts, Phys. Rev. D **49**, 4570 (1994).
[69] G. V. Fedotov *et al.*, Phys. Rev. C **79**, 015204 (2009).

[70] M. Ripani *et al.*, Phys. Rev. Lett. **91**, 022002 (2003).
[71] V. I. Mokeev *et al.*, Phys. Rev. C **80**, 045212 (2009).
[72] V. I. Mokeev *et al.*, Chin. Phys. C **33**, 1210 (2009).
[73] V. I. Mokeev, talk at *12th International Conference on Meson-Nucleon Physics and the Structure of the Nucleon*, May 31 -June 4 (2010), Williamsburg, VA.
[74] J. J. Kelly *et al.*, Phys. Rev. Lett. **95**, 102001 (2005).
[75] J. J. Kelly *et al.*, Phys. Rev. C **75**, 025201 (2007).
[76] V. V. Frolov *et al.*, Phys. Rev. Lett. **82**, 45 (1999).
[77] A.N. Villano *et al.*, Phys. Rev. C **80**, 035203 (2009).
[78] R. Thompson *et al.*, Phys. Rev. Lett. **86**, 1702 (2001).
[79] H. Denizli *et al.*, Phys. Rev. C **76**, 015204 (2007).
[80] C. S. Armstrong *et al.*, Phys. Rev. D **60**, 052004 (1999).
[81] M. M. Dalton *et al.*, Phys. Rev. C **80**, 015205 (2009).
[82] W. W. Ash, Phys. Lett. B **24**, 165 (1967).
[83] S. Stave *et al.*, Eur. Phys. J. A **30**, 471 (2006).
[84] N. F. Sparveris *et al.*, Phys. Lett. B **651**, 102 (2007).
[85] N. F. Sparveris *et al.*, Phys. Rev. Lett. **94**, 022003 (2005).
[86] T. Sato and T.-S. H. Lee, Phys. Rev. C **63**, 055201 (2001).
[87] V. Pascalutsa, M. Vanderhaeghen and S. N. Yang, Phys. Rep. **437**, 125 (2007).
[88] V. Pascalutsa and M. Vanderhaeghen, Phys. Rev. D **76**, 111501 (2007).
[89] S. Kamalov, S. N. Yang, D. Drechsel, O. Hanstein and L. Tiator, Phys. Rev. C **64**, 032201 (2001).
[90] A. Matsuyama, T. Sato and T.-S. H. Lee, Phys. Rep. **439**, 193 (2007).
[91] C. Alexandrou *et. al.*, Phys. Rev. D **77**, 085012 (2008).
[92] Z. P. Li, Phys. Rev. D **44**, 2841 (1991).
[93] Z. P. Li, V. Burkert and Zh. Li, Phys. Rev. D **46**, 70 (1992).
[94] O. Krehl *et al.*, Phys. Rev. C **62**, 025207 (2000).
[95] S. Capstick and B. D. Keister, Phys. Rev. D **51**, 3598 (1995).
[96] I. G. Aznauryan, Phys. Rev. C **76**, 025212 (2007).
[97] E. Pace, G. Salmé and S. Simula, Few Body Syst. Suppl. **10**, 407 (1999); E. Pace, G. Salmé, F. Cardarelli, and S. Simula, Nucl. Phys. **A666**, 33 (2000).
[98] V. M. Braun *et al.*, Phys. Rev. Lett. **103**, 072001 (2009).
[99] B. C. Liu and B. S. Zou, Phys. Rev. Lett. **96**, 142002 (2006).
[100] J. J. Xie, B. S. Zou and H. C. Chiang, Phys. Rev. C **77**, 015206 (2008).
[101] C. S. An and B. S. Zou, arXiv:0802.3996.
[102] N. Kaiser, P. B. Siegel and W. Weise, Phys. Lett. B **362**, 23 (1995).
[103] J. Nieves and E. Riuz Arriola, Phys. Rev. D **64**, 116008 (2001).
[104] T. Inoue, E. Oset and M. J. Vicente Vacas, Phys. Rev. C **65**, 035204 (2002).
[105] E. E. Kolomeitsev and M. F. M. Lutz, Phys. Lett. B **585**, 243 (2004).
[106] M. Warns, H. Schröder, W. Pfeil and H. Rollnik, Z.Phys. C **45**, 627 (1990).
[107] M. Aiello, M. M. Giannini and Santopinto, J. Phys. G **24**, 753 (1998).
[108] L. Tiator and M. Vanderhaeghen, Phys. Lett. B **672**, 344 (2009).
[109] B. Julia-Diaz, T.-S. H. Lee, A. Matsuyama and T. Sato, Phys. Rev. C **76**, 065201 (2007).
[110] J. Durand, B. Julia-Diaz, T.-S. H. Lee, B. Saghai and T. Sato, Phys. Rev. C **78**, 025204 (2008).
[111] H. Kamano, B. Julia-Diaz, T.-S. H. Lee, A. Matsuyama and T. Sato, Phys. Rev. C **79**, 025206 (2009).
[112] B. Julia-Diaz, T.-S. H. Lee, A. Matsuyama, T. Sato and L. C. Smith, Phys. Rev. C **77**, 045205 (2008).
[113] B. Julia-Diaz, H. Kamano, T.-S. H. Lee, A. Matsuyama, T. Sato and N. Suzuki, Phys. Rev. C **80**, 025207 (2009).
[114] N. Suzuki, T. Sato, and T.-S. H. Lee, Phys. Rev. C **79**, 025205 (2009).
[115] N. Suzuki, B. Julia-Diaz, H. Kamano, T.-S. H. Lee, A. Matsuyama and T. Sato, Phys. Rev. Lett. **104**, 042302 (2010).
[116] R. D. Young and T.-S. H. Lee, in preparation.
[117] I. Aznauryan *et al.*, arXiv:0707.1901.
[118] A. Höll *et. al.*, Phys. Rev. C **71**, 065204 (2005).
[119] G. Eichmann, R. Alkofer, I. C. Cloët, A. Krassnigg and C. D. Roberts, Phys. Rev. C **77** (2008) 042202(R).
[120] C. Alexandrou, arXiv:09064137.
[121] H. -W. Lin, S. D. Cohen, R. G. Edvards and D. G. Richards, Phys. Rev. D **78**, 114508 (2008).
[122] P. O. Bowman *et al.*, Phys. Rev. C **71**, 054507 (2005).
[123] M. S. Bhagwat, M. A. Pichowsky, C. D. Roberts and P. C. Tandy, Phys. Rev. C **68**, 015203 (2003).

Journal of Physics: Conference Series **299** (2011) 012009 doi:10.1088/1742-6596/299/1/012009

Transition to Perturbative QCD

R. Gilman[1,2], R.J. Holt[3] and P. Stoler[4]
[1]Rutgers University, Piscataway, New Jersey 08854, USA
[2]Jefferson Lab, Newport News, Virginia 23606, USA
[3]Argonne National Laboratory, Argonne, Illinois 60439, USA
[4]Rensselaer Polytechnic Institute, Troy, New York 12180, USA

Abstract. We review the Jefferson Lab program concerning the interplay between hadronic and underlying quark degrees of freedom in exclusive reactions. Much of the program was initially based on predictions from perturbative QCD (pQCD) concerning scaling of reaction cross sections, helicity conservation, and asymptotic behavior of form factors. Although much of the data do not follow simple pQCD expectations, some observables scale better than expected. Generally, the underlying dynamics are best understood with nonperturbative quark models, but the elastic deuteron form factors provide an example of the success of hadronic models to high momentum transfer.

1. Introduction

Determining the kinematic regime for transition from hadronic degrees of freedom to quark-gluon degrees of freedom has been a central goal of nuclear physics for the past several decades. More than 10 years ago, it was widely believed that observing cross sections that followed the constituent counting rules or measuring polarization observables consistent with hadron helicity conservation was sufficient to signal the transition region. The constituent counting rules, developed before quantum chromodynamics (QCD), are consistent with perturbative QCD (pQCD), but do not rely on being in the pQCD regime. In fact, it is now believed that the constituent counting rules could be a consequence of the Anti-deSitter Space/Conformal Field Theory (AdS/CFT) conjecture.

A number of observables are at least approximately consistent with constituent counting rules (CCR): the elastic form factor G_{Mp}, and cross sections for exclusive photopion production from the proton and neutron, two-body photodisintegration of the deuteron, and proton-proton elastic scattering. High-Q^2 electron-deuteron elastic scattering is consistent with an extension of the CCR, called reduced nuclear amplitudes (RNA) – but is also very well predicted by modern relativistic hadronic calculations. It is especially striking that reactions in simple nuclei obey the CCR for two reasons: (1) the nucleons in the deuteron are for the most part relatively far apart, whereas pQCD is a short-range process, and (2) there are a relatively large number of quarks in a simple nucleus, in principle requiring a large momentum transfer to each quark.

This agreement is likely misleading, since pQCD calculations have generally been unable to reproduce the magnitude of experimental cross sections or form factors. It appears that there are significant contributions of soft physics that increase cross sections while having little effect on the CCR behavior, although for nuclear reactions it has also been suggested that there are significant unaccounted contributions from hidden-color configurations in the wave function.[1]

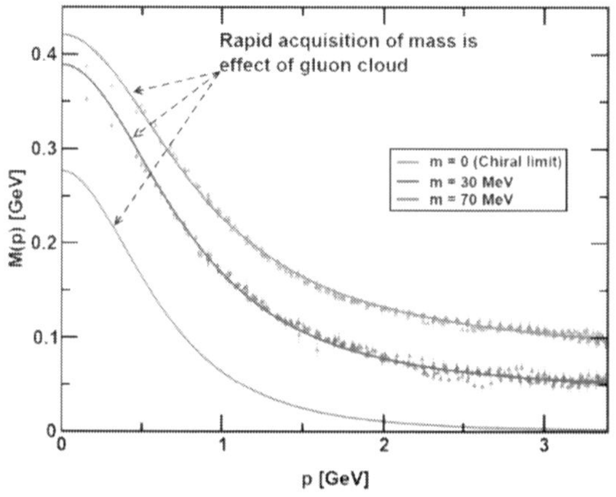

Figure 1. Dyson-Schwinger equation calculations of quark mass as a function of momentum compared to lattice QCD data [1].

Another way to think about the transition from hadronic degrees of freedom to partonic degrees of freedom is from the value of the quark momentum where the mass of the constituent quark becomes comparable with the mass of the current quark. The quark mass as a function of quark momentum has been calculated both from lattice gauge and Dyson-Schwinger approaches [2, 3]. The results are illustrated in Fig. 1. The curves represent three current quark masses, while the lattice calculations, of course, are performed for more massive quarks. From this figure, it appears that the quark momentum must be larger than about 2 GeV to begin to be in the perturbative regime, or more than 4 GeV^2 momentum transfer per quark. For the case of elastic scattering from the pion, one might expect perturbative QCD to become important at more than 16 GeV^2, well above the range of present data. One would require the 12 GeV upgrade at Jefferson Lab to perform this experiment up to a momentum transfer of only 6 GeV^2.

Furthermore, it now seems unlikely that hadron helicity conservation will be observed since it relies on quark helicity conservation and vanishing quark orbital angular momentum. We now know from a number of recent measurements, including proton form factors, the Sivers effect and Boer-Mulders asymmetries, that quark orbital angular momentum is ubiquitous. Studies of polarization in exclusive meson photoproduction and deuteron photodisintegration are also inconsistent with hadron-helicity conservation, supporting this idea, even though the cross sections in these reactions are consistent with the CCR.

Thus, as the approximate agreement of various processes with the constituent counting rules appears to be fortuitous, it remains a mystery why such good fortune is found in so many reactions. In the asymptotic pQCD limit, only the minimum *Fock state* valence quarks, exchanging the minimum number of gluons, participate in the reaction. The relative gluon-quark momenta are required to be high in all interaction vertexes so that the strong coupling constant α_S is relatively small. It seems instead that in the Jefferson Lab kinematic regime one in general has significant contributions from soft physics, but it also often appears that processes can be separated into soft and hard pQCD parts. For example, it has recently been shown for some reactions that the counting rule behavior is approximately reproduced in the generalized parton

[1] In hidden-color configurations, the deuteron is made up of two colored objects. These configurations can occur in the intermediate state in NN scattering, so theories with input from the measured NN force presumably already incorporate hidden-color effects.

distribution (GPD) / handbag picture, in which there is a hard perturbative scattering from a single quark in a hadron, with the momentum shared with other quarks through soft processes. The following sections summarize results of several of the more-studied reactions.

2. Hadron form factors

For mesons and baryons the leading pQCD form factor amplitudes are proportional to $1/Q^2$ and $1/Q^4$, respectively. In the following sections, we describe the pion, nucleon, and $N \to \Delta$ transition form factors.

2.1. Pion form factor

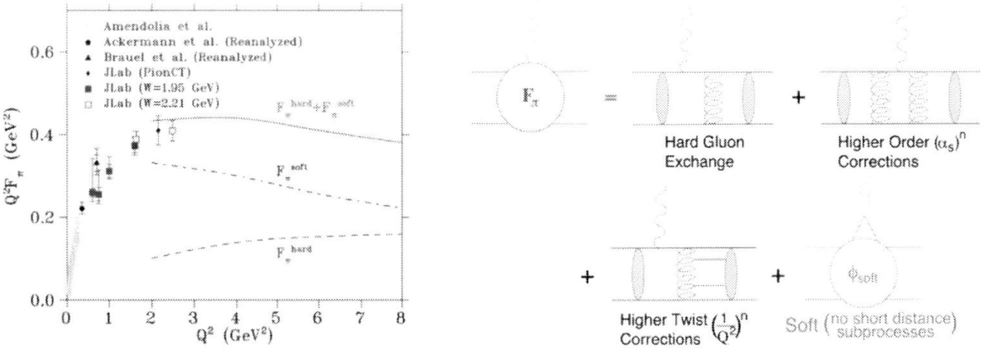

Figure 2. Left panel: Charged pion form factor. The dark and light filled squares are the result of the Hall C experiments [9, 10]. Right panel: Diagrams of lowest order pQCD and higher order hard processes which also contribute to the curves in the left panel.

Since the pion's valence quark structure consists of a single quark and antiquark, the lowest order pQCD mechanism requires only the exchange of a single gluon, with a high enough momentum relative to each of the quarks so that both quark-gluon vertices can be treated pertubatively. After the development of the SVZ QCD sum rules [4] there was considerable theoretical activity to obtain valence quark distributions and, from them, the leading order hadron form factors using pQCD [5]. Enthusiasm was kindled by the apparent agreement obtained with earlier measurements of the π^+ form factor at CESR in Cornell. However, this immediately became controversial. On the theoretical side it was pointed out that the $q\bar{q}$ valence quark distribution obtained by the sum rule method has peaks near the two physical limits of $x = 1$ and 0, implying that the active struck quark has either nearly all or none of the meson momentum. This situation leads to a large contribution from nonperturbative processes. The more physically reasonable asymptotic distribution, which peaks at $x = 1/2$, yields a much smaller F_{π^+} than experiment. On the experimental side several difficulties were pointed out. Since a free pion target does not exist, attempts to obtain F_{π^+} rely on the assumption that the proton sometimes consists of a neutron plus π^+, both of which are off-shell. Thus, experiments must measure cross sections at momentun transfer $t < 0$, separate out the longitudinal part of the cross section, to isolate the part of the cross section corresponding to directly knocking out the pion, and then extrapolate the results to the free pion pole, $t = m_\pi^2$, the kinematic point corresponding to the physical π mass, with an unbound $n\pi^+$ system. Model uncertainties and the very limited quality of the early π^+ data did not permit this.

Jefferson Lab undertook a major program in Hall C to obtain F_{π^+}. High quality data enabled clean extractions of the longitudinal cross sections using the Rosenbluth technique.

Improved models allowed the determination of the t-channel and non-t-channel contributions, and the magnitude of the pion form factor. Very precise measurements of F_{π^+} have been obtained for Q^2 up to 2 GeV^2, as shown in Fig. 2, along with the results of several calculations. The lowest order pQCD calculation using the asymptotic distribution function underestimates the result, but still contributes a significant fraction. Incrementally adding next-to-leading-order contributions involving more than the minimal gluon exchanges, and also $k_\perp$ components, incrementally reaches the experimental F_{π^+}. One may conclude [6] that although leading order pQCD cannot account for the form factor, it can be mostly accounted for by hard processes.[2] This is an important collateral finding for the eventual application of meson production for the study of GPDs, which requires hard perturbative treatments at the "handles" of the handbag. Also, dominance of the longitudinal cross section is essential for treating pion production in the framework of GPDs.

2.2. Nucleon elastic form factors

The hadronic current for nucleon elastic scattering is

$$J_\mu = \bar{u}(p_2)\left[\gamma_\mu F_1 + i\frac{\sigma_{\mu\nu}q^\nu}{2m}F_2\right]u(p_1) \tag{1}$$

where the Dirac form factor F_1 is the helicity conserving current distribution and the Pauli form factor F_2 is the helicity non-conserving current distribution. A linear combination of these form factors, the electric and magnetic form factors G_E and G_M, leads to simpler expressions for the cross section and polarization observables. The major finding of Jefferson Lab, discussed more elsewhere in this volume, is that the ratio G_E/G_M falls about linearly with Q^2, rather than being about constant.

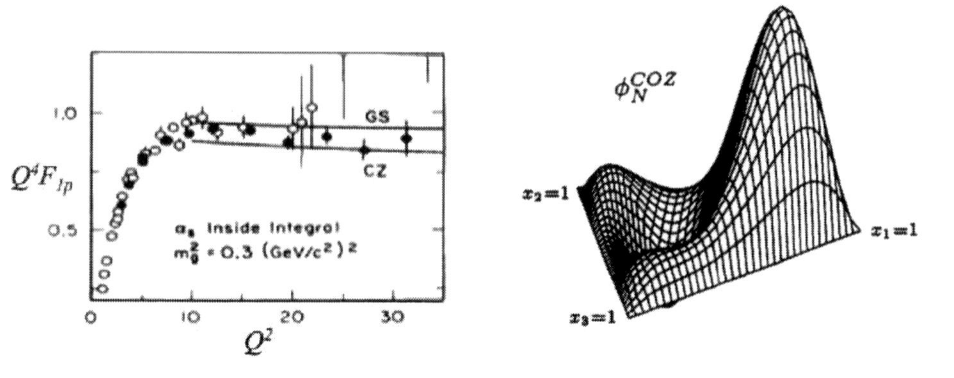

Figure 3. Left panel: Elastic proton helicity conserving form factor F_{1P}. Right panel: Valence quark distribution function $\Phi_{CZ}(x)$ obtained by Ref. [5] using SVZ sum rules.

2.2.1. Proton helicity conserving form factor F_{1p} By far the highest Q^2 exclusive data comes from the measurement at SLAC [11] of elastic ep scattering up to 36 GeV^2. Very high Q^2 is accessible by single arm electron scattering because elastic scattering is isolated in W. However,

[2] Of related interest is the $\pi^0\gamma^*\gamma$ form factor. It was long believed that this form factor was basically perturbative over the measured range, until this was called into question by recent BABAR data [7] that greatly exceed the perturbative prediction for $Q^2 > 15$ GeV^2. Objections exist to the correctness of this result [8], and the matter remains to be sorted out.

a single ep elastic cross section measurement cannot separate F_{1p} and F_{2p}. Figure 3 shows the result for F_{1p} assuming $G_{Ep} = G_{Mp}/\mu_p$,[3] and the results of a lowest order pQCD calculation [12], with a valence quark distribution $\Phi_{CZ}(x)$ obtained by the method of Ref. [5] using the SVZ sum rule, and a "running" strong coupling constant α_S. As in the case of the π^+ calculation, this distribution peaks where there is a large probability for the struck quark to have most of the proton momentum. If one repeats the calculation with the asymptotic distribution function, which peaks at values of x where all three quarks have an equal momentum fraction, the resulting curve accounts for a small fraction of the form factor. A more sophisticated formalism has been developed to include higher-order perturbative QCD resummation, or Sudakov form factor [13], which attempts to account for "strong radiative" effects.

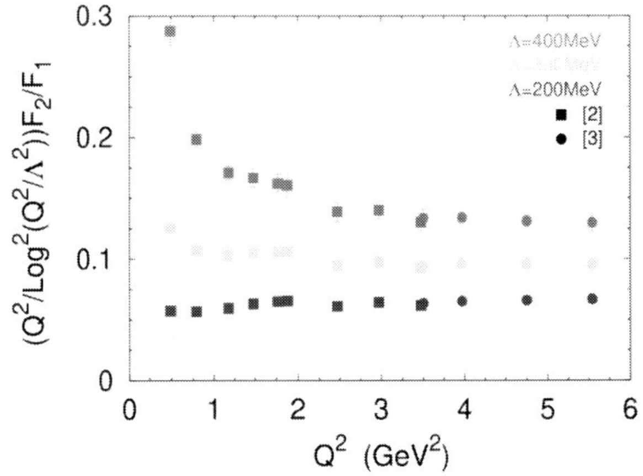

Figure 4. The Jefferson Lab Pauli form factor, plotted as $Q^2\log^2(Q^2/\Lambda^2)\,(F_{2P}/F_{1p})$ [15].

2.2.2. Proton helicity non conserving form factor F_{2p} The Pauli form factor F_{2p} is intrinsically helicity non-conserving and therefore would seem to violate one of the tenets of pQCD. More precisely, neglecting orbital angular momentum, pQCD predicts $F_{2p}/F_{1p} \propto 1/Q^2$. A major experimental finding of the Jefferson Lab program is that $F_{2p}/F_{1p} \propto 1/Q$ instead [14]. Reference [15] has pointed out that a pQCD framework can be valid in the case where overall helicity is not conserved but a spin flip comes from the inclusion of orbital angular momentum in the treatment of the form factor, in such a way that the helicity conservation can be retained at each vertex, while the orbital angular momentum of the initial and final state nucleons changes from $l = 1 \rightarrow l = 0$, or vice-versa. Indeed, a generalized power counting law can then be derived which agrees with the experimentally observed scaling. The assumption that the quarks have one unit of angular momentum in the initial state introduces a transverse term into the transition amplitude, schematically $T(x,y) \rightarrow T(x,k_{x\perp},y)$, as well as to the quark distribution amplitude for the initial state, schematically $\Psi(x) \rightarrow \Psi(x,k_{x\perp})$. Each $k_\perp$ adds a factor of $1/Q$ to the overall form factor compared to the leading order form factor. Finally, in order to avoid singularities coming from the regions of very small x, the Sudakov resummation is applied which basically cuts of the integral at $x \sim \Lambda^2/Q^2$, and adds a logarithmic factor to the form factor,

[3] The well known data obtained by Jefferson Lab during the past several years shows that this is not the case. Instead G_E^p/G_M^p decreases about linearly with Q^2, up to $Q^2 = 8.5$ GeV^2. The overall effect of a continued linear fall of G_E^p/G_M^p on the extraction of F_1^p is rather small at higher Q^2 and does not materially affect the conclusion of Fig. 3.

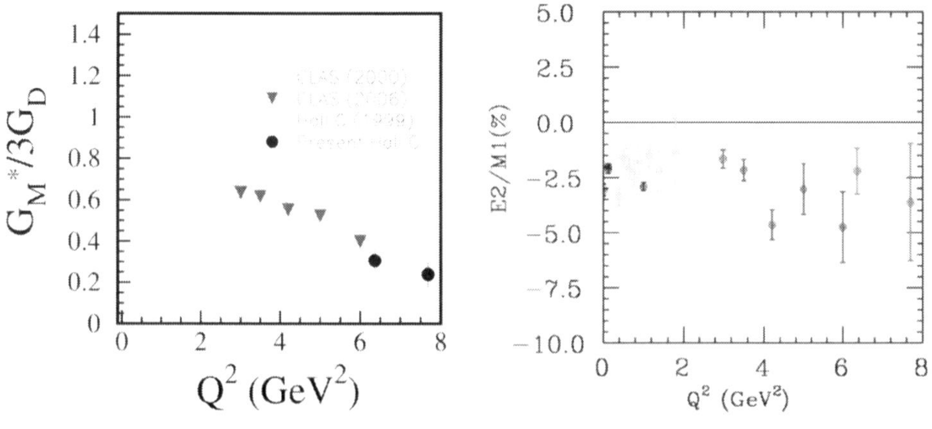

Figure 5. Results of Jefferson Lab measurements of the $p \to \Delta(1232)$ amplitudes. Left panel: G^*_M/G_D, from Ref. [17]. Right panel: R_{EM}. The open circles and dark solid points are Hall C data [16, 17]. The rest of the points are from Hall B [18, 19].

i.e. $F_{2p} \propto log(\Lambda^2/Q^2)/Q^6$. A very good fit to the recent Jefferson Lab data is obtained with $\Lambda \sim 400$ MeV.

2.3. Nucleon resonances

The lowest lying nucleon excited state is the $\Delta(1232)$, which has both spin and isospin quantum numbers 3/2. The Δ decays almost exclusively into a *p*-wave $N - \pi$ final state. It is relatively isolated from other resonances. It is also very strongly excited, and almost completely saturates the unitary circle in a pion scattering Argand plot. The $N \to \Delta$ transition has three electromagnetic multipoles — $M1$, $E2$ and $S1$, which are denoted *magnetic dipole, electric quadrupole* and *scalar dipole*, respectively. These are directly related to the CGLN [23] multipoles M_{1+}, E_{1+}, and S_{1+}, where the subscript 1+ denotes that the decay meson has an orbital angular momentum $l = 1$ and the total spin is $J = 1 + 1/2 = 3/2$.

For real photons (Q^2=0) the Δ is nearly a pure $M1$ excitation, and $R_{EM} \equiv E_{1+}/M_{1+}$ is very small (≈-2.5%). Early on this was explained in the framework of the SU6 CQM as a magnetic spin-flip excitation of one of the nucleon's quarks. At the asymptotic limit, $Q^2 \to \infty$, pQCD predicts the leading form factors should scale, just as in the elastic case, as $1/Q^4$. In addition, helicity conservation should be respected. For the $N \to \Delta$ the helicity amplitudes are related to the multipoles as $A_{1+} = 3E_{1+} + M_{1+}$ and $B_{l+} = 2(E_{1+} - M_{1+})$ where A_{1+} and B_{l+} are helicity conserving ($\Delta\Lambda$=0) and helicity non-conserving ($\Delta\Lambda$=2) amplitudes, respectively. Thus, near threshold, with $R_{EM} \sim 0$ one has $B_{l+} \sim -2A_{1+}$. However, for helicity to be conserved $B_{l+}/A_{l+} = 0$, so $E_{1+} = M_{1+}$ and $R_{EM} = +1$. Thus, the experimental result indicates that the helicity non-conserving amplitude is comparable to the helicity conserving amplitude, contradicting the requirements of pQCD.

The $N \to \Delta$ helicity conserving form factor can be defined in terms of hadronic currents as in Eq. (1). Since the Δ has spin 3/2 there are three form factors G^*_M, G^*_E and G^*_S. G^*_M, which is analogous to F_1, should also scale as $1/Q^4$. Earlier analysis of inclusive electron scattering data at SLAC [20, 21] suggested that the $p \to \Delta$ form factor is decreasing with Q^2 at a slope steeper than $1/Q^4$.

A major program was undertaken at Jefferson Lab to measure the details of the $p \to \Delta$ resonance over as large a range of Q^2 as possible to determine the evolution of the amplitudes from what is expected for a CQM toward what is expected from pQCD. Among the specific physics questions were whether G^*_M continues to fall anomalously fast as a function of Q^2, or begins to approach the scaling behavior, and whether R_{EM} remains very small and negative, or begins to turn positive, and asymptotically begins to approach +1. In addition to measurements of a few observables spanning a wide range of Q^2, one Hall A experiment [22] measured a complete set of recoil polarization observables to thoroughly determine amplitudes at a single $Q^2 = 1$ GeV^2.

Figure 5 indicates that neither G^*_M nor R_{EM} in the measured Q^2 region show indications of the behavior expected form pQCD. In the measured range, $R_{EM} \sim 0$ and G^*_M falls faster than the scaling prediction. Thus, the two primary signatures of the approach to pQCD are violated, up to $Q^2 \approx 8$ GeV^2, corresponding to a virtual photon wavelength that varies from near 1 fm down to about 1/20 fm.

Some theoretical arguments have been put forth to explain the fast falloff of G^*_M. Using the QCD sum rule approach, Ref. [24] argued that the leading $N \to \Delta$ form factor should be suppressed. The nucleon has a mixed symmetry valence quark distribution, $\Phi_{pS}+\Phi_{pA}$, while that of the Δ is purely symmetric $\Phi_{\Delta S}$. It was then found using the QCD sum rule technique that in the $P \to \Delta$ form factor $\langle\Phi_{pS}|T|\Phi_{\Delta S}\rangle$ and $\langle\Phi_{pA}|T|\Phi_{\Delta S}\rangle$ have opposite signs, and tend to cancel. Thus the G^*_M would be dominated by the sub-leading helicity non-conserving contribution, which should then decrease relative to the helicity conserving form factor, as observed. At some value in Q^2 the helicity non-conserving contribution would become small, and $G^*_M \to 1/Q^4$.

In another interesting case the transition amplitudes to the $S_{11}(1535)$, which has $J^\pi = 1/2^-$, were also measured in Hall B [25, 26] and Hall C [27] by observing the $p\eta$ decay channel. Since $J = 1/2$, there is an antisymmetric as well as symmetric distribution function (as for the proton). Ref. [24] finds no cancellation of various terms in the amplitude. Experimentally the helicity conserving $N \to S_{11}$ transition remains very strongly excited, at all Q^2, and shows signs of beginning to approach $1/Q^4$ at the highest Q^2 regions [20, 27].

3. Real compton scattering

Real Compton Scattering (RCS) probes the nucleon with incoming and outgoing real photons. As compared to pion photoproduction, there is a simpler final state, making it more likely that the GPD handbag mechanism describes RCS, and a slower expected energy dependence from the CCR, $ds/dt \sim s^{-6}$ *vs.* s^{-7}. But the cross sections are smaller and the experiment more difficult, as one must isolate the final-state photon from a much larger background of π^0 decay photons.

Cornell [33] found that cross sections at large c.m. angles approximately followed the CCR, for beam energies above about 3 GeV, though with poor statistical precision. The Jefferson Lab Hall A RCS experiment dramatically improved the quality of cross section data in the few GeV energy region, and also provided the first polarization point.[4] The cross section [34] shows that $ds/dt \sim s^{-8}$, a faster fall off than predicted by the CCR, even somewhat faster than predicted by Radyushkin [35] in a GPD approach. The longitudinal polarization transfer [28] shown in Fig. 6 indicates that the dynamics are consistent with predictions using the handbag approach, rather than pQCD.[5] Thus, RCS results to date support the handbag picture of underlying quark dynamics rather than the pQCD picture.

4. Pion photoproduction

Pion photoproduction would appear to be an ideal experiment to investigate the onset of perturbative physics, since it is known at low energies to be strongly influced by resonance

[4] A follow up experiment, 07-002, has taken additional polarization data in Hall C.

[5] It remains an unexplained oddity why, when polarization transfer on a quark is expected to follow the *KN* formula, the sum over all possible diagrams in the pQCD approach tends to give a result of opposite sign.

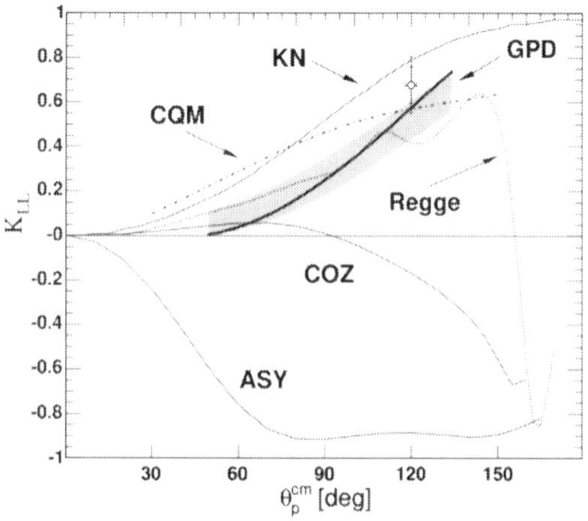

Figure 6. Longitudinal polarization transfer in RCS for $E_\gamma = 3.23$ GeV [28]. The Klein-Nishina (polarization transfer to a point-like object) prediction is indicated by "KN". The gray band indicates the handbag approach using GPDs [29]. A constituent quark model calculation [30] in the handbag approach is labelled "CQM". A Regge-exchange calculation [31] is indicated by "Regge". The labels "COZ" and "ASY" are for pQCD calculations [32] using the asymptotic (ASY) or Chernyak-Ogloblin-Zhitnitsky (COZ) distribution amplitudes.

production, but at high energies to show cross section scaling. Jefferson Lab has obtained precise cross sections [36] at beam energies significantly above 1 GeV; a small sample of the data, shown in Fig. 7, indicate that even in the scaling region the cross section does not exactly follow the scaling rules, but appears to oscillate. In the more studied case of pp elastic scattering, such oscillations have been attributed to either resonance (charm) thresholds, or to interference between long (Landshoff) and short (pQCD) range mechanisms. Similarly, recoil polarizations in $\gamma p \to p\pi^0$ show strong energy and angle dependences even above the resonance region [37]. In contrast to this, the $\gamma p \to K^+\Lambda^0$ reactions produces Λ's polarized in the direction of the photon spin [38], a simple behavior but one that is inconsistent with helicity conservation.

5. Few-body nuclei

Studies of the deuteron, particularly elastic scattering and photodisintegration, have been a primary source of information on the transition to pQCD in nuclei [39]. Ideally, as one probes the nucleus with increasing energy and momentum transfer, cross sections and polarization observables undergo a transition in their behavior, akin to a phase transition, that clearly signals the transition from a low energy, low momentum transfer hadronic regime to a high energy, high momentum transfer quark-gluon regime. Even with no clear signal in the data, one expects a transition in the ability to formulate a description in quark vs. hadronic degrees of freedom.

Observing the transition in exclusive reactions has been more problematic. Reactions such as quasi-free $d,^3$He$(e,e'p)$ and ed elastic scattering, measured to large momentum transfers, are generally well understood with hadronic theories [40, 41, 42, 47, 48, 49]. In inclusive reactions, one needs both large energy - more specifically large center of mass energy W - and large momentum transfer Q^2 to reach the deep inelastic regime. Thus, it makes sense that quasi-free scattering, with $W_{\gamma^* p} \approx m_p$, and elastic scattering, with $W_{\gamma^* d} = m_d$ for the deuteron as a whole, but $W^2_{\gamma^* p} < m_p^2$ for the struck nucleon, show little if any indication of quark behavior. Whatever the quark effects are, they are already largely incorporated into the hadronic theory.

To obtain high energy and momentum transfer, we turn to photodisintegration of the

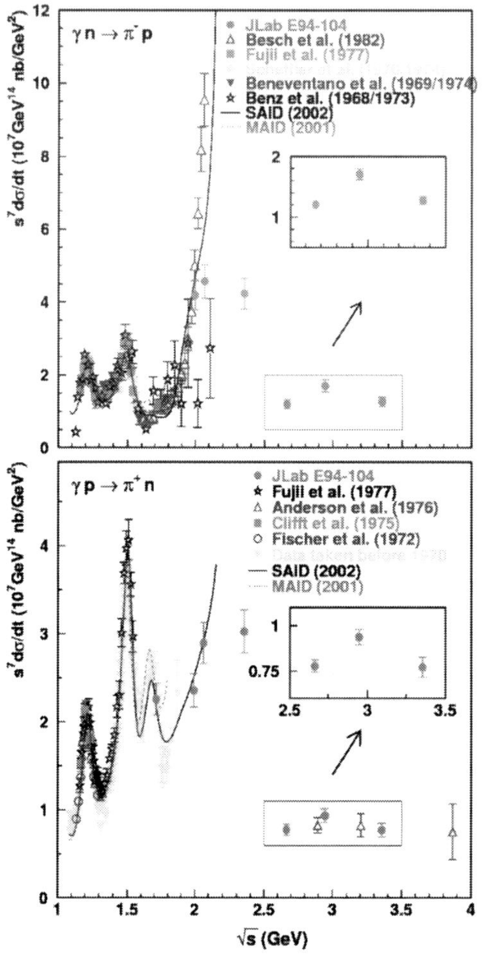

Figure 7. Comparison, taken from [36], of charged pion photoproduction from the nucleon, $\gamma p \to \pi^+ n$ and $\gamma n \to \pi^- p$, at $\theta_{c.m.} = 90°$.

deuteron. Above $E_\gamma = 1$ GeV, with $W^2 = s = 2E_\gamma m_d + m_d^2$, W is always above 2.7 GeV – for a proton target, the 1 GeV photon leads to $W > 1.66$ GeV. Also, at $\theta_{c.m.} = 90°$, we have a four momentum transfer $-t > 1$ GeV^2; $-t$ in real photo-reactions corresponds to Q^2 in virtual photon transfer. Thus, the kinematics for photodisintegration above 1 GeV roughly correspond to the kinematics for inclusive deep inelastic scattering.

5.1. Few-body nuclear form factors

Knowledge of *ed* elastic scattering has been vastly improved by experiments at Jefferson Lab [41, 42]; data have also been taken for the ^{3,4}He form factors. [50] The deuteron form factor should fall as Q^{10} at high momentum transfer, while the $A = 3$ ($A = 4$) form factors should fall as Q^{16} (Q^{22}). The cross section data [41] allow determination of the structure function $A(Q^2)$, a combination of the three form factors, which shows hints of possible scaling at the highest momentum transfers, as shown in Fig. 8.

The most complete study of the deuteron uses forward-angle cross sections for $A(Q^2)$, backward angle cross sections to determine the magnetic structure function $B(Q^2)$, and

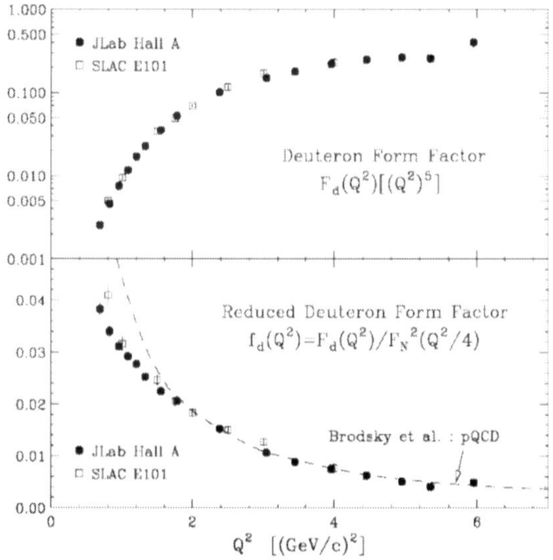

Figure 8. Top panel: Deuteron form factor times Q^{10}. The expected scaling behavior leads to $Q^{10}F_d$ being constant. Bottom panel: Deuteron form factor divided by nucleon form factor squared. The reduced nuclear amplitudes extension of pQCD predicts that the reduced form factor follows the dash line, which it does for $Q^2 \geq 2$ GeV^2 (taken from [41]).

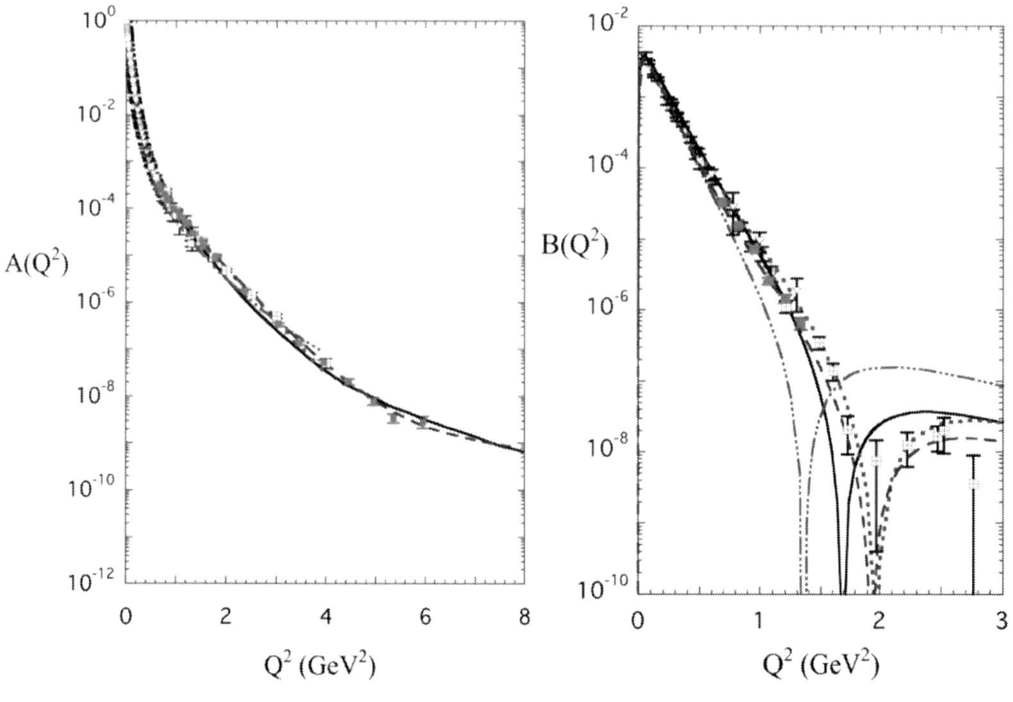

Figure 9. Relativistic hadronic calculations compared to *ed* scattering data for the structure functions *A* and *B*. Curves are from [43] (green dot dash), [44] (black solid), [45] (blue dash), and [46] (brown dot). Updated from [39].

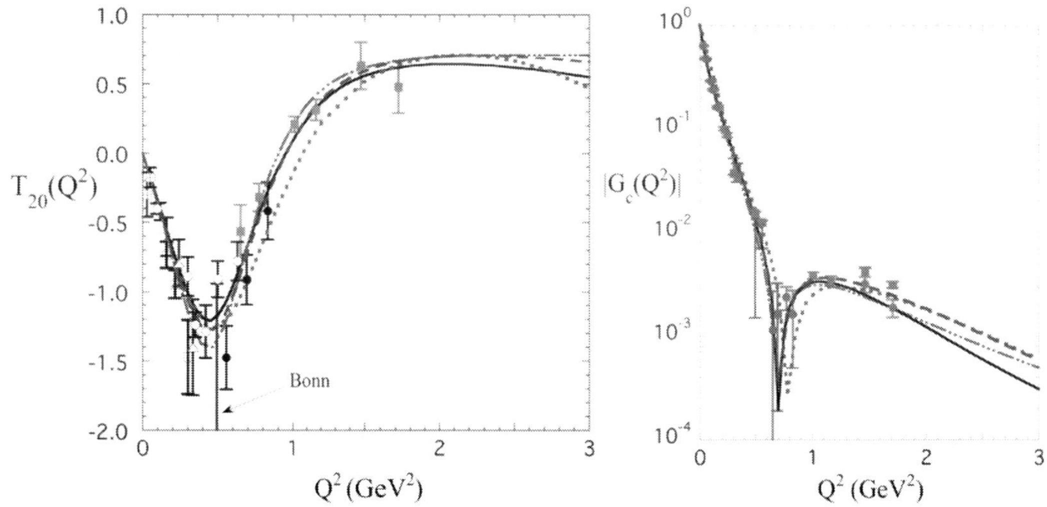

Figure 10. Relativistic hadronic calculations compared to *ed* scattering data for the tensor polarization t_{20} and the charge form factor G_C, extracted from A, B and t_{20}. More precisely, B only depends on the magnetic form factor, which is needed to extract G_C from A and t_{20}. Curves are from [43] (green dot dash), [44] (black solid), [45] (blue dash), and [46] (brown dot). Updated from [39].

polarization data [42] – or the individual form factors G_E, G_M, and G_Q extracted from the observables – as these are independent functions with differing sensitivity to the ingredients of the theory. Figures 9 and 10 show the results from four models selected because they all give a good account of the data for A and T_{20} (and hence also for G_C) up to $Q^2 \simeq 3$ GeV^2. One of these is an early calculation based on the quark compound bag model which uses hadronic degrees of freedom to describe the long-range physics [46]; the other three are models using only hadronic degrees of freedom. Two of these [44, 45] provide an excellent description of A to the highest Q^2, and illustrate the success of modern relativistic hadronic calculations at providing a good account of the data across the entire Q^2 range. It is the magnetic structure function B that is most difficult for theory to predict, but the minimum in B is also poorly determined by a few low statistics points, and most in need of an improved experiment. The essential ingredients in the best calculations are a complete treatment of relativistic effects along with an NN force based on data.

The hadronic calculations, particularly [44, 45], have superior predictive power, and are the preferred explanation in view of the fact that the pQCD estimates fail to predict the strength of the form factors by a huge factor, or the minimum in B. It seems that the hints of possible scaling shown in Fig. 8 are perhaps misleading and fortuitous, or reflect some important physics as yet not understood about the pQCD calculations. It is known that at the lowest momentum transfers measured, chiral perturbation theory, based on solving QCD with nucleons and pions, provides an excellent description of the data [51].

5.2. High-energy deuteron photodisintegration

Experiments [53, 54, 55, 56, 57, 58] at SLAC and Jefferson Lab have shown that the photodisintegration cross section for $E_\gamma > 1$ GeV and $p_T > 1.3$ GeV roughly follow the constituent counting prediction [59], $d\sigma/dt \propto s^{-11}$ at constant center of mass angle, with s the square of the center of mass energy; see Fig. 11. The behavior is amazingly good [60]; the

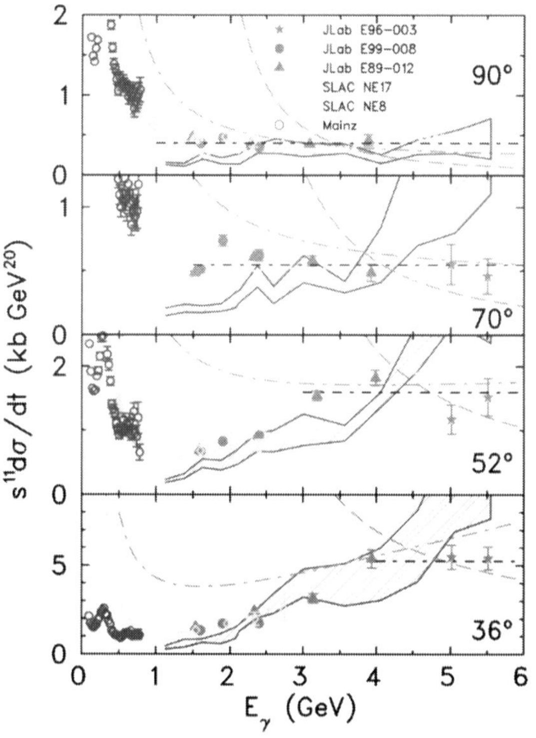

Figure 11. Scaling of the deuteron photodisintegration cross section is indicated by the flattening of the cross section at high energies. The extensive CLAS data set [58] is not shown (taken from [39]).

cross section has been measured to fall about a factor of $\approx$30,000 with an s^{-11} dependence, from 1 to 4 GeV at $\theta_{c.m.} = 90°$. This can be taken as an indication that quarks might be the appropriate underlying degrees of freedom, but it is not conclusive; simple models with hadronic degrees of freedom can lead to roughly the same behavior.

There is no satisfactory hadronic theory nor is there a prospect of a good conventional hadronic theory above 1 GeV photon energy [39]. There are too many possible resonance channels with poorly known amplitudes to have a reliable prediction. But there are no clean indications of any resonances in the data, and quark models, which automatically average over the resonances, are possible. Most existing quark models are based on the idea that the incoming photon is absorbed on a pair of quarks being exchanged between the two nucleons in the deuteron. Some models do approximate evaluations of this diagram [61, 62]; others evaluate it by relating the photodisintegration to nucleon-nucleon scattering [63, 64]. Finally, the quark-gluon string model [65] is based on the idea that the dominance of planar diagrams in QCD leads to the dominance of 3-quark exchange in photodisintegration, and evaluates the reaction using Regge phenomenology.

These models, and the underlying reaction dynamics, can be tested further with polarization observables. Only three high-energy experiments have been done. The linearly polarized photon asymmetry, Σ, was measured at Yerevan [66], up to 1.6 GeV at $\theta_{c.m.} = 90°$. The data show that above 1 GeV Σ is moderately sized, 0.2 or 0.3, and positive; the highest energy point indicates Σ might be increasing towards 1, but it is also only 1σ above 0.3. While it was previously believed that perturbative QCD leads to hadron helicity conservation [67], which leads to $\Sigma \to -1$, it is now understood that both chains of this argument have problems. Orbital angular momentum can lead to interesting spin effects, preventing hadron helicity conservation [68]. The limit of -1 relies on the photo-coupling being isoscalar; $\Sigma \rightarrow 1$ if the photo-coupling is isovector [69]. The data and its interpretation can be dramatically and easily improved by a more extensive measurement of the Σ asymmetry in Jefferson Lab Hall B.

Recoil proton polarizations have been measured in two high-energy experiments in Jefferson Lab Hall A. E89-019 [70] found that at $\theta_{c.m.} = 90°$ the induced polarization steadily decreases in magnitude, going from near -1 at 500 MeV to essentially 0 by 1 GeV. This contradicts the expected hadronic behavior, in which resonances lead to large, strongly energy-dependent induced polarizations. The transferred polarizations are moderate in size near 1 GeV, but appear to steadily decrease with energy at higher energies. The ensuing experiment, E00-007 [71], measured a 5-point angular distribution for $E_\gamma \approx 2$ GeV, from 37° to 110°. The polarizations have a smooth dependence with angle. The induced polarization p_y and transferred polarization $C_{x'}$ (transverse in the reaction plane) start out negative and moderately sized, but cross zero near $\theta_{c.m.} = 90°$ and are positive at larger angles. The transferred longitudinal polarization $C_{z'}$ starts out large and positive, decreasing to be zero near $\theta_{c.m.} = 110°$. From the hard rescattering model [72], if isovector photon coupling dominates, then the NN amplitude ϕ_5 dominates, which drives p_y and $C_{x'}$ to zero at 90° – these are each proportional to ϕ_5 multiplied by a sum of other amplitudes. It is interesting that the polarization observables Σ, p_y, and $C_{x'}$ all seem to indicate a dominantly isovector coupling.

5.3. Low-energy deuteron photodisintegration

Hadronic models have been much improved over the last several years, particularly due to the work of Schwamb and Arenhövel [73]. While various problems have been resolved, there remain some prominent disagreements, most notably as the energy increases above 300 MeV, the Δ resonance region, in the induced polarization p_y, which led to much discussion of exotic dibaryons in deuteron photodisintegration back in the 1970s and 1980s.[6] Two recent experiments have probed this region in more detail, to perhaps provide clues why the best hadronic calculations start to fail.

A Novosibirsk experiment [74] measured tensor polarizations for energies from 25 to 440 MeV. Generally, below pion production threshold there is little model dependence to the calculations, and there is an excellent prediction of the data. At energies near the Δ resonance, the most modern calculations of Schamb and Arenhövel improve the description of T_{20} and T_{22}, but hurt the description of T_{21}; it is not clear whether the discrepancy arises from uncertainties in the calculation or missing underlying dynamics.

Experiment 05-103 in Jefferson Lab Hall A [75] measured the recoil proton polarizations for energies from 280 – 360 MeV and angles from 20° – 110°. This is the energy region in which p_y starts to grow in magnitude, contradicting calculations; even the modern Schwamb and Arenhövel calculations show that the induced polarization tends to be small near $\theta_{c.m.} \approx 90°$.[7] The aim of the experiment was to provide a systematic set of polarization data to try to help identify the missing reaction dynamics. The experimental results clearly show the growth of the induced polarization with energy at 90°. For the polarization transfer observables, there is qualitative agreement but subtle quantitative differences between the data and the modern theory calculations. Generally, the older more phenomenological calculations of Schwamb and Arenhövel are in slightly better agreement with the data than the newer less phenomenological

[6] Another still unresolved notable problem from the same period is the induced neutron polarization at low energy.

[7] In the 1970s and early 1980s, this difference led to much consideration of dibaryons in deuteron photodisintegration, leading to much of the existing polarization data; while this explanation was eventually discarded, the missing reaction dynamics has yet to be identified.

calculations of Schwamb.

5.4. High-energy photodisintegration of ^3He

The underlying mechanisms of pn or deuteron photodisintegration can be illuminated by comparison with pp photodisintegration. Since there is no pp bound state, it is most natural to use ^{3}He, as with only one undetected neutron the reaction kinematics can be completely reconstructed and final state interactions are minimized. At low energies, pp disintegration is known to be small, which is understood to result from the two protons being largely in an $l = 0$ s state, coupled to total spin 0, with no net magnetic moment. At high energies, different ideas about the underlying reaction dynamics lead to a range of predictions for the pp disintegration cross section, from much larger than deuteron photodisintegration to much smaller [76, 77].

One important point in the photodisintegration of ^{3}He is that one can calculate the light cone momentum fraction of the undetected neutron. In general, $\alpha = (E - p_z)/m$; since α is a conserved quantity, we calculate $\alpha_n = \alpha_\gamma + \alpha_{^3He} - \alpha_{p_1} - \alpha_{p_2} = 0 + 3 - \alpha_{p_1} - \alpha_{p_2}$. The interesting thing is that α_n provides an independent check of whether the pp photodisintegration is long or short range; a short range process leads to a broader, flatter distribution centered near $\alpha_n = 1$. In addition, the asymmetry of this α_n distribution reflects how fast the cross section is falling with s. Thus, pp photodisintegration potentially has more information about the underlying dynamics than does deuteron photodisintegration.

An additional observation is that pp elastic scattering has prominent oscillations about the scaling prediction of energy dependence; in pn elastic scattering, oscillations are not so apparent. Thus, models which relate photodisintegration to NN scattering predict oscillations in the photodisintegration, compared to the scaling energy dependence.

An initial study of the disintegration of ^{3}He into a pp pair and n spectator [77] from Hall A at $\theta_{c.m.} = 90°$ and $E_\gamma = 0.8 - 4.6$ GeV shows that the cross section does approximately scale as expected, as s^{-11}, for photon energies above 2 GeV, but the scaled cross section is small, about 20 times smaller than the deuteron photodisintegration cross section. The small cross section results from cancellation in the pp amplitudes [78]. The decreased statistics resulting from the small cross section makes it difficult to determine if there are indeed oscillations as predicted, or to study the α_n distribution. More extensive data will soon be available from CLAS [79] at all angles, but only up to ≈1.5 GeV beam energy. A new experiment is needed to investigate the oscillations and α_n distribution.

These initial studies of ^{3}He photodisintegration are also yielding results for hard disintegration into a pd final state, data which are unpublished to date. Preliminary indications are that these cross sections scale as s^{-17} above about 1 GeV beam energy.

Thus, we have seen that several high-energy photoreactions exhibit cross section scaling behavior, though the onset of scaling occurs at very different kinematics for the different reactions. The limited spin measurements do not support helicity conservation. The underlying dynamics appear to be best explained as based on the quark-interchange mechanism, or evaluated using Regge theory techniques. At present we do not understand why different reaction channels appear to exhibit scaling starting at very different kinematics.

6. Summary

At the start of experiments at Jefferson Lab, a broad program of exclusive reactions was planned to understand the transition from hadronic to partonic degrees of freedom. Initially, it was anticipated that new high-energy data would be explained by perturbative QCD, with large momentum transferred to every quark in the nucleons in the reaction. The Jefferson Lab experimental program has allowed high-precision cross section data and polarization data to be obtained at large momentum transfer. These data along with other experimental and theoretical developments do not in general support a purely perturbative picture as had been expected. It appears that purely perturbative hard scattering is rarely the case. Instead, we have seen the role of orbital angular momentum of partons in hadrons, and the importance and separability of hard and soft physics in the handbag mechanism in intermediate energy reactions. The underlying dynamics typically involve one quark in the nucleon carrying much of the momentum transfer, sharing it with the other quarks through the wave function, rather than through hard momentum transfers. This is reflected in the cross sections and polarizations seen in reactions on the nucleon such as elastic scattering, real Compton scattering, and meson photoproduction. For systems with more than one nucleon, the handbag dominance leads to a related underlying picture of the dominance of quark-interchange mechanism, and the importance of quark orbital angular momentum obscures the phenomenon of color transparency.

Acknowledgments

The author acknowledges the support of the U.S. National Science Foundation grant PHY 06-52713 to Rutgers University, the U.S. Department of Energy, Office of Nuclear Physics, contract no. DE-AC02-06CH11357 to Argonne National Lab, and the U.S. Department of Energy contract No. DE-AC05-06OR23177 to Jefferson Science Associates, LLC. The U.S. Government retains a non-exclusive, paid-up, irrevocable, world-wide license to publish or reproduce this manuscript for U.S. Government purposes.

The authors thank S. Brodsky, G. Miller, M. Schwamb, and M. Strikman for many helpful discussions, and F. Gross and M. Sargsian for helpful discussions and a review of the manuscript.

References

[1] C. D. Roberts, Prog. Part. Nucl. Phys. **61**, 50 (2008).
[2] M. S. Bhagwat, M. A. Pichowsky, C. D. Roberts and P. C. Tandy, Phys. Rev. C **68**, 015203 (2003).
[3] P. O. Bowman, U. M. Heller, D. B. Leinweber and A. G. Williams, Phys. Proc. Suppl. **119**, 323 (2003).
[4] M. A. Shifman, A. I. Vainstein and V. I. Zakharov, Nucl. Phys. **B147**, 385 (1979), Nucl. Phys. **B147**, 448 (1979), Nucl. Phys. **B147**, 519 (1979).
[5] V. Chernyak and A. R. Zhitnitski, Phys. Rep. **112**, 173 (1983).
[6] T. Horn *et al.*, Phys. Rev. C **78**, 058201 (2008).
[7] B. Aubert *et al.*, Phys. Rev. D **80**, 052002 (2009).
[8] H. L. L. Roberts, C. D. Roberts, A. Bashir, L. X. Gutierrez-Guerrero and P. C. Tandy, Phys. Rev. C **82**, 065202 (2010).
[9] V. Tadavosian *et al*, Phys. Rev. C **75**, 055205 (2007).
[10] T. Horn *et al.*, Phys. Rev. Lett. **97**, 192001 (2006).
[11] R. Arnold *et al.*, Phys. Rev. Lett. **57**, 174 (1986).
[12] C. R. Ji, A. Sill and R. Lombard-Nelson, Phys. Rev. D **36**, 165 (1987).
[13] H. N. Li and G. Sterman, Nucl. Phys. **B381**, 129 (1992); H. N. Li, Phys. Rev. D **48**, 4243 (1993).
[14] M. K. Jones *et al.*, Phys. Rev. Lett. **84**, 1398 (2000); O. Gayou *et al.*, Phys. Rev. C **64**, 038202 (2001); O. Gayou *et al.*, Phys. Rev. Lett. **88**, 092301 (2002).
[15] A. V. Belitsky, X. Ji and F. Yuan, Phys. Rev. Lett. **91**, 092003 (2003).
[16] V. V. Frolov *et al.*, Phys. Rev. Lett. **82**, 45 (1999).
[17] A. Villano, Ph. D. thesis, Rensselaer Polytechnic Institute, to be published.
[18] K. Joo *et al.*, Phys. Rev. Lett. **88**, 122001 (2002).
[19] M. Ungaro *et al.*, Phys. Rev. Lett. **97**, 112003 (2006).
[20] P. Stoler, Phys. Rep. **226**, 103 (1993).
[21] L. M. Stuart *et al.*, Phys. Rev. D **58**, 032003 (1998).
[22] J. J. Kelly *et al.* Phys. Rev. Lett. **95**, 102001 (2005); Phys. Rev. C **75**, 025201 (2007).
[23] G. F. Chew, M. L. Goldberger, F. E. Low and Y. Nambu, Phys. Rev. **106**, 1345 (1957).
[24] C. E. Carlson and J. L. Poor, Phys. Rev. D **38**, 2758 (1988).
[25] R. Thompson *et al.*, Phys. Rev. Lett. **86**, 1702 (2001).
[26] H. Denizli *et al.*, Phys. Rev. C **76**, 015204 (2007).
[27] Jefferson Lab Experiment E01-002, M. Dalton; Ph. D. thesis, Univ. of Witwatersrand, Johannesburg, South Africa (to be published).
[28] D. J. Hamilton *et al.*, Phys. Rev. Lett. **94**, 242001 (2005).
[29] M. Diehl *et al.*, Phys. Rev. D **67**, 037502 (2003).

[30] G. A. Miller, Phys. Rev. **69**, 052201(R) (2004).
[31] F. Cano and J. M. Laget, Phys. Lett. B **551**, 317 (2003).
[32] T. Brooks and L. Dixon, Phys. Rev. D **62**, 114021 (2000).
[33] M. A. Shupe *et al.*, Phys. Rev. Lett. **19**, 1921 (1979).
[34] A. Danagoulian *et al.*, Phys. Rev. Lett. **98**, 152001 (2007).
[35] A. V. Radyushkin, Phys. Rev. D **58**, 114008 (1998).
[36] L. Zhu *et al.*, Phys. Rev. C **71**, 044603 (2005); Phys. Rev. Lett. **91**, 022003 (2003).
[37] K. Wijesooriya *et al.*, Phys. Rev. C **66**, 034614 (2002).
[38] http://cerncourier.com/cws/article/cern/30883; R. Bradford *et al.*, Phys. Rev. C **75**, 035205 (2006).
[39] R. Gilman and F. Gross, J. Phys. G **28**, R37 (2002).
[40] R. Gilman, Nucl. Phys. **A737**, 156-164 (2004).
[41] L. C. Alexa *et al.*, Phys. Rev. Lett. **82**, 1374 (1999).
[42] D. Abbott *et al.*, Phys. Rev. Lett. **84**, 5053 (2000).
[43] V. B. Mandelzweig and S. J. Wallace, Phys. Lett. B **197**, 469 (1987); S. J. Wallace and V. B. Mandelzweig, Nucl. Phys. **A503**, 673 (1989); D. R. Phillips, S. J. Wallace and N. K. Devine, Phys. Rev. C **58**, 2261 (1998).
[44] J. W. van Orden, N. Devine and F. Gross, Phys. Rev. Lett. **75**, 4369 (1995).
[45] Y. Huang and W. N. Polyzou, Phys. Rev. C **80**, 025503 (2009).
[46] H. Dijk and B. L. G. Bakker, Nucl. Phys. **A494**, 438 (1988).
[47] P. E. Ulmer *et al.*, Phys. Rev. Lett. **89**, 062301 (2002).
[48] F. Benmokhtar *et al.*, Phys. Rev. Lett. **94**, 082305 (2005)
[49] M. Rvachev *et al.*, Phys. Rev. Lett. **94**, 192302 (2005).
[50] Jefferson Lab Experiment E04-018, J. Gomez, A. Katramatou, G. Petratos *et al.*
[51] M. Pavón Valderrama, A. Nogga, E. Ruiz Arriola and D. R. Phillips, Eur. Phys. J. A **36**, 315 (2008).
[52] J. M. Laget, Few Body Syst. Supp. **15**, 171 (2003); Phys. Lett. B **609**, 49 (2005).
[53] J. Napolitano *et al.*, Phys. Rev. Lett. **61**, 2530 (1988); S. J. Freedman *et al.*, Phys. Rev. C **48**, 1864 (1993).
[54] J. E. Belz *et al.*, Phys. Rev. Lett. **74**, 646 (1995).
[55] C. Bochna *et al.*, Phys. Rev. Lett. **81**, 4576 (1998).
[56] E. C. Schulte *et al.*, Phys. Rev. Lett. **87**, 102302 (2001).
[57] E. C. Schulte *et al.*, Phys. Rev. C **66**, 042201R (2002).
[58] M. Mirazita *et al.*, Phys. Rev. C **70**, 014005 (2004).
[59] S. J. Brodsky and G. R. Farrar, Phys. Rev. Lett. **31**, 1153 (1973); V. A. Matveev, R. M. Muradyan and A. N. Tavkhelidze, Nuovo Cim. Lett. **7**, 719 (1973).
[60] P. Rossi *et al.*, Phys. Rev. Lett. **94**, 012301 (2005).
[61] S. J. Brodsky and J. R. Hiller, Phys. Rev. C **28**, 475 (1983).
[62] A. Radyushkin, private communication.
[63] L. L. Frankfurt, G. A. Miller, M. M. Sargsian and M. I. Strikman, Phys. Rev. Lett. **84**, 3045 (2000); Nucl. Phys. **A663**, 349 (2000).
[64] B. Julia-Diaz and T.-S. H. Lee, Mod. Phys. Lett. A **18**, 200 (2003).
[65] E. de Sanctis *et al.*, Few Body Syst. Suppl. **6**, 229 (1992); L. A. Kondratyuk *et al.*, Phys. Rev. C **48**, 2491 (1993); V. Yu. Grishina *et al.*, Eur. Phys. J. A **10**, 355 (2001).
[66] F. Adamian *et al.*, Eur. Phys. J. A **8**, 423 (2000).
[67] S. J. Brodsky and G. P. Lepage, Phys. Rev. D **24**, 2848 (1981).
[68] T. Gousset, B. Pire and J. P. Ralston, Phys. Rev. D **53**, 1202 (1996).
[69] V. Yu. Grishina *et al.*, Eur. Phys. J. A **18**, 207 (2003), Eur. Phys. J. A **19**, 117 (2004).
[70] K. Wijesooriya *et al.*, Phys. Rev. Lett. **86**, 2975 (2001).
[71] X. Jiang *et al.*, Phys. Rev. Lett. **98**, 182302 (2007).
[72] M. M. Sargsian, Phys. Lett. B **587**, 41 (2004).
[73] M. Schwamb and H. Arenhövel, Nucl. Phys. **A690**, 682 (2001), Nucl. Phys. **A696**, 556 (2001), Mod. Phys. Lett. A **18**, 190 (2003); M. Schwamb, Habilitation thesis, Johannes Gutenberg-Universität Mainz, 2006.
[74] I. Rachek *et al.*, Phys. Rev. Lett. **98**, 182303 (2007).
[75] J. Glister *et al.*, Phys. Lett. B **697**, 194 (2011).
[76] S. J. Brodsky *et al.*, Phys. Lett. B **578**, 69 (2003).
[77] I. Pomerantz *et al.*, Phys. Lett. B **684**, 106 (2010).
[78] C. Granados and M. M. Sargsian, Phys. Rev. C **80**, 014612 (2009).
[79] S. Niccolai *et al.*, Phys. Rev. C **70**, 064003 (2004).

Journal of Physics: Conference Series **299** (2011) 012010 doi:10.1088/1742-6596/299/1/012010

Short-Distance Structure of Nuclei

D. W. Higinbotham[1], E. Piasetzky[2], L. Weinstein[3] and S. A. Wood[1]

[1] Thomas Jefferson National Accelerator Facility, Newport News, Virginia 23606, USA
[2] Tel Aviv University, Tel Aviv 69978, Israel
[3] Old Dominion University, Norfolk, Virginia 23529, USA

Abstract. One of Jefferson Lab's original missions was to further our understanding of the short-distance structure of nuclei, in particular, to understand what happens when two or more nucleons within a nucleus have strongly overlapping wave-functions – a phenomena commonly referred to as short-range correlations. Herein, we review the results of the (e, e'), $(e, e'p)$ and $(e, e'pN)$ reactions that have been used at Jefferson Lab to probe this short-distance structure as well as provide an outlook for future experiments.

1. Introduction

The structure of nuclei is determined by the strong force with repulsion at short distances and attraction at moderate distances. This force, which binds the nucleons together while also keeping the structure from collapsing, makes the nucleus a fairly dilute system. This allows for calculations that treat the nucleus as a collection of hard objects in an average or mean field to describe many of the properties of nuclear matter; however, this simple picture breaks down when detailed features are studied. In particular, the short-distance part of the nucleon-nucleon potential can cause central, tensor and spin-orbit correlations between the nucleons in nuclei which deplete the low momentum states in nuclear systems [1].

Calculations indicate that these short-range correlations can lead to short-term local densities in the nucleus that are several times as high as the average nuclear density of 0.17 GeV/fm^3, and thus, comparable to densities predicted in neutron stars [2]. Isolating the signal of short-range correlated nucleons may therefore lead to a deeper understanding of cold, dense nuclear systems.

Directly identifying short-range correlations has been experimentally challenging. The low duty factors and the sub-GeV energies of previous facilities limited studies of short-range correlations to moderate momentum transfer kinematics. The Jefferson Lab accelerator and experimental facilities permit measurements at high energy and high luminosity with a continuous beam. These parameters have allowed measurements of the $A(e, e')$, $A(e, e'p)$ and $A(e, e'pN)$ reactions in kinematics above the quasi-elastic peak and at high momentum transfer. These kinematics are where the effects of competing mechanisms such as meson-exchange currents, isobar configurations, and final-state interactions are reduced and the signatures of short-range correlations become dominant.

2. Inclusive reactions

The inclusive $A(e, e')$ process is the simplest electromagnetic reaction for studying short-range correlations; though to use this reaction channel, it is important to be in kinematics where short-range correlations dominate over other reaction mechanisms such as meson-exchange currents or delta-isobar contributions. As first shown with the SLAC (e, e') data [3], this is done by requiring $x_B = Q^2/2m\omega > 1$ and $Q^2 > 1$ (GeV/c)^2 where m is the nucleon mass, ω is the energy transfer and Q^2 is the four-momentum transfer squared. The high x_B minimizes the inelastic scattering contributions, while high Q^2 minimizes meson-exchange currents. It was predicted that in these kinematics diagram b) shown in Fig. 1 would dominate the reaction [4]. For large Q^2 and $x_B > 1$ the value of x_B kinematically determines a minimum possible missing momentum; thus, by choosing x_B large enough, only contributions from nucleons with missing momentum above the Fermi sea, corresponding to large excitation energies, can contribute to the reaction.

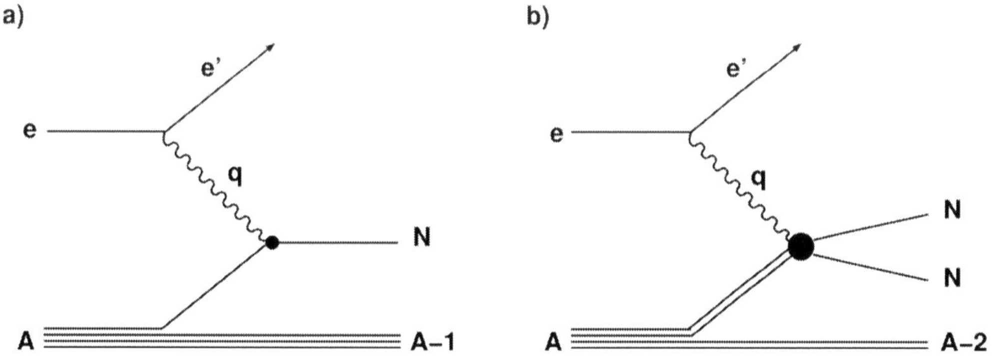

Figure 1. Shown are the dominant diagrams of the $A(e, e')$ reaction for $x > 1$ and $Q^2 > 1$ (GeV/c)^2. Figure a) shows single nucleon scattering while Figure b) shows scattering from a correlated initial-state pair.

For a given Q^2 range, the SLAC inclusive data was plotted as function of x_B for the weighted ratio of yields,

$$r(A, {}^2\mathrm{H}) = \frac{\mathcal{Y}(A)}{Z\sigma_{ep} + N\sigma_{en}} \frac{\sigma_{ep} + \sigma_{en}}{\mathcal{Y}({}^2\mathrm{H})} C^A_{\mathrm{rad}}, \tag{1}$$

where $\mathcal{Y}$ is the normalized yield, Z and N are the number of protons and neutrons in a nucleus A, σ_{eN} is the electron-nucleon cross section, and C^A_{rad} is the ratio of the radiative correction factors for A and ^{2}H. The SLAC data show for $Q^2 > 1$ (GeV/c)^2 and $x_B > 1.4$ a scaling in the data which was argued to be due to short-range correlations [3]. These events corresponded to nucleon momenta of about 250 – 300 MeV/c. For these measurements the statistics were rather limited and x_B was less than 2. Also, the deuteron data was measured in a different kinematics and comparisons to the other nuclei required nontrivial extrapolations.

In order to improve upon the SLAC (e, e') data, Jefferson Lab provided higher statistics data and was able to look for a second scaling region due to three-nucleon correlations. The three-nucleon correlation region was accessed by taking the ratios to ^{3}He data instead of deuterium. The data was taken with the CEBAF Large Acceptance Spectrometer (CLAS) [5, 6] and is shown in Fig. 2. The new data not only show scaling in the $1 < x_B < 2$ region, but also show, for the first time, a second scaling in the $x_B > 2$ region. This second scaling has been interpreted as being due to three-nucleon correlations and further strengthens the argument that the first scaling region is due to short-range two-nucleon correlations [7].

From these ratios, one can determine the probabilities of two-nucleon correlations in the various nuclei by using the relatively well understood deuteron. This is valid due to closure in the interaction of the knocked-out nucleons with nucleons not belonging to correlations and

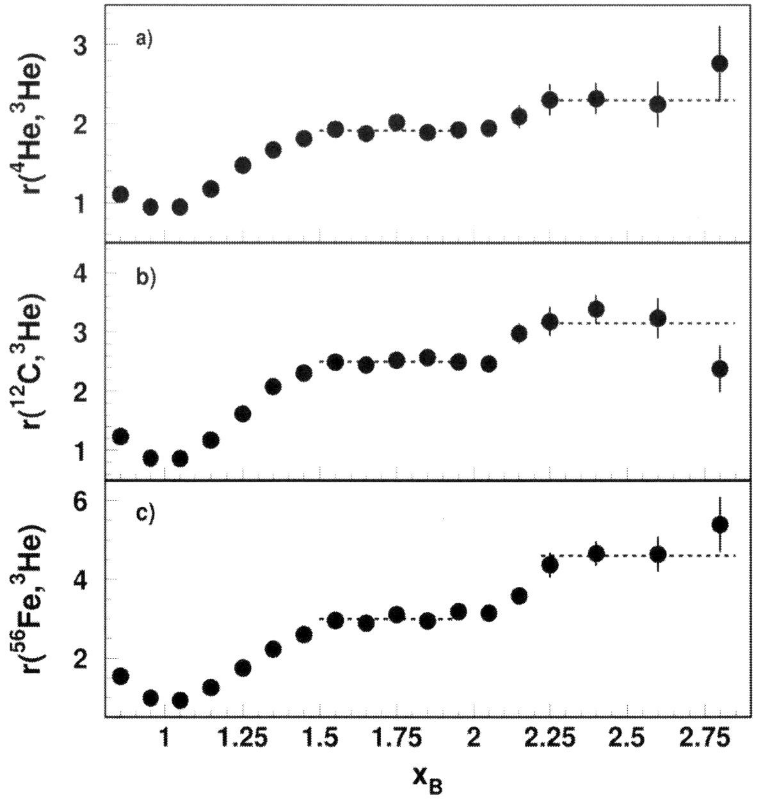

Figure 2. Weighted cross section ratios of (a) ^{4}He, (b) ^{12}C and (c) ^{56}Fe to ^{3}He as a function of x_B for $Q^2 > 1.4$ GeV^2. The horizontal dashed lines indicate the two-nucleon and three-nucleon scaling regions used to calculate the per-nucleon probabilities for two- and three-nucleon short-range correlations in nucleus A relative to ^{3}He. Reprinted with permission from K.S. Egiyan *et al.*, Phys. Rev. Lett. **96**, 082501 (2006). Copyright 2006 by the American Physical Society.

cancellation of final-state interactions within the correlations in the ratio [4]. The probability of a nucleon-nucleon correlation in deuterium is defined as the probability that a nucleon in deuterium has a momentum $p > 275$ MeV/c. These are momenta that have been shown in $\vec{D}(\vec{e}, e'p)$ asymmetry measurements to be dominated by highly correlated nucleons [8]. Thus by integrating the wave function, it has been found that the per-nucleon probability of a nucleon in deuterium, $a_2(^2\text{H})$, to be in a nucleon-nucleon correlation is 0.041 ± 0.008. The probability of correlations in deuterium versus ^{3}He, $a_2(^3\text{He}/^2\text{H})$ has been determined to be 1.97 ± 0.1 and thus $a_{2N}(^3\text{He}) = 0.08 \pm 0.016$. From this result, the general $a_{2N}(A)$ probabilities can be calculated from the scaling results for $1.15 < x_B < 2$ for $r(A/^3\text{He}) \times r(^3\text{He}/^2\text{H})$. To obtain the absolute probability of three-nucleon short-range correlations, a calculation of the Bochum group [9] using various potentials was used and gave an average value of $a_{3N}(^3\text{He}) = 0.18 \pm 0.06\%$. From this it was concluded that the $3N$ short-range correlations must be less then 1% [6]. These results are summarized in Table 1.

In Jefferson Lab's Hall C, inclusive data were also taken for $x_B > 1$ and Q^2 from 1–7 (GeV/c)^2. These data also showed, for initial nucleon momenta greater then 300 MeV/c, a scaling behavior consistent with short-range nucleon-nucleon correlations [10]. Recent calculations show that

Table 1. Shown are the average ratios, $r(A/^3\mathrm{He})$, for the scaling regions $1.5 < x_B < 2$ and $2.2 < x_B < 2.7$ along with the extracted absolute probabilities $a_{2N}(A)$ and $a_{3N}(A)$ that in a nucleus A a two- or three-nucleon short-range correlation is taking place at a given instant. Statistical and systematic errors have been combined in quadrature.

	$r(A/^3\mathrm{He})$ $1.5 < x_B < 2$	$a_{2N}(A)$ [%]	$r(A/^3\mathrm{He})$ $2.2 < x_B < 2.7$	$a_{3N}(A)$ [%]
^{3}He	1	8.0±1.6	1	0.18±0.06
^{4}He	1.93±0.03	15.4±3.2	2.33±0.13	0.42±0.14
^{12}C	2.49±0.15	19.8±4.4	3.18±0.27	0.56±0.21
^{56}Fe	2.98±0.18	23.9±5.3	4.63±0.33	0.83±0.27

these scaling regions can be used to obtain useful information about short-range correlations [12]. New high-precision data from Hall C on the $x_B > 1$ ratios should be available soon [11]

3. Proton knock-out

For proton knock-out experiments, it is useful to introduce the concept of an initial state momentum distribution of nucleons within the nucleus. This is done in a factorized approximation with a spectral function $S(\mathbf{p}, E)$ which relates the experimental missing momentum, $\mathbf{p}$, and missing energy, E to the measured $(e, e'p)$ cross sections by the relation:

$$\frac{d^6\sigma}{dp_e dp_p d\Omega_e d\Omega_p} = K\,\sigma_{eN} S(\mathbf{p}, E)\, T_A(Q^2), \tag{2}$$

where K is a kinematic factor, σ_{eN} is the elementary cross section, and the transparency, T_A, acts as the normalization factor between the experimental cross section and a theoretical momentum distribution. The transparency is often interpreted as the probability that a nucleon will be emitted from the nucleus without rescattering, though other effects are included in this factor, as it is experimentally just the scale parameter required to reconcile a theoretical momentum distribution with an experimental cross sections. Having theory calculate both the spectral function and the transparency allows us to further test our understanding of nuclear matter in a many-body system where exact solutions are not yet calculable.

In independent particle (IP) models, the strength of the spectral function is almost entirely limited to momenta and energies less than 250 MeV/c and 80 MeV, namely, the Fermi momentum and energy. In models built up from realistic nucleon-nucleon potentials, which include a repulsive core and medium range tensor component, the spectral function strength in the Fermi-gas region is reduced by about 20% [13, 14]. Measurements with reactions such as $(e, e'p)$ show a reduction of at least 30% [15, 16], which is consistent with the expectation that short-range correlations are an important part of the description of the nucleus. In the Hall A $^{16}\mathrm{O}(e, e'p)$ experiment, which measured cross sections and response functions for $25 < E < 120$ MeV and $p < 340$ MeV/c, it was found that calculations that included pion exchange currents, isobar currents and two-nucleon correlations were required to account for the shape of the data at high missing momentum, though the calculations still under estimated the measured cross section by a factor of two [17, 18].

Measurements of $A(e, e'p)$ were made in Jefferson Lab's Hall C to directly identify correlated strength rather than inferring it from the absence of strength predicted by independent particle models. Using the HMS and SOS spectrometers, protons knocked out of ^{12}C were detected in parallel kinematics with missing energy $E > 40$ MeV and missing momentum $k > 240$ MeV/c. The spectral function derived from this measurement, integrated over this range of missing energy and momentum, accounts for approximately 10% of the total strength of scattering from protons in ^{12}C [19]. This strength is consistent with or larger than the correlated strength predicted for this kinematic range from a Correlated Basis Function approach [13] or Green's function approach to calculating the spectral function [20]. These calculations both predict more of the correlated strength to be outside of the measured range, including at low missing energy and momentum where strength is indistinguishable from IP model proton knockout.

At Jefferson Lab, T_A has been measured for several nuclei [21, 22] and found to be essentially independent of proton energy for knocked-out protons with over 1 GeV of kinetic energy. The measured transparencies have been found to be somewhat higher than expected for a proton propagating through an independent particle model nucleus. For example, for ^{12}C the measured $T_A = 0.59$, while the prediction from an independent particle shell model is 0.55 [23]. Calculations with the Correlated Basis Function approach increase the transparency by about 10% due to short-range correlations and bring experiment and theory into much better agreement. This is interpreted as the short-range nucleon-nucleon causing a depletion in density near the point of interaction, giving a lower overall probability that is will interact as it leaves the nucleus.

Hall A measurements of the ^{3}He$(e,e'p)d$ and ^{3}He$(e,e'p)pn$ reactions were taken with a beam energy of 4.8 GeV, $Q^2 = 1.5$ (GeV/c)^2 in $x_B = 1$ kinematics. When this experiment was proposed, it was expected that these kinematics would cleanly show short-range correlations at missing momenta greater then 300 MeV/c. What was observed was a much greater strength in the high missing momentum region then expected [24, 25] as shown in Figure 3. The three-body break-up (3bbu) has been integrated over missing energy so that it could be plotted with the two-body break-up (2bbu) and the strengths of the two reactions compared.

The unexpected strength was explained as an interference between correlations in the initial state and final-state interaction [26], where neither effect alone could explain the observed cross section. Subsequently calculations by Ciofi degli Atti and Kaptari showed that a parameter free calculation which included a realistic wave function [27], *i.e.* one in which correlations are built in, and final-state interactions calculated with the generalized Eikonal approximation could explain the data [28, 29, 30].

4. Triple coincidence reactions

While the $A(e,e')$ and $A(e,e'p)$ data shown in Sections 2 and 3 clearly suggest strong local correlations, it is the new exclusive data that confirms that the inclusive scaling is indeed due to short-range correlations. Also, with exclusive reactions, it possible to determine types and abundances of nucleon pairs involved in correlations. Such experiments where two nucleons are detected were proposed in the 1960's [31], but only with the advent of high luminosity and high energy machines were clean kinematics (*i.e.* $Q^2 > 1$ GeV^2 and $x_B > 1$) obtainable.

Experimentally, a high-momentum, small de Broglie wavelength probe can knock out one nucleon of a nucleus while leaving the residual nucleus nearly unaffected; however, if the nucleon being struck is part of an initial-state pair, the high relative momentum of the pair will cause the correlated nucleon to recoil and be ejected [4, 32]. Figure 4 shows an illustration of a high-energy electron knocking out two nucleons from a nucleus.

Historically, the interpretation of triple-coincidence experimental data at low Q^2 in terms of short-range correlations has been complicated by contributions from meson-exchange currents, isobar configurations and final-state interactions [33, 34, 35, 36]. The kinematics for the measurements described herein were chosen to minimize these effects. For example, at high Q^2, meson-exchange contributions decrease as $1/Q^2$ relative to plane-wave impulse contributions and relative to those due to correlations [37, 38]. Large Q^2 and x_B also drastically reduce

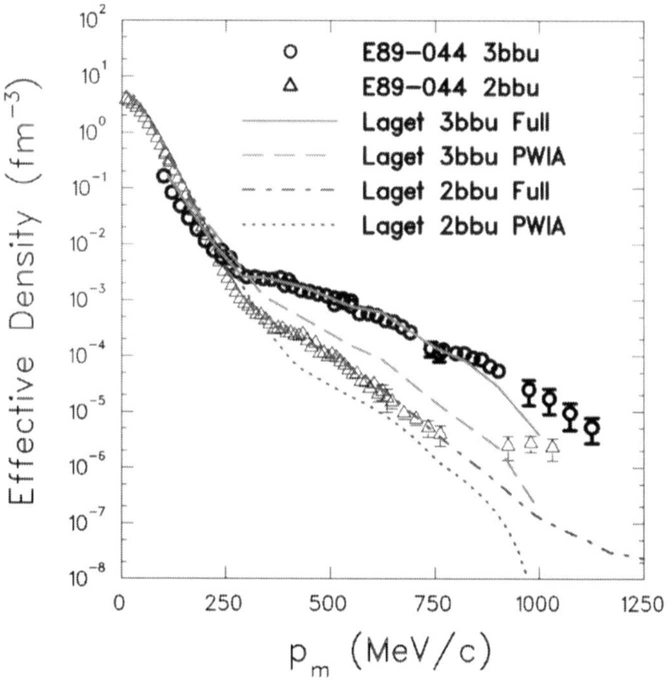

Figure 3. Proton effective momentum density distributions in ^{3}He extracted from ^{3}He$(e, e'p)pn$ three-body break-up (3bbu) is shown as the open black circles and the ^{3}He$(e, e'p)d$ two-body break-up (2bbu) is shown as open black triangles. The three-body break-up (3bbu) integration covers E_M from threshold to 140 MeV. The results are compared to calculations from J.-M. Laget [26] which explain the dominance of the continuum cross section at large missing momentum as a strong interference between short-range correlations and final-state interactions. Reprinted with permission from F. Benmokhtar *et al*, Phys. Rev. Lett. **94**, 082305 (2005). Copyright 2005 by the American Physical Society.

isobar currents contributions [39, 40]. Final-state interactions are not small, but, in the chosen kinematics, predominately restricted to be within the correlated pair [2].

The high luminosity Hall A triple coincidence experiment used an incident electron beam of 4.627 GeV and the two Hall A high-resolution spectrometers (HRS) [41] to identify the ^{12}C$(e, e'p)$ reaction. Scattered electrons were detected in the left HRS at a central scattering angle (momentum) of 19.5° (3.724 GeV/c). This corresponds to the quasi-free knockout of a single proton with transferred three-momentum $|\vec{q}| = 1.65$ GeV/c, transferred energy $\omega = 0.865$ GeV, $Q^2 = 2$ (GeV/c)^2, and $x_B = 1.2$. Knocked-out protons were detected using the right HRS which was set at 3 different combinations of central angle and momentum: 40.1° & 1.45 GeV/c, 35.8° & 1.42 GeV/c, and 32.0° & 1.36 GeV/c. These kinematic settings correspond to median missing-momentum values $p_{miss} = 0.35$, 0.45 and 0.55 GeV/c, respectively, covering the p_{miss} range of 300 – 600 MeV/c.

A large-acceptance spectrometer, BigBite, was used to detect recoiling protons in ^{12}C$(e, e'pp)$ events. The BigBite spectrometer consists of a large-acceptance, non-focusing dipole magnet [42] and a custom detector package. For this measurement, the magnet was located at an angle of 99° and 1.1 m from the target with a resulting angular acceptance of about 96 msr and a nominal momentum acceptance from 0.25 GeV/c to 0.9 GeV/c. Located immediately behind BigBite was a large neutron array matched to BigBite's acceptance. This made it possible to

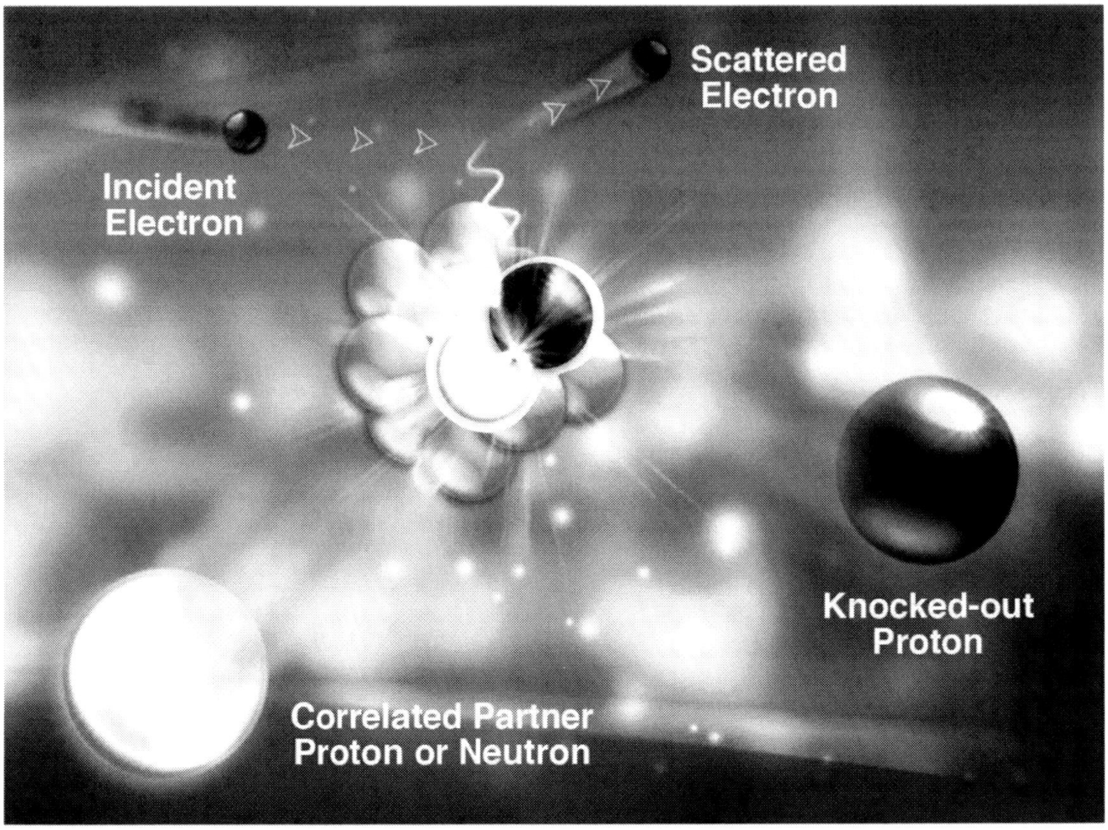

Figure 4. Illustration showing the $(e, e'pN)$ reaction. The incident electron couples to a nucleon-nucleon pair via a virtual photon. In the final state, the scattered electron and struck nucleon are detected along with the correlated nucleon that is ejected from the nucleus. Illustration courtesy of Joanna Griffin and inspired by Anna Shneor.

simultaneously detect the $^{12}C(e, e'pp)$ and $^{12}C(e, e'pn)$ reactions.

Figure 5 shows the cosine of the angle, γ, between the missing momentum ($\vec{p}_{miss}$) and the recoiling proton detected in BigBite ($\vec{p}_{rec}$) for the highest p_{miss} setting of 550 MeV/c [43]. Also shown in Fig. 5 is the angular correlation for the random background as defined by a time window offset from the coincidence peak. The back-to-back peak of the real triple coincidence events is demonstrated clearly. The curve is a result of a simulation of the scattering off a moving pair having a center-of-mass momentum width of 0.136 GeV/c. That width was extracted from the data and is consistent with the width measured in the (p, ppn) experiment at Brookhaven National Lab [44] and well as with a theoretical calculation based on the convolution of two independent single particle momentum distributions [45].

The measured ratio of ^{12}C$(e, e'pp)$ to ^{12}C$(e, e'p)$ events is given by the ratio of events in the background-subtracted time-of-flight peak to those in the ^{12}C$(e, e'p)$ spectra. The measured ratios are limited by the finite acceptance of BigBite. The center of mass momentum distribution obtained from the measured angular correlation is used to account for this finite acceptance; the resulting extrapolated ratios are shown in Fig. 6.

To compare the $(e, e'pn)$ to $(e, e'p)$ yields, a similar procedure as above was followed. With neutrons, momentum determination was done with time-of-flight and the neutron detection

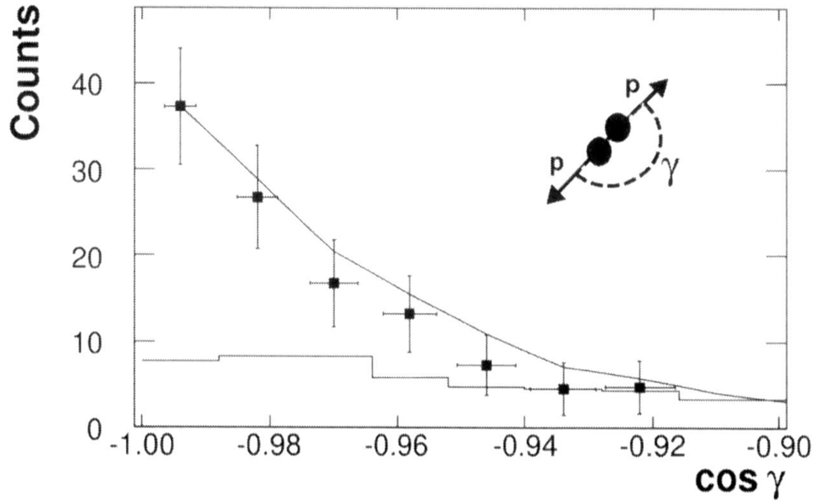

Figure 5. For the ^{12}C$(e, e'pp)$ reaction at $Q^2 > 1$ (GeV/c)^2, the distribution of the cosine of the opening angle between the $\vec{p}_{miss}$ and $\vec{p}_{rec}$ for a $p_{miss} = 0.55$ GeV/c. The histogram shows the distribution of random events. The curve is a simulation of the scattering off a moving pair with a width of 0.136 GeV/c for the pair center of mass momentum. Reprinted with permission from R. Shneor *et al.*, Phys. Rev. Lett. **99**, 072501 (2007). Copyright 2007 by the American Physical Society.

efficiency was only about 40%. This results in a poorer signal to background ratio than the $(e, e'pp)$ measurement and also a larger uncertainty. Taking into account the finite acceptance of the neutron detector and the neutron detection efficiency, it was found that $96 \pm 22\%$ of the $(e, e'p)$ events with a missing momentum above 300 MeV/c had a recoiling neutron. This result agrees with the hadron beam measurement of the $(p, 2pn)/(p, 2p)$ reaction in which $92 \pm 18\%$ of the $(p, 2p)$ events with a missing momentum above the Fermi momentum of 275 MeV/c were found to have a single recoiling neutron carrying the momentum [46].

Since the recoiling proton ^{12}C$(e, e'pp)$ and neutron ^{12}C$(e, e'pn)$ data were collected simultaneously with detection systems covering nearly identical solid angles, the ratio of ^{12}C$(e, e'pn)/^{12}$C$(e, e'pp)$ could be directly determined with many of the systematic factors needed to compare the rates of the ^{12}C$(e, e'pp)$ and ^{12}C$(e, e'pn)$ reactions canceling out. Correcting only for detector efficiencies, a ratio of 8.1 ± 2.2 was determined. To estimate the effect of final-state interactions (*i.e.*, reactions that happen after the initial scattering), the attenuation of the recoiling protons and neutrons was assumed to be almost equal. In this case, the only correction related to final-state interactions of the measured ^{12}C$(e, e'pn)$ to ^{12}C$(e, e'pp)$ ratio is due to single charge exchange. For the experiment, the neutron to proton single charge exchange dominates proton to neutron single charge exchange since the $(e, e'pn)$ rate is about an order of magnitude larger than the $(e, e'pp)$ rate and thus decreases the measured ^{12}C$(e, e'pn)/^{12}$C$(e, e'pp)$ ratio. Using Glauber approximation [47], it was estimated this effect was 11%. Taking this into account, the corrected experimental ratio for ^{12}C$(e, e'pn)/^{12}$C$(e, e'pp)$ is 9.0 ± 2.5.

To deduce the ratio of pn to pp correlated pairs in the ground state of ^{12}C, the measured ^{12}C$(e, e'pn)$ to ^{12}C$(e, e'pp)$ ratio was used. Since the experiment triggered only on forward $(e, e'p)$ events, the probability of detecting pp pairs was twice that of pn pairs; thus, we conclude that the ratio of pn/pp pairs in the ^{12}C ground state is 18 ± 5 as shown in Fig. 7. This result

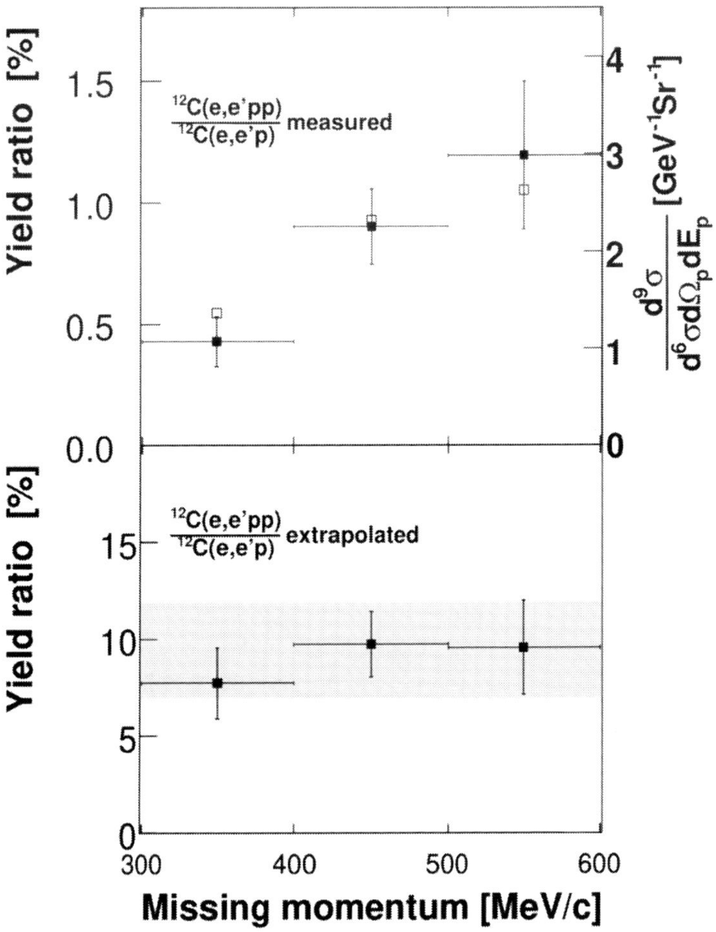

Figure 6. The measured and extrapolated ratios of yields for the $^{12}C(e,e'pp)$ and the $^{12}C(e,e'p)$ reactions. The shaded area represents a band of $\pm 2\sigma$ uncertainty in the width of the center-of-mass momentum of the pair. Reprinted with permission from R. Shneor *et al.*, Phys. Rev. Lett. **99**, 072501 (2007). Copyright 2007 by the American Physical Society.

is consistent with the $(p, 2pn)$ data [46]. Since both the probe and the kinematics of these two experiment were very different, it furthers the interpretation of the process as being due to scattering off a correlated pair of nucleons [48].

The small ratio of pp/np has been explained by several theoretical groups [50, 51, 52]. These calculations clearly show that the measured pp/np ratio is expected and is a clear indication of the nucleon-nucleon tensor force at the probed distances and relative momenta of the nucleons in a short-range two-nucleon correlation.

Instead of directly knocking out one nucleon of the correlated pair, it is also possible in the three-body system to knock out the third nucleon and observe the decay of the spectator correlated pair. This was done in the $^3He(e,e'pp)n$ using 2.2 and 4.4 GeV electron beams and detecting the scattered electron and ejected protons in CLAS over a wide kinematic range [53]. When all three final state nucleons have momenta greater than 250 MeV/c, the reaction is dominated by events where two nucleons each have less than 20% of the transferred energy and

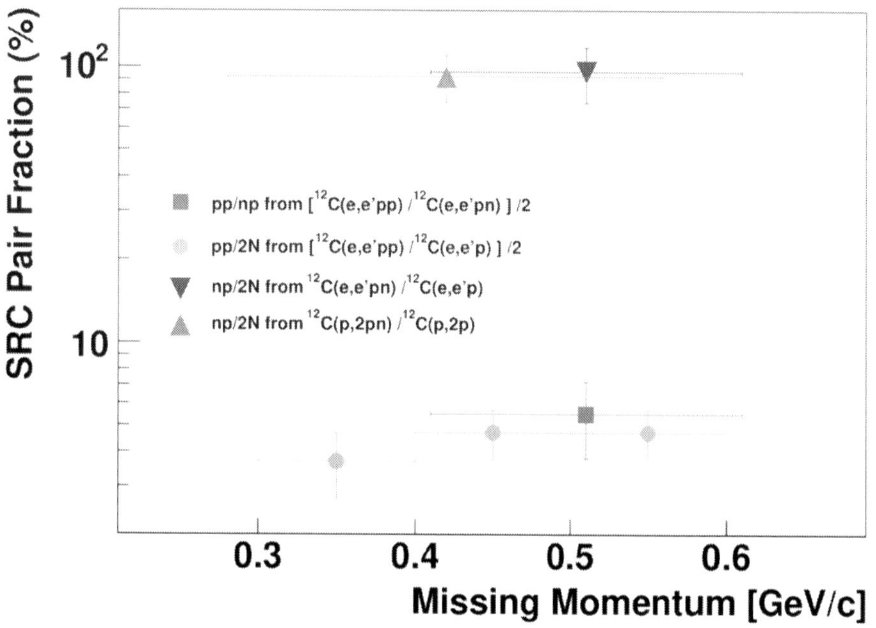

Figure 7. The fractions of correlated pair combinations in carbon as obtained from the $(e, e'pp)$ to $(e, e'pn)$ reactions [49], as well as from previous $(p, 2pn)$ data [46]. Reprinted with permission from R. Subedi *et al.*, Science **320**, 1476 (2008).

the third 'leading' nucleon has the remainder. Final state interactions of the leading nucleon are suppressed by requiring that it has perpendicular momentum with respect to $\vec{q}$ of less than 300 MeV/c. In these cases the two other nucleons (the pn or pp pair) are predominantly back-to-back and have very little total momentum in the momentum transfer direction. The relative pair momentum is also mostly isotropic. This indicates that the nucleon-nucleon pair is correlated and is a spectator to the virtual photon absorption. This means that the measured relative and total pair momenta, $\vec{p}_{rel} = (\vec{p}_1 - \vec{p}_2)/2$ and $\vec{p}_{tot} = \vec{p}_1 + \vec{p}_2$ are closely related to the initial momenta of the correlated pair. As shown in Fig. 8, the pair relative momentum peaks at about 300–400 MeV/c and the pair total momentum peaks at about 300 MeV/c. The pp and pn momentum distributions are very similar. The advantage of this approach is that there is no contribution from meson-exchange currents or isobar configurations since the virtual photon does not interact with the correlated pair. However, because the continuum interaction of the correlated pair is very strong and reduces the calculated cross section by a factor of about ten, this reaction is very difficult to calculate precisely. Diagrammatic calculations by Laget are in qualitative agreement with the measured momentum distributions, but only after including the effects of the continuum interaction.

5. Short range correlations and the EMC effect

One of the outstanding questions in nuclear physics is whether free nucleons and bound nucleons have the same quark structure. The first significant evidence that they differ is that the per nucleon deep-inelastic electron-scattering cross section ratio of heavy to light nuclei is not equal to one. This phenomenon, discovered by the European Muon Collaboration, became known as the EMC effect. It has generated considerable experimental and theoretical interest over the

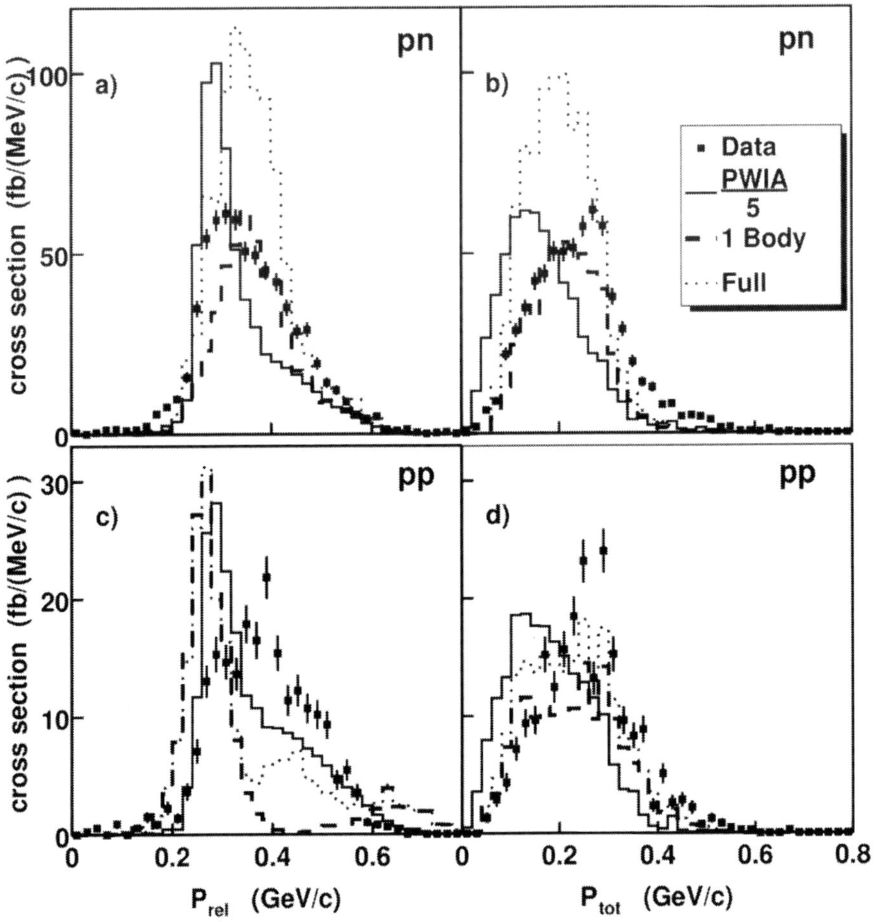

Figure 8. Figure a) shows lab frame cross section versus *pn* pair relative momentum. Points show the data, solid histogram shows the PWIA calculation times $\frac{1}{5}$, dashed histogram shows Laget's one-body calculation [54, 55, 56], dotted histogram shows Laget's full calculation; b) the same for total momentum; c), d) the same for *pp* pairs. Reprinted with permission from R. A. Niyazov *et al.*, Phys. Rev. Lett. **92**, 052303 (2004). Copyright 2004 by the American Physical Society.

years, though there no generally excepted explanation for the effect [57, 58]. In general, short-range correlation and the EMC effect were considered completely different phenomena; yet why shouldn't the quark distributions in two nucleons that are close together be affected or modified in some way?

Quantitatively comparing the magnitude of the EMC effect and the SRC probabilities in different nuclei shows that the two effects are remarkably correlated [59] (see Fig. 9). This strongly supports the idea that nuclear short-range correlations and the deep inelastic EMC effect are related phenomena. While theorists have debated the EMC effect for decades, this phenomenological observation is the first experimental clue of the cause of the EMC effect. In addition, this result allows experimentalists for the first time to extract the quark structure of a free neutron from proton and deuteron data without a theoretical model and should create renewed interest in understanding the short-range part of the nuclear wave-function.

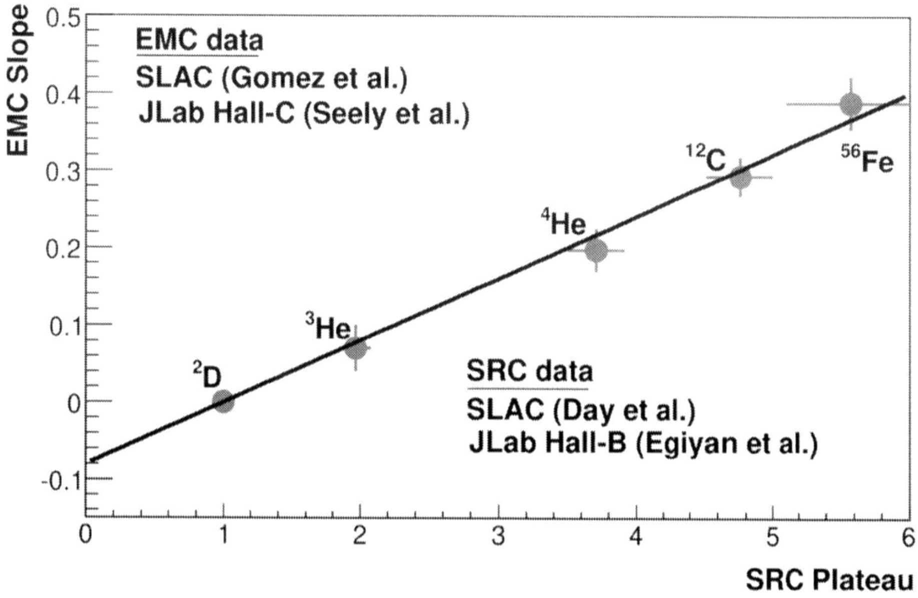

Figure 9. The absolute value of the slope of the EMC effect for Bjorken x_B values from 0.3 to 0.7 is plotted versus the relative magnitude of the short-range correlation strength as compared to deuterium (*i.e.*, the height of the plateau in the ratio of cross sections of nucleus A to deuterium for $1.5 < x_B < 2$).

This phenomenological linking of the EMC effect and short-range correlations should greatly increase our understanding of the EMC effect, nucleon modification in nuclei (if any), and the quark structure of free neutrons. It also has implications for the physics of neutron stars and for the search for new physics.

6. Summary and outlook

From these new results, one can get a good general picture of two-nucleon correlations that take place within the nucleus. From (e, e') we learn the probability of correlations in various nuclear systems. From $(e, e'p)$ we learn the probability of independent single particle knock-out and about the high-momentum tail in the extracted momentum distributions. From $^3\mathrm{He}(e, e'pp)n$ we learn the about shape of the distorted correlated pair momentum distributions and from $^{12}\mathrm{C}(e, e'pN)$ we learn the relative importance of pp, pn and nn. Putting this all together, we learn that in ^{12}C, 60–70% of the nucleons within the nucleus act like independent particles, while approximately 20% have a short-range correlation with another nucleus and that these partner nucleons almost always have the opposite isospin due to short-range tensor correlations. The remaining fraction of nucleons within the nucleus are thought to be in long-range correlations.

The association of the small $(e, e'pp)/(e, e'pn)$ ratio with the dominance of the nucleon-nucleon short-range tensor force leads naturally to the quest to increase the missing momentum and to look for pairs which are even closer to each other, at distances that are dominated by the repulsive core. In particular, the observed dominance of np to pp correlations is predicted to decrease as missing momentum is increased and the nucleon-nucleon short-range tensor force gives way to the short range repulsive force [50, 51, 52, 60]. This momentum dependence is also expected to be more pronounced in light nuclei, motivating an upcoming Hall A triple coincidence experiment, similar to the ^{12}C$(e,e'pN)$ measurement, which will map out the np to pp ratio over the range $400 < p_{miss} < 850$ MeV/c using ^{4}He.

Further (e,e') $x_B > 1$ inclusive measurements are also planned. In the near term, Hall A will make a comparison of symmetric and neutron rich isotopes of Calcium to probe the isospin dependence of $2N$ and $3N$ correlations. This experiment will also make a detailed study of the onset of scaling due to $3N$ correlations will be made from the Q^2 dependence of $x_B > 2$ cross sections. After the 12 GeV upgrade, Hall C will measure the (e,e') reaction in $x_B > 1$ kinematics and in the deep inelastic scattering region of $W > 2$ GeV. This will allow an investigation of quarks which are being shared by more than one nucleon [61, 62].

Beyond the approved experiments, future facility upgrades could include a large acceptance detector to detect backward recoil particles in coincidence with forward angle spectrometers. Such a detector system would allow broader studies of $(e,e'pN)$ reactions and the detection of the products of $3N$ or $N\Delta$ correlations.

Acknowledgments
Many thanks to theorists Mark Strikman and Misak Sargian for their invaluable help in teaching us the physics of correlations; and well as thanks to theorists Claudio Ciofi degli Atti, Leonid Frankfurt, Jean-Marc Laget, Rocco Schiavilla, and many others for thought-provoking discussions and help in understanding the results. Thanks to experimentalists Stepan Stepanya, Shalev Gilad, William "Bill" Bertozzi, John Watson, John Arrington, Donal Day, and many others who have helped push forward the short-range correlations program at Jefferson Lab. But most of all, our thanks goes to the people who received their Ph.D.'s doing the research and without whom these exciting results would not have been possible. In Hall A, thank to Nilanga Liyanage for his work on the E89-003 ^{16}O$(e,e'p)$ experiment; to Fatiha Benmokhtar and Marat Rvachev for their work on the E89-044 ^{3}He$(e,e'p)$ experiment; and to Peter Monaghan, Ran Shneor, and Ramesh Subedi for their work on the E01-015 ^{12}C$(e,e'pN)$ experiment. In Hall B, thanks to Natalia Dashyan for her work on the CLAS (e,e') analysis and Rustam Niyazov for his work on the CLAS ^{3}He$(e,e'pp)n$ analysis. In Hall C, thanks to Ben Clasie, Dipangkar Dutta, Dave McKee, Kristoff Normand, Daniela Rohe, and Derek van Westrum for their work on nuclear transparency and spectral function measurements; and Nadia Fomin for her high-precision work on $x_B > 1$ reactions. And finally, we remember our deceased friend Kim Egiyan, who, shorty before his death, found the signatures for short-range correlations that he had spent his life searching for.

References
[1] A. Rios, A. Polls and W. H. Dickhoff, Phys. Rev. C **79**, 064308 (2009).
[2] L. Frankfurt, M. Sargsian and M. Strikman, Int. J. Mod. Phys. A **23**, 2991 (2008).
[3] L. L. Frankfurt, M. I. Strikman, D. B. Day and M. Sargsian, Phys. Rev. C **48**, 2451 (1993).
[4] L. L. Frankfurt and M. I. Strikman, Phys. Rept. **76**, 215 (1981).
[5] K. S. Egiyan *et al.*, Phys. Rev. C **68**, 014313 (2003).
[6] K. S. Egiyan *et al.*, Phys. Rev. Lett. **96**, 082501 (2006).
[7] M. Strikman, CERN Cour. **45N9**, 37 (2005).
[8] I. Passchier *et al.*, Phys. Rev. Lett. **88**, 102302 (2002).
[9] A. Nogga *et al.*, Phys. Rev. C **67**, 034004 (2003).
[10] J. Arrington *et al.*, Phys. Rev. Lett. **82**, 2056 (1999).
[11] N. Fomin, arXiv:0812.2144.
[12] C. Ciofi degli Atti and C. B. Mezzetti, Phys. Rev. C **79**, 051302 (2009).
[13] O. Benhar, A. Barocini and S. Fantoni, Nucl. Phys. **A505**, 267 (1989).
[14] O. Benhar, N. Farina, H. Nakamura, M. Sakuda and R. Seki, Phys. Rev. D **72**, 053005 (2005).
[15] L. Lapikás, Nucl. Phys. **A553**, 297 (1993).
[16] J. J. Kelly, Adv. Nucl. Phys. **23**, 75 (1996).

[17] N. Liyanage *et al.*, Phys. Rev. Lett. **86**, 5670 (2001).
[18] K. G. Fissum *et al.*, Phys. Rev. C **70**, 034606 (2004).
[19] D. Rohe *et al.*, Phys. Rev. Lett. **93**, 182501 (2004).
[20] H. Muther, G. Knehr and A. Polls, Phys. Rev. C **52**, 2955 (1995).
[21] D. Abbott *et al.*, Phys. Rev. Lett. **80**, 5072 (1998).
[22] K. Garrow *et al.*, Phys. Rev. C **66**, 044613 (2002).
[23] D. Rohe *et al.*, Phys. Rev. C **72**, 054602 (2005).
[24] F. Benmokhtar *et al.*, Phys. Rev. Lett. **94**, 082305 (2005).
[25] M. M. Rvachev *et al.*, Phys. Rev. Lett. **94**, 192302 (2005).
[26] J. M. Laget, Phys. Lett. B **609**, 49 (2005).
[27] R. B. Wiringa, V. G. J. Stoks and R. Schiavilla, Phys. Rev. C **51**, 38 (1995).
[28] C. Ciofi degli Atti and L. P. Kaptari, Phys. Rev. C **71**, 024005 (2005).
[29] C. Ciofi degli Atti and L. P. Kaptari, Phys. Rev. Lett. **95**, 052502 (2005).
[30] C. Ciofi delgi Atti and L. P. Kaptari, Phys. Rev. Lett. **100**, 122301 (2008).
[31] D. U. L. Yu, Annals of Physics **38**, 392 (1966).
[32] L. L. Frankfurt and M. I. Strikman, Phys. Rept. **160**, 235 (1988).
[33] L. J. H. M. Kester *et al.*, Phys. Rev. Lett. **74**, 1712 (1995).
[34] K. I. Blomqvist *et al.*, Phys. Lett. B **421**, 71 (1998).
[35] C. J. G. Onderwater *et al.*, Phys. Rev. Lett. **81**, 2213 (1998).
[36] D. L. Groep *et al.*, Phys. Rev. C **63**, 014005 (2001).
[37] R. G. Arnold *et al.*, Phys. Rev. C **42**, 1 (1990).
[38] J. M. Laget, Phys. Lett. B **199**, 493 (1987).
[39] L. L. Frankfurt, M. M. Sargsian and M. I. Strikman, Phys. Rev. C **56**, 1124 (1997).
[40] M. M. Sargsian, Int. J. Mod. Phys. E **10**, 405 (2001).
[41] J. Alcorn *et al.*, Nucl. Instrum. Meth. A **522**, 294 (2004).
[42] D. J. J. de Lange *et al.*, Nucl. Instrum. Meth. A **412**, 254 (1998).
[43] R. Shneor *et al.*, Phys. Rev. Lett. **99**, 072501 (2007).
[44] A. Tang *et al.*, Phys. Rev. Lett. **90**, 042301 (2003).
[45] C. Ciofi degli Atti and S. Simula, Phys. Rev. C **53**, 1689 (1996).
[46] E. Piasetzky *et al.*, Phys. Rev. Lett. **97**, 162504 (2006).
[47] I. Mardor, Y. Mardor, E. Piasetzky, J. Alster and M. M. Sargsian, Phys. Rev. C **46**, 761 (1992).
[48] D. Higinbotham, E. Piasetzky and M. Strikman, CERN Cour. **49N1**, 22 (2009).
[49] R. Subedi *et al.*, Science **320**, 1476 (2008).
[50] R. Schiavilla, R. B. Wiringa, S. C. Pieper and J. Carlson, Phys. Rev. Lett. **98**, 132501 (2007).
[51] M. M. Sargsian, T. V. Abrahamyan, M. I. Strikman and L. L. Frankfurt, Phys. Rev. C **71**, 044615 (2005).
[52] M. Alvioli, C. Ciofi degli Atti and H. Morita, Phys. Rev. Lett. **100**, 162503 (2008).
[53] R. A. Niyazov *et al.*, Phys. Rev. Lett. **92**, 052303 (2004).
[54] J. M. Laget, J. Phys. G **14**, 1445 (1988).
[55] G. Audit *et al.*, Nucl. Phys. **A614**, 461 (1997).
[56] J.-M. Laget, Phys. Rev. C **35**, 832 (1987).
[57] P. Norton, Rep. Prog. Phys. **66**, 1253 (2003).
[58] D. F. Geesaman, K. Saito and A. W. Thomas, Ann. Rev. Nucl. Part. Sci. **45**, 337 (1995).
[59] L. B. Weinstein *et al.*, Phys. Rev. Lett. **106**, 052301 (2011).
[60] R. B. Wiringa, R. Schiavilla, S. C. Pieper and J. Carlson, Phys. Rev. C **78**, 021001 (2008).
[61] M. M. Sargsian *et al.*, J. Phys. G **29**, R1 (2003).
[62] F. Froemel and S. Leupold, Phys. Rev. C **76**, 035207 (2007).

Journal of Physics: Conference Series **299** (2011) 012011 doi:10.1088/1742-6596/299/1/012011

Medium Modifications of Hadron Properties and Partonic Processes

W. K. Brooks[1], **S. Strauch**[2] **and K. Tsushima**[3]

[1] Universidad Técnica Federico Santa María, Valparaíso, Chile
[2] University of South Carolina, Columbia, South Carolina 29208, USA
[3] University of Adelaide, South Australia 5005, Australia

Abstract. Chiral symmetry is one of the most fundamental symmetries in QCD. It is closely connected to hadron properties in the nuclear medium *via* the reduction of the quark condensate $< \bar{q}q >$, manifesting the partial restoration of chiral symmetry. To better understand this important issue, a number of Jefferson Lab experiments over the past decade have focused on understanding properties of mesons and nucleons in the nuclear medium, often benefiting from the high polarization and luminosity of the CEBAF accelerator. In particular, a novel, accurate, polarization transfer measurement technique revealed for the first time a strong indication that the bound proton electromagnetic form factors in ^{4}He may be modified compared to those in the vacuum. Second, the photoproduction of vector mesons on various nuclei has been measured *via* their decay to e^+e^- to study possible in-medium effects on the properties of the ρ meson. In this experiment, no significant mass shift and some broadening consistent with expected collisional broadening for the ρ meson has been observed, providing tight constraints on model calculations. Finally, processes involving in-medium parton propagation have been studied. The medium modifications of the quark fragmentation functions have been extracted with much higher statistical accuracy than previously possible.

1. Introduction

One of the most exciting topics in nuclear physics is to study how the properties of hadrons are modified by the nuclear environment, and how such modifications are related to the properties of nuclei and nuclear phenomena. Since nucleons and mesons are made of quarks, antiquarks and gluons, one expects their internal structure to be modified when placed in a nuclear medium or atomic nuclei [1]. From the QCD point of view, this is primarily the manifestation of modifications of the quark condensate in the nuclear medium, which is closely associated with the partial restoration of chiral symmetry. Similarly, medium modifications of partonic processes are directly tied to the gluon density of the medium. The modern view of the nucleus must inescapably include these ingredients, which come directly from the properties of the QCD Lagrangian.

There is little doubt that, at sufficiently high nuclear density and/or temperature, quarks and gluons are the correct degrees of freedom. By contrast, the general success of conventional nuclear physics (with *effective* interactions) indicates that nucleons and mesons provide a good starting point for describing a nucleus at low energy. Therefore, a consistent nuclear theory describing the transition from nucleon and meson degrees of freedom to quarks and gluons is truly required to describe nuclei and nuclear matter over a wide range of density and temperature. Of course, theoretically, lattice QCD simulations may eventually give reliable information on the density and temperature dependence of hadron properties in the nuclear medium. However, current simulations have just started to study the 2-body, nucleon-nucleon [2] and nucleon-hyperon interactions [3], which is still very far from what is needed for the description of a finite nucleus. At present, although one is forced to rely on some models, it is a very important step to understand the main features of nuclear phenomena and the structure of finite nuclei based on the quark and gluon degrees of freedom [4].

We know that explicit quark degrees of freedom are certainly necessary to understand deep-inelastic scattering (DIS) at momentum transfers of several GeV. In particular, the nuclear EMC effect [5] has suggested that it is vital to include some dynamics beyond the conventional nucleon-meson treatment of nuclear physics to explain the nuclear structure function data [6, 7]. This leads to the extraction of nuclear parton, or bound-nucleon parton distributions [8, 9]. In-medium structure functions were studied at Jefferson Lab in an EMC-type experiment in heavy nuclei, emphasizing the large-x region [10, 11], and recently also in few-body nuclei [12, 13]. Another series of Jefferson Lab experiments focus on the longitudinal-transverse separation of in-medium structure functions [14, 15].

Furthermore, the search for evidence of some modification of nucleon properties in medium has recently been extended to the nucleon electromagnetic form factors in polarized $(\vec{e}, e'\vec{p})$ scattering experiments on ^{16}O [16, 17] and ^{4}He [18, 19, 20, 21, 22] nuclei at MAMI and Jefferson Lab. These experiments observed the double ratio of proton-recoil polarization transfer coefficients in the scattering off nuclei with respect to the elastic $^4\mathrm{He}(\vec{e}, e'\vec{p})$ reaction at four-momentum transfers squared of several 100 (MeV/c)^2 to a few (GeV/c)^2. The results from ^{4}He strongly hint at the need to include medium modifications of the proton electromagnetic form factors. The recent Coulomb sum rule experiment at Jefferson Lab [23] is a precision measurement of response functions in quasi-elastic electron scattering and will provide important data on the nucleon charge properties in the nuclear medium, providing important constraints on the electric form factor of the proton.

The photoproduction of vector mesons on various nuclei has been studied to search for possible in-medium modifications of the vector-meson spectral functions. The vector mesons, ρ, ω, and ϕ, are observed *via* their decay to e^+e^-, in order to reduce the effects of final state interactions in the nucleus [24, 25, 26]. However, in these experiments mesons measured were moving relatively fast. In-medium modifications at lower momenta have not yet been observed experimentally. One can expect that the quarks inside the fast moving mesons do not have enough time to adjust to the surrounding nuclear medium. Thus, further experiments with nearly stopped meson kinematics are certainly required to draw more concrete conclusions.

Another interesting class of medium modifications consists of the changes in fundamental partonic processes that are embedded in the medium. The well-established features of quark fragmentation in vacuum are strongly modified in the cold nuclear medium, a topic best addressed in semi-inclusive DIS on nuclei. A second effect is the alteration of the process whereby a highly virtual quark radiates gluons, leading to an increased rate of energy loss. Precise characterization of these effects in the cold medium, in addition to giving new insight into QCD, may also aid interpretation of observables in the hot medium. These medium modifications of fundamental partonic vacuum processes are now beginning to be understood theoretically after more than two decades of investigation.

In this article, the medium modifications of nucleon structure, electromagnetic form factors and structure functions are focused on in Section 2, while the modifications of meson structure, masses and widths are focused on in Section 3. Section 4 treats the medium modification of partonic processes, discussing modification of fragmentation functions and medium-induced quark energy loss. An overview of experiments at Jefferson Lab, dedicated to the study of medium modifications of both nucleon and meson properties, is given in Table 1.

Table 1. Jefferson Lab experiments to study medium modifications.

In-medium property	Reaction	Target	Exp. proposals
Form factors	$(\vec{e}, e'\vec{p})$	^{1}H, ^{2}H, ^{4}He, ^{16}O	[27, 19, 20, 16]
Nucleon charge	(e, e')	^{4}He, ^{12}C, ^{56}Fe, ^{208}Pb	[23]
Structure function F_2	(e, e')	^{1}H, ^{2}H, ^{3}He, ^{4}He, Be, C, Al, Fe, Cu, Au	[10, 11, 12]
Structure functions F_2, σ_L/σ_T	(e, e')	^{1}H, ^{2}H, C, Al, Fe	[14, 15]
Vector-meson spectral functions	(γ, e^+e^-)	^{2}H, ^{12}C, Fe-Ti	[24]
Fragmentation functions, Δp_T	$(e, e'X)$	^{2}H, C, Al, Fe, ^{120}Sn, Pb	[28]

2. Nucleon structure modification

2.1. Modification of electromagnetic form factors

2.1.1. In-medium form factors Whether the nucleon changes its fundamental properties while embedded in a nuclear medium has been a long-standing question in nuclear physics, attracting experimental and theoretical attention. QCD is established as the theory of the strong nuclear force but the degrees of freedom observed in nature, hadrons and nuclei, are different from those appearing in the QCD Lagrangian, quarks and gluons. There are no calculations available for nuclei within the QCD framework. Nuclei are effectively and well described as clusters of protons and neutrons held together by a strong, long-range force mediated by meson exchange [29].

Distinguishing possible changes in the structure of nucleons embedded in a nucleus from more conventional many-body effects like meson exchange currents (MEC), isobar configurations (IC) or final state interactions (FSI) is only possible within the context of a model. Therefore, interpretation of an experimental signature as an indication of nuclear medium modifications of the form factors is better motivated if this results in a more economical description of the nuclear many-body system. In this context, a calculation by Lu *et al.* [30], using a quark-meson coupling (QMC) model, suggests a measurable deviation from the free-space electromagnetic form factor over the four-momentum transfer squared Q^2 range $0 < Q^2 < 2.5$ (GeV/c)^2. Similar measurable effects have been calculated in a light-front constituent quark model by Frank *et al.* [31], a modified Skyrme model by Yakshiev *et al.* [32], a chiral quark soliton model (CQS) by Smith and Miller [33], and the Nambu-Jona-Lasinio model of Horikawa and Bentz [34]. Medium modifications of nucleon properties in nuclear matter and finite nuclei have been also discussed by Wen and Shen [35]. Furthermore, the nuclear modification of axial form factors [36] also may be measured.

The connection between the modifications induced by the nuclear medium of the nucleon form factors and of the deep inelastic structure functions was discussed by Liuti [37] using the concept of generalized parton distributions (GPDs). Guzey *et al.* [38] have studied incoherent deeply virtual Compton scattering (DVCS) on ^{4}He in the ^{4}He$(e, e'\gamma p)X$ reaction, which probes medium modifications of the bound nucleon GPDs and elastic form factors. They have also investigated medium modification of the quark contribution to the spin sum rule [39]. The relation between the medium modified form factors and structure functions was also discussed in Ref. [40] in the framework of quark-hadron duality, where the size of the medium modification in the former was used to place constraints on models of the nuclear EMC effect which assume a large deformation of the intrinsic structure of the nucleon in medium.

The best experimental constraints on the changes in the electromagnetic form factors come from the analysis of y-scaling data. For example, in the iron nucleus (Fe) the nucleon root-mean-square radius cannot vary by more than 3% [41]. However, in the kinematic range covered by this y-scaling analysis, the eN cross section is predominantly magnetic, so this limit applies essentially to G_M. (As the electric and magnetic form factors contribute typically in the ratio 1:3 the corresponding limit on G_E would be nearer 10%.) For the QMC model, the calculated increase in the root-mean-square radius of the magnetic form factors is less than 0.8% at ρ_0 [30]. For the electric form factors the best experimental limit seems to come from the Coulomb sum-rule, where a variation bigger than 4% would be excluded [42]. This is similar in size to the variations calculated in the QMC model (*e.g.*, 5.5% for G_E^p at ρ_0) [30] and not sufficient to reject them.

2.1.2. Medium modifications from recoil polarization experiments One of the most intuitive methods to investigate the properties of nucleons inside nuclei is quasi-elastic scattering off nuclei. Since the charge and magnetic responses of a single nucleon are quite well studied from elastic scattering experiments, measuring the same response from quasi-elastic scattering off nuclei and comparing with a single nucleon is likely to shed light on the question. The polarization transfer ratio in elastic $\vec{e}p$ scattering, P'_x/P'_z, is directly proportional to the ratio of the electric and magnetic form factors of the proton [43]; here P'_x and P'_z are the polarization transfer observables transverse and longitudinal to the momentum transfer direction. When such measurements are performed on a nuclear target in quasi-elastic kinematics, the polarization transfer observables are sensitive to the form factor ratio of the proton embedded in the nuclear medium. Experimental results for the polarization transfer ratio are conveniently expressed in terms of the polarization double ratio

$$R = \frac{(P'_x/P'_z)_A}{(P'_x/P'_z)_{^1\mathrm{H}}}, \tag{1}$$

in order to emphasize differences between the in-medium and free values. Here the polarization transfer ratio for the quasi-elastic proton knockout $A(\vec{e}, e'\vec{p})$ reaction is normalized to the hydrogen polarization transfer ratio measured in the identical setting. Such a double ratio cancels nearly all experimental systematic uncertainties.

Experiment E89-033 was the first to measure the polarization transfer in a complex nucleus, ^{16}O [17]. The results are consistent with predictions of relativistic calculations based on the free proton from factor with an experimental uncertainty of about 18%. Earlier polarization transfer experiments have studied nuclear medium effects in deuterium [44, 45, 46] at the Mainz microtron (MAMI) and MIT-Bates. More recently, polarization transfer data on ^{2}H were measured in Jefferson Lab experiment E89-028 [47]. The data are shown in Fig. 1 and compared with a model calculation by Arenhövel, which includes final state interactions, meson exchange currents, and isobar configurations, as well as relativistic contributions (RC) of leading order in p/m_N to the kinematic wave function boost and to the nucleon current. Arenhövel's calculation describes the ^{2}H data well. Within statistical uncertainties, no evidence of medium modifications is found. As the sampled nuclear density is small and the bound proton is nearly on-shell in the kinematics of this experiment, it is not surprising that there are no indications for medium modifications of the proton electromagnetic form factors in the ^{2}H data.

One might expect to find larger medium effects in ^{4}He, with its significantly higher density. Although estimates of the many-body effects in ^{4}He may be more difficult than in ^{2}H, calculations for ^{4}He indicate they are small [48]. The first $^4\mathrm{He}(\vec{e}, e'\vec{p})^3\mathrm{H}$ polarization transfer measurements were performed at MAMI at $Q^2 = 0.4$ (GeV/c)^2 [18] and at Jefferson Lab Hall A at $Q^2 = 0.5$, 1.0, 1.6, and 2.6 (GeV/c)^2, E93-049 [21]. Experiment E03-104 extended these measurements with two high-precision points at $Q^2 = 0.8$ and 1.3 (GeV/c)^2 [22]. All these data were taken in quasi-elastic kinematics at low missing momentum with symmetry about the three-momentum transfer direction to minimize conventional many-body effects in the reaction. The missing-mass technique was used to identify ^{3}H in the final state.

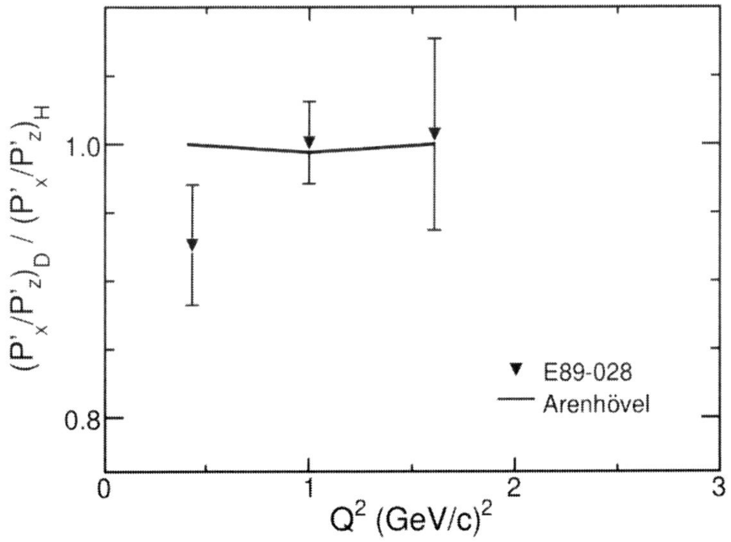

Figure 1. Bound-to-free polarization transfer double ratio R for $^2\mathrm{H}(\vec{e}, e'\vec{p})n$ (triangles) from [47] as a function of Q^2. The curve shows results of a calculation by Arenhövel; see [47].

The ^{4}He polarization transfer double ratio is shown in Figure 2. The new data from E03-104 [22] are consistent with the previous data from E93-049 [21] and MAMI [18]. The polarization transfer ratio (P'_x/P'_z) in the $(\vec{e}, e'\vec{p})$ reaction on helium is significantly different from those on deuterium or hydrogen. The helium data are compared with results of a relativistic distorted-wave impulse approximation (RDWIA) calculation by the Madrid group [49, 50] (dotted curve). In this model FSI are incorporated using relativistic optical potentials that distort the final nucleon wave function. MEC are not explicitly included in the Madrid calculation. Predictions by Meucci *et al.* [51] show that the two-body current (the seagull diagram) effects are generally small (less than 3% close to zero missing momenta) and visible only at high missing momenta. It can be seen that the Madrid RDWIA calculation (dotted curve) overestimates the data by about 6%. The calculations shown uses the Coulomb gauge, the *cc1* and *cc2* current operators as defined in [52], and the MRW optical potential of [53]. The *cc1* current operator gives lower values of R than the *cc2* operator. In general, various choices for, *e.g.*, spinor distortions, current operators, and relativistic corrections affect the theoretical predictions by $\leq 3\%$ within the RDWIA model, and presently cannot explain the disagreement between the data and the RDWIA calculation. We note that these relativistic calculations provide good descriptions of, *e.g.*, the induced polarizations as measured at Bates in the $^{12}\mathrm{C}(\mathrm{e},\mathrm{e}'\vec{p})$ reaction [54, 50] and of A_{TL} in $^{16}\mathrm{O}(e, e'p)$ as previously measured at Jefferson Lab [55]. After including the density-dependent medium-modified form factors from the QMC [30] or CQS [33] models in the RDWIA calculation (solid and dashed curves), good agreement with the polarization transfer data is obtained.

This agreement has been interpreted as possible evidence of proton medium modifications [21]. This interpretation is based on the description of the data in a particular model in terms of medium modifications of nucleon form factors and requires excellent control of the reaction mechanisms such as meson exchange currents, isobar configurations, and final state interactions. In fact, there is an alternative interpretation of the observed suppression of the polarization transfer ratio within a more traditional calculation by Schiavilla *et al.* [56] (shaded band). The latter calculation uses free nucleon form factors and explicitly includes MEC effects

Journal of Physics: Conference Series **299** (2011) 012011 doi:10.1088/1742-6596/299/1/012011

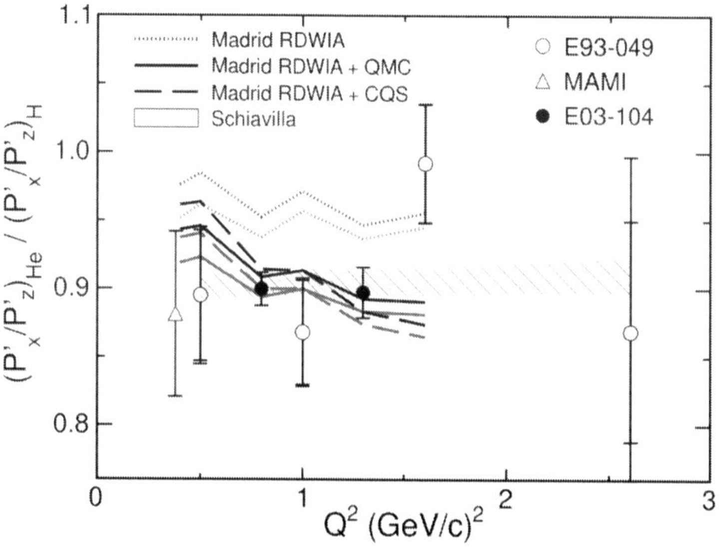

Figure 2. ^{4}He$(\vec{e}, e'\vec{p})^3$H polarization transfer double ratio R as a function of Q^2 from Mainz [18] and Jefferson Lab experiments E93-049 [21] (open symbols) and E03-104 [22] (filled circles). The data are compared to calculations from Schiavilla *et al.* [56] and the Madrid group [49, 50] using the *cc1* (lower set of curves) and *cc2* (upper set of curves) current operators. In-medium form factors from the QMC [30] (solid curves) and CQS [33] (dashed curves) models were used in two of the Madrid calculations. Not shown are a relativistic Glauber model calculation by the Ghent group [57] and results from Laget [48] which give both a value of $R \approx 1$.

which suppress R by almost 4%. The FSIs are treated within the optical potentials framework and include both a spin-dependent and spin-independent charge-exchange term; the spin-orbit terms, however, are not well constrained by data. In the model of Schiavilla *et al.*, the final state interaction effects suppress R by an additional 6% bringing this calculation also in good agreement with the data within the statistical uncertainties associated with the Monte Carlo technique in this calculation. It should be noted that charge-exchange terms are not taken into account in the Madrid RDWIA calculation. The difference in the modeling of final state interactions is the origin of the major part of the difference between the results of the calculations by Udias *et al.* [49, 50] and Schiavilla *et al.* [56] for the polarization observables.

Effects from final state interactions can be studied experimentally with the induced polarization, P_y, which is a direct measure of final state interactions. Induced-polarization data were taken simultaneously to the polarization transfer data. Figure 3 shows the data for P_y. The induced polarization is small at the low missing momenta in this reaction. The sizable systematic uncertainties are due to possible instrumental asymmetries. Dedicated data have been taken during E03-104 to study these and work is underway to significantly reduce the systematic uncertainties in P_y in the final analysis. The data are compared with the results of the calculations from the Madrid group and Schiavilla *et al.* at missing momenta of about zero. To facilitate this comparison, the data have been corrected for the spectrometer acceptance. The preliminary data suggest that the induced polarization (and thus the final state interaction) is underestimated in the MRW optical potential in the model of the Madrid group and overestimated in the model of Schiavilla *et al.* Note that the charge-exchange terms, particularly the spin-dependent one, gives the largest contribution to P_y in the Schiavilla *et al.*

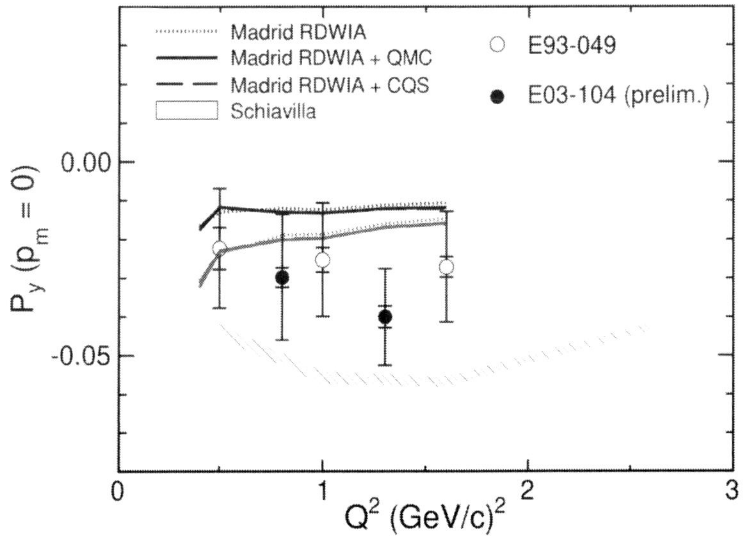

Figure 3. Induced polarization data from Jefferson Lab experiment E93-049 [21] along with preliminary results from experiment E03-104 [58]. The data are compared to calculations from the Madrid group [49, 50] and Schiavilla *et al.* [56] as in Fig. 2. Note that the comparison is made for missing momentum $p_m \approx 0$; the experimental data have been corrected for the spectrometer acceptance for this comparison.

calculation. The induced polarization proves to be sensitive to the choice of optical potential allowing this observable to be used to constrain theoretical models of FSI. A comparison of the model calculations in Figure 2 and Figure 3 shows that the in-medium form factors mostly affect the ratio of polarization transfer observables and to a lesser extent the induced polarization. It is a great advantage of E03-104 to have access to both the polarization transfer and the induced polarization.

In summary, polarization transfer in the quasi-elastic $(e, e'p)$ reaction is sensitive to possible medium modifications of the bound-nucleon form factor, while at the same time largely insensitive to other reaction mechanisms. Currently, the $^4\mathrm{He}(\vec{e}, e'\vec{p})^3\mathrm{H}$ polarization transfer data can be well described by either the inclusion of medium-modified form factors or strong charge-exchange FSI in the models. The analysis of the new high-precision induced-polarization data from Jefferson Lab Hall A has recently been finalized and the results provide now a more stringent test of these calculations [59].

2.1.3. Coulomb sum rule Experimental constraints on possible proton medium modifications are available for the electric form factor G_E from tests of the Coulomb sum rule (CSR) in quasi-elastic electron scattering off nuclei. This sum rule states that at large enough three-momentum transfer q the integration of the charge response $R_L(q, \omega)$ of a nucleus over the full range of energy loss ω, excluding the elastic peak at the lower limit, should be equal to the total charge Z of the nucleus [60]

$$S_L(q) = \frac{1}{Z} \int_{0^+}^{\infty} \frac{R_L(q, \omega)}{\tilde{G}_E^2} d\omega \,, \qquad (2)$$

where $\tilde{G}_E = (G_E^p + (N/Z)G_E^n)\zeta$ takes into account the nucleon charge form factors and a relativistic correction, ζ. Pauli blocking at very small momentum transfer as well as short- and long-range correlations at medium and larger momentum transfer result in a quenching of the Coulomb sum rule. However, at sufficiently high momentum transfer, only short range correlation effects remain. They have been estimated by various theoretical calculations using different NN forces and found to be responsible for at most 10% quenching of the CSR integral. As a result, any further quenching of this quantity at sufficiently high momentum transfer may indicate the possibility of modified properties of the nucleon inside the nucleus [61].

Starting from this quite simple idea, for the past 20 years, various laboratories around the world performed experiments in diverse conditions, but the final conclusion is still controversial: on ^{4}He, good agreement between theory and experiment is obtained when using free-nucleon form factors [62] and similar results were obtained on ^{12}C and ^{56}Fe [42]. However, a re-examination of the extraction of the longitudinal and transverse response functions on medium-weight and heavy nuclei (^{40}Ca, ^{48}Ca, ^{56}Fe, ^{197}Au, ^{208}Pb and ^{238}U) in the framework of the Effective Momentum Approximation showed quenching of the Coulomb sum rule [60, 61]. This quenching corresponds to a relative change of the proton charge radius of $13 \pm 4\%$ and it was interpreted as an indication for a change of the nucleon properties inside the nuclear medium.

Jefferson Lab Experiment E05-110 is a precision test of the Coulomb sum rule [23]. The experiment measured the inclusive electron scattering cross sections off ^{4}He, ^{12}C, ^{56}Fe, and ^{208}Pb in the quasi-elastic region and will make it possible to study A- or density-dependent effects of the CSR. The measurement covered a wide range of momentum transfers, 0.55 GeV/c $\leq q \leq$ 0.9 GeV/c, and thereby expanded the rather limited coverage of previous experiments. This measurement will provide data on the saturation or quenching of the Coulomb sum rule in a kinematic region where a clean access of the nucleon charge properties in the nuclear medium is plausible, free from Pauli blocking and long range correlations. Also, in this region short-range correlations are found to explain less than 10% reduction from the naive Coulomb sum rule. The kinematic range of the experiment overlaps with previous experiments at the low-q side and it provides a unique opportunity to investigate the q evolution of the Coulomb sum from the nucleonic to sub-nucleonic scale. Improved control of systematic uncertainties is possible through the measurement of quasi-elastic scattering cross sections at four different angles, providing four virtual photon polarization values ϵ at almost equal spacing. Two additional ϵ settings will provide a tool to study the effect of the Coulomb corrections. Also, the use of two different thicknesses for ^{208}Pb target will help to reduce any systematic errors from the radiative corrections. Data taking was completed in early 2008 and the analysis of the data, particularly the extraction of the longitudinal and transverse response functions, is underway.

2.2. Modification of unpolarized structure functions

The European Muon Collaboration (EMC) measured the ratios of structure functions of iron and deuterium in deep-inelastic scattering and found a reduction of the structure function of a nucleus compared to that of a free nucleon at intermediate values of Bjorken x, $0.3 < x < 0.6$. These measurements provided the first clear evidence that the quark structure of nucleons and nuclei were significantly different. Indeed, the nucleon bound in a nucleus carries less momentum than in free space. However, the specific causes of the modifications observed in nuclear structure functions have not yet been identified with certainty [63]. Although the simple kinematic effects of binding energy and Fermi motion do not account for the EMC effect [63], they do play a significant role at large x and are for those kinematics important in understanding the modification of the nuclear quark distributions [64]. A detailed study of these effects, however, is hindered by limited data. Most of the data are available only for heavy nuclei, yet few-body nuclei have the advantage that exact calculations are available. Particularly, uncertainties in the effect of binding due to uncertainties in the nuclear structure are reduced. Calculations also predict large differences in both the magnitude and x dependence of the EMC effect in ^{3}He and ^{4}He, whereas the effect in various heavy nuclei show the same x dependence. The study of few-body nuclei can thus help to shed more light on the EMC effect. Data are furthermore sparse above $x = 0.8$, where binding and other short-distance effects should dominate. The F_2 structure function in inclusive electron scattering off nuclei at large Q^2 has been measured at Jefferson Lab Hall C in a series of experiments: E89-008 [10], E02-019 [11], and most recently in E03-103 [12]. The latter will provide better constraints on the effect of binding in the large-x region.

Jefferson Lab Experiment E03-103 provides a precision measurement of the EMC effect in both few-body nuclei and heavy nuclei. The experiment ran in 2004 in Hall C and measured inclusive electron scattering from ^{1}H, ^{2}H, ^{3}He, and ^{4}He as well as Be, C, Al, Cu, and Au over a broad range of x ($0.3 < x < 1.0$) up to $Q^2 \approx 8.0$ GeV^2. The ^{3}He structure function was measured at x larger than 0.5 for the first time. Results from E03-103 along with previous SLAC results are shown in Fig. 4 [13]. The Jefferson Lab data are in good agreement with these previous

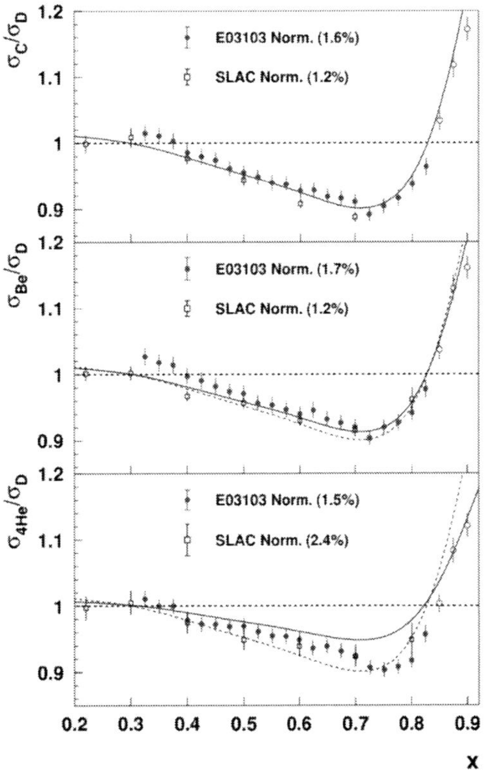

Figure 4. EMC cross-section ratios for ^{12}C (top), ^{9}Be, and ^{4}He from E03-103 [13] and SLAC [65]. The ^{9}Be results include a correction for the neutron excess (see [13]). Filled (open) circles denote W^2 above (below) 4 GeV^2. The solid curve is the A-dependent fit to the SLAC data, while the dashed curve is the fit to ^{12}C. Normalization uncertainties are shown in parentheses for both measurements. Figure taken from [13].

results and improve significantly the precision of the EMC ratio at larger x. The findings of E03-103 indicate that the nuclear dependence of the cross section is comparable for ^{4}He, ^{9}Be, and ^{12}C (dashed curve); whereas the observed nuclear effects are clearly smaller for ^{3}He (not shown). The data thus do not support previous A-dependent fits to the EMC effect nor fits dependent on the *average* nuclear density. The average nuclear density is relatively low in ^{9}Be. The results rather suggest that the nuclear dependence of the quark distributions may depend on the *local* nuclear environment. The better coverage at larger x was achieved by relaxing the

Journal of Physics: Conference Series **299** (2011) 012011 doi:10.1088/1742-6596/299/1/012011

usual W^2 cut for deep inelastic scattering from 4 GeV^2 down to below 2 GeV^2 after previous data showed that the structure function scales also at lower values of W (quark-hadron duality) [66]. Specific data were taken to verify this assumption. Data from E02-019 will further allow a detailed study of the scaling of nuclear structure functions in the $x > 1$ region.

Several experiments in Hall C measured the longitudinal-transverse separated structure functions F_2 and the ratio $R = \sigma_L/\sigma_T$ from nuclear targets at $Q^2 < 4$ (GeV/c)^2 [14, 15]. The data on R for hydrogen and deuterium have been published [67], and the data on the heavier targets are being analyzed. When fully available, these data will greatly increase our understanding of nuclear effects.

2.3. *Modification of polarized structure functions*

The well-known EMC effect has fascinated a generation of physicists and has stimulated much theoretical work. There is little question that the explanation of this phenomenon involves the modification of hadron structure in the medium, although the details of the descriptions vary. It is then to be expected that there may be an analogous change in the *spin* structure of the nucleon as well. Detailed calculations support this conjecture [68] and suggest that the observable modification is frequently larger than the unpolarized EMC effect.

Letters of intent have proposed measurements at 6 GeV [69, 70], however, the best measurement will be performed with the upgraded 12 GeV Jefferson Lab accelerator due to improved kinematic reach and better control of systematic uncertainties. The foundation that provides confidence that these studies can be realized is the large and successful Jefferson Lab program of polarized target studies on ^{1}H, ^{2}H, and ^{3}He described elsewhere in this document.

The method requires polarizing a nuclear target that is chosen to satisfy the following conditions:

- a large fraction of the overall nuclear polarization is concentrated on one nucleon within the nucleus;
- reliable calculations can be performed of the degree of polarization for the highly polarized nucleon;
- sufficient nuclear size to provide a medium environment representative of larger nuclei, while still limiting the dilution factor.

One example of a nucleus that satisfies these conditions is ^{7}Li, for which detailed nuclear structure calculations exist [71, 72]. Asymmetry measurements can constrain the value of the medium-modified g_1 to the few-percent level over the range of interest in x_B from 0.05 to more than 0.5.

3. Meson structure modification

The masses of hadrons are created dynamically and are much larger than the summed masses of their constituents. The generation of hadronic masses is connected to spontaneous breaking of chiral symmetry and in-medium hadron properties, such as their masses and widths, are expected to change with chiral symmetry restoration. Various theoretical models predict an at least partial restoration of chiral symmetry at high temperatures and/or densities. By using effective chiral Lagrangians with a suitable incorporation of the scaling property of QCD, Brown and Rho [1] derived approximative in-medium scaling laws and predict for the ratio of vector-meson masses at nuclear-matter density, ρ_0, compared to their free values $m^*(\rho_0)/m \approx 0.80$. Using QCD sum rules, Hatsuda and Lee [73] obtained the density dependence of this ratio, $m^*(\rho)/m \approx 1 - \alpha(\rho/\rho_0)$, with $\alpha = 0.18 \pm 0.06$. Models based on nuclear many-body effects predict a broadening in the width of the ρ meson with increasing density. This prediction is based on the assumption that many-body excitations may be present with the same quantum numbers and can mix with the hadronic states [74, 75, 76, 77, 78, 79]. Furthermore, due to the uncertainty of the coupling constants as a function of density, the branching ratios are expected to change in the nuclear medium and also distort the invariant mass spectrum of the resonance [80].

First experimental studies of the properties of vector mesons as a function of temperature and density were done in relativistic and ultra-relativistic heavy-ion reactions. The CERES collaboration at CERN performed pioneering investigations of possible ρ-meson medium modifications through e^+e^- pair production in nucleus-nucleus collisions. A di-lepton yield exceeding expectations from hadron decays has been observed in the mass region 0.2 to 0.6 GeV/c^2. An upgraded CERES experiment with improved mass resolution confirmed these results, and found in addition a substantial in-medium broadening of the ρ spectral function over a density dependent shift of the ρ pole mass [81]. The result of $\mu^+\mu^-$ measurements in In-In collisions in the NA60 experiment at CERN has also shown a considerable broadening (doubling of the width) of the ρ spectrum, while no shift in the mass was observed [82, 83, 84]. In these heavy-ion reactions the temperature and/or density is varied and they proceed far from equilibrium under difficult to separate reaction mechanisms. This makes an interpretation of these results in terms of the chiral symmetry structure of the vacuum very challenging.

Medium modifications of mesons at zero temperature and normal nuclear matter density are experimentally accessible in photonuclear or elementary hadronic reactions on heavy nuclei. The best approach, free of final state interactions, is the study of the leptonic decay channel of these mesons. An observation of a medium-modified vector meson invariant mass decrease ($\alpha = 0.09 \pm 0.002$) has been claimed by a KEK-PS collaboration in an experiment where 12 GeV protons were incident on nuclear targets, and the e^+e^- pairs were detected [85, 86, 87]. The Crystal Barrel/TAPS collaboration has reported a 14% downward shift in the mass of the ω, where the analysis focused on the $\pi^0\gamma$ decay of *low-momentum* ω mesons photoproduced on a nuclear target [88]. However, in a recent re-analysis of these data the earlier claim of an in-medium mass shift of the ω could not be confirmed [89]. It is predicted that the ω potential in nuclear matter is attractive and possible to form nuclear bound state [90, 91, 92], if it is produced nearly with zero momentum.

The Jefferson Lab Experiment E01-112 studied the photoproduction of light vector mesons (ρ, ω, and ϕ) on deuterium and the heavier targets of carbon, titanium, and iron [24] with the CEBAF Large Acceptance Spectrometer (CLAS) [93]. The photon beam from the Hall B photon tagging facility covered an energy range from about 0.6 to 3.8 GeV. The CLAS has an excellent electron/pion discrimination, which allowed detection of the vector mesons *via* their rare leptonic decay in the e^+e^- channel and thus avoiding the complications from final state interactions of the hadronic decay channels.

The experimental e^+e^- invariant mass distribution includes the spectral distribution from the decay of the vector mesons ρ, ω, and ϕ. It also contains background from other physical processes, for example meson decays into γe^+e^- or the Bethe-Heitler process. Detailed simulations showed that from all background processes, which lead to correlated e^+e^- pairs, only the Dalitz decay of the ω mesons contributes appreciably in the e^+e^- mass region of interest. To simulate each physics process, a realistic model was employed and corrected for the CLAS acceptance. The events were generated using a code based on a semi-classical Boltzmann-Uehling-Uhlenbeck (BUU) transport model developed by the Giessen group that treats photon-nucleus reactions as a two-step process [94]. In the first step, the incoming photons react with a single nucleon, taking into account various nuclear effects, *e.g.* shadowing, Fermi motion, collisional broadening, Pauli blocking, and nuclear binding. Then, in the second step, the produced particles are propagated explicitly through the nucleus allowing for final state interactions, governed by the semi-classical BUU transport equations. A rather complete treatment of the e^+e^- pair production from γA reactions at Jefferson Lab energies using this code can be found in Ref. [95].

Another source of background for the e^+e^- mass distributions, the combinatorial background, are events where the electron and the positron originate in uncorrelated processes within the same 2 ns wide CEBAF beam bucket. The most salient feature of the uncorrelated sources is that they produce same-charge lepton pairs as well as oppositely charged pairs. This is also true for the measurement of opposite-sign pions or muons for which the combinatorial method has also been used in the past [96, 97]. This method has also been used in the extraction of resonance signals [98] and proton femtoscopy of eA interactions [99]. The combinatorial background is determined by an event-mixing technique. The electrons of a given event are combined with positrons of another event, as the two samples of electrons and positrons are completely uncorrelated. This produces the phase-space distribution of the combinatorial background. This distribution was then normalized to the number of expected opposite-charge pairs, which unambiguously can be determined from same-charge pairs.

The experimental e^+e^- mass distributions are shown in Fig. 5 after subtraction of the combinatorial background. Also shown are BUU model calculations for various e^+e^- channels, fitted to the data, which describe the data very well. The narrow distributions of the ϕ and ω meson, including its Dalitz decay, are readily normalized to the data. The subtraction of these distributions results in the experimental spectra of the ρ mass shown in Fig. 6. The shape of the spectral function is well approximated by a Breit-Wigner function times a factor $1/\mu^3$, where μ is the mass of the e^+e^- pair [26]. Experimental values of the in-medium ρ mass and width are then obtained from fits to the data (solid curves). The in-medium widths of the ρ for the nuclear targets are slightly larger than the free value, *e.g.* $\Gamma = 218 \pm 15$ MeV for the Fe-Ti target, but are well understood as collisional broadening [100] as modeled in the BUU calculations. They are not compatible with the doubling of the ρ width reported by NA60 [82].

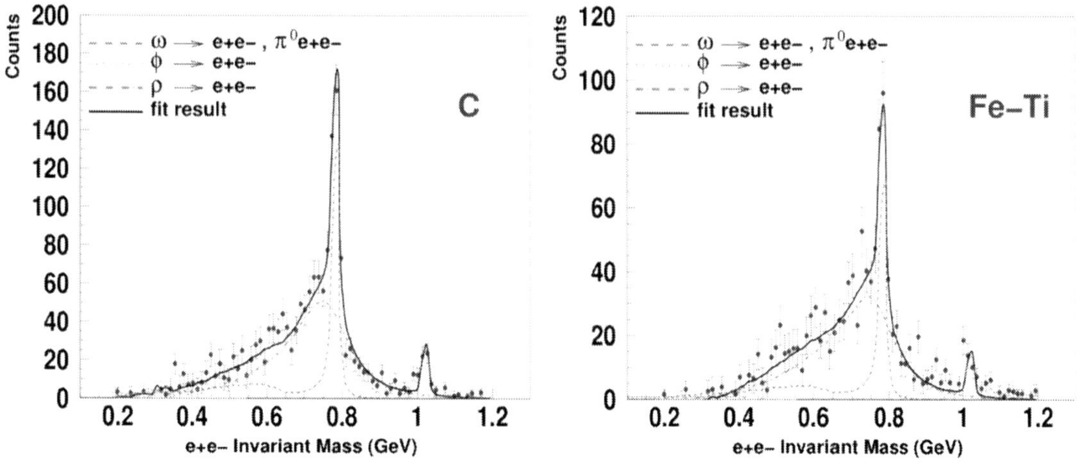

Figure 5. Experimental e^+e^- invariant mass distributions obtained for carbon (left panel) and Fe-Ti (right panel) after subtraction of the combinatorial background. The curves are calculations by the BUU model [101, 102] for various e^+e^- channels. Figures are from [25].

The observed in-medium ρ masses are consistent with the free value. The relative ρ-mass shift for the Fe-Ti target and ρ momenta ranging from 0.8 to 3.0 GeV is $\alpha = 0.02 \pm 0.02$. This is consistent with the predictions of no significant mass shift by the calculations of Ref. [77, 78] and those of Ref. [94, 103] at ρ vector meson momenta > 1 GeV. The total systematic uncertainty for the measured α due to various sources is estimated to be $\Delta\alpha = \pm 0.01$ [26]. The result from

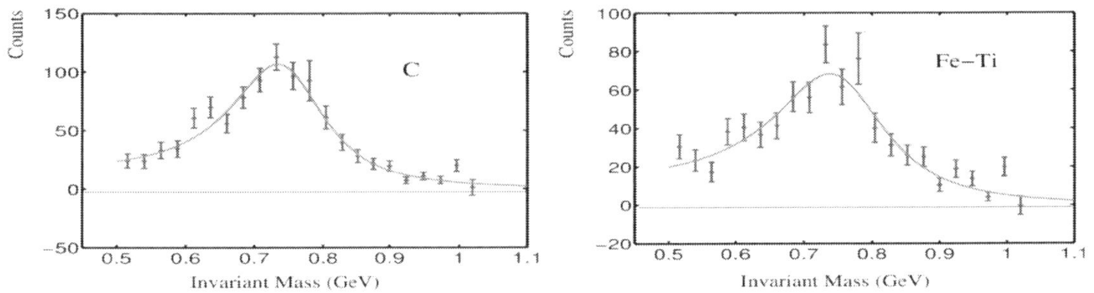

Figure 6. Extracted ρ mass spectra from the CLAS g7 run data for the C (left), and Fe-Ti (right panel) targets. The curves show fits of a Breit-Wigner function times $1/\mu^3$, where μ is the e^+e^- invariant mass. Figures are from [25].

the CLAS g7 run sets an upper limit of $\alpha = 0.04$ with a 95% confidence level. It does not favor the prediction of Refs. [1] and [73] for a 20% mass shift and $\alpha = 0.16 \pm 0.06$ respectively, and is significantly different from other similar experiments [85, 86, 87], where $\alpha = 0.092 \pm 0.002$, with no broadening in the width of the ρ meson. The g7 results are consistent with the result of a recent re-analysis of ω meson photoproduction data [89] which found broadening, but no mass shift, in the ω signal.

The extracted experimental ρ mass spectrum with the unique characteristic of electromagnetic interactions in both the production and decay is well described by the ρ functional form obtained from the exact calculations given in Refs. [95, 104, 105] with no modification beyond standard nuclear many-body effects. Recently, the dilepton spectra were calculated by Riek *et al.* [106, 107] within the same effective hadronic model that describes NA60 data [108] and found to be in good agreement with the CLAS data. With the availability of more sophisticated theoretical models and improved analysis techniques, future experiments with higher statistics are expected to make a conclusive statement about the momentum dependence of the in-medium modifications and the nature of the QCD vacuum.

4. Modification of partonic processes

A tremendous amount of progress has been achieved over the past decade in understanding the modification of partonic processes in the medium. New experiments at Fermilab, DESY, and Jefferson Lab have uncovered a wealth of new information on partonic processes in the *cold* nuclear medium through the Drell-Yan process and through semi-exclusive DIS. Experiments at RHIC have generated a strong interest in the same partonic processes in the *hot* medium through the observation of jet quenching, and these observations will soon be tested with much higher energies at the LHC [109]. The Jefferson Lab experiments [28, 110] in particular promise to uniquely measure spacetime properties of QCD inaccessible to any other experiments, with sufficient luminosity and kinematic reach to achieve a detailed understanding of the processes by which hadrons are formed and of the propagation of quarks both in the medium and in vacuum.

A primary aim of the measurements in the cold medium is to explicate the fundamental QCD processes involved in parton propagation and hadron formation. Particularly in the space-time domain, the well-understood properties of nuclei can be exploited to extract quantities such as the free quark lifetime (the so-called production time) and hadron formation times. In the process of confronting the data with model calculations, the detailed mechanisms of hadron formation can also be constrained and characterized. As a secondary benefit, the insight gained by these studies can in principle be used to better understand the data from the relativistic heavy ion collisions. Much theoretical work is needed to actually accomplish this, however, there are examples of efforts to describe both the hot and cold matter within the same theoretical language [111, 112].

The following discussion centers around two primary interests. First, the modification of fragmentation functions is discussed. The primary observable, the hadronic multiplicity ratio, is a measure of the modification of the fragmentation functions. Through modeling it can be related to hadron formation lengths and fragmentation mechanisms. Second, the topic of quark energy loss is discussed. Here the primary observable is the broadening of the distributions in hadron transverse momentum. It is expected that the production time can be estimated from this variable within certain kinematic regions. Further, this observable can be connected to a variety of important and interrelated topics, such as partonic multiple scattering in the medium, energy loss *via* gluon bremsstrahlung, and jet quenching in relativistic heavy ion collisions. Recent reviews, with a more complete coverage of theoretical approaches, may be found in Refs. [113, 114].

4.1. Fragmentation functions

In the simplest picture, fragmentation functions $D_h^q(z)$ describe the probability that an energetic quark q or antiquark $\bar{q}$ evolves into a specific hadron h, where the ratio of the hadron energy to the quark energy is $z = E_h/E_q$. They enter into the expression for the cross section for semi-inclusive electron scattering in combination with the parton distribution functions. A direct measure of the nuclear medium modification of the fragmentation functions is given by the *hadronic multiplicity ratio* R_M^h:

$$R_M^h(z,\nu,Q^2,p_T^2,\phi) = \frac{\left(\frac{N_h(z,\nu,Q^2,p_T^2,\phi)}{N_e^{DIS}(\nu)}\right)_A}{\left(\frac{N_h(z,\nu,Q^2,p_T^2,\phi)}{N_e^{DIS}(\nu)}\right)_D}, \tag{3}$$

where the numerator pertains to larger nucleus A, the denominator pertains to the deuterium nucleus D, and the variables not previously defined are p_T^2, the hadron momentum transverse to the virtual photon direction squared, and ϕ, the azimuthal angle around the virtual photon direction of the electron-hadron scattering plane. This observable, from the parton model perspective, is determined by the parton distribution functions and the fragmentation functions in the two nuclei. Experimentally it is now known that the nuclear medium induces a strong z dependence on R_M^h, demonstrating that it is indeed the fragmentation functions that are modified, in addition to the much smaller changes in the parton distribution functions that are associated with the EMC effect.

While there were historical measurements of semi-inclusive hadron production on nuclear targets prior to the 1990's, the first measurements that included identification of the produced hadrons were those of the HERMES experiment at DESY. These pioneering measurements established the basic features of the multiplicity ratio:

- suppression of R_M^h at high z and at low ν that systematically increases for larger nuclei;
- an enhancement of R_M^h at high p_T^2 in analogy with the Cronin effect seen in $p-A$ collisions;
- R_M^h depends on the hadron species produced.

The Jefferson Lab 5 GeV experiment, while more limited in kinematic reach than the HERMES experiment, has been able to observe qualitatively new behavior due to the much higher Jefferson Lab luminosity and the capacity for accommodating solid targets, thus probing

Journal of Physics: Conference Series **299** (2011) 012011 doi:10.1088/1742-6596/299/1/012011

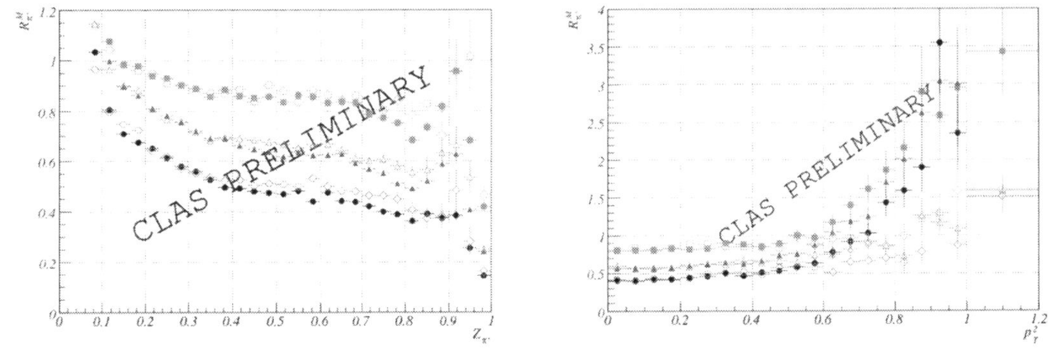

Figure 7. Preliminary data for the CLAS hadronic multiplicity ratio measurement as a function of z_{π^+} (upper plot) and p_T^2 (lower plot). The data shown are for positive pions for carbon (squares), iron (triangles), and lead (circles) targets normalized to a deuterium target. The data with filled (open) symbols correspond to the range $\nu = 2.2 - 3.2$ $(3.2 - 3.7)$ GeV and $Q^2 = 1.0 - 1.3$ GeV^2. The errors shown are statistical only. See Refs. [115, 116] for more discussion of the preliminary results for positive pions, and Ref. [117] for preliminary results for neutral kaons.

the largest nuclei. Because the Jefferson Lab data have two orders of magnitude more integrated luminosity, it is possible to bin the data in up to three kinematic variables while preserving good statistical accuracy. Thus, a fuller exploration of the kinematic dependence of R_M^h is possible. A second consequence of the higher luminosity is access to production of rarer hadrons, such as K^0, η, and potentially η', as well as exploratory studies with baryons such as protons, Λs and Σs. This complements the HERMES data set, in which the RICH detector was able to identify K^+ and K^- as well as antiprotons; both experiments have access to the three pion charge states, which provides a good cross-check to validate the consistency of the two data sets. Figure 7 shows a small subset of the preliminary Jefferson Lab data for R_M^h for positive pions. It can be seen from this figure that the basic dependencies established by the HERMES data are well reproduced by these lower energy data (suppression at high z, enhancement at high p_T^2), and are extended to the heaviest nuclei and to three-fold differential distributions (ν, Q^2, and z in the upper panel, and ν, Q^2, and p_T^2 in the lower panel). In addition to these studies for positive pions, preliminary results for neutral kaons can be found in Ref. [117], and studies are underway for neutral pions and η [118], negative pions, protons, and positive kaons [119], and Λ and Σ baryons. These types of studies lay the groundwork for future experiments at high luminosity facilities [110, 120].

To interpret these data and extract the physics of interest, a model is needed. Many models have been constructed to describe the HERMES results, and a number of these are now being used to address the new Jefferson Lab data. Existing models fall into three classes. First, there are models which assume that hadron formation only takes place outside the nucleus, and then use a pQCD framework to describe the interaction of a moving parton with the medium. In this picture, the behavior of R_M^h is due entirely to the stimulated emission of gluons resulting from partonic multiple scattering. The second class of models invokes hadron formation within the medium as the primary cause of the behavior of R_M^h, such that the semi-classical interaction of a forming hadron (or 'prehadron') removes flux from the observed final state channel through hadronic interactions. The third class of model has emerged only recently; it retains ingredients from the other two approaches, but treats the interaction in a

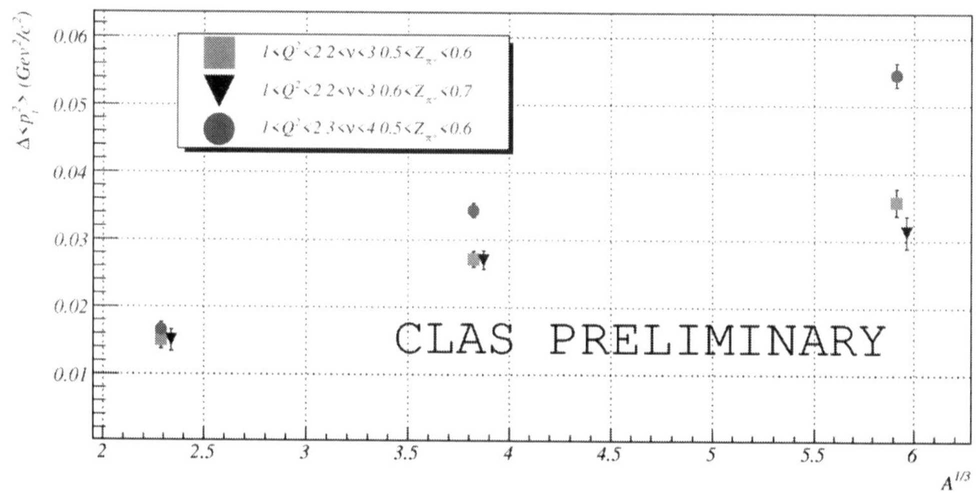

Figure 8. Mass number dependence of p_T-broadening in nuclear DIS. CLAS preliminary data for positive pions for carbon, iron, and lead targets three-fold differential in Q^2, ν, and z_{π^+} as shown in the legend. The errors shown are statistical only. See Refs. [115, 116] for more discussion of the preliminary results for positive pions, and Ref. [117] for preliminary results for neutral kaons.

fully quantum-mechanical framework. In this last approach an interference is observed between hadron formation inside the medium and outside the medium, and this gives rise to a non-trivial modification of R_M^h.

It is likely that the stringent constraints imposed by the new Jefferson Lab data will soon clarify what the essential model ingredients are. In particular, it is crucial to establish the relative importance of the two processes of gluon bremsstrahlung and prehadron interactions in describing the kinematic and flavor dependencies of R_M^h, and to understand the role of quantum interference in these processes. At that point it will be feasible to fully analyze the data from the 5 GeV data and from the planned 12 GeV Jefferson Lab Experiment E12-06-117, extracting hadron formation lengths from a variety of produced hadrons and constraining the mechanisms involved in their formation.

4.2. Quark energy loss

Quarks passing through a strongly-interacting medium are expected to lose energy through the medium-stimulated emission of gluons. This topic underwent intensive theoretical study by a number of groups beginning in the 1990's (see [121, 113]). Experimentally this topic was a focus of the Fermilab Drell-Yan experiments [122] as well as a phenomenon invoked to explain the jet quenching observed at RHIC. Extraction of a reliable value for the quark energy loss has proven to be an elusive goal, and estimated values from both theoretical and experimental work have varied by more than one order of magnitude.

While obtaining a quantitative value for quark energy loss is still a challenge, there are closely related parameters that are now accessible through experiment. Principal among these is the *transport coefficient* $\hat{q}$, which quantifies the transverse momentum accumulated by the quark per unit path length. This parameter or its equivalent appears in every calculation of quark energy loss for cold and hot matter and thus measuring it is of high interest. Using the BDMPS formalism [123, 124] for the following discussion, the transport coefficient is related to the mean transverse momentum acquired by quark q along a trajectory of length L:

$$< p_{T,q}^2 >= \hat{q}L, \tag{4}$$

and the radiative energy loss is given by

$$-\frac{dE}{dz} = \frac{1}{4}\alpha_s N_c \hat{q} L, \tag{5}$$

where α_s is the strong coupling constant and N_c is the number of colors. These equations, while specific to one theoretical approach, demonstrate the close connection between energy loss, p_T broadening, and the transport coefficient.

As with the quark energy loss, there is an order of magnitude spread among the theoretical and experimental estimates for $\hat{q}$. For example, an estimate from pQCD is the following:

$$\hat{q} = \frac{4\pi^2 \alpha_s C_R}{N_c^2 - 1} \rho x G(x, Q^2), \tag{6}$$

where C_R is the SU(3) color charge of the parton, equal to $(N_c^2 - 1)/2N_c$ and N_c for quarks and gluons respectively, x is the Bjorken scaling variable, ρ is the nuclear medium density, and G is the gluon density in the nucleon. The connection between gluon density and $\hat{q}$ is a consistent theme among the various theoretical approaches.

The new precision Jefferson Lab and HERMES data from the past decade will provide a definitive measurement of unprecedented precision for the value of $\hat{q}$ in cold nuclear matter. In this case, the experimentally measured hadron observable $\Delta p_{T,h}^2$ is very closely related to $\hat{q}$ and rather minor assumptions are required to derive it from the data.

A related exercise is the extraction of the lifetime of the free quark, referred to as the 'production time' τ_p, from the data. Although heuristic estimates of this quantity can be found, it has never been measured experimentally. While an exact form is not known, its general dependence on ν and z is expected to be of the form

$$\tau_p \approx z(1 - z)\nu, \tag{7}$$

based on energy conservation and special relativity.

The Jefferson Lab data suggest that a unique window exists for measuring τ_p where the production length τ_p/c is comparable to nuclear dimensions. In this regime, interactions with the nucleus allow a determination of τ_p, while at higher ν the quark hadronizes outside the nucleus and no information on its lifetime can be gained. The method of extraction makes use of the p_T broadening, which occurs during the phase in which the quark is propagating as a colored object. Once it undergoes color neutralization and becomes a prehadron, its interactions with the nucleus contribute much less broadening per unit path length. Thus, when the quark's color is neutralized, the broadening essentially stops. This is manifested experimentally by a non-linear saturation in the relationship between Δp_T and the nuclear size $\approx A^{\frac{1}{3}}$.

Figure 8 shows a small subset of the preliminary Jefferson Lab 5 GeV data for three different three-dimensional bins in Q^2, ν, and z. On the vertical axis is the experimentally observed p_T broadening, $\Delta p_{T,h}^2$. On the horizontal axis is $A^{\frac{1}{3}}$ which is proportional to the average medium thickness for the three nuclei shown. It is seen that these data exhibit the non-linear saturation described above, in varying degrees. To interpret these in terms of quark degrees of freedom one must correct the hadron's $\Delta p_{T,h}^2$ to the value of the quark-level $\Delta p_{T,q}^2$, typically using a simple kinematic multiplier. If the prescription of Domdey *et al.* [125] is used for this, one finds that the data shown in this figure follow the behavior expected from Eq. (7) above. A detailed study of the full data set is underway, but it is already evident that there is qualitative confirmation of the physical picture and a clear path to extracting the free quark lifetime τ_p from these data.

The determination of τ_p provides a precise value of L in Eq. (4), and the value of the quark-level $\Delta p^2_{T,q}$ gives the quantity on the left-hand side of this equation, thus determining $\hat{q}$. This can then be used to estimate the quark energy loss through Eq. (5).

In principle, the quark energy loss can also be determined from a direct measurement using the same data and comparing the energy spectra of leading hadrons emerging from nuclei of different thicknesses. Careful consideration of Fermi motion and nuclear pion production are required to extract a direct estimate. An evaluation of these possibilities is also underway.

5. Summary

In this article we have reviewed the tremendous successes of Jefferson Lab experiments that aimed at the understanding of in-medium modifications of hadron properties and quark interactions over the last decade. Particularly, we have highlighted three experiments: First, the ^{4}He$(\vec{e}, e'\vec{p})^3$H polarization transfer measurements revealed possible medium modifications of the bound proton electromagnetic form factors. Second, the study of vector meson photoproduction on nuclei has observed no mass shift of the ρ meson in the nuclear medium and a width consistent with expected collision broadening. These results put tight constrains on models of in-medium hadron properties. Finally, in-medium modification of fragmentation function ratios extracted in Jefferson Lab with high statistics and accuracy, has opened new experimental possibilities that connect to the future Jefferson Lab 12 GeV upgrade. The data analysis of some more recent experiments to study the Coulomb sum rule or the nuclear EMC effect are underway and exciting results from those are expected to come out soon.

Acknowledgments

Notice: Authored by Jefferson Science Associates, LLC under U.S. DOE Contract No. DE-AC05-06OR23177. The U.S. Government retains a non-exclusive, paid-up, irrevocable, world-wide license to publish or reproduce this manuscript for U.S. Government purposes. We would like to thank C. Djalali and J. Arrington for useful discussions about this work. S.S. acknowledges support from the U.S. National Science Foundation, NSF PHY-0856010. W.K.B. acknowledges support from CONICYT grant number 1080564.

References

[1] G. E. Brown and M. Rho, Phys. Rev. Lett. **66**, 2720 (1991).
[2] S. R. Beane, P. F. Bedaque, K. Orginos and M. J. Savage, Phys. Rev. Lett. **97**, 012001 (2006); N. Ishii, S. Aoki and T. Hatsuda, Phys. Rev. Lett. **99**, 022001 (2007).
[3] S. R. Beane, P. F. Bedaque, T. C. Luu, K. Orginos, E. Pallante, A. Parreno and M. J. Savage, Nucl. Phys. **A794**, 62 (2007); H. Nemura, N. Ishii, S. Aoki and T. Hatsuda, Phys. Lett. B **673**, 136 (2009).
[4] K. Saito, K. Tsushima and A. W. Thomas, Prog. Part. Nucl. Phys. **58**, 1 (2007).
[5] R.P. Bickerstaff and A.W. Thomas, J. Phys. G **15**, 1523 (1989); M. Arneodo, Phys. Rep. **240**, 301 (1994); D. F. Geesaman, K. Saito and A. W. Thomas, Annu. Rev. Nucl. Part. Sci. **45**, 337 (1995); G. Piller and W. Weise, Phys. Rep. **330**, 1 (2000).
[6] G.A. Miller and J.R. Smith, Phys. Rev. C **65**, 015211 (2002); Phys. Rev. C **66**, 049903(E) (2002); J.R. Smith and G.A. Miller, Phys. Rev. C **65**, 055206 (2002).
[7] J.R. Smith and G.A. Miller, Phys. Rev. Lett. **91**, 212301 (2003).
[8] K. J. Eskola, V. J. Kolhinen and C. A. Salgado, Eur. Phys. J. C **9**, 61 (1999); K. J. Eskola, V. J. Kolhinen and P. V. Ruuskanen, Nucl. Phys. **B535**, 351 (1998).
[9] M. Hirai, S. Kumano and M. Miyama, Phys. Rev. D **64**, 034003 (2001).
[10] Jefferson Lab Experiment E89-008, D. Day and B. Filippone, co-spokespersons.
[11] Jefferson Lab Experiment E02-019, A. Lung, J. Arrington, D. Day and B. Filippone, co-spokespersons.
[12] Jefferson Lab Experiment E03-103, J. Arrington and D. Gaskell, co-spokespersons.
[13] J. Seely *et al.*, Phys. Rev. Lett. **103**, 202301 (2009).
[14] Jefferson Lab Experiment E99-118, A. Bruell, J. Dunne and C. Keppel, co-spokespersons.

[15] Jefferson Lab Experiment E04-001, A. Bodek and C. Keppel, co-spokespersons.
[16] Jefferson Lab Experiment E89-033, C. Chang, C. Glashausser, S. Nanda and P. Rutt, co-spokespersons.
[17] S. Malov *et al.*, Phys. Rev. C **62**, 057302 (2000).
[18] S. Dieterich *et al.*, Phys. Lett. B **500**, 47 (2001).
[19] Jefferson Lab Experiment E93-049, R. Ent and P. Ulmer, co-spokespersons.
[20] Jefferson Lab Experiment E03-104, R. Ent, R. Ransome, S. Strauch and P. Ulmer, co-spokespersons.
[21] S. Strauch *et al.*, Phys. Rev. Lett. **91**, 052301 (2003).
[22] M. Paolone *et al.*, Phys. Rev. Lett. **105**, 072001 (2010).
[23] Jefferson Lab Experiment E05-110, J.-P. Chen, S. Choi and Z.-E. Meziani, co-spokespersons.
[24] Jefferson Lab Experiment E01-112, C. Djalali, M. Kossov and D. Weygand, co-spokespersons.
[25] R. Nasseripour *et al.*, Phys. Rev. Lett. **99** 262302 (2007).
[26] M. H. Wood, *et al.*, Phys. Rev. C **78** 015201 (2008).
[27] Jefferson Lab Experiment E89-028, J. Finn and P. Ulmer, co-spokespersons.
[28] Jefferson Lab Experiment E02-104, W. Brooks, spokesperson.
[29] S. A. Moszkowski and B. L. Scott, Ann. Phys. (Paris) **11**, 65 (1960).
[30] D. H. Lu *et al.*, Phys. Lett. B **417**, 217 (1998); Phys. Rev. C **60**, 068201 (1999).
[31] M. R. Frank, B. K. Jennings and G .A. Miller, Phys. Rev. C **54**, 920 (1996).
[32] U. T. Yakhshiev, U.-G. Meissner, and A. Wirzba, Eur. Phys. J. A **16**, 569 (2003).
[33] J. R. Smith and G. A. Miller, Phys. Rev. C **70**, 065205 (2004).
[34] T. Horikawa and W. Bentz, Nucl. Phys. **A762**, 102 (2005).
[35] W. Wen and H. Shen, Phys. Rev. C **77**, 065204 (2008).
[36] D. H. Lu, A. W. Thomas and K. Tsushima, arXiv:nucl-th/0112001; K. Tsushima, H.-C. Kim and K. Saito, Phys. Rev. C **70**, 038501 (2004).
[37] S. Liuti, arXiv:hep-ph/0601125v2.
[38] V. Guzey, A. W. Thomas and K. Tsushima, Phys. Lett. B **673**, 9 (2009).
[39] V. Guzey, A. W. Thomas and K. Tsushima, Phys. Rev. C **79**, 055205 (2009).
[40] W. Melnitchouk, K. Tsushima and A. W. Thomas, Eur. Phys. J. A **14**, 105 (2002).
[41] I. Sick, in *Proc. Int. Conf. on Weak and Electromagnetic Interactions in Nuclei*, ed. H. Klapdor (Springer-Verlag, Berlin, 1986) p. 415.
[42] J. Jourdan, Phys. Lett. B **353**, 189 (1995).
[43] R. G. Arnold, C. E. Carlson and F. Gross, Phys. Rev. C **23**, 363 (1981).
[44] D. Eyl *et al.*, Z. Phys. A **352**, 211 (1995).
[45] B. D. Milbrath and J. McIntyre *et al.*, Phys. Rev. Lett. **80**, 452 (1998).
[46] D. H. Barkhuff *et al.*, Phys. Lett. B **470**, 39 (1999).
[47] B. Hu *et al.*, Phys. Rev. C **73**, 064004 (2006).
[48] J. M. Laget, Nucl. Phys. **A579**, 333 (1994).
[49] J. M. Udias *et al.*, Phys. Rev. Lett. **83**, 5451 (1999); J. A. Caballero, T. W. Donnelly, E. Moya de Guerra and J. M. Udias, Nucl. Phys. **A632**, 323 (1998).
[50] J. M. Udias and J. R. Vignote, Phys. Rev. C **62**, 034302 (2000).
[51] A. Meucci, C. Giusti and F. D. Pacati, Phys. Rev. C **66**, 034610 (2002).
[52] T. de Forest, Nucl. Phys. **A392**, 232 (1983).
[53] J. A. McNeil, L. Ray and S. J. Wallace, Phys. Rev. C **27**, 2123 (1983).
[54] R. J. Woo *et al.*, Phys. Rev. Lett. **80**, 456 (1998).
[55] J. Gao *et al.*, Phys. Rev. Lett. **84**, 3265 (2000).
[56] R. Schiavilla *et al.*, Phys. Rev. Lett. **94**, 072303 (2005).
[57] P. Lava *et al.*, Phys. Rev. C **71**, 014605 (2005).
[58] S. P. Malace, M. Paolone and S. Strauch, AIP Conf. Proc. **1056**, 141 (2008).
[59] S. P. Malace *et al.*, Phys. Rev. Lett. **106**, 052501 (2011).
[60] J. Morgenstern and Z.-E. Meziani, Phys. Lett. B **515**, 269 (2001).
[61] K. Saito, K. Tsushima and A. W. Thomas, Phys. Lett. B **465**, 27 (1999).
[62] J. Carlson, J. Jourdan, R. Schiavilla and I. Sick, Phys. Lett. B **553**, 191 (2003).
[63] G.A. Miller, Eur. Phys. J. A **31**, 578 (2007).
[64] J. Arrington, J. Phys. Conf. Ser. **69**, 012024 (2007).
[65] J. Gomez *et al.*, Phys. Rev. D **49**, 4348 (1994).
[66] J. Arrington *el al.*, Phys. Rev. C **73**, 035205 (2006).
[67] V. Tvaskis *et al.*, Phys. Rev. Lett. **98**, 142301 (2007).
[68] I. C. Cloet, W. Bentz, A. W. Thomas, Phys. Rev. Lett. **95**, 052302 (2005); Phys. Lett. B **642**, 210 (2006).
[69] Jefferson Lab Letter of Intent LOI-06-003, V. Dharmawardane, contact person.
[70] Jefferson Lab Letter of Intent LOI-06-004, X. Zheng and A. Deur, contact persons.

[71] O. A. Rondon, Phys. Rev. C **60**, 035201 (1999).
[72] K. Saito, M. Ueda, K. Tsushima and A. W. Thomas, Nucl. Phys. **A705**, 119 (2002).
[73] T. Hatsuda and S. H. Lee, Phys. Rev. C **46**, 34 (1992).
[74] M. Herrman *et al.*, Nucl. Phys. **A545**, 267c (1992).
[75] R. Rapp *et al.*, Nucl. Phys. **A617**, 472 (1997).
[76] W. Peters, M. Post, H. Lenske, S. Leupold and U. Mosel, Nucl. Phys. **A632**, 109 (1997).
[77] M. Urban *et al.*, Nucl. Phys. **A673**, 357 (2000).
[78] D. Cabrera *et al.*, Nucl. Phys. **A705**, 90 (2002).
[79] M. F. M. Lutz *et al.*, Nucl. Phys. **A706**, 431 (2002).
[80] F. Eichstaedt *et al.*, nucl-th/07040154.
[81] D. Adamova *et al.*, nucl-ex/0611022.
[82] R. Arnaldi *et al.*, Phys. Rev. Lett. **96**, 162302 (2006).
[83] S. Damjanovic *et al.*, Eur. Phys. J. C **49**, 235 (2007).
[84] S. Damjanovic *et al.*, Nucl. Phys. **A783**, 327 (2007).
[85] R. Muto *et al.*, J. Phys. G: Nucl. Part. Phys. **30**, 1023 (2004).
[86] R. Muto *et al.*, Phys. Rev. Lett. **98**, 042501 (2007).
[87] M. Naruki *et al.*, Phys. Rev. Lett. **96**, 092301 (2006).
[88] D. Trnka *et al.*, Phys. Rev. Lett. **94**, 192303 (2005).
[89] M. Nanova *et al.*, arXiv:1005.5694 [nucl-ex].
[90] K. Tsushima, D. H. Lu, A. W. Thomas and K. Saito, Phys. Lett. B **443**, 26 (1998).
[91] R. S. Hayano, S. Hirenzaki and A. Gillitzer, Eur. Phys. J. A **6**, 99 (1999).
[92] K. Saito, K. Tsushima, D. H. Lu and A. W. Thomas, Phys. Rev. C **59**, 1203 (1999).
[93] B. A. Mecking *et al.*, Nucl. Instr. Meth. A **503**, 513 (2003).
[94] P. Muehlich, T. Falter, C. Greiner, J. Lehr, M. Post and U. Mosel, Phys. Rev. C **67**, 024605 (2003).
[95] M. Effenberger, E.L. Bratkovskaya and U. Mosel, Phys. Rev. C **60**, 044614 (1999).
[96] G. Jancso *et al.*, Nucl. Phys. **B124**, 1 (1977).
[97] D. Jouan *et al.*, Internal Report, Institut de Physique Nucleaire d'Orsay, IPNO-DR-02.015 (2002).
[98] D. Drijard, H. G. Fischer and T. Nakada, Nucl. Inst. Meth. **225**, 367 (1984).
[99] A. V. Stavinsky *et al.*, Phys. Rev. Lett. **93**, 192301 (2004).
[100] D. V. Bugg, Nucl. Phys. **B88**, 381 (1975).
[101] M. Effenberger and U. Mosel, Phys. Rev. C **62**, 014605 (2000).
[102] M. Effenberger, E. L. Bratkovskaya, W. Cassing and U. Mosel, Phys. Rev. C **60**, 027601 (1999).
[103] M. Post, S. Leupold and U. Mosel, Nucl. Phys. **A741**, 81 (2004). Phys. Rev. C **63**, 015204 (2001).
[104] G.-Q. Li, C. M. Ko, G. E. Brown and H. Sorge, Nucl. Phys. **A611**, 539 (1996).
[105] H. B. O'Connell *et al.*, Prog. Part. Nucl. Phys. **39**, 201 (1997).
[106] F. Riek, R. Rapp, T.-S.H. Lee, Yongseok Oh, Phys. Lett. B **677**, 116 (2009).
[107] F. Riek, R. Rapp, Yongseok Oh, and T.-S.H. Lee, Phys. Rev. C **82**, 015202 (2010).
[108] H. van Hees and R. Rapp, Nucl. Phys. A **806**, 339 (2008).
[109] W. K. Brooks, AIP Conf. Proc. **1265**, 276 (2010).
[110] Jefferson Lab Experiment E12-06-117, W. Brooks, G. Gilfoyle, K. Hafidi, K. Hicks, M. Holtrop, K. Joo, G. Niculescu and M. Wood, co-spokespersons.
[111] A. Majumder and B. Müller, Phys. Rev. C **77**, 054903 (2008).
[112] A. Majumder, R. J. Fries and B. Müller, Phys. Rev. C **77**, 065209 (2008).
[113] A. Accardi, F. Arleo, W. K. Brooks, D. D'Enterria, and V. Muccifora, Riv. Nuov. Cim. **09**, 439 (2009).
[114] A. Majumder and M. van Leeuwen, arXiv:1002.2206v3 [hep-ph].
[115] H. Hakobyan, PhD thesis, Yerevan State University (2008).
[116] W. K. Brooks and H. Hakobyan, AIP Conf. Proc. **1056**, 215 (2008), Nucl. Phys. **A830**, 361c (2009), AIP Conf. Proc. **1265**, 230 (2010).
[117] K. Hicks and A. Daniel, arXiv:0907.5362v1 [nucl-ex].
[118] T. Mineeva, W. K. Brooks, K. Joo and M. Ungaro, Bull. Amer. Phys. Soc. **54**, Number 10, http://meetings.aps.org/link/BAPS.2009.HAW.EF.12 (2009).
[119] R. Dupre and K. Hafidi, Bull. Amer. Phys. Soc. **55**, Number 1, http://meetings.aps.org/link/BAPS.2010.APR.S8.9 (2010).
[120] W. K. Brooks, PoS **DIS2010**, 258 (2010).
[121] R. Baier, D. Schiff and B. G. Zakharov, Annu. Rev. Nucl. Part. Sci. **50**, 37 (2000).
[122] M. J. Leitch *et al.*, Phys. Rev. Lett. **84**, 3256 (2000).
[123] R. Baier, Y. L. Dokshitzer, A. H. Mueller, S. Peigne and D. Schiff, Nucl. Phys. **B484**, 265 (1997).
[124] R. Baier, Y. L. Dokshitzer, A. H. Mueller and D. Schiff, Nucl. Phys. **B531**, 403 (1998).
[125] S. Domdey, D. Grunewald, B. Z. Kopeliovich and H. J. Pirner, Nucl. Phys. **A825**, 200 (2009).

Journal of Physics: Conference Series **299** (2011) 012012 doi:10.1088/1742-6596/299/1/012012

Searches for Physics Beyond the Standard Model

Julie Roche[1,2], **Willem T. H. van Oers**[3,4] **and Ross D. Young**[5,6]

[1] Ohio University, Athens, Ohio 45071, USA
[2] Jefferson Lab, Newport News, Virginia 23606, USA
[3] University of Manitoba, Winnipeg, MB, Canada R3T 2N2
[4] TRIUMF, 4004 Wesbrook Mall, Vancouver, BC, Canada V6T 2A3
[5] University of Adelaide, South Australia 5005, Australia
[6] Argonne National Laboratory, Argonne, Illinois 60439, USA

Abstract. Jefferson Lab has now demonstrated ablility to test the fundamental symmetries of nature, and thereby probe for new physics beyond the Standard Model. Here we review the tremendous advances in precision parity-violation measurements with CEBAF that enable searches for new physics. This has been demonstrated with a determination of the weak charge of the proton, which is found to be in agreement with the prediction of the standard electroweak theory, and at a precision that rules out relevant new physics to the TeV scale. We also review the planned future experiments which aim to further test the electroweak theory at Jefferson Lab, including a further improvement on the proton weak charge, an ultra-precise Møller measurement, and a probe of the axial quark charges in PVDIS.

1. Outline

Our contemporary understanding of physics at the smallest of scales is concisely described by the Standard Model of nuclear and particle physics. The study of the strong interaction component of this Standard Model is the principal focus of Jefferson Lab — as described throughout this volume. In this chapter we review measurements at Jefferson Lab that are outside the realm of strong-interaction phenomena and contribute to ongoing programs in the search for new physics beyond the Standard Model.

The study of new physics is driven by the perpetual search for a more fundamental understanding of the constituents and forces of nature. The current formulation of the Standard Model is built upon three generations of fermions, with interactions mediated by the exchange of gauge bosons. This model has had tremendous success in describing a wide range of physical phenomena. Nevertheless, the reported discovery of neutrino oscillations [1] is evidence of the need for a new theory beyond the Standard Model. In the near future, in light of the starting up of the much-anticipated Large Hadron Collider (LHC), we are anticipating further discoveries which will guide the development of this new theory. Through the tremendous amount of energy available, the LHC has the capacity to directly produce new particles which lie beyond the classification of the Standard Model.

In contrast to high-energy colliders, an alternative strategy to search for new physics is to perform precision measurements of interactions that can reliably be predicted within the Standard Model [2, 3, 4, 5]. While not directly producing new particles, any deviation from the prediction of the Standard Model could provide a signature of these new forms of matter. For example, the potential impact of supersymmetry on low-energy observables has recently been reviewed [6]. It is this low-energy domain where Jefferson Lab can contribute to this complimentary approach to searching for new physical phenomena.

As stated above, Jefferson Lab is primarily involved in the study of strong-interaction phenomena. There has been a large program dedicated to resolving the strange-quark content of the vector form factors of the nucleon [7, 8, 9, 10, 11, 12, 13]. These measurements have made use of the weak interaction as a tool to probe this strange-quark content [14, 15]. Details on the strange-quark measurements have been described elsewhere in this volume — here, we review how these measurements have also been used to test the probe itself [16].

In Section 2 we introduce the neutral-current sector of the electroweak theory and the predicted running of the weak mixing angle. In Section 3, the advances in experimental techniques that have paved the way for precision electroweak tests are reviewed. The extraction of the weak charge of the proton from parity-violating electron scattering measurements and the associated hadronic backgrounds are discussed in Section 4. In Section 5 we show the latest global analysis of the weak charges of the quarks and the implications for new physics. In Section 6 we review ongoing and future planned experiments at Jefferson Lab which will provide further tests of the low-energy weak interaction, in particular, those of Q-weak, PVDIS and the 11 GeV Møller programs.

2. Running of the weak mixing angle

The electroweak component of the Standard Model unifies the interactions of electromagnetism and the weak nuclear force into a single framework described by an SU(2)×U(1) gauge field theory. Left-hand fermion currents transform as SU(2) doublets under weak isospin, with a coupling strength conventionally denoted by g. In addition, both left- and right-hand currents carry a weak hypercharge quantum number, which couples to a U(1) gauge field, with strength g'. In the Standard Model, a scalar doublet is also coupled to these gauge groups. In the vacuum of the theory, this field acquires a nonvanishing expectation value through the Higgs mechanism and spontaneously breaks the full gauge symmetry down to just a single U(1) phase — corresponding to the standard electromagnetic theory. For each of the broken symmetries, the associated gauge bosons, the $W^\pm$ and Z, acquire finite masses by coupling to the Higgs field. While this theory has continued to perform remarkably well over many years of experimental tests, there is compelling evidence that a new discovery is imminent. This anticipation is driven by recent observations of neutrino oscillations, outstanding issues with modern cosmology; and the LHC, which promises a definitive result on the existence of the Higgs boson.

For the purposes of our discussion, the electroweak theory is characterised by three parameters — one coupling constant for each of the two gauge sectors, plus the mass scale of spontaneous symmetry breaking. The first of these is just the usual coupling constant of quantum electrodynamics, α. The next best-measured quantity is the μ-decay coupling constant $G_\mu = 1.166367(5) \times 10^{-5}\,\mathrm{GeV}^{-2}$ [17, 18, 19], which normalizes the corresponding Fermi 4-point interaction. In terms of the electroweak theory, this quantity essentially defines a ratio of coupling constant to mass scale. The final input typically used to define the theory is the mass of the Z boson, $M_Z = 91.1876(21)\,\mathrm{GeV}$ [20].

These three observables constrain the relevant component of the electroweak theory, with any other dependencies becoming predictions. In particular, the Weinberg angle is one such quantity, defined in terms of the measured coupling constants. While there are several renormalization schemes commonly considered [17], we work with the modified minimal subtraction ($\overline{\mathtt{MS}}$) scheme, where at the Z mass $\hat{s}_Z^2 \equiv \sin^2\hat{\theta}_W(M_Z) = 0.23120$. We note that defining the theory at the Z scale involves evolution of both α and G_μ to this scale. Given that the running of the couplings are determined by the theory, the evolution of $\sin^2\theta_W$ describes a consise and instructive prediction of the theory, as displayed in Figure 1. This running has now been confirmed in atomic physics measurements [2] (see also the latest theory update [21]), PV Møller scattering

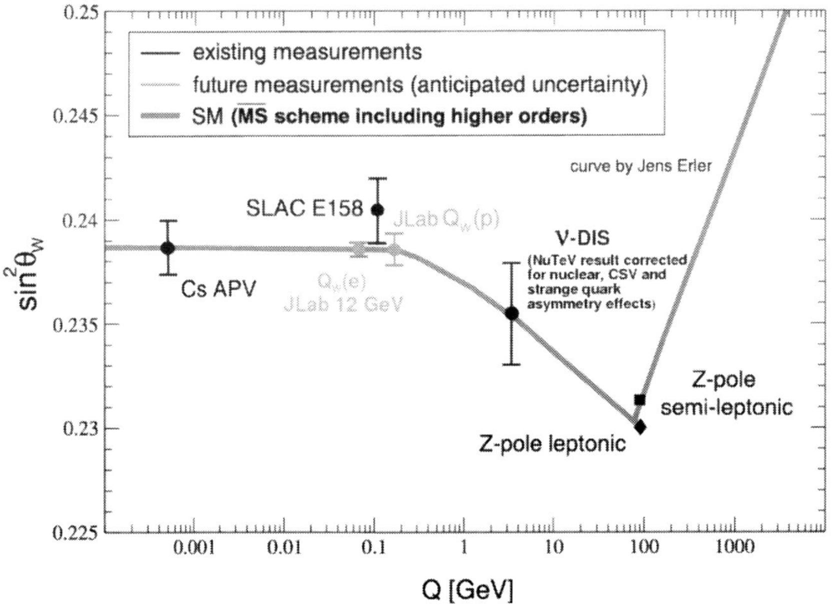

Figure 1. Calculated running of the weak mixing angle in the Standard Model, as defined in the modified minimal subtraction scheme [24]. The uncertainty in the predicted running is given by the thickness of the (blue) curve. The black points with error bars show the existing data, while the pink points with error bars (with arbitrarily chosen ordinates) refer to the future Jefferson Lab Q-weak and Møller measurements discussed in this chapter. The existing measurements are from atomic parity violation [2, 21]; SLAC E-158 Møller scattering [5]; NuTeV deep inelastic neutrino scattering [3, 25]; and the LEP and SLC Z-pole asymmetries [20].

[4] and latest collider results from the Tevatron [22, 23] (not shown in the figure).

The neutrino– and antineutrino–nucleus scattering measurements reported by NuTeV indicated a deviation from the Standard Model prediction [3]. Charge symmetry breaking in quark distributions has been recognised to resolve a portion of this discrepancy [26]. Further, it has recently been reported that the isospin dependence of the partons' medium modifications, induced by a non-vanishing isovector-vector ("ρ"-like) mean field, induces a substantial correction [27]. A reassessment of the associated hadronic and nuclear corrections renders the NuTeV measurement consistent with the Standard Model prediction [25], as seen in Figure 1.

Beyond confirming the predicted running of the couplings within the electroweak theory, these measurements are also sensitive to any new contact interaction in these relevant channels, and hence place limits on the possible forms of new physics. This is just the approach we wish to exploit here in precision tests at Jefferson Lab.

3. Experimental advances in precision parity-violating electron scattering

Parity-violating electron scattering (PVES) was first used to study the electroweak theory in the 1970's, where Prescott *et al.* pioneered the experimental technique [28] that makes it possible to perform a precision test of the Standard Model. Out of the fifteen or so parity-violating electron scattering experiments performed or planned world-wide, half of them have used or will be using the CEBAF beam. The stability of that beam as well as the accumulated operational know-how

Table 1. Survey of parity-violating electron scattering experiments. (Updated from [29].)

Experiment	Target	Physics	A_{RL}
		Completed Experiments	
SLAC E122 [28]	D	Weak mixing angle	10^{-4}
Mainz [30]	^{9}Be	New physics	10^{-5}
MIT-Bates [31]	^{12}C	New physics	10^{-6}
SAMPLE (MIT-Bates) [32]	H, D	Strange form factor	10^{-5}
HAPPEX (JLab) [7, 8, 9, 10, 11]	H, He	Strange form factor	10^{-6}
G^0 (JLab) [12]	H, D	Strange form factor	10^{-5}
PVA4 (Mainz) [33, 34, 35]	H	Strange form factor	10^{-5}
Møller (SLAC) [5]	e	New physics	10^{-7}
		Upcoming Experiments	
HAPPEX (JLab)	H	Strange form factor	10^{-6}
PREX (JLab)	Pb	Neutron radius	10^{-7}
PVDIS (JLab)	D	Vector-axial coupling of quarks	10^{-4}
Q-weak (JLab)	H	New physics	10^{-7}
Møller (JLab)	e	New physics	10^{-8}

in tuning the polarized electron source for these types of experiments are keys to the feasibility of Standard Model tests at Jefferson Lab. Table 1 gives an historical listing of the world-wide experiments, as well as the reaction studied, the main goals of the experiments and the order of magnitude of the measured asymmetry.

Completed PVES experiments (including HAPPEX and G0) have already produced results that test the Standard Model. The principal goal of these experiments was to measure the strange form factors of the nucleon, as extensively described in an earlier chapter of this volume. Here we summarize the major improvements in the PVES technique since the pioneering work of Prescott *et al.*, that now makes it possible to perform high-precision tests of the Standard Model.

In PVES, one measures the fractional difference $\mathcal{E}^{exp}$ of the scattered flux from an unpolarized target for an incident beam of longitudinally polarized electrons. Under the assumption that the solid angle and luminosity is the same for the two helicity states, this difference in the scattered flux is related to the analyzing power, A_{RL}, the physical observable (see Equation (9)), as follows:

$$\mathcal{E}^{exp} = \frac{N^+ - N^-}{N^+ + N^-} = P_e A_{RL} \tag{1}$$

where P_e is the polarization of the incoming electron beam. The statistical precision of the measurement is given by

$$\sigma_{stat}(A_{RL}) = \frac{1}{\sqrt{N} P_e} \tag{2}$$

where $N \equiv N^+ + N^-$ is the total number of scattered events detected.

3.1. Achieving statistical precision

The striking feature of A_{RL} is its small size (ranging from 10^{-4} to 10^{-9}). In order to obtain a statistically meaningful measurement (of the order of a few percent) within a reasonable time frame, both the beam polarization and the rate of detected events need to be large. At CEBAF, polarized electrons are photo-emitted from a GaAS semi-conductor cathode excited by circularly

Table 2. Some of the key features of each CEBAF photo-injector [36].

	First photo-injector	Second photo-injector	Third photo-injector
Date	Feb. 1995-Jan. 1998	Feb. 1998-June 1999	July 1999 to present
Charge lifetime (C)	< 10	~ 100	200
Polarization	$\sim 35\%$	70-75%	>80%
GaAs surface	bulk	strained	super-lattice

polarized laser light of energy matching the semi-conductor band gap. Since the beginning of operations, the CEBAF polarized electron source has been upgraded three times [36]. Each upgrade has provided an increase in both the polarization and the cathode operational lifetime (see Table 2). The design of the injector itself changed with two goals in mind: understanding and reducing the underlying causes of photo-cathode quantum efficiency changes during beam delivery, and, constructing a source with the highest possible transmission from the cathode to the end users. These modifications have been so successful that CEBAF can deliver 200μA continuous wave at full energy, limited only by the available radio-frequency power. The high average beam current can be delivered for uninterrupted periods of months.

In parallel to the source design modifications, the polarization of the beam also increased over time as better and better engineered photo-cathodes were used. Because the beam is delivered in three different halls, its polarization can be measured by five different electron polarimeters based on Compton, Møller and Mott scattering located throughout the accelerator. A special measurement campaign [37] offered a unique opportunity to compare the polarizations measured by these very different polarimeters. A Wien filter in the injector region was used to vary the electron spin orientation at the polarimeter site such that each polarimeter could report the actual magnitude of the polarization. This study allowed a statistically significant check of the estimation of the respective analyzing powers and the systematic precision of the polarimeters. The absolute measurements of the beam polarization at CEBAF were determined with a precision of the order of 1%.

Given these features of the CEBAF machine, high luminosity PVES experiments are being performed. Both the HAPPEX and G0 experiments ran with instantaneous luminosities of the order of $2 \times 10^{38}\,\mathrm{cm}^{-2}\,\mathrm{s}^{-1}$. The Q-weak experiment will run at an instantaneous luminosity of $1.7 \times 10^{39}\,\mathrm{cm}^{-2}\,\mathrm{s}^{-1}$. In consequence, the target system must accommodate large power deposition in the liquid with minimal global yield reduction and density fluctuations. The target heat exchanger for the G0 experiment removed 0.6 kW from the liquid while the Q-weak one will handle 2.5 kW. Experimental tests and computational fluid dynamic simulations are used to optimize the target design. Careful studies of the G0 LH2 target [38] have shown that under nominal running conditions the target density fluctuations increased the asymmetry distribution's RMS width at most by 2%, and reduced the net yield by at most by 1.5%. Both effects have negligible impact on the precision of the PVES results.

3.2. Controlling helicity correlated false asymmetries

Given these properties of the CEBAF accelerator, high current measurements are achieving sub-part per million statistical precision routinely with matching normalizations of the electron beam polarization and target thickness. Other sources of potential systematic errors are the helicity correlated modulations in the beam parameters, such as the incident beam current, the energy, the polarization, and beam size, profiles (intensity and polarization), position, and direction [39]. Helicity correlated modulations in these beam parameters lead to false asymmetries corrupting the physics asymmetry to be measured. The apparent asymmetry $\mathcal{E}^{meas}$ can be written in terms

Journal of Physics: Conference Series **299** (2011) 012012 doi:10.1088/1742-6596/299/1/012012

Table 3. Average beam asymmetries under polarization reversal in intensity, energy and beam differences in horizontal and vertical position already achieved (G0 and HAPPEX II) or planned Q-weak.

	G0	HAPPEX II	Q-weak
A_Q [ppm]	-0.3	0.4	<0.1
A_E [ppb]	19	0.2	<1
Δx [nm]	6	0.5	<2
$\Delta x'$ [nrad]	2	-0.2	<30
Δy [nm]	8	1.7	<2
$\Delta y'$ [nrad]	3	0.2	<30

of the physics asymmetry $\mathcal{E}^{exp}$ and the helicity correlated false asymmetries:

$$\mathcal{E}^{meas} = \mathcal{E}^{exp} + \mathcal{E}^{q} + \sum_i \alpha_i \Delta X_i \tag{3}$$

where $\mathcal{E}^q$ is the asymmetry due to helicity correlated modulations in the integrated beam current and the helicity correlated modulations in all other beam parameters.

The approach followed in all precision parity-violating experiments is to minimize the unwanted modulations in the beam parameters, to design and construct a main detection system that has as much as possible cylindrical symmetry around the incident electron beam axis (or left–right symmetry in the case of the HAPPEX experiments), to measure the remaining sensitivities of the detection system by modulating the beam in a set of control measurements in order to measure the effects of the potential beam helicity correlated differences, and to correct for these effects when necessary.

Most of the unwanted helicity correlated modulations in the beam parameters originate in the polarized electron source. The electrons of the two helicity states are produced by shining circularly polarized laser light on the strained GaAs photocathode. The helicity of the polarized laser light is defined by the Pockels cell. Gradients in the quantum efficiency coupled to the position of the laser beam spot on the GaAs crystal result in helicity correlated modulations of the electron beam parameters. Great care is being applied to configuring and aligning the polarized source optics. Proper tuning of the electron beam through the various components of the accelerator can reduce the beam emittance and unwanted helicity correlated beam parameters. There are two further options for reducing helicity correlated changes in the beam parameters and in particular those due to energy changes: the insertion of a half wave plate in the laser beam and the use of a Wien filter (both to change the helicity of the electron beam). Thus with two laser polarizations, there are in principle eight different combinations of helicity state of the polarized laser light to the helicity state of the electron beam on target.

A further technique used to reduce the effects of helicity correlated changes in the beam parameters is the use of feedback systems during data taking. Table 3 gives a tabulation of average beam asymmetries under helicity reversal obtained in the G0 and HAPPEX experiments and projected for the Q-weak experiment. It appears that the latter are within reach for the Q-weak experiment presently being prepared for data taking starting in 2010.

4. Electroweak form factors

4.1. Born asymmetry

In physical measurements, the electron scattering asymmetries probe the full electroweak structure of the target. For a proton, this means resolving its hadronic structure in order to

Journal of Physics: Conference Series **299** (2011) 012012 doi:10.1088/1742-6596/299/1/012012

Table 4. Z-boson couplings to vector and axial-vector fermion currents.

Fermion	g_V^f	g_A^f
e	$-1+4\sin^2\theta_W$	1
u	$1-\frac{8}{3}\sin^2\theta_W$	-1
d	$-1+\frac{4}{3}\sin^2\theta_W$	1
s	$-1+\frac{4}{3}\sin^2\theta_W$	1

extract the underlying electroweak parameters. Here we review how this is achieved in practice. We first outline the nucleon matrix elements which are probed in the PVES program. The usual electromagnetic Dirac and Pauli form factors of the nucleon are defined by the matrix element

$$\langle N(P')|J_\mu(0)|N(P)\rangle = \bar{u}_N(P')\left[\gamma_\mu F_1(q^2) + i\frac{\sigma_{\mu\nu}q^\nu}{2M}F_2(q^2)\right]u_N(P)\,, \tag{4}$$

where the momentum transfer between in (P) and out (P') nucleon states is given by $q = P' - P$. These form factors measure the response of the nucleon to the electromagnetic current

$$J_\mu = \frac{2}{3}\bar{u}\gamma_\mu u - \frac{1}{3}\bar{d}\gamma_\mu d - \frac{1}{3}\bar{s}\gamma_\mu s\,. \tag{5}$$

The form factors are commonly reexpressed in terms of the Sachs electric and magnetic form factors [40] (with $Q^2 = -q^2$)

$$G_E(Q^2) = F_1(Q^2) - \frac{Q^2}{4M^2}F_2(Q^2), \quad G_M(Q^2) = F_1(Q^2) + F_2(Q^2)\,. \tag{6}$$

The vanishing momentum transfer limit of these form factors, $G_E(Q^2=0)$ and $G_M(Q^2=0)$, are respectively, the total electric charge and magnetic moment of the nucleon. The neutral weak vector current can be expressed as

$$J_\mu^Z = g_V^u\bar{u}\gamma_\mu u + g_V^d\bar{d}\gamma_\mu d + g_V^s\bar{s}\gamma_\mu s \tag{7}$$

where the (tree-level) neutral-current charges of the quarks are listed in Table 4. Analogous to the electromagnetic case, we can define the neutral-current form factors

$$\langle N(P')|J_\mu^Z|N(P)\rangle = \bar{u}_N(p')\left[\gamma_\mu F_1^Z(q^2) + i\frac{\sigma_{\mu\nu}q^\nu}{2M}F_2^Z(q^2)\right]u_N(p)\,, \tag{8}$$

with similarly defined Sachs forms, G_E^Z and G_M^Z. Of particular interest in the present study is the normalization of the neutral-current electric form factor, $G_E^Z(Q^2=0) \equiv g_E^Z$, which describes the *weak charge* of the nucleon. At tree level, the proton and neutron weak charges are given by $1 - 4\sin^2\theta_W$ and -1, respectively.

These neutral-current form factors are probed by measuring the analyzing power between the elastic scattering of pure right- and left-hand polarised electrons. The leading Born contribution to the proton analyzing power can be written as

$$A_{RL}^p = \frac{\sigma_R - \sigma_L}{\sigma_R + \sigma_L} = -\frac{G_\mu Q^2}{4\pi\alpha\sqrt{2}}\frac{g_A^e\left(\epsilon G_E^p G_E^{Z(p)} + \tau G_M^p G_M^{Z(p)}\right) + g_V^e\sqrt{1-\epsilon^2}\sqrt{\tau(1+\tau)}G_M^p G_A^p}{\epsilon {G_E^p}^2 + \tau {G_M^p}^2}\,, \tag{9}$$

Journal of Physics: Conference Series **299** (2011) 012012 doi:10.1088/1742-6596/299/1/012012

with the relevant kinematic factors given by $\tau = |Q^2|/4M_p^2$ and $\epsilon = \left[1 + 2(1+\tau)\tan^2\theta/2\right]^{-1}$, with θ the scattering angle in the laboratory frame. We refer the reader to Ref. [41] for an extensive review and further details on scattering from other targets.

To illustrate the extraction of the weak charge term, we consider the forward angle and small momentum transfer limit. In this kinematic limit, the analyzing power is seen to behave like

$$\begin{aligned} A_{RL}^p &= -\frac{G_\mu Q^2}{4\pi\alpha\sqrt{2}}\left\{g_A^e g_E^{Z(p)} + Q^2\frac{g_A^e}{12M_p^2}\left[3\mu_p\mu_p^Z + g_E^{Z(p)}\left(2M_p^2(\langle r_E^2\rangle_p - \langle r_E^2\rangle_p^Z) - 3\mu_p^2\right)\right]\right. \\ &\quad \left. +|Q|\theta\frac{g_V^e}{2M_p}\mu_p G_A^p + \mathcal{O}\left(|Q|^3, \theta^2\right)\right\}. \end{aligned} \tag{10}$$

The electric radii are defined by $\langle r_E^2\rangle_p = -\frac{6}{G_E(0)}\frac{dG_E(Q^2)}{dQ^2}\Big|_{Q^2=0}$. From Eq. (10) we see that the small-Q^2 and forward angle analyzing power projects onto the weak charge of the proton. As Q^2 increases one becomes sensitive to the electroweak hadronic structure of the nucleon — this is precisely the structure that is being investigated by the strangeness program. As the scattering angle moves away from the forward angle limit, measurements have increased sensitivity to the axial form factor (which encompasses the anapole contribution).

4.2. *Beyond the Born term*

As discussed in Section 2, the precision of contemporary parity-violation measurements are sensitive to the scale evolution of the electroweak theory. We therefore require knowledge of the corrections to the tree-level asymmetry. The first step in resolving the electron-proton scattering asymmetry is to evolve the electroweak interactions to the appropriate ($Q^2 \sim 0$) scale, as illustrated in Figure 1. Most important for the present discussion are the effective 4-point electron–quark interactions, defined by

$$\mathcal{L}_{NC} = \frac{G_\mu}{\sqrt{2}}\sum_q \left[C_{1q}\bar{e}\gamma^\mu\gamma_5 e\,\bar{q}\gamma_\mu q + C_{2q}\bar{e}\gamma^\mu e\,\bar{q}\gamma_\mu\gamma_5 q\right]. \tag{11}$$

Using this framework involving the renormalised effective interactions, the products of the tree-level charges used in expressions above are replaced by these defined couplings, $g_A^e g_V^q \to C_{1q}$ and $g_V^e g_A^q \to C_{2q}$. We will now refer to the C_{1q} parameters to denote the renormalised weak charges of the quarks.

The necessary radiative corrections which evolve these interactions were first evaluated in the early 1980's [42, 43, 44]. The latest precision determination of these radiative corrections have been reported by Erler, Kurylov and Ramsey-Musolf [45, 24], with the effective couplings given by [17]

$$C_{1u} + C_{1d} = 0.1528 \pm 0.0001 \tag{12}$$
$$C_{1u} - C_{1d} = -0.5298 \pm 0.0003 \tag{13}$$
$$C_{2u} + C_{2d} = -0.0095 \tag{14}$$
$$C_{2u} - C_{2d} = -0.0623 \pm 0.0005. \tag{15}$$

The uncertainties incorporate both the experimental precision in the Z-pole measurements and the precision in the scale evolution within the Standard Model.

In elastic electromagnetic scattering, the influence of hadronic structure in the two-photon exhange diagram has been demonstrated [46, 47] to play an important role in resolving the apparent discrepancy between Rosenbluth and polarization transfer measurements [48, 49] — see [50] for a recent review. The importance of hadronic structure in the computation of the two-photon effects has led to efforts to understand these effects in the analysis of parity violation measurements [51, 52, 53, 54]. A detailed analysis of these effects has recently been reported in [55]. Based on the same hadronic model used in the purely electrogmatic case [56], corrections are typically found to be less than 1% in magnitude.

Dispersion calculations of the inelastic contributions to the γZ-box, in the forward limit, suggest a somewhat larger correction [57, 58]. At the kinematics of the Q-weak experiment (discussed below in Section 6), these contributions give a 6.0–8.1% correction to $Q_W^p (= -2(2C_{1u} + C_{1d}))$, with the range bounded by proton structure function data [58]. This uncertainty of $\pm 1\%$ imposes a rather marginal shift in the total precision when added in quadrature to the nominal 4% measurement of Q-weak.

4.3. Hadronic form factors

In the extraction of the weak charge, other required inputs of the analyzing power, Eq. (9), are also the form factors of the proton. These involve both electromagnetic and neutral-current vector form factors and the neutral-current axial form factor. The term involving the axial form factor is also sensitive to a correction described by the nucleon anapole form factor [59, 60].

The dominant backgrounds in the forward scattering limit are the vector form factors of the proton, as highlighted by the Q^4 term in the expansion (10). This contribution involves both the standard electromagnetic form factors and the neutral current form factors. The electromagnetic form factors are well constrained through extensive study — see Refs. [61, 62] for recent reviews. In the analysis to be discussed, the simple parameterization of Kelly [63, 64] has been implemented. As a test, results results have been found to show negligible sensitivity to the use of an alternative parameterization, as given by Friedrich and Walcher [65].

While the electromagnetic form factors are themselves well known, the dominant uncertainty in the extraction of the weak charge lies in the neutral current form factors. For instance, under the assumption of charge symmetry, one can rewrite the proton neutral current form factors as

$$G_{E,M}^{Z(p)} = \left(2g_V^u + g_V^d\right) G_{E,M}^p + \left(2g_V^d + g_V^u\right) G_{E,M}^n + \left(g_V^u + g_V^d + g_V^s\right) G_{E,M}^s \,, \tag{16}$$

where we see the standard proton and neutron form factors together with the strangeness form factor. This is the separation that provides the isolation of strangeness, and we refer to the earlier chapter for details. From the point of view of a Standard Model test, the G^s term can simply be regarded as a sum over unknown hadronic structure in $G^{Z(p)}$ — which can dominantly be thought of as a sum of strangeness and charge symmetry violation.

As mentioned above, the axial form factor contribution is subject to a correction from a nucleon anapole moment. Being a nonperturbative hadronic phenomenon, we keep this contribution separate from the radiative corrections described in the previous section. Keeping the anapole contribution as explicit, we write the *effective* axial form factor as

$$g_V^e \tilde{G}_A^N = 2\left\{(C_{2u} - C_{2d})G_A^{(3)}\tau_3 + \sqrt{3}(C_{2u} + C_{2d})G_A^{(8)} + (C_{2u} + 2C_{2d})G_A^{(s)}\right\} + g_V^e A_{\rm ana}^N \tag{17}$$

with $\tau_3 = 1(-1)$ for the proton (neutron). The single-quark radiative corrections have been absorbed into the substitution $g_V^e g_A^q \to C_{2q}$. The normalisation of these axial form factors are defined such that at $Q^2 = 0$ these correspond to the moments of the quark helicity structure functions

$$G_A^{(3)}(0) = \frac{1}{2}(\Delta u - \Delta d)\,, \tag{18}$$

$$G_A^{(8)}(0) = \frac{1}{2\sqrt{3}}(\Delta u + \Delta d - 2\Delta s)\,, \tag{19}$$

$$G_A^{(s)}(0) = \Delta s. \tag{20}$$

These moments are reasonably well determined experimentally. The octet isoscalar axial charge, $G_A^{(8)}(0) = (3F - D)/2\sqrt{3}$, can be estimated by assuming SU(3)-flavour symmetry in hyperon β-decays, producing $a_8 = 3F - D = 0.58 \pm 0.03$ [66]. The strange axial charge $G_A^{(s)}(Q^2 = 0) = a_s$ can be estimated by combining this estimate with measurements of the first moment of the g_1^p structure function, where an analysis by the SMC Collaboration has determined $a_s = -0.07 \pm 0.04$ [67]. An estimate of the violations of SU(3)-flavour symmetry indicates that the corrections to a_8 could be of order 20% [68] — suggesting a more conservative uncertainty on this quantity $a_8 = 0.58 \pm 0.03 \pm 0.12$. Further, the use of SU(3) gives an estimate of a_s that is highly correlated with a_8 [67], this uncertainty in a_8 is reflected by $a_s = -0.07 \pm 0.04 \mp 0.05$[1].

With the standard axial form factors known to reasonable precision, it appears that these PVES measurements could also provide improved limits on the C_{2q} parameters — particularly in backward angle scattering, where ϵ in Eq. (9) becomes small. Unfortunately the anapole term, $A_{\rm ana}^N$, is unique to these measurements and therefore cannot be experimentally constrained independently of the terms involving C_{2q}. Theoretical estimates suggest this term is relatively small [59, 60, 71, 72, 73] relative to best experimental constraints [74]. In the interest of minimizing the theoretical bias, the whole effective axial form factor is fit as an independent term, without attempting to distiguish between the axial charges and the anapole contribution.

Neutrino scattering measurements suggest the Q^2 dependence of the axial form factors can be well described by a dipole form, with mass parameter of the order of $\sim 1\,\text{GeV}$ [75]. For the small Q^2 range reported here, this form is used to parameterise the *effective* axial form factors — *assuming* the anapole term follows the same Q^2 dependence.

5. Weak charges of the quarks

5.1. Global analysis

As discussed in Sec. 4.2, the effective lepton–quark contact interactions are known within the Standard Model to high precision. Experimentally, there have only been limited measurements of the weak neutral current in the low-energy domain. The most precise measurement on these interactions to date comes from the atomic parity violation measurements on cesium [2] — see [76] for a recent review on these types of experiments. The interpretation of these precision measurements as Standard Model tests are subject to the theoretical precision with which the atomic wave functions can be described, we refer to [77] for a review. The latest theoretical update on the cesium weak charge extraction provides $Q_w^{\rm Cs} \equiv -376C_{1u} - 422C_{1d} = -73.16(29)_{\rm exp}(20)_{\rm th}$ [21], which is to be compared with the Standard Model prediction $-73.15(4)$ [17]. The observed agreement is a fantastic result for the Standard Model, where the precision rules out new physics to a characteristic energy scale in the multi-TeV range (as we will elaborate on below).

This large nucleus has of course almost equal number of up and down quarks, and thereby primarily constrains new physics in the isoscalar $(u + d)$ channel. For many years, the best measurement of the weak neutral current in the isovector $(u - d)$ channel dates back to the pioneering work of Prescott *et al.* [28], with a weak charge extraction giving $2C_{1u} - C_{1d} = -0.90(18)$ — compared to the Standard Model prediction $-0.7183(5)$. While this work played a significant role in the establishment of the electroweak theory, the constraints on new physics are limited to $\sim 400\,\text{GeV}$, which is below the scale of modern direct searches [20].

The weak charge of the proton, as detailed above, is sensitive to the combination $2C_{1u} + C_{1d}$, which also provides sufficient orthogonality to the primarily isoscalar combination probed by cesium. Measurements of the parity-violating elastic scattering asymmetries have now been

[1] We note that semi-inclusive measurements at HERMES suggest that a_8 and a_s are both somewhat smaller [69, 70] — though limited by a finite domain of Bjorken x over which the structure function can be integrated.

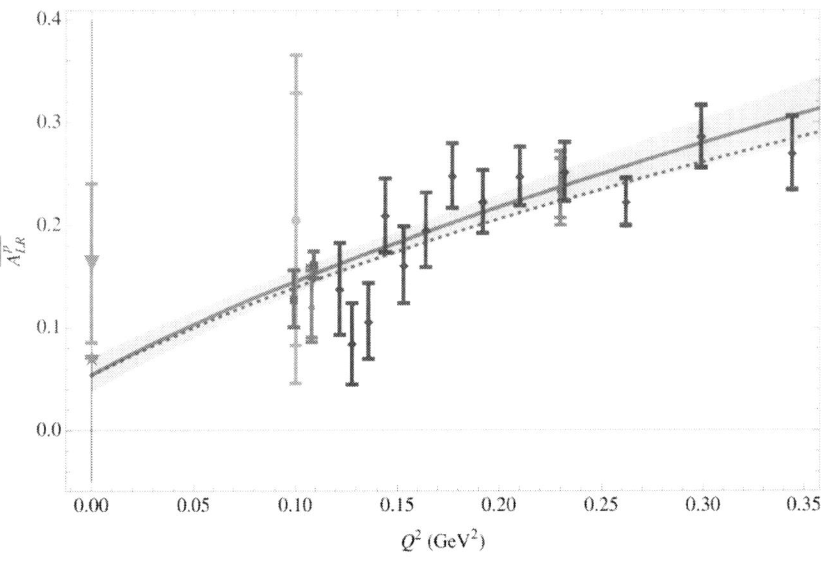

Figure 2. The measured parity-violating analyzing power on a proton target, extrapolated to the forward-angle limit using the analysis of all world data on PVES. The points have been normalised such that the extrapolation to $Q^2 = 0$ measures the proton's weak charge. The previous experimental knowledge (within uncertainties on the neutron weak charge) is shown by the triangular data point, and the Standard Model prediction by the star. The solid curve and shaded region indicate, respectively, the best fit and 1-σ bound, based upon the global fit to all electroweak data. The dotted curve shows the resulting fit if one incorporates the theoretical estimates [60] of the anapole form factors of the nucleon.

carried out by a number of experiments: SAMPLE at MIT-Bates [32]; PVA4 at Mainz [33, 34, 35]; and G0 [12] and HAPPEX [7, 8, 9, 11, 13] at Jefferson Lab. The range of kinematic coverage of these measurements, together with the known elastic form factors, provides enough information to constrain the asymmetry (9) to enable a relatively precise determination of the proton's weak charge [16].

In addition to the proton measurements, there is additional constraint on the electroweak structure from deuteron measurements by SAMPLE [78] and helium-4 by HAPPEX [10, 13] — though these are somewhat less significant in the extraction of the proton weak charge.

The range of proton measurements is displayed in Figure 2, where we have projected onto the forward angle limit (as explained below) and divided through by the leading factor in (10) such that the axis projects onto the weak charge of the proton. In terms of an extrapolation to $Q^2 \to 0$, we note that the G0 experiment [12] dominates the kinematic coverage and the best single precision point is offered by the HAPPEX result at $Q^2 \sim 0.11\,\mathrm{GeV}^2$ [13] — both being experiments that have run at Jefferson Lab.

Because each measurement has been performed at various scattering angles, the data points displayed in Fig. 2 have been rotated to the forward-angle limit using the global fit of this analysis, with the outer error bar on the data points indicating the uncertainty arising from the $\theta \to 0$ extrapolation. The dominant source of uncertainty in this $\theta \to 0$ extrapolation lies in the determination of the anapole form factor of the nucleon. The experimentally-constrained uncertainty on the anapole form factor is relatively large compared to estimates based upon chiral effective field theory supplemented with a vector meson dominance model [60]. Further constraining our fits to this theoretical estimate yields the dotted curve in Fig. 2, where the

Journal of Physics: Conference Series **299** (2011) 012012 doi:10.1088/1742-6596/299/1/012012

discrepancy with the experimentally determined fit is always less than one standard deviation — and yields negligible impact on the weak charge extraction.

The fit curve and uncertainty band is the result of the full global fits, using all the new electroweak data [16] and all the other earlier measurements [79, 80], which are dominated by the cesium result. We see a significant improvement in the previous knowledge of the protons weak charge, these measurements provide a new result of $Q^p_{\rm weak} = -2(2C_{1u} + C_{1d}) = 0.054 \pm 0.017$ — to be compared with the combined knowledge of the earlier measurements of 0.163 ± 0.077. This new improvement is displayed in Fig. 3, where the other experimental knowledge is summarised by the grey ellipse [79, 80]. The green filled ellipse shows the combined constraint provided by the latest electron scattering measurements. As outlined, these are dominated by measurements on a proton target, which corresponds to the more precisely determined direction of the ellipse. The helium measurements have also produced a constraint on the isoscalar combination which is comparable in precision to the early MIT-Bates measurement on carbon [31] — yet far from the precision attained in the atomic measurements.

The new result, combining all these measurement, is observed to be in tremendous agreement with the predictions of the Standard Model.

5.2. *Current limits on physics beyond the Standard Model*

While the increase in precision offered by the new constraint is evident in Fig. 3, it is useful to report the significance of this constraint relative to potential deviations from physics beyond the Standard Model. While there could be many possible types of new physics that could generate such deviations, it is most convenient to simply parameterise a new effective contact interaction [45],

$$\mathcal{L}^{eq}_{\rm NP} = \frac{g^2}{4\Lambda^2}\bar{e}\gamma_\mu\gamma_5 e \sum_q h^q_V \bar{q}\gamma^\mu q \,. \tag{21}$$

This essentially describes the resultant contact interaction surviving upon integrating out any of the potential new high-energy degrees of freedom. For further discussion on possible extensions, see [6, 45].

This contact interaction is characterised by an energy scale, Λ, and coupling strength, g. The isospin dependence of this new interaction can be parameterised in terms of an angle θ_h, with $h^u_V = \cos\theta_h$ and $h^d_V = \sin\theta_h$. This definition simply defines the quadrature sum of the independent isospin combinations to be unity, and then fixes the overall coupling to be characterised by g. The value of g is then dependent upon the type of new physics interaction. This would range from something small for a new weakly-coupled Z', to roughly unity for supersymmetric scenarios, up to a somewhat larger value for strongly-interacting composite models. For comparison, the Standard Model neutral-current effective interaction, with this parameterisation would correspond to $\Lambda \simeq 91\,{\rm GeV}$, $g \simeq 1/3$ and $\theta_h \simeq 5\pi/3$.

The new bounds on possible new physics interactions, characterised by Λ/g, are shown in Fig. 4, where new physics is ruled out at the 95% CL below the curve. Whereas earlier data had constrained a lower bound on relevant new physics $\Lambda/g > 0.4\,{\rm TeV}$, the new PVES measurements lift this limit to almost $1\,{\rm TeV}$, which is comparable to the direct searches at Fermilab [81]. Further, we see that in the phase space relevant to the isoscalar combination, as probed in atomic parity violation, the relevant scale is ~ 4–$5\,{\rm TeV}$.

Note that this limit applies to a generic class of quark-lepton parity-violating contact interactions, arising from integrating out all heavy, new-physics degrees of freedom. The possibility of cancellations between conspiring new physics at lower energies cannot be ruled out by such a low-energy analysis — yet results can constrain such a cancellation for any new-physics scenarios. We further note that the determined limit involves the ratio Λ/g, and hence a new physics discovery at a relatively low energy scale must be weakly coupled.

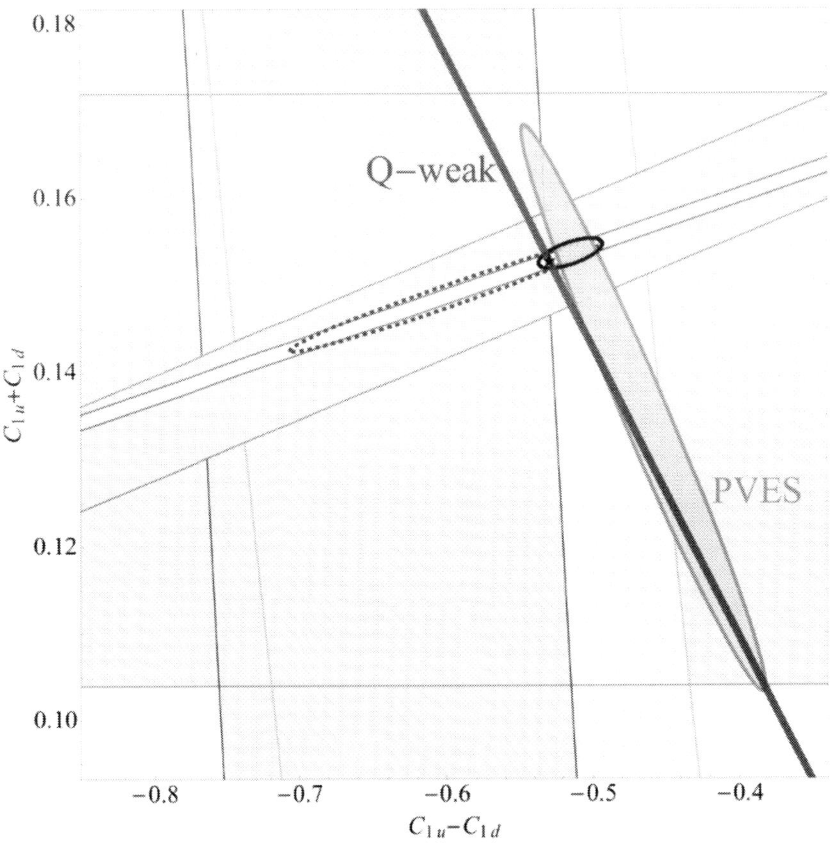

Figure 3. Experimental constraints on the effective neutral weak couplings. The dotted contour displays the experimental limits (68% CL) from all earlier sources, as dominated by the precision measurement on Cesium (narrow horizontal band). The filled ellipse denotes the additional constraint (at 1-σ) provided by recent high-precision PVES scattering measurements from SAMPLE, PVA4, G0 and HAPPEX. The total constraint (68% CL) from combining all measurements is displayed by the black contour. The future precision projected for the Q-weak experiment is shown by the narrow solid band, assuming agreement with the Standard Model is observed.

6. Future outlook

6.1. Q-weak experiment: precision measurement of the proton's weak charge

The Q-weak experiment is to carry out the first precision measurement of the weak charge of the proton, $Q_W^p = 1 - 4\sin^2\theta_W$, at Jefferson Lab, building on the technical advances that have been made in the laboratory's world-leading parity-violation program discussed above and using the results of earlier measurements to constrain hadronic corrections. The experiment is a high-precision measurement of the parity-violating longitudinal analyzing power (after dividing the measured asymmetry by the electron beam's polarization) in electron-proton elastic scattering at $Q^2 = 0.026\,(\mathrm{GeV/c})^2$ employing approximately 180 μA of 85% polarized beam on a 0.35 m long liquid hydrogen target. The experiment will determine the weak charge of the proton with about 4% combined statistical and systematic errors.

In the absence of physics beyond the Standard Model, the experiment will provide an approximately 0.3% measurement of $\sin^2\theta_W$, which will make it the most precise stand alone measurement of the weak mixing angle at low Q^2, and in combination with other parity-violation

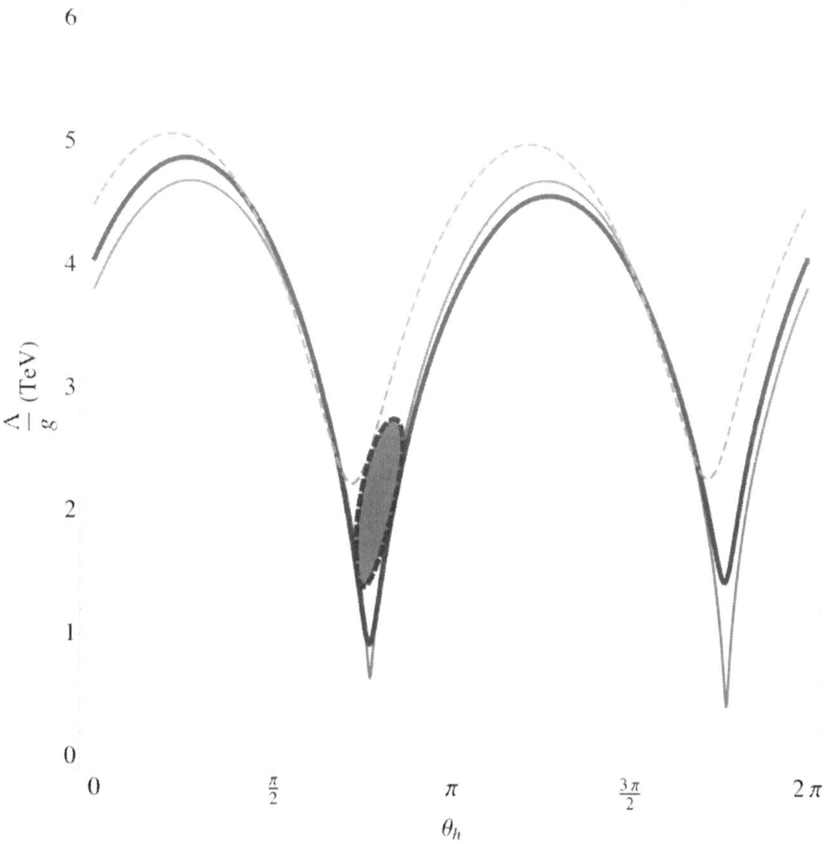

Figure 4. Constraints on new physics beyond the Standard Model in terms of the mass scale versus the isospin dependence. New physics is ruled out at the 95% CL below the curves. The thin (red) curve shows the limits based on the earlier measurements, corresponding to the grey ellipse of Fig. 3. The thick (blue) curve shows the new limits based on the red contour in Fig 3. The dashed curve displays the reach of the future proton weak charge measurement to be performed at Jefferson Lab, *assuming* agreement with the Standard Model is observed. The solid-filled region displays the possible discovery potential of Q-weak (also at 95% CL), *assuming* central values agree with current measurements.

measurements, a high precision determination of the weak couplings to the up and down quarks, improving on the present knowledge, see Figure 3. The measurement of Q_W^p will be performed with statistical and systematic errors significantly smaller than of the existing low Q^2 data. Any significant deviation from the Standard Model prediction for $\sin^2\theta_W$ at low Q^2 would be a signal of new physics, whereas agreement would place new and significant constraints on possible extensions of the Standard Model.

The Standard Model makes a firm prediction for Q_W^p, based on the "running" of the weak mixing angle or $\sin^2\theta_W$, from the Z pole down to low energies as shown in Figure 1. The average of the precise measurements at the Z pole makes it possible to calibrate the curve at one particular energy. But note that the most precise leptonic and semi-leptonic measurements at the Z pole differ by 3σ, a difference so far without definitive explanation. The shape of the curve away from this point is a prediction of the Standard Model. Consequently, to test this prediction one needs measurements away from the Z pole. At the present there exist several low

Q^2 precise determinations of $\sin^2\theta_W$: from an atomic parity-violation measurement on ^{133}Cs [2], and from a parity-violating electron-electron (Møller) scattering measurement [5] (E158 at SLAC).

It is to be noted that electroweak radiative corrections affect the proton and electron weak charges rather differently. In addition to the effect from the "running" of $\sin^2\theta_W$, there is a relatively large WW box graph contribution to the weak charge of the proton that does not appear in the case of the electron. This contribution compensates numerically for a large part the effect of the running of $\sin^2\theta_W$ for the proton, which is not the case for the electron. In addition there are two-boson exchange (γ and Z) contribution corrections [55] with intermediate states described by nucleons and inelastic excitations (see Sec. 4.2 for further discussion).

The experiment began commissioning and data taking in 2010 and will continue until 2012 before full statistics will be reached.

6.2. An ultra precise measurement of the weak mixing angle using Møller scattering

As commented above, the theory of electroweak interactions in the Standard Model has survived stringent experimental tests over more than three decades. Consistency has been achieved at the 0.1% level, where electroweak radiative corrections involving the top quark, the massive vector bosons (Z and $W^{\pm}$), and the Higgs boson become manifest. All experimental data to date, with direct access to a center of mass energy of 200 GeV in e^+e^- collisions, and approaching 1 TeV in the hard scattering of partons in $p\bar{p}$ collisions are in agreement with theoretical predictions. Nevertheless, compelling theoretical arguments, input from cosmological observations, and the discovery of neutrino mass strongly motivate the exploration of the multi-TeV scale *via* direct searches at colliders, as well as *via* ultra-precise electroweak measurements at low energy.

The Møller experiment is to measure the parity-violating analyzing power in the scattering of longitudinally polarized electrons off electrons of a liquid hydrogen target, using an 11 GeV beam in Hall A of the upgraded CEBAF of Jefferson Lab. The measurement will improve on the existing result of the same quantity from the E158 experiment at SLAC by more than a factor of five in fractional precision (Figure 1). The precision (about ± 0.00025) is comparable to the precision reached in the two better measurements at the Z pole. As stated above the latter two measurements of $\sin^2\theta_W$ differ by three standard deviations (see Figure 5). While the world average is consistent with other electroweak measurements and constraints on the Higgs boson mass, choosing one or the other central value ruins this consistency. Consequently, an independent determination of $\sin^2\theta_W$ with the same level of precision and interpretability is needed to resolve this issue. As described in the previous section, new neutral current interactions are best parameterized model-independently at low energies by effective four-fermion interactions (as in Eq. (21)) *via* the quantity Λ/g, where g indicates a coupling strength and Λ the scale of the new physics. The Møller experiment will reach a sensitivity in Λ/g of 7 TeV, which then constitutes the most sensitive probe of new flavor and CP conserving neutral current interactions in the leptonic sector until the advent of a linear collider and a neutrino factory. The 11 GeV Møller experiment and the Q-weak experiment form a set of complementary experiments that are each unique and will test the Standard Model with great precision.

6.3. Measurements of parity-violating deep inelastic scattering

Parity-violating deep inelastic scattering from the deuteron will provide constraints on physics beyond the Standard Model, in a similar spirit to the Q-weak and Møller experiments. Importantly, this program will be sensitive to the axial hadronic currents, unlike the latter two experiments. There are two deep inelastic scattering experiments at Jefferson Lab: a scheduled 6 GeV experiment and a proposed 11 GeV experiment (PVDIS). The first one will measure the parity-violating longitudinal analyzing power in electron-deuteron deep inelastic scattering at $Q^2 = 1.11$ and $1.90\,(\mathrm{GeV/c})^2$ at $x \simeq 0.3$. The combination of the two measurements will

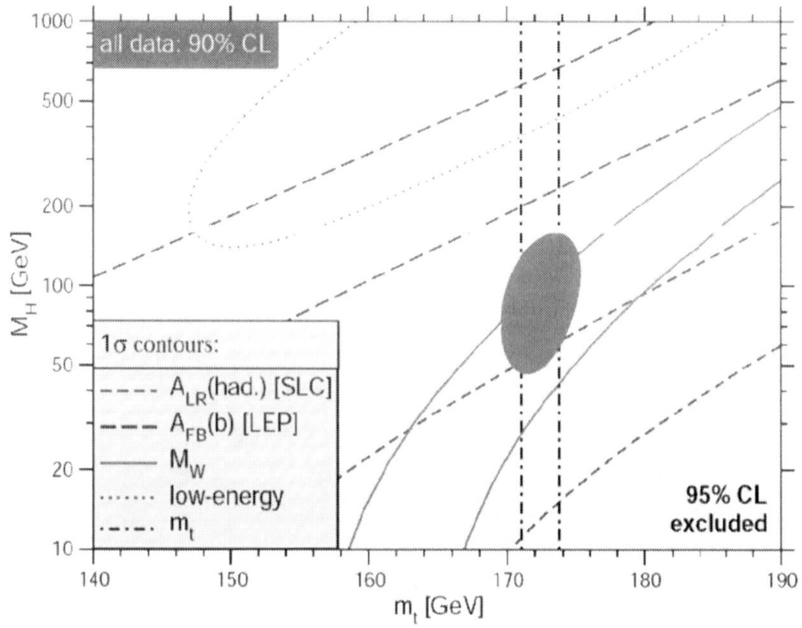

Figure 5. 1-σ bands from various precision measurements. The red filled ellipse is a 90% C.L. contour of all precision electroweak data. The shaded yellow region is excluded *via* direct searches at colliders — including the latest exclusion band from the Tevatron [82]. The strongest constraint on the width of the purple dotted contour from current low energy measurements is already now obtained from the SLAC Møllerexperiment [5], while its shape and location on the plot is mostly due to the NuTeV result [3]. With the targeted precision, the 11 GeV Møller experiment will provide a constraint of similar slope and precision as both the Z-pole measurements at SLC and LEP — and the potential to resolve the disrepancy between these results.

provide a constraint on higher-twist effects in parity-violating deep inelastic scattering at a level of $2.8\%/Q^2$. In turn, the experiment will provide a determination of the combination of the vector–axial-vector weak couplings $(2C_{2u} - C_{2d})$. The proposed 11 GeV experiment will measure the parity-violating analyzing power in electron-deuteron deep inelastic scattering (with additional measurements on a hydrogen target) with a precision of about 0.5% over the range $0.3 < x < 0.7$ and with a large dynamic range of Q^2 values. As already mentioned above, the contributions from higher-twist operators need to be delineated, although in the measured analyzing power some higher-twist contributions will cancel. Another complication is the presence of charge symmetry breaking of the quark distributions. At present there is paucity of precise information on charge symmetry breaking of the quark distributions and thus will have to be determined in the experiment. The 11 GeV experiment requires very high luminosity combined with a large azimuthal acceptance, a large solenoidal spectrometer with fast counting techniques for tracking, Cherenkov and calorimeter detectors, and polarimetry at the level of 0.5%.

7. Summary

It should be apparent from what has been described in this chapter that Jefferson Lab is fullfilling a prominent role in elucidating electroweak physics in the low energy regime. It will assume a unique role, fully complementary to the LHC, by performing a series of precision electron parity violating experiments (in preparation or planned for the 12 GeV upgrade) which are designed to test the current Standard Model and decipher the signatures of new physics.

Acknowledgments

The authors would like to acknowledge the many stimulating discussions with colleagues in this subfield of electroweak physics. This work was supported by DOE contract DE-AC02-06CH11357, under which UChicago Argonne, LLC, operates Argonne National Laboratory.

References

[1] M. H. Ahn *et al.*, Phys. Rev. Lett. **90**, 041801 (2003).
[2] S. C. Bennett and C. E. Wieman, Phys. Rev. Lett. **82**, 2484 (1999).
[3] G. P. Zeller *et al.*, Phys. Rev. Lett. **88**, 091802 (2002).
[4] P. L. Anthony *et al.*, Phys. Rev. Lett. **92**, 181602 (2004).
[5] P. L. Anthony *et al.*, Phys. Rev. Lett. **95**, 081601 (2005).
[6] M. J. Ramsey-Musolf and S. Su, Phys. Rept. **456**, 1 (2008).
[7] K. A. Aniol *et al.*, Phys. Rev. Lett. **82**, 1096 (1999).
[8] K. A. Aniol *et al.*, Phys. Lett. B **509**, 211 (2001).
[9] K. A. Aniol *et al.*, Phys. Rev. C **69**, 065501 (2004).
[10] K. A. Aniol *et al.*, Phys. Rev. Lett. **96**, 022003 (2006).
[11] K. A. Aniol *et al.*, Phys. Lett. B **635**, 275 (2006).
[12] D. S. Armstrong *et al.*, Phys. Rev. Lett. **95**, 092001 (2005).
[13] A. Acha *et al.*, Phys. Rev. Lett. **98**, 032301 (2007).
[14] D. B. Kaplan and A. Manohar, Nucl. Phys. **B310**, 527 (1988).
[15] D. H. Beck, Phys. Rev. D **39**, 3248 (1989).
[16] R. D. Young, R. D. Carlini, A. W. Thomas and J. Roche, Phys. Rev. Lett. **99**, 122003 (2007).
[17] C. Amsler *et al.*, Phys. Lett. B **667**, 1 (2008).
[18] D. B. Chitwood *et al.*, Phys. Rev. Lett. **99**, 032001 (2007).
[19] A. Barczyk *et al.*, Phys. Lett. B **663**, 172 (2008).
[20] S. Schael *et al.*, Phys. Rept. **427**, 257 (2006).
[21] S. G. Porsev, K. Beloy and A. Derevianko, Phys. Rev. Lett. **102**, 181601 (2009).
[22] D. E. Acosta *et al.*, Phys. Rev. D **71**, 052002 (2005).
[23] V. M. Abazov *et al.*, Phys. Rev. Lett. **101**, 191801 (2008).
[24] J. Erler and M. J. Ramsey-Musolf, Phys. Rev. D **72**, 073003 (2005).
[25] W. Bentz, I. C. Cloët, J. T. Londergan and A. W. Thomas, Phys. Lett. B **693**, 462 (2010).
[26] J. T. Londergan, Nucl. Phys. Proc. Suppl. **141**, 68 (2005).
[27] I. C. Cloët, W. Bentz and A. W. Thomas, Phys. Rev. Lett. **102**, 252301 (2009).
[28] C. Y. Prescott *et al.*, Phys. Lett. B **84**, 524 (1979).
[29] K. S. Kumar and P. A. Souder, Prog. Part. Nucl. Phys. **45**, S333 (2000).
[30] W. Heil *et al.*, Nucl. Phys. **B327**, 1 (1989).
[31] P. A. Souder *et al.*, Phys. Rev. Lett. **65**, 694 (1990).
[32] D. T. Spayde *et al.*, Phys. Lett. B **583**, 79 (2004).
[33] F. Maas *et al.*, Phys. Rev. Lett. **93**, 022002 (2004).
[34] F. E. Maas *et al.*, Phys. Rev. Lett. **94**, 152001 (2005).
[35] S. Baunack *et al.*, Phys. Rev. Lett. **102**, 151803 (2009).
[36] C. K. Sinclair *et al.*, Phys. Rev. ST Accel. Beams **10**, 023501 (2007).
[37] J. M. Grames *et al.*, Phys. Rev. ST Accel. Beams **7**, 042802 (2004).
[38] S. D. Covrig *et al.*, Nucl. Instrum. Meth. A **551**, 218 (2005).
[39] K. D. Paschke, Eur. Phys. J. A **32**, 549 (2007).
[40] F. J. Ernst, R. G. Sachs and K. C. Wali, Phys. Rev. **119**, 1105 (1960).
[41] M. J. Musolf *et al.*, Phys. Rept. **239**, 1 (1994).
[42] C. H. Llewellyn Smith and J. F. Wheater, Phys. Lett. B **105**, 486 (1981).
[43] W. J. Marciano and A. Sirlin, Phys. Rev. D **27**, 552 (1983).
[44] W. J. Marciano and A. Sirlin, Phys. Rev. D **29**, 75 (1984).
[45] J. Erler, A. Kurylov and M. J. Ramsey-Musolf, Phys. Rev. D **68**, 016006 (2003).
[46] P. G. Blunden, W. Melnitchouk and J. A. Tjon, Phys. Rev. Lett. **91**, 142304 (2003).
[47] P. A. M. Guichon and M. Vanderhaeghen, Phys. Rev. Lett. **91**, 142303 (2003).
[48] M. K. Jones *et al.*, Phys. Rev. Lett. **84**, 1398 (2000).

[49] O. Gayou *et al.*, Phys. Rev. Lett. **88**, 092301 (2002).
[50] C. E. Carlson and M. Vanderhaeghen, Ann. Rev. Nucl. Part. Sci. **57**, 171 (2007).
[51] A. V. Afanasev and C. E. Carlson, Phys. Rev. Lett. **94**, 212301 (2005).
[52] J. Arrington and I. Sick, Phys. Rev. C **76**, 035201 (2007).
[53] H. Q. Zhou, C. W. Kao and S. N. Yang, Phys. Rev. Lett. **99**, 262001 (2007).
[54] J. A. Tjon and W. Melnitchouk, Phys. Rev. Lett. **100**, 082003 (2008).
[55] J. A. Tjon, P. G. Blunden and W. Melnitchouk, Phys. Rev. C **79**, 055201 (2009).
[56] P. G. Blunden, W. Melnitchouk and J. A. Tjon, Phys. Rev. C **72**, 034612 (2005).
[57] M. Gorchtein and C. J. Horowitz, Phys. Rev. Lett. **102**, 091806 (2009).
[58] A. Sibirtsev, P. G. Blunden, W. Melnitchouk and A. W. Thomas, Phys. Rev. D **82**, 013011 (2010).
[59] W. C. Haxton, E. M. Henley and M. J. Musolf, Phys. Rev. Lett. **63**, 949 (1989).
[60] S.-L. Zhu, S. J. Puglia, B. R. Holstein and M. J. Ramsey-Musolf, Phys. Rev. D **62**, 033008 (2000).
[61] C. F. Perdrisat, V. Punjabi and M. Vanderhaeghen, Prog. Part. Nucl. Phys. **59**, 694 (2007).
[62] J. Arrington, C. D. Roberts and J. M. Zanotti, J. Phys. G **34**, S23 (2007).
[63] J. J. Kelly, Phys. Rev. C **70**, 068202 (2004).
[64] J. J. Kelly, Phys. Rev. C **66**, 065203 (2002).
[65] J. Friedrich and T. Walcher, Eur. Phys. J. **A17**, 607 (2003).
[66] B. W. Filippone and X. Ji, Adv. Nucl. Phys. **26**, 1 (2001).
[67] D. Adams *et al.*, Phys. Rev. D **56**, 5330 (1997).
[68] S.-L. Zhu, S. Puglia and M. J. Ramsey-Musolf, Phys. Rev. D **63**, 034002 (2001).
[69] A. Airapetian *et al.*, Phys. Rev. D **71**, 012003 (2005).
[70] A. Airapetian *et al.*, Phys. Lett. B **666**, 446 (2008).
[71] D. O. Riska, Nucl. Phys. **A678**, 79 (2000).
[72] C. M. Maekawa, J. S. Veiga and U. van Kolck, Phys. Lett. B **488**, 167 (2000).
[73] C. M. Maekawa and U. van Kolck, Phys. Lett. B **478**, 73 (2000).
[74] R. D. Young, J. Roche, R. D. Carlini and A. W. Thomas, Phys. Rev. Lett. **97**, 102002 (2006).
[75] V. Bernard, L. Elouadrhiri and U. G. Meissner, J. Phys. G **28**, R1 (2002).
[76] J. A. Behr and G. Gwinner, J. Phys. G **36**, 033101 (2009).
[77] J. S. M. Ginges and V. V. Flambaum, Phys. Rept. **397**, 63 (2004).
[78] T. M. Ito *et al.*, Phys. Rev. Lett. **92**, 102003 (2004).
[79] W. M. Yao *et al.*, J. Phys. G **33**, 1 (2006).
[80] J. Erler and M. J. Ramsey-Musolf, Prog. Part. Nucl. Phys. **54**, 351 (2005).
[81] T. Aaltonen *et al.*, Phys. Rev. Lett. **102**, 091805 (2009).
[82] T. Aaltonen *et al.*, Phys. Rev. Lett. **104**, 061802 (2010).

Journal of Physics: Conference Series **299** (2011) 012013 doi:10.1088/1742-6596/299/1/012013

Hypernuclear Spectroscopy

F. Garibaldi[1], O. Hashimoto[2], J. J. LeRose[3], P. Markowitz[5], S. N. Nakamura[2], J. Reinhold[5], and L. Tang[3,4]

[1] Istituto Nazionale di Fisica Nucleare and Istituto Superiore di Sanitá, I-00161 Roma, Italy
[2] Tohoku University, Sendai, 980-8578, Japan
[3] Jefferson Lab, Newport News, Virginia 23606, USA
[4] Hampton University, Hampton, Virginia 23668, USA
[5] Florida International University, Miami, Florida 33199, USA

Abstract. A program of hypernuclear spectroscopy experiments encompassing many hypernuclei has been undertaken in both Halls A and C using complimentary approaches. Spectra with sub-MeV resolution have been obtained for ${}^{9}_{\Lambda}$Li, ${}^{12}_{\Lambda}$B, and ${}^{16}_{\Lambda}$N in Hall A, while results from Hall C include ${}^{7}_{\Lambda}$He, ${}^{12}_{\Lambda}$B, and ${}^{28}_{\Lambda}$Al with new data still under analysis for ${}^{7}_{\Lambda}$He, ${}^{9}_{\Lambda}$Li, ${}^{10}_{\Lambda}$Be, ${}^{12}_{\Lambda}$B and ${}^{52}_{\Lambda}$V. High resolution and high precision in the determination of the single Λ binding energy at various shell levels has been the key success of these experiments using the (e,e′K^{+}) reaction to produce Λ hypernuclei.

1. Introduction

Since the first observation of hypernuclei in 1953 [1], spectroscopic investigation of hypernuclei has been considered to be a unique method to provide invaluable information on many-body hadronic systems by utilizing the new degree of freedom – "strangeness". This impurity can be used as a probe to study both the structure and properties of baryons in the nuclear medium and the structure of nuclei as baryonic many-body systems [2, 3, 4]. Without Pauli blocking, the Λ can penetrate into the nuclear interior and form deeply bound hypernuclear states. Therefore, new nuclear structures or unknown properties of the baryonic interaction, which cannot be seen from the investigation of ordinary nuclei with conventional probes, may manifest themselves in hypernuclei, providing indispensable information on the flavor SU(3) basis for baryonic systems.

To create a unified description of the baryonic interaction within the flavor SU(3) basis, one must understand baryonic interactions beyond nucleon-nucleon (NN) interactions, such as hyperon-nucleon (YN) and hyperon-hyperon (YY) interactions. Spectroscopic investigation of Λ hypernuclei, a nuclear many-body system containing one Λ particle, provides a unique and currently the only practical tool to study the ΛN interaction, since direct ΛN scattering experiments are technically difficult. Since the Λ decays only weakly and has a relatively long lifetime, Λ hypernuclei feature narrow states commonly described by coupling of low-lying core states to a Λ in low level shell states (s, p, ...) with widths ranging from a few to $\sim$ 100 keV depending on the decay channels (weak decay, EM transitions, and nucleon emissions or breakups at high excitation levels). This makes spectroscopic studies possible.

One novel feature of the ΛN interaction is that the Λ has isospin 0. This prevents one-pion exchange (OPE) due to the isospin conservation requirement for the strong interaction. This means that the OPE which is the long range component in the NN interaction is missing in the

ΛN interaction. This makes the study of the ΛN interaction more attractive in exploring the short-range components in the strong interaction.

However, going from spectroscopic investigation to a full understanding of the Λ hypernuclear structure and ΛN interaction is a complicated process which requires solving the baryonic many-body problem. This process is well outlined and summarized in the introduction section in the most recent review article written by Hashimoto and Tamura [5], which says: "Assuming that a Λ hypernuclear wavefunction can be decomposed into a core nucleus and a Λ hyperon, the hypernuclear Hamiltonian is expressed as

$$H = H_{\mathrm{CoreNucleus}} + t_\Lambda + \sum v_{\Lambda\mathrm{N}}^{\mathrm{effective}}, \tag{1}$$

where $H_{\mathrm{CoreNucleus}}$ is the Hamiltonian for the core nucleus, t_Λ is the kinetic energy of the Λ hyperon and $v_{\Lambda\mathrm{N}}^{\mathrm{effective}}$ describes the effective ΛN interaction. The effective interaction can be constructed *via* a G-matrix calculation, starting from the two-body interactions in free space. One-boson exchange models such as the Nijmegen [7] and Jülich interactions [8] (which were constructed by extending NN interaction models on the basis of flavor SU(3) symmetry with scarce ΛN and ΣN scattering data used to adjust parameters) are widely used to describe the elementary two-body interactions.

Analytical forms of the effective potentials are often given in the form of a three-range Gaussian,

$$V_{\Lambda\mathrm{N}}(r) = \sum_i (a_i + b_i k_f + c_i k_f^2)\exp(-r^2/\beta_i^2) \text{ [8]}. \tag{2}$$

A wide variety of hypernuclear properties such as level structure and reaction cross sections are calculated using this potential and compared directly with experimental data. These calculations are reasonably reliable, in part, because the ΛN interaction is much weaker than the NN interaction and no anti-symmetrization against the nucleon is required. It is to be noted that few-body hypernuclear systems with $A \leq 5$ can be calculated directly from the two-body interactions, in which the calculated binding energies of ground and excited states are compared precisely with experimental data. In such studies, information on Λ hypernuclear structure obtained by spectroscopic studies plays an essential role in testing and improving YN interaction models.

There is also a phenomenological approach to the effective interaction [9, 10] in which the ΛN effective interaction in p-shell Λ hypernuclei is written in the form

$$V_{\Lambda\mathrm{N}}(r) = V_0(r) + V_\sigma(r)\boldsymbol{s}_\Lambda\boldsymbol{s}_\mathrm{N} + V_\Lambda(r)\boldsymbol{l}_{\Lambda\mathrm{N}}\boldsymbol{s}_\Lambda + V_\mathrm{N}(r)\boldsymbol{l}_{\Lambda\mathrm{N}}\boldsymbol{s}_\mathrm{N} + V_T(r)S_{12} \tag{3}$$

where $S_{12} = 3(\boldsymbol{\sigma}_\Lambda\hat{\boldsymbol{r}})(\boldsymbol{\sigma}_\mathrm{N}\hat{\boldsymbol{r}})$ - $\boldsymbol{\sigma}_\Lambda\boldsymbol{\sigma}_\mathrm{N}$. Low-lying level energies of p-shell hypernuclei can be described with radial integrals over the $s_\Lambda\, p_N$ wave function for each of the five terms in Eq. (3). These integrals, denoted as $\bar{V}$, Δ, S_Λ, S_N and T, can be determined from p-shell Λ hypernuclear data, then compared with theoretical predictions given by free ΛN interactions through G-matrix calculation."

This passage outlines the general approach applied in hypernuclear physics with which one attempts to understand the baryonic interactions beyond nucleons. On the other hand, it also shows that experimental precision in spectroscopic studies is crucially important, which is the main reason for the development of the Jefferson Lab hypernuclear spectroscopy program as well as the high precision γ spectroscopy program developed at KEK-PS/BNL-AGS in the last decade.

Traditionally, hypernuclei have been studied *via* their production using secondary mesonic beams, π or K, through various strangeness producing reactions, such as (K^-, π^-), (π^+, K^+), *etc.* For the reaction mass spectroscopy, the secondary nature of the initial beam forces a limit on the energy resolution for the observed hypernuclear states. Resolution has been limited to above

1.5 MeV (FWHM). The low-lying states were difficult to observe clearly and their corresponding binding energies could not be accurately extracted. Although the high precision Λ spectroscopy experiments at these facilities have been able to provide precision level spacing information for many light hypernuclei, there is no determination of the binding energy for the ground states. Therefore, high precision mass spectroscopy still has an important and irreplaceable role in providing valuable information for finding the correct models.

The Jefferson Lab accelerator provides a unique opportunity to produce hypernuclei using the $(e,e'K^+)$ reaction with high precision. The high energy resolution is achieved by the combination of (1) precision high intensity beam; (2) high resolution of both scattering angles and momenta in spectrometers for scattered electrons and kaons; and (3) thin target foils which minimize target straggling effects. It was anticipated that the energy resolution for the Jefferson Lab hypernuclear spectroscopy program could be at the sub-MeV level with the ultimate goal of 300–400 keV (FWHM). The absolute mass scale depends on the method and technique applied in the $(e,e'K^+)$ kinematics calibration which determines the level of precision in determination of binding energy of the observed hypernuclear states.

To compensate the small cross sections in electroproduction, not only high beam intensity is needed but also both the scattered electrons and kaons must be detected at small forward angles. This makes the experimental design with double spectrometers difficult. The issues are associated with (1) the level of virtual photon flux, (2) the solid angle acceptance and (3) the short-lived kaon survival rate. The most obvious technical challenges for both the experiment and the data analysis are from the extremely high particle singles rates.

Other than the attractive high precision, electroproduction of hypernuclei brings in additional new features to the overall investigation of hypernuclei. First of all, electroproduction with the $(e,e'K^+)$ reaction produces a Λ from a proton in the nucleus creating a nuclear core state with a proton hole to which the Λ couples. This can produce either mirror hypernuclei to those produced by the hadronic reactions (K^-, π^-) and (π^+, K^+) or states with different isospin. In other words, electroproduction provides neutron rich hypernuclei which are suitable candidates to investigate the Λ-Σ coupling contribution in the effective potential as well as the contribution from the ΛNN three-body force with two π exchanges. Furthermore, the large 3-momentum transfer electroproduction *via* absorption of a virtual photon promotes large spin-flip transition amplitudes from the initial nuclear to the final hypernuclear states while the non-spin-flip amplitude remains non-negligible. Thus, deeply bound hypernuclear states with both natural and unnatural parities may simultaneously appear. This provides rich and new spectroscopy complementary to that produced by the hadronic reactions.

Over the last decade, two independent hypernuclear experimental programs in Halls A [11] and C [12] were developed and successfully carried out. In the following sections, the programs from both halls, their applied technique, the achievements in physics results, and future prospects will be described.

2. The Hall C experimental technique and development – three distinct phases

After the first production of $^{3,4}_{\Lambda}H$ hypernuclei with electron beams was confirmed in Hall C [13], $(e,e'K^+)$ hypernuclear spectroscopy was initiated by the Hall C Collaboration at Jefferson Lab. Beginning in 2000 when the Hall C Collaboration successfully carried out the first spectroscopy experiment (E89-009/HNSS) at Jefferson Lab with electroproduction, the program in Hall C has gone through three phases with two major spectrometer upgrades, HKS and HES, funded by JSPS (Japanese Society for Promotion of Science, 16GS0201, 12002001) through the Tohoku University group, as well as modification of the experimental technique in tagging the scattered electrons (see illustration in Fig. 1). These upgrades and the introduction of new techniques continued to improve energy resolution and precision in determination of absolute mass, to increase the physics production yield rate, and to widen the mass range of the produced hypernuclei in spectroscopic studies in a variety of nuclear shell levels.

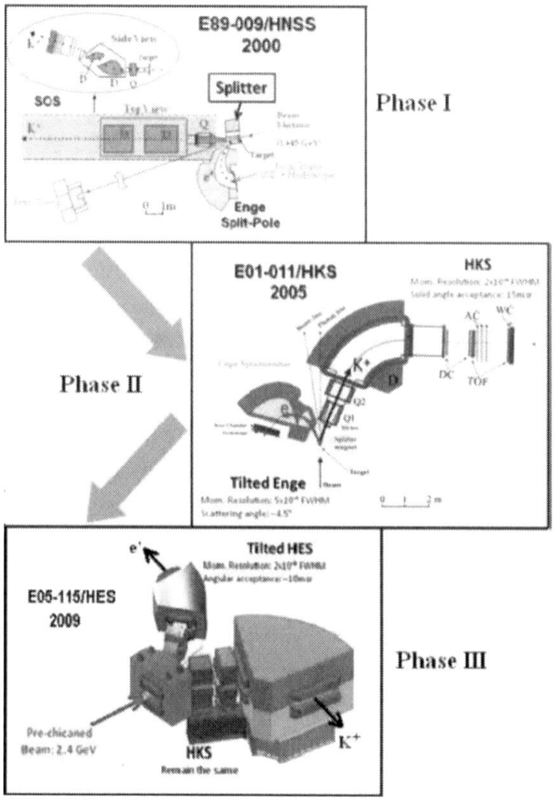

Figure 1. Schematic illustration of the experimental setup, technique and equipment upgrades for the three Hall C hypernuclear spectroscopy experiments: E89-009, E01-011, and E05-115/E08-002.

2.1. A common splitter magnet

The central element of the Hall C experiment is the charge separation splitter magnet (SPL). With the target located at the SPLs upstream effective field boundary (EFB), this dipole magnet bends the oppositely charged particles away from the beam in opposite directions. This common splitter magnet technique allows the subsequent spectrometers for forward peaked electrons and K^+ mesons to be located at angles sufficiently large with respect to the beam line. Therefore, it is easier to locate the double spectrometer system to detect extremely forward-angled particles. Furthermore, a single common splitter magnet immediately behind the target makes larger solid angle acceptance, which is important to compensate low yield due to the small production cross sections with an electromagnetic probe. The overall kinematics and the spectrometer optics thus depend on the design and application of this magnet.

Introduction of the splitter magnet configuration presents unavoidable challenges to the design of the experiment. First of all, the unused primary beam that passes through the SPL is deflected. Therefore, this magnet is simultaneously handling the beam, the scattered electrons, and the reaction kaons at the same time. The beam must be carefully handled to avoid a serious radiation problem. From the view point of kinematics, it couples the beam and the two spectrometer systems together in a fixed kinematics space. It is impossible to perform an absolute calibration for each individual system separately. A special calibration method is necessary to ensure the high quality of the spectra and high accuracy in determining the Λ binding energy.

Secondly, the magnetic field right after the target can also direct the background particles near zero degrees into the spectrometers causing high particle singles rates and high accidental coincidence rates. In the electron arm, this background comes from bremsstrahlung and Møller scattered electrons with extremely high rates at very forward angles. They are the sources of the accidental background in the hypernuclear spectra obtained in the Hall C experiments. Positrons from pair production at near zero degrees can come into the kaon spectrometer. In general, they are not in the kaon spectrometer solid angle acceptance. However, some of them hit the dense materials in the exit area of the spectrometer and can create subsequent showers with low energy electrons going toward the detector system aided by the fringe field of the last dipole magnet. This can cause a high particle flux leading to high singles rates as well as jeopardizing the tracking efficiency and position precision of the focal plane wire chambers. This is particularly serious for heavy targets such as ^{52}Cr used in 2009 for E05-115.

2.2. *Kinematics considerations and calibration*

The central virtual photon energy was chosen to be 1.5 GeV, *i.e.* ω = E–E$'$ = 1.5 GeV, at which the elementary Λ production cross section is at a plateau. At small forward angles, the corresponding central momentum of the kaon arm (SOS for the experiment in 2000, HKS for the runs in the 2005 and 2009 experiments) was at 1.2 GeV/c. Fixing the virtual photon energy and the kaon momentum, the beam energy is determined by the spectrometer used for the electron arm. In the Phase I (HNSS/2000) and II (HKS/2005) experiments the Enge Split-Pole spectrometer was used to tag the scattered electron at zero degrees in HNSS/2000 and at about 4.5° in the HKS/2005 experiments. The central momentum of the Enge was chosen to be 0.34 GeV/c, thus the beam energy was then 1.85 GeV. For the 2009 HKS-HES experiment, the introduction of the newly developed HES increased the electron central momentum to 0.84 GeV/c, therefore the beam energy was increased to 2.34 GeV.

The relatively lower energy kinematics made it easy to design and construct short orbit spectrometers (SOS and later HKS) which maximized the survival rate against the short lifetime of kaons ($c\tau \sim 3.6$ m) and allowed the use of conventional spectrometer techniques for kaon detection. The conventional detector techniques, like time-of-flight, aerogel/water Čerenkov detectors and conventional drift chambers, are able to identify and momentum-analyze the 1.2 GeV/c kaons.

With this kinematics design, the physics yield rate depends on the solid angle and momentum acceptances of the two spectrometers. Figure 2 illustrates the kinematic acceptance using the 2005 HKS setup as an example. The large momentum acceptances of both the electron and kaon arms allow allocation of the events of free Λ and Σ^0 production from the CH2 target and events from the production of various hypernuclei and hypernuclear states from different nuclear targets in a single kinematics space without changing any setting of the spectrometers. The events of the ground state of $^{12}_{\Lambda}$B hypernuclei are illustrated in the figure. The dash lines show the events with central angles. The distributions of Λ and Σ^0 in Fig. 2 are spread due to the recoil angles of the produced hyperons. This spread becomes smaller for heavier systems, *i.e.* hypernuclei with mass A > 6 are less affected by uncertainty from finite angular resolution. Figure 2 also shows the reconstructed one-dimensional mass spectrum for the correlated events. The shaded area is due to events from accidental e$'$ and K$^+$ coincident events which is the main background of the Hall C experiments. The distribution away from the Λ and Σ^0 peaks and above the accidental background is due to the quasi-free hyperon production from C in the CH2 target.

With angular resolution and central angle precision at a level of 2–3 mr (r.m.s.), the dominant

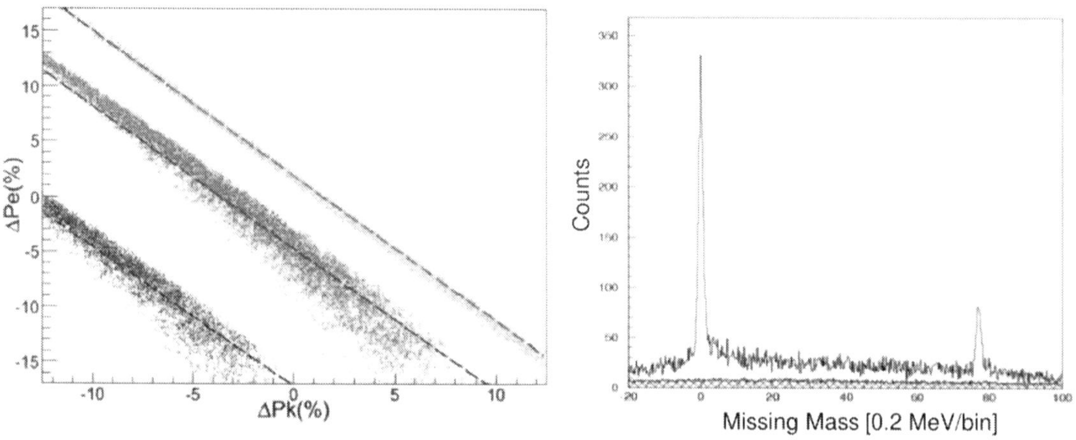

Figure 2. Illustrative view of the kinematics acceptance of the Hall C experiments and the free Λ and Σ^0 spectrum from the p(e,e$'$K$^+$)Λ(Σ^0) reaction.

factors crucial to the precision of the absolute reconstructed missing masses are the precise knowledge of the absolute central beam energy and the absolute central momenta of the two spectrometers shown as 0% in Fig. 2 for ΔPe$'$ and ΔPk. These three kinematic parameters should be precisely calibrated, however, it should be noted that there remain only two degrees of freedom for a fixed missing mass.

The large acceptance of the kaon and electron spectrometers that simultaneously includes both Λ and Σ^0 production is a major advantage of this setup. The masses of the Λ and Σ^0 are sufficiently well known that an absolute energy calibration is possible. This is impossible for the (π, K) or (K, π) reactions. The Λ and Σ^0 peaks can also be used to calibrate the kinematical parameter combination since the right combination will give the correct masses as well as the narrowest peak widths for both the Λ and the Σ^0. The resultant kinematic precision (a part of the systematic error in the Λ binding energy) depends on the statistics of the Λ/Σ^0 events and the knowledge of the line shapes which are affected by the optics tune and the radiative tail. The later can be obtained from detailed simulation studies. For the three conducted experiments, E89-009/HNSS, E01-011/HKS, and E05-115/HES, the kinematic accuracy continuously improved due to the dramatic increase of the physics yield rate with each upgrade. It was at the $\sim$ 300 keV level for HNSS. It was then improved to $\lesssim$100 keV for HKS. The HES data are currently under analysis and the accuracy is anticipated $\lesssim$50 keV.

2.3. Optics optimization

Because of the application of the common splitter magnet, the optics of the two spectrometers can only be optimized by using well defined physical events from the (e,e$'$K$^+$) reaction. These events include those from Λ and Σ^0 production. However, due to their light recoil mass, even small angular uncertainties can cause large momentum uncertainties. Furthermore, as illustrated by Fig. 2, events from these two processes do not fully cover the entire momentum acceptances. Therefore, events from well observed hypernuclear states were also selected in the process of the optics optimization. Hypernuclei with A $>$ 6 have heavy recoil masses so that the momentum uncertainty from these events is very small even with large angular errors. The quality of the optics in the two dimensional space in Fig. 2 could be translated into the width of the observed peak in the one dimensional mass spectrum. Therefore, a χ^2 was defined for each selected event for its reconstructed mass from momentum matrices with respect to the observed mean in the spectrum. The elements of the matrices were optimized using a non-linear least χ^2 process using the summed χ^2 from events selected from various well observed hypernuclear states and different production target masses. The basic requirement is a simultaneous minimization of the width of all the observed peaks including those not involved in the event selection. Iteration is required between this optics optimization and the kinematics calibration until both reach stable results.

A blind analysis of simulation data was carried out to evaluate the systematic errors originating from the optics optimization processes. Calibration data as from Λ and Σ^0 elementary production and data from physics states with arbitrary masses were generated by a Monte Carlo simulation with various statistics following the conditions of the actual experiment. The same level of background was also added accordingly. The arbitrarily assumed masses of states in the simulation were hidden and the same optimization process was applied to these simulated data.

The obtained result was then compared to the assumed masses of the states. The signal/accidental background (S/A) ratio is certainly the major factor that limits the achievable optics quality. The optics qualities like uniformity in the kinematics space, absolute missing mass (or Λ binding energy) accuracy as a function of statistics and S/A ratio, and the actual line shape of a single particle state affected by the tuning process were evaluated by this blind analysis of the simulated events.

2.4. Phase I Experiment: E89-009/HNSS (2000)

This was an important pioneering experiment that proved the feasibility of high precision spectroscopy using the CEBAF electron beam. The basic layout is shown in Fig. 1. In the experiment scattered electrons were tagged by an existing Enge split-pole spectrometer at zero degrees to maximize the virtual photon flux. However, bremsstrahlung and Møller scattered electrons are also highly peaked near zero degrees thus the electron arm singles rate was extremely high. When using 22 mg/cm^2 C as the production target and applying 0.6 μA beam current which was the maximum usable luminosity, the electron arm single rate reached 200 MHz. With the Enge split-pole spectrometer, position information at the focal plane (FP) was enough to give the required resolution and thus tracking was not necessary. Since electrons from various processes are all highly peaked at $\sim 0°$, neglecting the angular information of the electrons resulted in only $\sim$ 2 mr angular uncertainty. To measure the FP position under such a high rate, a special silicon-strip-detector (SSD) system was developed by the Tohoku-Houston collaboration [14] and successfully applied.

The existing Short-Orbit-Spectrometer (SOS) in Hall C was used as the kaon spectrometer. Its short orbit is an advantage in the detection of short lived particles such as the kaon but the solid angle acceptance with the front splitter magnet was limited to $\approx$ 4.5 msr. In addition, its momentum resolution was 6×10^{-4} (FWHM) which was the major limitation to the overall experimental energy resolution. The existing SOS detector system, including TOF, aerogel and Lucite Čerenkov detectors, and shower detectors, was utilized for kaon identification. A detailed description of the experimental technique as well as the spectrometers and detector systems can be found in Ref. [15].

The most important achievement of this experiment was proving the feasibility of hypernuclear spectroscopy with an electron beam. The obtained missing mass spectrum for $^{12}_{\Lambda}$B demonstrated the first sub-MeV hypernuclear reaction spectroscopy with a resolution of 800 keV (FWHM) [16] (see Fig. 3), a factor of two better than the best resolution achieved by the (π^+, K$^+$) experiments at KEK. However, the production yield rate was low, ~0.9 counts/hour from the ground state doublets of $^{12}_{\Lambda}$B, and the accidental background was high with a S/A ratio of ~1:1 at the peak from the ground state doublets. Two major peaks corresponding to Λs in s and p orbits were clearly seen with the sub-MeV resolution achieved for the first time in hypernuclear reaction spectroscopy.

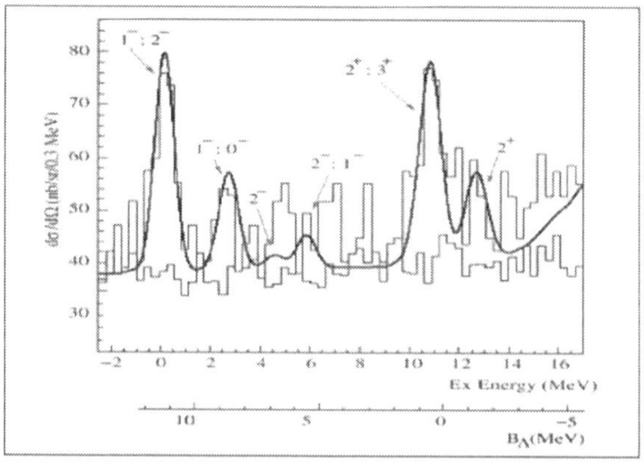

Figure 3. Missing mass spectrum of ${}^{12}_{\Lambda}$B obtained by E89-009 (HNSS) [16].

2.5. Phase II Experiment: E01-011/HKS (2005) with the first upgrade

The major equipment upgrade, inspired by the first successful experiment, was the design and construction of the High resolution Kaon Spectrometer, HKS. The HKS consists of normal conducting Q-Q-D magnets and this new spectrometer featured three times better momentum resolution, 2×10^{-4} (FWHM), and three times larger solid angle acceptance, 15 msr, with the front splitter magnet. Centered at a reaction angle of 7° it covers a reaction angle range from 1° to 13°.

Another essential improvement in the experimental technique was the optimized configuration of the electron spectrometer, "tilt method". The Enge split-pole spectrometer, which was used in the first phase experiment, was tilted off the splitter's dispersion plane by an angle of 7.75° and aimed at a virtual target point below the dispersion plane by a small vertical shift ($\sim$ 5.6 cm). This created an optical acceptance away from scattering angles below 4°, so that bremsstrahlung and Møller electrons were blocked by collimators and the magnet iron reducing the electron singles rate by a factor of 5×10^{-5} while reducing the integrated virtual photon flux acceptance by only a factor of 2×10^{-2}. By the combination of increasing the luminosity and increasing the HKS solid angle acceptance, the physics yield rate was increased by almost a factor of 10.

Since the kaon arm singles rate also increased significantly, a more sophisticated detector system was designed and built. This new system included two sets of wire chambers for tracking, three layers of timing detectors for TOF, three layers of Aerogel Čerenkov detectors for pion and positron rejection, and two layers of water Čerenkov detectors for proton rejection. These detectors are segmented and grouped with sophisticated trigger modules with an FPGA (TUL-8040) to avoid accidental vetoes of kaons due to the high background single particle rates and to the reduced background from events out of the spectrometer optics by using the right combination of segmented counters. By introducing the "tilt method", angles and momentum of scattered electrons must be fully reconstructed using the measured focal plane parameters from trajectory tracking. Therefore, a new wire chamber with a honeycomb-cell structure and timing counters was designed and constructed. Due to the significant improvement of the resolution of the kaon arm, the energy resolution was improved to the level of ~500 keV (FWHM).

With these upgrades, in 2005 the E01-011/HKS experiment was successfully carried out. It aimed to study beyond p-shell hypernuclei and obtained an ${}^{28}_{\Lambda}$Al spectrum for the first time, observing states corresponding to the Λ in s, p, and d shells coupling to low lying nuclear core states. In addition, spectra from two additional hypernuclei, ${}^{7}_{\Lambda}$He and ${}^{12}_{\Lambda}$B, were obtained, from a short test run on light hypernuclei and the necessary calibration data, respectively. Discussion of the obtained spectra is given in a later section on the physics outlook and a detailed description of the experimental setup can be found in Ref. [17].

2.6. *Phase III Experiment: E05-115/HES (2009) with the second upgrade*

For the kinematics of the second upgrade, the photon energy ω was kept at about 1.5 GeV while the beam energy E and the energy of the scattered electrons E′ were increased by the same amount. With the higher energy E (2.34 GeV) and E′ (0.84 GeV), the background from Bremsstrahlung and Møller scattering are more forward and thus we can make the electron spectrometer tilt angle smaller where the virtual photon flux factor Γ is larger. The upgrade replaced the Enge Split-Pole spectrometer with a new spectrometer, the High resolution Electron Spectrometer (HES). In addition, a newly designed and constructed splitter magnet (SPL) replaced the old one in order to better handle higher energy beams and scattered electrons (see schematic view in Fig. 1). These new magnets were designed and constructed by the Tohoku University group as HKS.

By optimized matching between the HKS and the HES, including the SPL magnet with the tilt method used successfully in the second generation experiment, E01-011, the solid angle acceptances are increased, therefore, the total integrated virtual photon flux (virtual photon tagging efficiency) increases by a factor of ~10. However, the new splitter magnet, being larger than the previous one, reduced the HKS solid angle acceptance by a factor of two. So, the new system was expected to increase the overall physics yield rate by another factor of five.

With this new upgrade, the experiment E05-115/E08-002 tried to study hypernuclear spectra over a wide mass range, including the medium heavy hypernucleus $^{52}_{\Lambda}$V, and a variety of p-shell hypernuclei $^{7}_{\Lambda}$He, $^{9}_{\Lambda}$Li, $^{10}_{\Lambda}$Be, and $^{12}_{\Lambda}$B. With the improved momentum resolution by the HES for the electron arm, the overall resolution is expected to reach 400 keV (FWHM) or better, a further improvement from the 2005 HKS experiment. The data taking for the experiment was successfully carried out in 2009 and proved the yield rate increase factor using the p(e,e′K$^+$)Λ,Σ^0 reactions. With better resolution and three times higher statistics, the p-shell hypernuclei including the previously studied hypernuclei, $^{7}_{\Lambda}$He and $^{12}_{\Lambda}$B. Their level structures, especially the core excited states will be studied in detail. The first medium heavy hypenucleus $^{52}_{\Lambda}$V, produced by an electron beam will give precious information about the single particle energy of the Λ, core-configuration mixing and ls splitting. A detailed shell model calculation with DWIA was recently performed for the ^{52}Cr(e,e′K$^+$)$^{52}_{\Lambda}$V reaction [18].

3. The Hall A experimental technique and setup

The E94-107 experiment in Hall A [11] started a systematic study of high-resolution hypernuclear spectroscopy on 1p shell nuclei, specifically ^{9}Be, ^{12}C, ^{16}O. Moreover a very useful (e,e′K) experiment on hydrogen was possible at the same kinematics allowing the study the elementary reaction at the same kinematics as the hypernuclear spectroscopy experiments. The results on ^{12}C and ^{16}O have been published [19, 20] and preliminary results are available for ^{9}Be and hydrogen.

Hall A at Jefferson Lab is well suited to perform (e, e′K$^+$) experiments. Scattered electrons can be detected in the HRS electron arm while coincident kaons are detected in the HRS hadron arm [21]. Kinematics for the first experiment, utilizing a ^{12}C target, were set to particle detection at 6° for both electrons and kaons, incident beam energy of 3.77 GeV, virtual photon energy of 2.2 GeV, scattered electron momentum of 1.56 GeV/c, kaon momentum of 1.96 GeV/c, and $Q^2 = 0.079$ GeV^2. This choice was driven by several factors. Since the cross section depends strongly on Q^2 (through the virtual photon flux as determined by the electron kinematics), to get reasonable counting rates the measurements have to be made at low Q^2. Hence, the electron scattering angle must be small, and the kaon angle must be close to the virtual photon direction in order to minimize the momentum transfer. However, to keep a reasonable kaon survival fraction the kaon momentum must be fairly high. A superconducting septum magnet was added to each HRS to allow detection of particles at central scattering angles as small as 6°. This new spectrometer configuration (Septum + HRS) provides a general purpose device that extends the HRS features to small scattering angles while preserving the spectrometer optical performance [22]. Funding for the septum magnet pair was provided by Istituto Nazionale di Fisica Nucleare, INFN.

Good energy resolution together with a low level of background is mandatory for these experiments. Energy resolution depends on the momentum resolution of the HRS spectrometers, on the straggling and energy loss in the target, and on the beam energy spread. A momentum resolution of the system (Septum + HRS magnets) of $\delta p/p = 10^{-4}$ (FWHM) and an incident beam energy spread as small as 6×10^{-5} (FWHM) are necessary to be able to get an excitation energy resolution of 600 keV or less. To reduce the background level in produced spectra, a very efficient PID system is necessary for unambiguous kaon identification. For all these reasons significant modifications were needed to the Hall A spectrometers setup.

3.1. Septum magnets

A pair of superconducting septum magnets, each providing a 6.5° horizontal bend was introduced into the HRS spectrometer configuration. A very nice feature of this septum magnet setup was that the two arms were essentially independent and could be tuned and optimized separately. Due to their small bend angle and relatively short length (80 cm) the septum magnets made only a small perturbation on the standard HRS optics that was easily corrected by a small modification of the settings of the three quadrupoles in each arm. The quality and exact character of the optics transformation tensor were measured with a series of elastic scattering measurements using a 2 GeV electron beam on C and Ta targets. Measurements were also made using a sieve-like mask in front of each spectrometer to optimize and calibrate the angular reconstruction. The results of the calibration and optimization effort are illustrated in Fig. 4. For details of the design and construction of the septum magnets see Ref. [23].

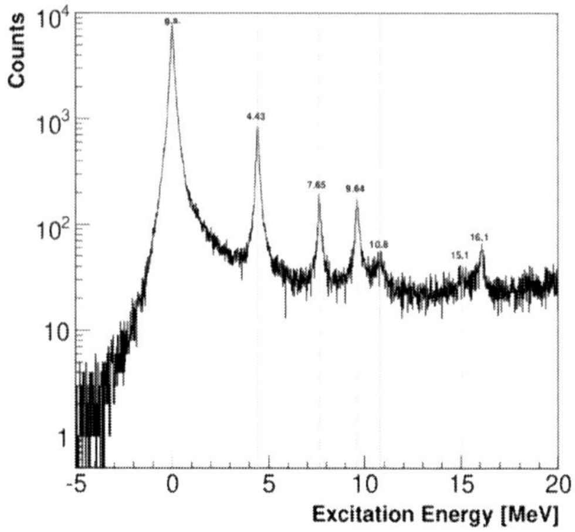

Figure 4. Elastic ^{12}C scattering spectrum as seen in one arm of the HRS + Septum configuration after optimization. The width of all the peaks, elastic and inelastic, is $\leq 10^{-4}$ (FWHM).

3.2. Waterfall target

The waterfall target system provides a target for experiments on ^{16}O. Using a waterfall for oxygen experiments has many advantages. Pure oxygen is difficult to handle, as it is highly reactive. The use of other oxygen compounds requires additional measurements to subtract the non-oxygen background, whereas the hydrogen in water can be used for calibration purposes. The technique of using continuously flowing water as an electron scattering target was first developed by Voegler and Friedrich [24], and later refined by Garibaldi *et al.* [25].

3.3. Detectors

The detector packages of the two spectrometers are designed to perform various functions in the characterization of charged particles passing through the spectrometer. These include: providing a trigger to activate the data-acquisition electronics, collecting tracking information (position and direction), precise timing for time of flight measurements and coincidence determination, and identification of the scattered particles. The timing information is provided from scintillators, as well as the main trigger. The particle identification is obtained from a variety of Čerenkov type detectors (aerogel and gas) and lead-glass shower counters. A pair of VDCs provides tracking information. The main part of the detector package in the two spectrometers (trigger scintillators and VDCs) is identical. For details of the tracking see Ref. [21].

3.3.1. Time of flight (TOF) The long path from the target to the HRS focal plane (25 m) allows accurate time-of-flight identification in coincidence experiments if the accidental rate is low. After correcting for differences in trajectory lengths, a TOF resolution of $\sim$0.5 ns (σ) is obtained. The time-of-flight between the S1 and S2 planes is also used to measure the speed of particles, β, with a resolution of 7% (σ).

3.3.2. Shower counters Two layers of shower detectors [26] are installed in each HRS. The blocks in both layers in HRS-L and in the first layer in HRS-R are oriented perpendicular to the particle tracks. In the second layer of HRS-R, the blocks are parallel to the tracks. Typical pion rejection ratios of 500:1 are achieved using two-dimensional cuts of the energy deposited in the front layer versus the total energy deposited.

3.3.3. Gas Čerenkov A gas Čerenkov detector filled with CO2 at atmospheric pressure [27] is mounted between the trigger scintillator planes S1 and S2. The detector allows electron identification with 99% efficiency and has a threshold for pions of 4.8 GeV/c. The detector has ten spherical mirrors with 80 cm focal length, each viewed by a PMT (Burle 8854). The light-weight mirrors were developed at INFN [28]. The length of the particle path in the gas radiator is 130 cm (80 cm) for the gas Čerenkov in the HRS-R(L), leading to an average of about twelve (seven) photoelectrons. The combination of the gas Čerenkov and shower detectors provides a pion suppression above 2 GeV/c of a factor of 2×105, with a 98% efficiency for electron selection in the HRS-R.

3.3.4. Aerogel Čerenkov detectors There are two aerogel Čerenkov counters available with different indices of refraction, which allow a clean separation of pions, kaons and protons over the full momentum range of the HRS spectrometers. The aerogel is continuously flushed with dry CO_2 gas. The two counters (A1 and A2) are diffusion-type aerogel counters. The 6 cm thick aerogel radiator used in A1 has a refractive index of 1.015, giving a threshold of 2.84 (0.803) GeV/c for kaons (pions). The average number of photoelectrons for GeV electrons in A1 is $\sim$8. The 9 cm thick aerogel radiator used in A2 has a refractive index of 1.055, giving a threshold of 1.55 (2.94) GeV/c for kaons (protons). Trigger logic is used to require that A1 not fire (*e.g.*, rejecting pions) but that A2 does fire (requiring kaons). Rejection factors of 70:1 for rejecting pions and $> 60:1$ for protons were achieved using the aerogels in the hardware trigger.

3.3.5. Ring imaging Čerenkov detector (RICH) The need for an unambiguous identification of kaons has driven the design, construction, and installation of a Ring Imaging Čerenkov (RICH) detector in the hadron HRS focal plane detector package. The layout of the RICH is conceptually identical to the ALICE HMPID design [29]. It uses a proximity focusing geometry, a CsI photocathode, and a 15 mm thick liquid perfluorohexane radiator. A detailed description of the layout and the performance of the RICH detector can be found in Ref. [30]. The fundamental role of the RICH in identifying the kaons is shown in Fig. 5. In the left panel the unfilled timing spectrum of coincidences between the electron and the hadron spectrometers, obtained by selecting for kaons using the two threshold aerogel counters, shows a barely visible kaon signal with a dominant contribution from misidentified pions and protons. The flat part of this spectrum is given by random coincidences. The 2 ns structure is a reflection of the pulse structure of the electron beam. The filled spectrum and its exploded version (right panel), is obtained by adding the RICH to the kaon selection. Here, all contributions from pions and protons completely vanish. The crucial role of the RICH detector can be understood also from Fig. 6. A background free spectrum shows up after RICH cuts. The core excited state peaks clearly emerge from background.

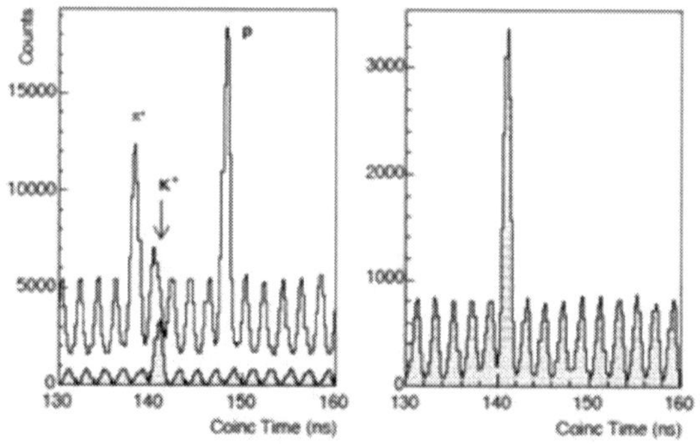

Figure 5. Hadron plus electron arm coincidence time spectra. Left panel: the unfilled histograms are obtained by selecting kaons with only the threshold Aerogel Čerenkov detectors. Filled historgram (expanded in the the right panel) also includes the RICH kaon selection. The remaining contamination is due to accidental (e,e′) and (e,K^+) coincidences.

4. Physics outlook from both the Hall A and C experiments

During a decade of effort, the Jefferson Lab hypernuclear program has been fruitful, even though the (e,e′K^+) hypernuclear spectroscopy has proven to be very challenging technically for both performing the experiments and analyzing the data in order to achieve high precision spectroscopy.

4.1. The elementary process, Λ and Σ^0 electroproduction

The study of p(e,e′K^+)Λ/Σ^0 is important not only for the understanding of strangeness electroproduction but also for absolute missing mass calibration of the spectrometer systems by using the well known Λ and Σ^0 masses (see Fig. 7 from the 2009 HKS experiment). Due to the lack of a neutron target, an absolute mass calibration with hyperon production is impossible for the (π^+, K^+) and (K^-, π^-) reactions.

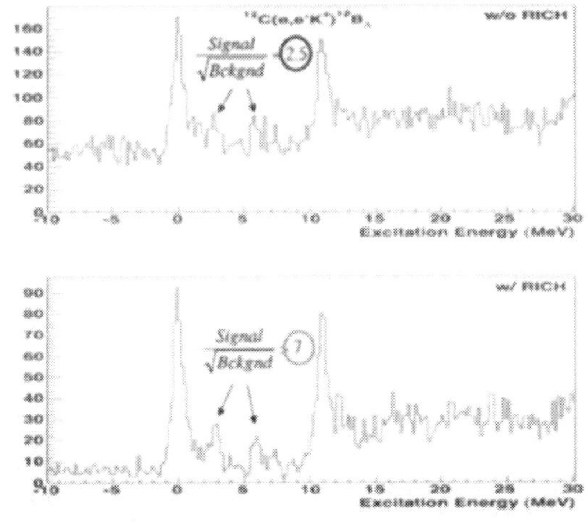

Figure 6. The missing mas spectra for ^{12}C(e,e$'$K$^+$)$^{12}_{\Lambda}$B reaction with and without the RICH cuts.

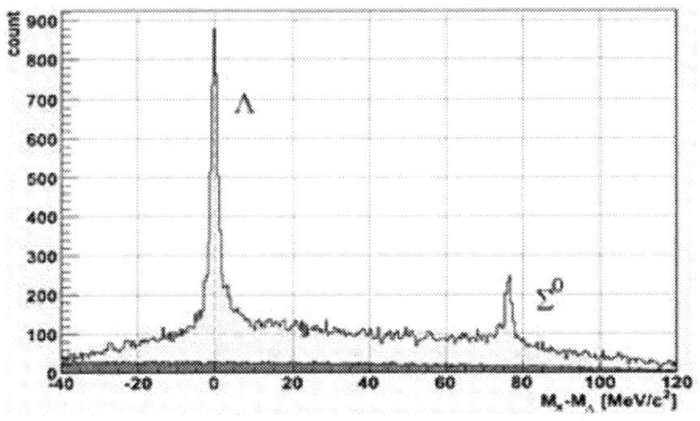

Figure 7. Missing mass spectrum for hypersons measured by HKS-HES system with CH2 target in E05-115. Accidental coincidence background is shown under the sharp hyperon peaks originating from quasi-free hyperon production from carbon.

Electroproduction of Λs at very forward angles ($Q^2 \sim 0$) is important to provide reference data to isobar models which give non-consistent predictions on the Λ production cross section at forward angles. Water and CH2 target data provide this information though the CH2 data need to be extrapolated to the point where the hydrogen escape is zero in order to deduce the cross section. High precision data on the elementary electroproduction of hyperons will be obtained with a water fall target by E07-012 scheduled to run in Hall A in 2012.

The p(e,e$'$K$^+$)Y reaction (where Y is either a Λ or Σ^0 hyperon) has also already been measured as part of the Hall A program. The measurements meet several goals. The cross section for the elementary reaction serves as input for the models of hypernuclear cross sections. The separated Λ and Σ^0 mass peaks calibrate the excitation energy spectra. And finally, the measured Λ and Σ^0 cross sections are inherently interesting.

The ratio of Λ:Σ^0 cross sections is known to drop rapidly with increasing Q^2. Photoproduction measurements at this invariant mass W = 2.2 GeV have ratios at forward angles of 0.7, while electroproduction at $Q^2 = 1$ GeV^2 sees a $\Lambda : \Sigma^0$ ratio of 0.1. (It should be noted that the most forward angle available for the photoproduction data is 25.8° or $\cos\theta_{CM} = 0.9$.) The data were taken with both a 4 cm cryogenic hydrogen target and the thin waterfall foil. The kinematics have $Q^2 = 0.07$ GeV^2, W= 2.2 GeV, and $\theta = 6°$. The results are shown in Fig. 8. Plotted is the differential cross section versus center-of-mass angle, with both the hypernuclear measurements (at $Q^2 = 0.07$ GeV^2) and the LEPS and CLAS photoproduction results ($Q^2 = 0$). Also plotted are model calculations such as the Saclay-Lyon, Regge-plus-resonance approach, and Kaon-MAID.

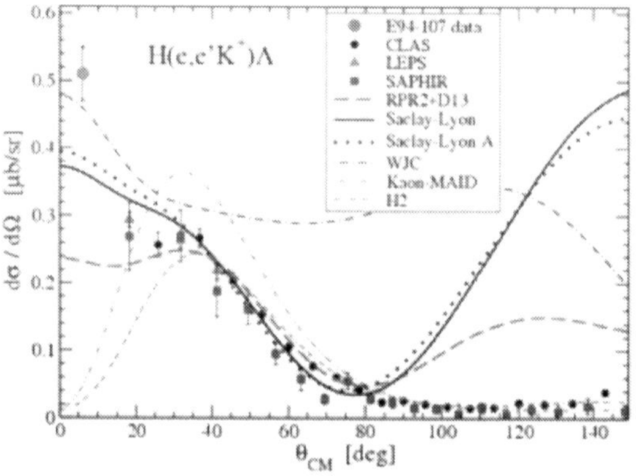

Figure 8. E94-107 elementary kaon electoprodution data along with various other data sets and calculations.

As can be seen, the data suggest that not only do the present models fail to describe the data over the full angular range, but that the cross section rises at the forward angles. The failure of existing models to describe the data suggests the reaction mechanisms may be incomplete. The large magnitude of the cross section, coupled with the fact that this is electroproduction, raises the possibility that longitudinal contributions may obscure the comparison to (transverse) photoproduction data. Although there are no measurements of the longitudinal response at this low Q^2, the entirety of the data on the kaon longitudinal response indicates that it is everywhere less than half the transverse response, and that the interference responses (longitudinal-transverse and transverse-transverse) are approximately an order of magnitude less than the transverse. Assuming that the longitudinal response is 50% of the transverse response, the measured cross section at this value of ϵ would mean that the extracted transverse cross section drops by about 25%.

4.2. $^7_\Lambda$He *hypernucleus*

A recent cluster calculation gives a precise prediction of the $^7_\Lambda$He mass as a system of α + n + n + Λ [31]. It has been recognized in the analysis of the A=7 hypernuclear isotriplet, $^7_\Lambda$He, $^7_\Lambda$Li*(T=1) and $^7_\Lambda$Be, that it is quite important to study the charge symmetry breaking (CSB) effect of the ΛN interaction which is necessary to explain mass difference between $^4_\Lambda$H and $^4_\Lambda$He. However, the statistics of the existing $^7_\Lambda$He data from emulsion are too poor to give any reliable conclusion.

Preliminary analysis of the results of E01-011 [32] shows a clear peak of the $^7_\Lambda$He ground state with enough statistics for the first time. The obtained binding energy was compared with results from the cluster calculation. Surprisingly, inclusion of a CSB term in the ΛN interaction makes the discrepancy between experiment and theory worse, though the CSB term is essential for A=4 hypernuclei. This indicates that current understanding of the CSB effect in the ΛN interaction potential is still imperfect and further systematic study is necessary.

Journal of Physics: Conference Series **299** (2011) 012013 doi:10.1088/1742-6596/299/1/012013

4.3. ^{12}C(e,e′K$^+$)$^{12}_{\Lambda}$B

$^{12}_{\Lambda}$C serves as a mass reference system in the (π^+, K$^+$) and (K$^-$, π^-) hypernuclear reactions so that very high statistics data have been accumulated since the ground state doublet and states with Λ in the p-shell make prominent peaks with sufficient cross sections. Therefore, its mirror hypernucleus, $^{12}_{\Lambda}$B, which is expected to have the same features as $^{12}_{\Lambda}$C, can be used as a reference system for hypernuclear (e,e′K$^+$) spectroscopy though the (e,e′K$^+$) reaction needs no emulsion mass reference, since the well known hyperon masses can be used.

The high resolution achievable in the (e,e′K$^+$) reaction enables us to compare the hypernuclear iso-doublet $^{12}_{\Lambda}$B and $^{12}_{\Lambda}$C to extract the iso-spin dependent term of ΛN interaction, *i.e.* the CSB effect. The $^{12}_{\Lambda}$C spectrum obtained by E336 at KEK-PS (Fig. 9 top) [33], $^{12}_{\Lambda}$B spectrum obtained by E94-107 at Jefferson Lab Hall A (Fig. 9 middle) [19] and $^{12}_{\Lambda}$B spectrum obtained by E01-011 in Hall C (Fig. 9 bottom) [34] are compared. One can recognize that the global structures of A=12 hypernuclei are similar and the $^{12}_{\Lambda}$B excitation energies are consistent between the Hall A and Hall C data. Thanks to the high resolution of (e,e′K$^+$) hypernuclear spectroscopy, the states with Λs coupled to low-lying core excited states were clearly observed for $^{12}_{\Lambda}$B as well as the prominent s_{Λ} and p_{Λ} peaks.

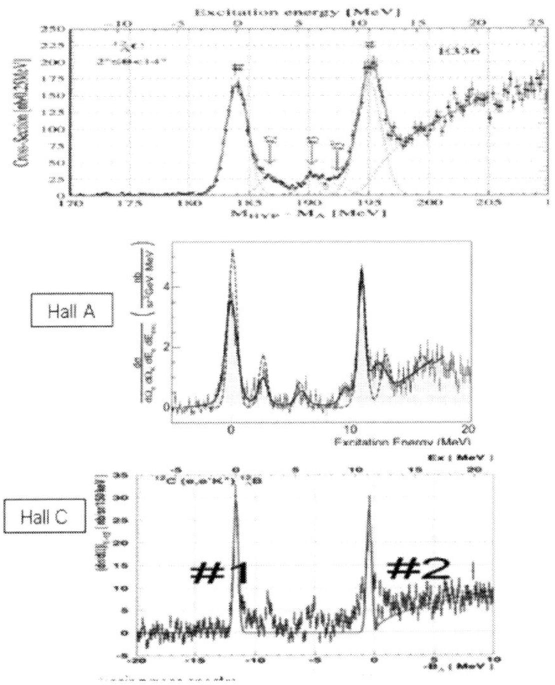

Figure 9. Missing mass spectra of $^{12}_{\Lambda}$B and $^{12}_{\Lambda}$C [19, 33, 34].

The Hall A spectrum features excellent signal-to-noise ratio due to the high performance RICH detector and the Hall C spectrum shows improved resolution through the introduction of the newly developed HKS. With the improved resolution, the positions of the two excited-state peaks and the existence of additional strength outside the major peaks are much more clearly defined than in the (π^+, K$^+$) spectrum [35]. In particular, the position of the first-excited peak is in agreement with γ-ray data. Detailed analysis and comparison of A=12 hypernuclear spectra is now underway.

Journal of Physics: Conference Series **299** (2011) 012013 doi:10.1088/1742-6596/299/1/012013

4.4. ^{9}Be(e,e'K$^+$)$^9_\Lambda$Li

There are still some unresolved problems in the spectroscopy of hypernuclei in the lower part of p-shell. The spectra of $^{10}_\Lambda$B (ground state doublet splitting) and the $^{11}_\Lambda$B (energy of the J=1/2$^+$ member of the first excited doublet) as measured in precise (K$^-$, π^- γ) and (π^+, K$^+$ γ) experiments are inconsistent with the standard shell model description of p-shell hypernuclei.

The electroproduction of a $^9_\Lambda$Li hypernucleus from a ^{9}Be target can hopefully shed new light on this problem. In this case the ground state doublet and two excited doublets of $^9_\Lambda$Li (all lying below the strong neutron decay threshold) are produced with comparable cross sections. In addition, splitting of the ground-state doublet and the second excited-state doublet are predicted to be large enough to be detected ($\sim$ 500 keV), while the first excited state doublet is predicted to be almost degenerate. In an electroproduction experiment with good energy resolution five hypernuclear states below the strong particle decay threshold can, in principle, be resolved in the $^9_\Lambda$Li hypernucleus.

The results of the analysis of the ^{9}Be(e,e'K$^+$)$^9_\Lambda$Li, reported here, are still preliminary. In Fig. 10, the theoretical curve is superimposed on the fitted data. The theoretical cross sections are obtained using the Saclay-Lyon (SLA) model for the elementary p(e,e'K$^+$)Λ reaction. Shell-model wave functions are determined using a parametrization of the ΛN interaction that fits the precise γ-ray hypernuclear spectra of $^7_\Lambda$Li. A fit with three peaks is guided by the three main structures predicted by the theory. The fit procedure is described elsewhere [19]. An attempt has been made to also disentangle the different contributions coming from the multiple levels which calls for a fit with five peaks, as guided by the theoretical model.

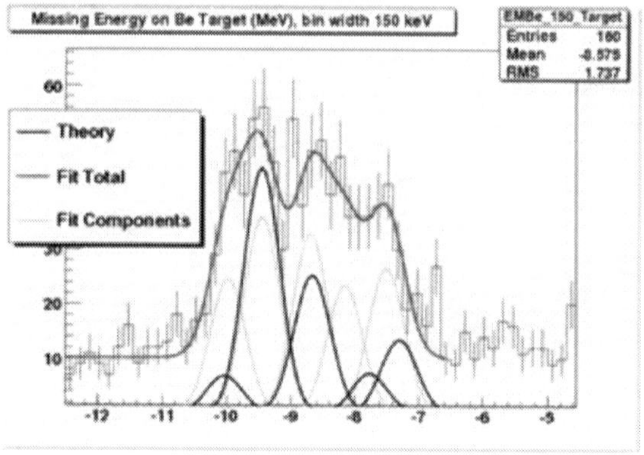

Figure 10. ^{9}Be(e,e'K$^+$)$^9_\Lambda$Li missing mass spectrum fitted with five gaussians. Also shown are the theory predictions for this spectrum.

As mentioned previously the results are still preliminary. At first look it seems evident a disagreement exists between experiment and theory for the strength of the second and third doublets. This can be due both to the elementary part of the reaction and to the nuclear part (spectroscopic factors).

4.5. ^{16}O(e,e'K$^+$)$^{16}_\Lambda$N

The ^{16}O(e,e'K$^+$)$^{16}_\Lambda$N reaction was measured in Hall A. The $^{16}_\Lambda$N final state is particularly interesting both because it has not been measured previously and because it allows a comparison to the mirror nucleus, $^{16}_\Lambda$O whose bound energy levels were measured both with germanium detectors (allowing very precise determinations of the excitation spectra) [36] and with a variety of hadronic reactions, *e.g.*, (K$^-$, π^-) [37], (π^+, K$^+$) [33] and (K$^-_{\rm stop}$, π^-) [38] (allowing moderate determination of the ground state energy).

The presence of the hydrogen has many advantages. In particular, it permits a calibration of the missing-mass scale and thus an accurate measurement of the Λ binding energy in the hypernucleus. Moreover, an interesting measurement of the (e,e′K$^+$) cross section on a proton in a kinematical region never previously measured was possible.

The kinematics for the experiment were almost the same as for the ^{12}C running: particles were detected at 6° (both electrons and kaons), an incident beam energy of 3.66 GeV, virtual photon energy of 2.2 GeV, scattered electron momentum of 1.45 GeV/c, kaon momentum of 1.96 GeV/c, and $Q^2 = 0.079$ GeV^2.

Figure 11, left panel, shows the $^{16}_{\Lambda}$N excitation-energy spectrum as obtained after kaon selection with aerogel detectors and RICH without background subtraction. The electron kaon random coincidence contribution evaluated in a large timing widow and normalized is superimposed on the spectrum. Again, the uniquely low backgrounds can be seen. The shaded region shows the (e,e′)⊗(e,K$^+$) random-coincidence background. The large broad peak observed at around 30 MeV is the mis-reconstructed binding energy due to the contribution of the hydrogen in the waterfall when the oxygen mass is used for the target mass in constructing the missing mass. The excitation region of the $^{16}_{\Lambda}$N production with background subtraction is shown on the right side of the figure.

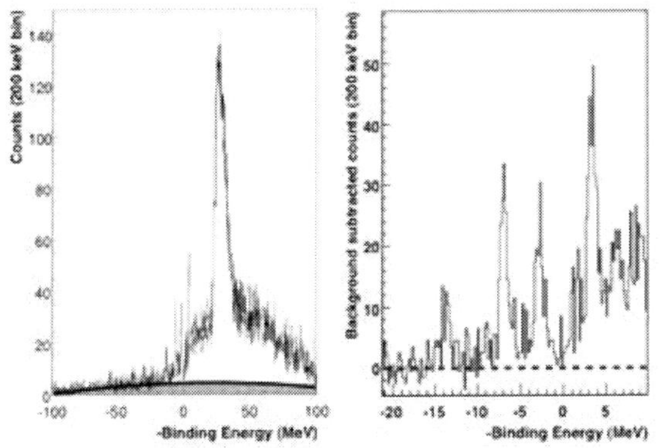

Figure 11. The $^{16}_{\Lambda}$N excitation-energy spectrum as obtained after kaon selection with aerogel detectors and RICH without background substraction.

Figure 12 shows the six-fold differential cross section expressed in nb/(sr^2 GeV MeV). Similarly to the ^{12}C target, the background was evaluated by fitting the data obtained for random coincidences in a large timing window and subtracted, resulting in no residual background in the region on the left of the ground state. The fit to the data has been made using Voigt functions. Four isolated peaks are observed. The ground state peak gives a Λ separation energy of $B_\Lambda =$ 13.76 ± 0.16 MeV for the 1^- member of the ground-state doublet in $^{16}_{\Lambda}$N. Three more peaks are observed at binding energies of 6.93, 2.84, and 3.34 MeV.

The theoretical cross sections were obtained in the framework of the distorted-wave impulse approximation (DWIA) [39] using the Saclay-Lyon (SLA) model [40] for the elementary p(e,e′K$^+$)Λ reaction. The ground state of ^{16}O is assumed to be a simple closed shell and the shell-model wave functions for $^{16}_{\Lambda}$N are computed in a simple particle-hole model space. As in the case of $^{12}_{\Lambda}$B, the energy of the first-excited peak (a doublet of 1^- and 2^- states) agrees much better with the γ-ray data than does the data from hadronic reactions. This gives confidence in the energies extracted for the third and fourth peaks. The largest discrepancy between theory and experiment is in the position of the fourth peak. The excitation energies of the positive-parity states depend on the spacing of the p_Λ and s_Λ single-particle energies which

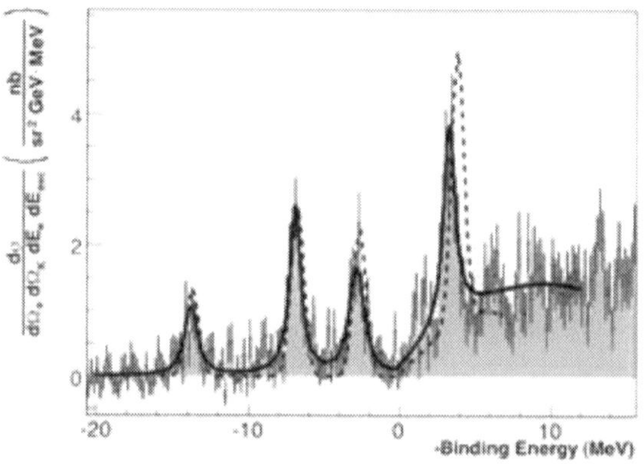

Figure 12. The ^{16}O(e,e′K$^+$)$^{16}_{\Lambda}$N spectrum from Hall A expressed in nb/(sr^2 GeV MeV).

have to be extrapolated from $^{13}_{\Lambda}$C and could therefore be uncertain by several hundred keV. The gap between the third and fourth peaks is predicted to be slightly larger (6.5 MeV) than the underlying separation (6.324 MeV) of the p-hole states in ^{15}N, in contrast to the observed splitting of 6.18 MeV. In fact, the first clear indication that the spin-orbit splitting of the p_Λ orbits is very small came from the observation that the separation between the 0^+ states is very close to the underlying core-state separation.

The Λ separation energy of $B_\Lambda = 13.76 \pm 0.16$ MeV obtained for the first peak is important. There are few emulsion events for the heavier p-shell hypernuclei and these events tend to have ambiguous interpretations. The reactions involving the production of a Λ from a neutron are more difficult to normalize. For example, the (π^+,K$^+$) data are normalized to 10.80 MeV for the B_Λ of $^{12}_{\Lambda}$C which differs considerably from the much better determined value of 11.37 ± 0.06 MeV for the mirror hypernucleus $^{12}_{\Lambda}$B. This difference is often cited as evidence for charge-dependent effects. However, the (K^-_{stop},π^-) reaction gives 11.38 ± 0.09 MeV for $^{12}_{\Lambda}$C [38]. Within errors, the binding energy and the excited levels of the mirror hypernuclei $^{16}_{\Lambda}$O and $^{16}_{\Lambda}$N (this experiment) are in agreement, giving no strong evidence of charge-dependent effects. Details and references can be found in Ref. [20].

4.6. ^{28}Si(e,e′K$^+$)$^{28}_{\Lambda}$Al

The ^{28}Al hypernucleus was studied with an isotopically enriched ^{28}Si target by E01-011 in Hall C. This is the first beyond p-shell hypernucleus studied with an electron beam and the developed technique to realize the ^{28}Al spectrum opened a door to medium heavy hypernuclear spectroscopy by the (e,e′K$^+$) reaction in the E05-115 experiment. The mirror hypernucleus ^{28}Si was studied at KEK-PS with a natural Si target [41]. The global structures of the spectra look similar, but the binding energies of the ground states are apparently different and the excitation energy of the p_Λ is much larger than that of ^{28}Si (see comparison in Fig. 13). Careful investigations to understand this difference are now underway experimentally as well as theoretically.

4.7. Other spectra to be analyzed – $^{10}_{\Lambda}$Be and $^{52}_{\Lambda}$V

The main target of the third generation experiment, E05-115, in Hall C is isotopically enriched ^{52}Cr. By the ^{52}Cr(e,e′K$^+$)$^{52}_{\Lambda}$V reaction, a spectroscopic study of a medium heavy hypernucleus was carried out for the first time with an electron beam. Precise single particle energies of $^{52}_{\Lambda}$V are expected to provide important constraints to mean field theories and a systematic study of the Λ's single particle energy will reveal the distinguishability of hadrons deep inside

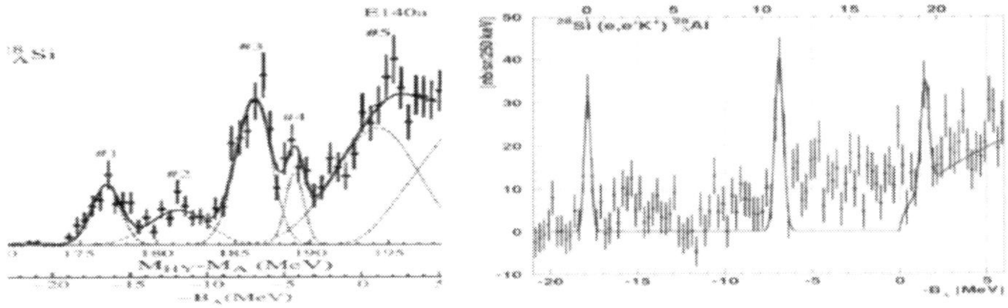

Figure 13. Missing mass spectra of $^{28}_{\Lambda}$Si [40] and $^{28}_{\Lambda}$Al (preliminary from Hall C E01-011).

of a nucleus. The precise structure of $^{52}_{\Lambda}$V will shed light on the nature (core-configuration mixing or l-s splitting of the ΛN force) of high-l Λ states which were observed in medium heavy hypernuclei with the (π^+, K$^+$) reaction [42]. The experiment is technically challenging since the high electromagnetic background causes a high multiple hit rate in the detectors. Development of a new tracking code to handle high multi-hit events is now underway.

In addition, the p-shell hypernucleus $^{10}_{\Lambda}$Be was studied with an enriched $^{10}_{\Lambda}$B target. As mentioned previously, the naive charge symmetry breaking term in the ΛN interaction cannot explain the masses of the A=7 iso-triplet hypernuclei. The CSB term can be modified to explain the A=4 and A=7 systems by simultaneously introducing another degree of freedom, but the constructed CSB interaction should predict the masses of A=10 hypernuclei which can be treated as an iso-doublet $\alpha + \alpha + N + \Lambda$ by the same technique as the A=7 system in the framework of a cluster model. Therefore the results of E05-115's $^{10}_{\Lambda}$Be spectroscopy are anxiously awaited.

5. Future prospects

In 2008, a high-intensity proton machine, J-PARC (Japan Proton Accelerator Research Complex), started operations. One of the major research subjects at J-PARC is hypernuclear spectroscopy with high intensity kaon beams. New results on S = -2 hypernuclear spectroscopy and high-precision γ-ray spectroscopy of S = -1 hypernuclei will be obtained at J-PARC. Hypernuclear spectroscopy with electron beams is complementary to research subjects at J-PARC. Therefore, combining research outputs from J-PARC and Jefferson Lab will make great progress in this research field in the next decade.

5.1. Experiments before the 12 GeV era

A proposal (E07-012) [43] has been approved by the Jefferson Lab PAC to study the angular dependence of the ^{16}O(e,e$'$K$^+$)$^{16}_{\Lambda}$N and p(e,e$'$K$^+$)Λ reactions. In the performed Jefferson Lab hypernuclear electroproduction experiments (E89-009, E94-107, E01-011, and E05-015) the K$^+$ mesons are detected at very small (few degrees) laboratory scattering angels (measured with respect to virtual photon momentum). This region of kaon scattering angles is not covered, unfortunately, by recent photo and electroproduction data on the elementary production process from CLAS [44], SAPHIR [45], and LEPS [46] Collaborations. The angular dependences of the cross section at small angles and at photon energies $E_\gamma \sim 2$ GeV predicted by different models for the K$^+$ electromagnetic production (see Fig. 8) differ drastically and this lack of relevant information about the elementary process makes an interpretation of obtained hypernuclear spectra difficult. In addition, the ratio of the hypernuclear (calculated in DWIA) and elementary cross section measured at the same kinematics should be almost model independent at very forward kaon scattering angles. The ratio therefore contains direct information on the target and hypernuclear structure, production mechanisms and, possibly on the modification of the dynamics of the p(e,e$'$K$^+$) process in the nuclear environment.

The Hall A experimental setup with septum magnets, the waterfall target, and virtual photons with energy $E_\gamma = 2$ GeV has a unique opportunity to solve the above mentioned problems. We will measure:

- The electroproduction cross section on the proton in H_2O at lab kaon scattering angles, ${}^{lab}\theta_{ke} = 8.5°$ and 11° (${}^{lab}\theta_{k\gamma} = 4°$ and 7°), which together with our previous measurements for ${}^{lab}\theta_{ke} = 6°$ (${}^{lab}\theta_{k\gamma} = 2°$) will cover the angular region missing in the CLAS and SAPHIR data. New precise data will clearly discriminate between various models of photo and electroproduction of strangeness, such as Saclay-Lyon and Kaon-MAID.
- The angular dependence of the hypernuclear cross section (HN) on ^{16}O will be determined simultaneously. These data and, especially, the ratio of HN to the elementary cross section will give new valuable information on hypernuclear structure (including spin assignment of produced hypernuclear states), reaction mechanisms and, even possibly the modification of the dynamics of the $(e,e'K^+)$ process in the nuclear medium.

5.2. Decay pion spectroscopy

So far, the $(e,e'K^+)$ hypernuclear study has been developed in missing mass spectroscopy. The production mechanism and structure of hypernuclei have been studied by this method. As an addition, a new experimental technique, hypernucleus decay pion spectroscopy, was proposed as E10-001 [47].

Utilizing the relatively weak electromagnetic probe and high momentum transfer in the reaction kinematics, "hot hypernuclei" (*i.e.* at highly excited states) or hypernuclei with Λ produced directly in the s-shell then coupling to the highly excited deep particle hole states of the nuclear core (*i.e.* states with Λ coupled to high lying core states) are produced in the continuum part of the overall hypernuclear production from one common nuclear target. This type of hypernucleus has high spin and some of them have masses well above the nucleon/nucleus emission threshold. The strong nuclear breakup process which produces various hyper-fragments (*i.e.* light hypernuclei) will take place as seen in emulsion or bubble chamber experiments. The selectivity is expected to also follow Young's scheme [48]. Therefore, the high continuum production from hypernuclear electroproduction becomes a rich source of light hypernuclei. After a cascade ("cool down") process to the ground state, light hypernuclei decay *via* weak decay modes in which the negatively charge pion decay channel is expected to have above a 10% rate for all p-shell hypernuclei. The light hypernucleus stops easily in the target due to the low release energy.

Decaying at rest, the pions from the two body decay channel have monochromatic momenta which contain characteristic information about the ground state of the decayed hypernuclei *via* the well known structures (to a few or below keV level) of normal nuclei involved in the two-body decay. Therefore, precisely measured pion momentum spectroscopy will provide important information about the light hypernuclei.

Wide ranging physics can be learned from the pion spectroscopy. For example (but not limited to), with resolution < 170 keV/c FWHM the ground state binding energy for a variety of simultaneously produced light hypernuclei can be determined with a precision better than 20 keV. This is particularly important to reconfirm the values obtained from emulsion data analysis and to resolve the puzzle in the CSB. Many pairs of mirror hypernuclei can be obtained at the same time. From pion spectroscopy for a specific hypernucleus the spin-parity, which is still missing for many light hypernuclei, of the initial hypernucleus can be determined. This pion decay may prove to be an unique tool in studying extremely neutron rich hypernuclei at or near drip line, such as ${}^{6-8}_{\Lambda}H$ (the heavy hyper-hydrogen).

A pioneering run to prove the principle of this new technique will be carried out simultaneously with E07-012 in Hall A. Future experiments will be proposed and planned based

Journal of Physics: Conference Series **299** (2011) 012013 doi:10.1088/1742-6596/299/1/012013

realistically on the outcome of this test run. The success of this test run is not only important for establishing an excellent hypernuclear program at Jefferson Lab based on this technique but also opens doors for future studies using high intensity mesonic beams and further technical development for new physics aspects which are not yet accessible, such as pion decay asymmetry with respect to the hypernuclear polarization which pin-points the exact formation mechanism of a hypernucleus and the deep interior structure of the nucleus through the production rate of specific light hypernuclei (such as ${}^{4}_{\Lambda}$H) from different parental hypernuclear high-lying states [49].

5.3. Future hypernuclear spectroscopy with electron beams at Jefferson Lab
After the CEBAF 12 GeV upgrade, the Jefferson Lab Hypernuclear Collaboration which consists of the previous Hall A and Hall C Hypernuclear Collaborations will pursue a wide mass range of hypernuclear study. One direction of the (e,e$'$K$^+$) reaction spectroscopy is to study heavier hypernuclear systems to extract Λ's single particle energies to study hadron behavior deep inside of the nucleus. Heavier targets result in severe background and thus the experiments need to control the high background. In order to accumulate reasonable statistics in a given amount of beamtime, large acceptance and the high-resolution short-orbit spectrometers, HKS and HES, will be powerful tools for such a study.

Detailed studies of relatively light hypernuclei for which the signal-to-noise ratio is more important than the statistics could be studied by using the HRS with a higher beam energy. The clean kaon selection capability of the RICH detector will play a significant role in such programs. In a decade of electroproduction of hypernuclei studies, we have developed many tools and techniques, and thus we can choose the right tools depending on the physics program. Though its feasibility is not yet proven, the new decay pion hypernuclear spectroscopy is an interesting research subject. It might break through a limit of the (e,e$'$K$^+$) hypernuclear spectroscopy which has not yet been combined with the study of the decay products. For example, hypernuclear γ-ray spectroscopy is a very powerful research technique which can be combined with (π,K) and (K,π) reactions, however, the Ge γ-ray detector cannot work in a high-rate electron background. If decay pion spectroscopy is proven to be feasible, a new large solid-angle spectrometer (HπS) which would concurrently serve as a splitter magnet can be introduced with the existing HKS and HES. Such spectrometers could provide data for standard (e,e$'$K$^+$) spectroscopy, hypernuclear decay spectrum, and triple-coincidence information.

6. Summary

Hypernuclear spectroscopy by electromagnetic probes has been established at Jefferson Lab (Hall A and Hall C) showing that the experimental knowledge coming from hypernuclear studies by hadronic probes can be greatly improved. In a decade, a wide variety of hypernuclei, ${}^{7}_{\Lambda}$He, ${}^{9}_{\Lambda}$Li, ${}^{10}_{\Lambda}$Be, ${}^{12}_{\Lambda}$B, ${}^{16}_{\Lambda}$N, ${}^{28}_{\Lambda}$Al and ${}^{52}_{\Lambda}$V, have been studied by the (e,e$'$K$^+$) reaction. Large momentum transfer, strong spin flip, the excitation of both natural and unnatural parity states, and the creation of new species of hypernculei not otherwise available are the characteristics of electromagnetic hypernuclear spectroscopy. Moreover, high energy resolution and precision measurements, also characteristic of the electromagnetic probe, make these experiments unique and crucial tools for this field. High accuracy determination of the Λ binding energy for different hypernuclei has been achieved and much more is expected from the new line of research by pionic decay experiments. Systematic studies of mirror hypernuclei could shed light on charge dependence in hypernuclei and the underlying YN interaction.

These experiments were and are challenging. They have required many novel technologies and new detectors to be built in both Halls A and C. These detectors and the new technologies have served in the past and will serve in the future for other important experiments at Jefferson Lab. For these reasons, in the decade to come, the Hypernuclear Collaboration of Jefferson Lab is expected to play a key role in this research field in the framework of collaborations with other existing or new facilities. The recent upgrade of MAMI-C has enabled it to produce hypernuclei with an electron beam and J-PARC soon starts to deliver high-intensity π and K beams. In the exploration of hypernuclear physics, the next decade looks very promising.

References

[1] M. Danysz and J. Pniewski, Phil. Mag. **44**, 348 (1953).
[2] B. Povh, Prog. Part. Nucl. Phys. **18**, 183 (1987).
[3] R. Chrien and C. Dover, Ann. Rev. Nucl. Part. Sci. **39**, 113 (1989).
[4] H. Bando, T. Motoba and J. Žofka, Int. J. Modern Phys. **21**, 4021 (1990).
[5] O. Hashimoto and H. Tamura, Prog. Part. Nucl. Phys. **57**, 564 (2006).
[6] M. M. Nagels *et al.*, Phys Rev. D **15**, 2547 (1977); Phys. Rev. D **20**, 1633 (1979); P. M. Maessen *et al.*, Phys. Rev. C **40**, 2226 (1989); Th. A. Rijken, V. G. J. Stokes and Y. Yamamoto, Phys. Rev.C **59**, 21 (1999).
[7] B. Holzenkamp *et al.*, Nucl. Phys. **A500**, 458 (1989); K. Holinde, Nucl. Phys. **A547**, 245c (1992).
[8] Y. Yamamoto *et al.*, Prog. Theor. Phys. Suppl. **118**, 361 (1994).
[9] R. H. Dalitz and A. Gal, Ann. Phys. **116**, 167 (1978).
[10] D. J. Millener, A. Gal, C. B. Dover and R. H. Dalitz, Phys. Rev. C **31**, 499 (1985).
[11] Jefferson Lab Experiment E94-107, *High resolution 1p-shell hypernuclear spectroscopy*, F. Garibaldi, S. Frullani, P. Markowitz and J. LeRose, spokespersons.
[12] Jefferson Lab Experiment E89-009, *Investigation of the spin dependence of the ΛN effective interaction in the p shell*, E. Hungerford, R. Chrien and L. Tang, spokespersons; Jefferson Lab Experiment E01-011, *Spectroscopic study of Λ hypernuclei up to medium-heavy mass region through the $(e,e'K^+)$ reaction*, O. Hashimoto, S. N. Nakamura, L. Tang and J. Reinhold, spokespersons; Jefferson Lab Experiments E05-115 and E08-002, *Spectroscopic investigation of Λ hypernuclei in the wide mass region using the $(e,e'K^+)$ reaction*, O. Hashimoto, S. N. Nakamura, L. Tang and J. Reinhold, spokespersons.
[13] F.Dohrmann *et al.*, Phys. Rev. Let. **93**, 242501 (2004).
[14] T. Miyoshi *et al.*, Nucl. Instr. and Meth. A **496**, 362 (2003).
[15] E89-009 Doctoral Dissertations: L. Yuan (Hampton Univ.), T. Miyoshi (Tohoku Univ.), and M. Sasour (Univ. of Houston).
[16] T. Miyoshi *et al.*, Phys. Rev. Lett. **90**, 232502 (2003), L. Yuan *et al.*, Phys. Rev. C **73**, 044607 (2006).
[17] E01-011 Doctoral Dissertations: Y. Okayasu (Tohoku Univ.), V. Rodriguez (Univ. Of Houston), T. Seva (Univ. of Zagreb), A. Matsumura (Tohoku Univ.), P. Baturin (Florida Int. Univ.).
[18] T. Motoba *et al.*, *Strangeness in nuclear and hadronic systems – SENDAI08*, World Scientific, ISBN981-4277-60-6, 178 (2010).
[19] M. Iodice *et al.*, Phys. Rev. Lett. **99**, 052501 (2007).
[20] F. Cusanno *et al.*, Phys. Rev. Lett. **103**, 202501 (2009).
[21] J. Alcorn *et al.*, Nucl. Instrum. Methods A **522**, 294 (2004).
[22] G. M. Urciuoli *et al.*, Nucl. Phys. **A691**, 43 (2001).
[23] P. Brindza *et al.*, IEEE Transactions on Applied Superconductivity **11**, 1594 (2001).
[24] N. Voegler *et al.*, Nucl. Instr. and Meth. **198**, 293 (1982).
[25] F. Garibaldi *et al.*, Nucl. Instr. and Meth. A **314**, 1 (1992).
[26] A. Ketikyan *et al.*, *hallaweb.jlab.org/equipment/detectors/detectors.*
[27] M. Iodice *et al.*, Nucl. Instr. and Meth. A **411**, 223 (1998).
[28] E. Cisbani *et al.*, Nucl. Instr. and Meth. A **496**, 305 (2003).
[29] CERN/LHCC 98-19, ALICE TDR 1, 14 August 1998.
[30] M. Iodice *et al.*, Nucl. Instr. And Meth. A **553**, 231 (2005).
[31] E. Hiyama, Y. Yamamoto, T. Motoba and M. Kamimura, Phys. Rev. C **80**, 054321 (2009).
[32] O. Hashimoto *et al.*, Nucl. Phys. **A835**, 121 (2010).
[33] O. Hashimoto *et al.*, Nucl. Phys. **A639**, 93 (1998).
[34] HKS Collaboration, to be published.
[35] Y. Ma, Nucl. Phys. **A835**, 422 (2010); H. Tamura, Nucl. Phys. **A835**, 3 (2010).
[36] H. Akikawa *et al.*, Phys. Rev. Lett. **88**, 082501 (2002).
[37] R. E. Chrien *et al.*, Phys. Lett. B **89**, 31 (1979).
[38] H. Tamura *et al.*, Prog. Theo. Phys. Suppl. **117**, 1 (1994).
[39] M. Sotona and S. Frullani, Prog. Theor. Phys. Suppl. **117**, 151 (1994).
[40] T. Mizutani, C. Fayard, G.-H. Lamot and B. Saghai, Phys. Rev. C **58**, 75 (1998).
[41] T. Hasegawa *et al.*, Phys. Rev. C **53**, 1210 (1996).
[42] H. Hotch *et al.*, Phys. Rev. C **64**, 044302 (2001).

[43] Jefferson Lab Experiment E07-012, *The angular dependence of* $^{16}O(e,e'K^+)^{16}_{\Lambda}N$ *and* $H(e,e'K^+)$, F. Garibaldi *et al.*, spokespersons.

[44] R. Bradford *et al.*, Phys. Rev. C **73**, 035201 (2006).

[45] K. H. Glander *et al.*, Eur. Phys. J. A **19**, 251 (2005).

[46] R. G. T. Zegers *et al.*, Phys. Rev. Lett. **91**, 092001 (2003).

[47] Jefferson Lab Proposal PR10-001, *Study of light hypernuclei by pionic decay at Jefferson Lab*, L. Tang, L. Yuan, A. Margaryan, S. N. Nakamura, J. Reinhold, J. LeRose and F. Garibaldi.

[48] Yu. F. Smirnov, Proc. Int. Conf. Clustering Phenomena in Nuclei, Bochum 1969, IAEA Vienna, p. 153 (1969).

[49] L. Majling and J. Žofka, Czech. J. Phys. **41**, 1083 (1991).

Journal of Physics: Conference Series **299** (2011) 012014 doi:10.1088/1742-6596/299/1/012014

The Jefferson Lab Free Electron Laser Program

Stephen V. Benson, David Douglas, George R. Neil, and Michelle D. Shinn

Jefferson Lab, Newport News, Virginia 23606 USA

Abstract. The development of superconducting radio frequency (SRF) accelerator technology for CEBAF's nuclear physics program set the stage for its application to a number of other efforts. We describe below the development of the a major advance in Free Electron Laser (FEL) Technology based on SRF linacs. The Jefferson Lab efforts achieved three orders of magnitude increase in the power delivered by FELs and firmly established the viability of energy recovering linac technology. We describe the details of the physics and engineering challenges addressed to accomplish this effort and then discuss some of the applications performed using the light source. We conclude with a look at the planned directions for the program in the future.

1. Introduction

The development of reliable and cost effective SRF accelerators as demonstrated by the operation of the CEBAF accelerator in 1994 [1] set the stage for an application to high average power Free Electron Lasers. It had been realized early in CEBAF construction that the system could be utilized for an FEL [2] but it wasn't clear in the early stage of this study what the specific advantage would be over other systems or what the goals of such a program might be. FELs had been considered as candidates for high average power almost from their inception and the decade that followed was a wild ride of great hopes driven by large infusions of cash from Strategic Defense Initiative (SDI) interest. Ultimately these efforts in the US lapsed into a number of low level applied research studies with little to show from the nearly \$1B investment [3]. At that time there was continuing improvement in the performance of solid state lasers with optical parametric amplifiers extending to longer wavelengths and beginning to answer needs for laboratory sources of tunable coherent infrared radiation. Yet there was a need for capabilities that could not be satisfied by existing or near-term foreseeable lasers: short pulses of high energy in high-repetition-rate pulse trains, tunable throughout the infrared region. These needs spanned basic research in condensed matter, atomic and molecular physics and, if powers up to 100 kW could be produced, industrial applications. Support for such an effort was eventually obtained through the Office of Naval Research establishing a development effort for a defense system against cruise missiles that continues today. Initial studies [4] considered use of the CEBAF accelerator but the fear of compromise of the nuclear physics program led to the construction of a dedicated FEL facility utilizing not only SRF technology but applying energy recovery [5] for improvement of performance, and efficient, cost-effective scaling to higher powers.

The article below summarizes the successful development of a superconducting, energy-recovering Free Electron Laser which became the highest power FEL in the world. It also established a new technology, Energy Recovering Linacs (ERLs), as a basis for development of research light sources for the 21st century and other applications. We follow the physics background with a discussion of the measured performance of the laser and then highlight a few of the user activities which leverage the performance of this unique light source. We will conclude with a discussion of the legacy and prospects for application of this development to future light sources.

2. FEL background

FELs are generally acknowledged as being re-invented by J. M. J. Madey based on the relativistic extension of earlier work of Mott and Phillips on the ubitron tube [6, 7]. The FEL utilizes a relativistic electron bunch propagating through a periodic magnetic field (wiggler or undulator) that radiates synchrotron radiation up-shifted in frequency by one Doppler shift and one Lorentz contraction of the wiggler wavelength. As a result the emitted wavelength is inversely proportional to the square of the electron beam energy (Fig. 1). Generation of 1 micron light is feasible with electron beam energies around 100 MeV. The electric field of the initial spontaneous emission combined with the wiggler field works back on the electron longitudinal distribution and causes bunching of the electrons at the optical wavelength. In this way the emitted light establishes coherence through feedback.

In an oscillator configuration mirrors are used to re-inject the previously produced light to subsequent fresh electron bunches and the intensity builds until non-linear effects take over at saturation. About 1% of the electron beam power can be extracted as optical radiation. A portion of the light is out-coupled from the optical cavity *e.g.*, through a partially transmitting mirror, and transported by mirrors to laboratories where the light may be utilized or studied. The remaining 99% of the electron beam power may be simply dumped in a water cooled copper block or, as discussed below in more detail, sent back through the accelerator 180 degrees out of the radiofrequency (rf) phase to decelerate the beam and recover the beam energy. A number of references on the physics and technology of FELs exist to which we refer the interested reader [8, 9].

Prior to the construction of the IR Demo FEL, all linac-based FELs had used pulsed machines (even the FEL on the Stanford Superconducting Accelerator - SCA because of cavity heating limitations due to electron multipacting) [10]. As a consequence they all operated at low duty factors of 10^{-3} or so [11, 12]. During the time the FEL was on, significant power was produced but low duty factor kept the true average power down and significantly limited the stability that could be achieved. Nonetheless, several US efforts [10, 11, 12] as well as several important groups outside the US [13, 14] utilized IR FEL output from RF accelerators for R&D. Although successful in delivering a number of interesting studies, independent reviews [15] remained unimpressed and FELs were relegated primarily to novelty status.

The temporal characteristics of the accelerator output dictate the temporal qualities of the laser so to achieve continuous operation of an FEL (as a continuous train of short pulses accelerated by continuous wave (CW) RF fields) it is essential to obtain a high quality CW electron source. Luckily around this time the development of Cs:GaAs photo-cathodes was deemed sufficiently mature that a DC photogun could be chosen, with temporal control obtained from a mode-locked drive laser [16]. The development of such an injector was crucial to achieving the performance of the IR Demo FEL, and its progeny, the IR/UV Upgrade. Even the present development of X-ray FELs still rests on the capability of photoinjectors initiating the process [17]. To further discuss the details of the injector system goes beyond the scope of this paper but we refer the reader to injector and gun discussions elsewhere [18].

Once the initial bunch is formed and accelerated to an energy where space charge forces no longer predominate (~10 MeV) the beam can be accelerated in standard CEBAF superconducting cavities to the energy desired for short wavelength lasing. The maintenance of electron beam quality and control of high currents is not so simple and thus our story of the accelerator system begins.

3. Accelerator developments leading to CW energy recovery

The emergence of two paradigm shifts set the stage for the evolution of FEL operation to high average power. The first addressed the FEL process itself, and was the recognition that high average powers would be most readily produced through the use of a high repetition rate drive beam rather than one with high instantaneous power. Room-temperature accelerator drivers for FELs are based on the use of extremely high single-bunch charges and high peak currents at the FEL with attendant extremely high peak laser powers. This approach is limiting because of both instabilities in the drive beam itself (due to its high instantaneous intensity) and because of beam quality limitations imposed by space charge forces and the response of the accelerator to the very large transients associated with pulsing the intense beam off and on.

The move to low peak, high average power operation was thus directly enabled by the advent of SRF technology. CW SRF accelerators for nuclear physics such as the HEPL linac, the Illinois microtron, and later, CEBAF, produce continuous streams of individual bunches, individual packets of charge, at very high repetition rate (hundreds of MHz or a few GHz) typically accelerating a bunch in every RF period. Early work in this area at HEPL [20] quickly led to the recognition that this advantage could be applied to FEL drivers. Beams with lower instantaneous currents would be of extremely high quality; so high, in fact, that the high peak current required by the FEL could be produced by beams with quite modest single bunch charge. Moreover, given the capability of SRF systems to be on all the time that is, to operate CW the average current of such a system could in fact compete with, or exceed, that available from a conventional pulsed accelerator. Success of initial CEBAF operation showed that a very high power electron drive beam of exceptional quality would be available in an SRF based system.

The second paradigm shift involved the recognition that both the efficiency and operability of a high power FEL system would be limited unless novel methods evolved for power management

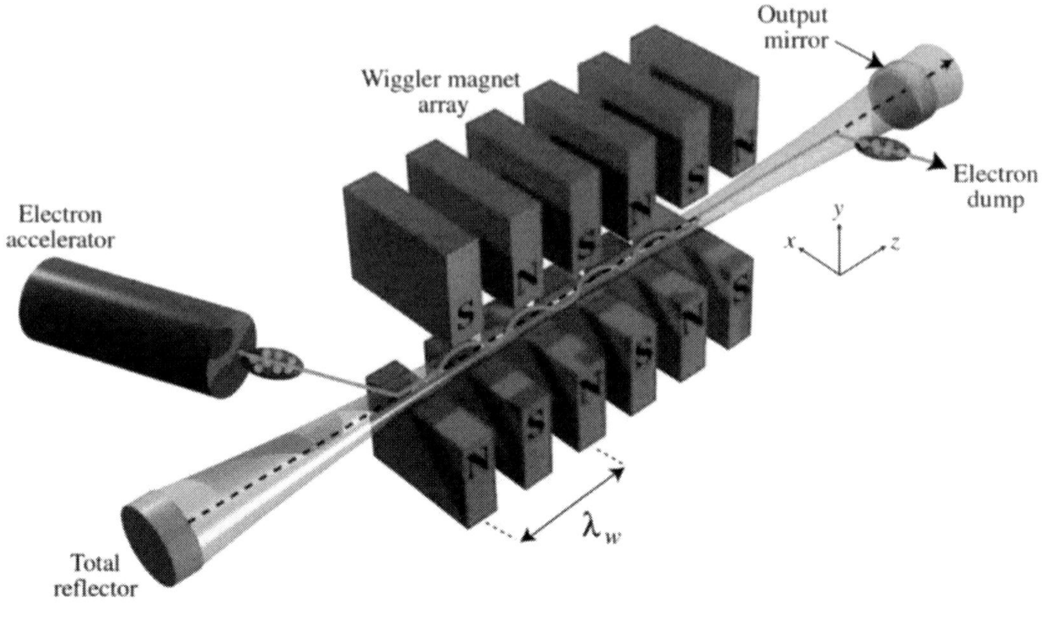

Figure 1. The FEL interaction region where bunches of relativistic electrons from the accelerator enter the wiggler simultaneously with light pulses which are amplified provided they are near the resonant wavelength.

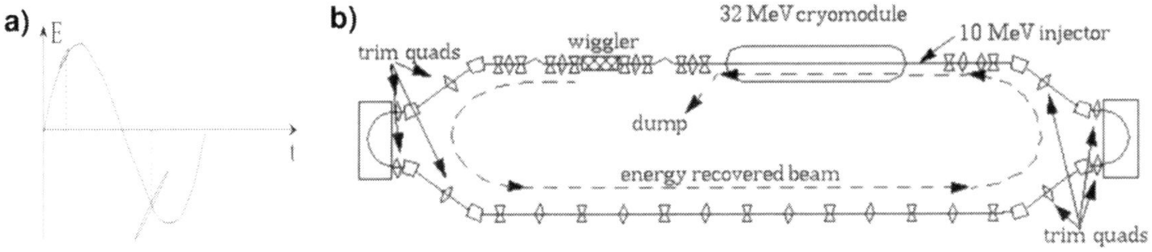

Figure 2. a) shows the timing of a beam bunch relative to the RF waveform and b) shows the accelerator geometry for energy recovery.

of the electron drive beam. FELs transform only a few percent of drive beam power into light. Kilowatt class FELs therefore require drive beams of hundreds of kilowatts of power and generate exhaust beams with similar power levels! Thus, conventional SRF linac drivers for FELs require impressive RF and wall plug power. They also demand radiation control of very high power waste beams. The expense of such a facility would be very limiting unless a viable alternative to conventional linac architectures was found. Such an enabling technology was found in energy recovery. In an effort to overcome precisely these issues, Tigner [5] had earlier proposed the notion of re-using the electron beam after it was exploited by experimentalists to put RF power back in the accelerator. Notionally, after acceleration and experimental use, the beam would be returned to the linac exactly a half-RF-period out of phase and decelerated to low energy. The power extracted from the beam would be reused for acceleration of subsequent beam. The timing of the beam with respect to the RF waveform, and a notional machine concept using this method are shown in Fig. 2. Multiple groups recognized over the next two decades the value of this technique, with an MIT group [19] performing recovery experiments with the Bates linac in the early 1980s. Even more importantly, the Stanford HEPL laboratory realizing that an SRF linac was ideally suited for this application because the absence of accelerating cavity ohmic wall losses would allow an essentially power free generation of extremely high current, high power, high quality electron beams [20].

The synergy between the success of initial CEBAF CW operation and these paradigm shifts served to further motivate efforts to pursue an FEL driven by an SRF energy recovering linac (or ERL). By the early 1990s, it had become apparent that SRF ERL drivers could be used to provide beams ideally suited for FELs and as study of this topic evolved, increasingly subtle details of a system design emerged. Three key interrelated requirements specific to CW operation quickly became apparent which have allowed the CW implementation of energy recovery and generation of extremely high FEL output powers. First of these was the emergence of a clear process of beam control through the accelerator (phase space matching) that would provide an appropriate drive beam to the FEL and recover the power from it. Secondly, the interaction of the beam with its environment and with itself must be managed to handle a menagerie of instabilities and beam-quality-degrading effects from beam currents and intensities some 25 times higher than CEBAF. Finally, essentially loss-free transport was required to handle the unprecedented beam intensities.

Solution of these demanding requirements was achieved using a novel combination of pre-existing hardware (from CEBAF) and beamline designs (largely from MIT-Bates) in a process that was truly bricolage [21]. The resulting system the IR Demo was initially operated CW in the spring of 1998 as a 50 kW single-pass linac and used in early summer 1998 to generate 155 W of CW light. Upon completion of the recirculation transport in the summer of that year, the machine was fully commissioned, acting as a true ERL, generating a drive beam of higher power than that available using only the installed RF. CW FEL performance was extended to well over 1 kW by the following summer. The follow-on system (the IR Upgrade) was a scaled-up version of the Demo; it came on line in 2003 and produced 14.3 kW CW light in the fall of 2006. It later added terahertz output ports extracting 100W in broadband femtosecond pulses and a UV FEL (UV Upgrade) producing over 100 W output. The physics considerations that went into the design of these machines are presented next.

3.1. Beam control throughout the accelerator

High power beams generated by modern CW SRF accelerators must be well controlled so as to avoid damaging levels of beam loss, typically around 1 W/m, about 10^{-6} of the total beam power. Considerable care is therefore required in the design, construction and operation of FEL driver accelerators. Beam sizes must be carefully controlled throughout the system; this requires the use of numerous focusing magnets, quadrupoles, adjusted so as to provide smooth transitions between the various regions of the machine. When properly executed and operated, control of the core beam will also provide management of halo, or beam tails, and background at large amplitude that can lead to significant beam loss.

Even more important than beam size is the control of beam energy and timing, so called longitudinal phase space management. FELs require high peak current to lase. Notionally, one might therefore try to generate and accelerate a high charge short pulse of electrons. However, such an approach would be doomed to failure by the self-fields of the electron bunch: at low energy a short, intense pulse of electrons is rapidly and severely degraded by the action of the internal repulsive forces of the charge on itself. A key advance in the design of the IR Demo was the recognition that a process of longitudinal phase space management could successfully produce, accelerate, deliver, and recover the beam needed for a high power FEL by first generating a temporally long bunch of electrons with a small energy spread. This bunch, by virtue of its length and consequentially low spatial density, would be insensitive to degradation from internal repulsive forces (the so-called space charge effects), and could be accelerated to high energy without loss of beam quality.

Acceleration occurs on the rising part of the wave form, resulting in a phase-energy correlation across the bunch, a chirp that can later be used to compress the bunch length and generate the peak current needed by the FEL. The head of the beam is at lower energy than the centroid, which is in turn at lower energy than the tail (Fig. 2a). The transport system need only be designed to bunch the chirped beam, that is, to ensure that the low energy components of the beam traverse a longer trajectory than the higher allowing the rear of the pulse to catch up with the front. The resulting short bunch can be utilized by the FEL to generate an intense optical beam; this process not only removes energy from the electrons but also smears out the beam energy spectrum resulting in a very large energy spread exhaust beam after the FEL.

A second key design advance enabling CW operation is appropriately introduced at this point of our discussion. If simply recovered, the large energy spread exhaust beam would be unmanageable when decelerated to low energy. However, the initial bunch compression process can be executed in reverse to provide energy compression during energy recovery. The large energy spread/very short bunch after the FEL would, during transport back to the linac, be decompressed, reintroducing the time-energy correlation that existed at the end of the linac after acceleration. By proper choice of transport system path length, this bunch could be re-injected exactly out of phase with the accelerated beam, resulting in recovery of beam power to the linac. Proper choice of time-energy correlation (momentum compaction) then assures that the exhaust energy, at each time, will exactly match the energy variation along the RF waveform. After recovery, at the end of the linac, the electron beam is nearly mono-energetic and can be losslessly transported into a dump. Though not novel (energy compression systems had been operated in accelerator systems prior to this time) the coupling of energy compression to the energy recovery process allows CW operation that would otherwise be precluded by beam loss

Journal of Physics: Conference Series **299** (2011) 012014 doi:10.1088/1742-6596/299/1/012014

at low energy. The process is depicted schematically in Fig. 3.

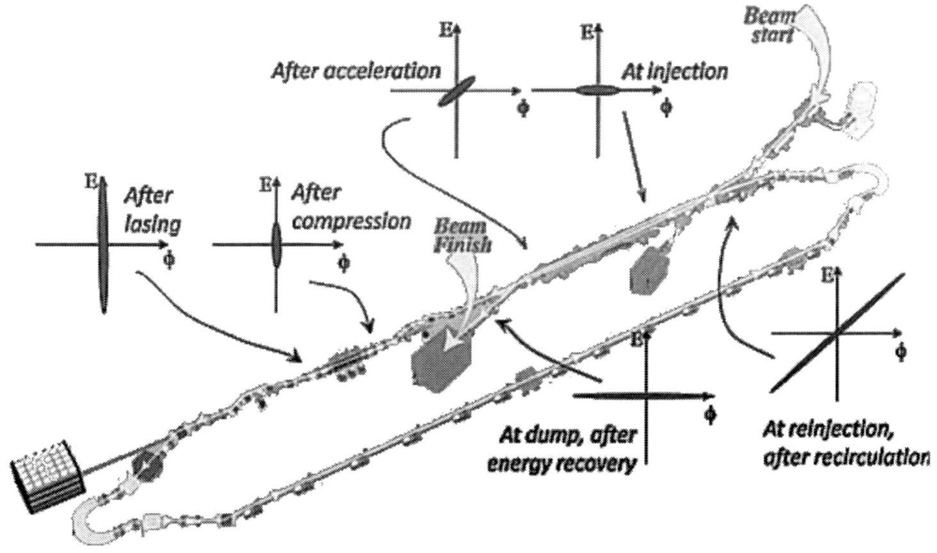

Figure 3. Schematic of longitudinal phase space management.

Extrapolation to very high powers requires use of a large energy acceptance transport system. In addition, nonlinear effects must be included such as curvature of the RF waveform and compensation for variations in bending and focusing during beam transport to the wiggler and back to energy recovery. Methods to accomplish this were confirmed for FEL-like parameters at MIT-Bates in the 1980s. The Jefferson Lab FEL linacs were designed as clones of the Bates recirculator and have repeatedly demonstrated very-low-loss operation ($< 10^{-6}$ loss). Without energy recovery, ~4.5 kW of power is needed to operate each cavity while accelerating 1 mA of beam; with energy recovery, only 2 kW is needed, regardless of current.

3.2. Effects due to high beam intensity
The unprecedented CW beam intensity required by high power FEL operation stimulated concern about numerous current-driven phenomena during design and operation of the Jefferson Lab FEL drivers. A critical issue in the design of CEBAF had been the potential impact of beam break-up (BBU), a beam-intensity-driven instability that had plagued earlier generations of recirculated linacs. SRF cavities resonate at a spectrum of frequencies, not just the RF fundamental, and under certain circumstances the parasitic "higher order modes" (HOMs) can significantly influence beam behavior. In particular, if the timing of the bunch train is commensurate with the resonant frequency of an HOM, the beam can potentially put power into the mode, causing the mode to drive the beam in a manner further exciting the mode. This feedback loop can readily go unstable, causing beam loss and limiting the operating current of the accelerator.

Evaluation of this effect for the initial design indicated that it, like CEBAF, would not be susceptible to unstable behavior because of the care exercised during initial cavity design. Early experimental studies of this effect [22] stimulated a comprehensive investigation of the instability. As a result, accelerator designers are now essentially free to choose the instability threshold current [23] with well-defined engineering design criteria.

Journal of Physics: Conference Series **299** (2011) 012014 doi:10.1088/1742-6596/299/1/012014

Extrapolation to higher beam currents has required investigation and management of a number of beam-intensity-driven effects, in addition to the BBU instability. These include interactions of the bunch with itself, such as coherent synchrotron radiation (CSR), and with the accelerator environment. The latter includes the interaction of the beam electromagnetic fields (wakes) with various resonant and conductive devices in the system, resulting in localized heating. These studies point to an emergent understanding of power-flow management in high-power FEL systems. CSR-driven heating of FEL optical cavity mirrors, RF and resistive wall heating of beam-line components, and wake- and HOM-driven power deposition along the accelerator all represent potential performance limitations for the next generation of accelerator driver. Experience with the Jefferson Lab FEL drivers has shown that proper management of these effects is required for successful operation.

3.3. Management of beam loss, halo, and irregularities

Even after appropriate handling of the core beam and addressing effects related to high intensity, there are significant obstacles to high-power beam operation. Figure 4 shows a viewer image of the beam; the complexity of the spatial distribution is obvious. Historically, such structure has been attributed solely to complex nonlinear processes arising during beam formation and transport, but tests have shown the sources of halo, large amplitude beam components of intensity too low to be easily observed, but too high to be neglected, are typically quite prosaic. They include:

- drive laser light scattered over the active area of the photocathode (ameliorated by anodizing much of the active area),
- finite time response of the cathode (leading to momentum tails which are managed by the large acceptance of the transport system), and
- imperfect gating of the drive laser pulse train, leading to so-called "ghost pulses" of low intensity but well beyond the previously cited loss rate tolerance.

Experience with the Jefferson Lab drivers has shown that the halo tends to propagate independently of the core beam. Appropriate measures may thus be taken to manage it so that it is transported through the system with acceptably limited losses. Operationally, we typically measure loss patterns around the machine, and adjust focusing elements to which the core beam is insensitive but the halo responds strongly so as to alleviate excessive localized loss. We have thereby successfully achieved reliable and repeatable operation at beam powers at 1 MW levels with little loss ($< 10^{-6}$) and essentially no activation of accelerator components.

4. Lasing: IR Demo, IR Upgrade, THz, and UV output

4.1. IR Demo

Though the FEL group had great hopes for a high power FEL, one thing that cannot be stressed enough was the uncertainties in our expectations for this first machine dubbed the IR Demo. No machine like this had ever been built. The average design current for the IR Demo was 25 times that of CEBAF. The design of the FEL systems actually started with the formation of a team that worked on the design of an industrial UV laser system in 1995. The chief design principles of the IR Demo were to be as conservative and inexpensive as possible. A major issue was the uncertainty in the electron beam parameters. No one had used a photocathode DC gun with a free electron laser before or characterized injector performance at the desired charge so we had no good benchmark to use for the electron beam properties. We therefore assumed a factor-of-two safety margin over simulated values on the electron beam quality numbers, *i.e.* the transverse emittance and the bunch length were twice what the simulation codes predicted. The maximum operating energy of the accelerator would be 41 MeV based on cryomodule performance and the current would be 5 mA in the form of 135 pC bunches at 37.425 MHz. For the wiggler we chose

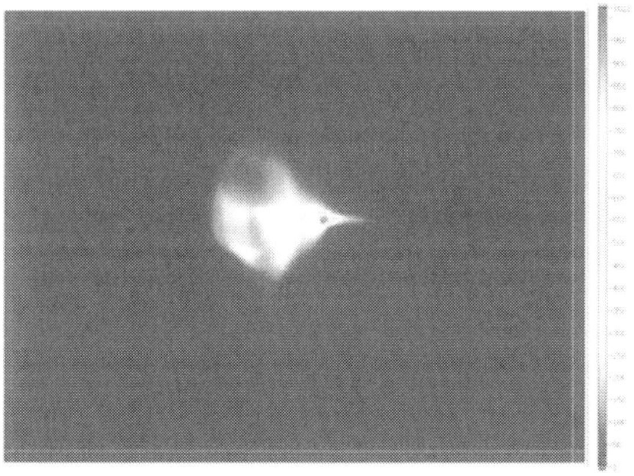

Figure 4. Beam viewer image of electron beam. The complexity of the beam structure and therefore underlying beam dynamics is clearly evident.

a proven design from the Argonne Advanced Photon Source. With a 41 MeV beam energy and a design wavelength of 3 microns we needed a fairly short wiggler period. On the other hand we were very worried about beam loss at the entrance to the wiggler so we wanted a very large gap (making achievement of the required field harder). We therefore settled on a design with a 2.8 cm period and an rms field strength of 2.7 kG. The calculated small signal gain was at least 40%, sufficient gain margin to get over 1 kW at 3.2 microns assuming perfect optical cavity mirrors. The final wiggler is shown in Fig. 5.

Figure 5. The IR Demo wiggler mounted on an optical bench along with four of the red matching quadrupoles. Above the wiggler one can also see three of the viewers used to align the electron beam to the optical mode.

As the machine was commissioned we started to see payoffs from our conservative design choices. The wiggler vendor had provided us with a wiggler that could achieve the specified field with a gap much larger than the design gap. We therefore had the capability of closing the gap and using a field 50% larger than the design. Another pleasant surprise was that the beam energy from the accelerator could be as high as 48 MeV instead of 41 MeV. This meant that we could use the higher field and electron beam energy to get to the design wavelength. This increases the gain and power. To save commissioning time, we started up with only half of the design charge. This still allowed us to reach 5 mA if the repetition rate was increased to 74.85 MHz. We were surprised to find that the measured beam parameters matched the simulation values fairly closely. When all these factors were taken into account, the projected laser gain increased to about 70% per pass. With a cavity loss of 10% this provided a very comfortable gain margin for lasing.

Another essential tool in getting the laser started was the alignment system. With the help of Dick Oepts from the Felix FEL project in the Netherlands we developed a simple but very reliable method of aligning the optical cavity with both the wiggler magnetic axis and the electron beam. It is worth pointing out that the optical mode and the electron beam in the wiggler are both about the diameter of strands of spaghetti. These three-foot long strands have to be overlapped to a fraction of their diameter and one can only see them in a few locations. Nevertheless, the alignment system allowed us to routinely align the electron beam and optical cavity so that very little subsequent adjustments had to be carried out.

Because of both the alignment system and the large gain margin, the laser lased surprisingly easily when the wiggler was first installed (we had left the wiggler off the beamline until we were convinced we could actually get beam through the small gap in the wiggler vacuum chamber). Initial lasing was with 42 MeV beam at 5 microns on June 15, 1998 using our so-called "first light" optics. These only had an output coupling of 3%. We quickly ramped up to over 150 W of power in a configuration in which the beam was dumped at full energy in a straight-ahead dump (see figure 3 for a drawing of the IR Demo layout). We then changed out the output coupler for a 10% output coupler and achieved over 300 W on July 27 [24]. In straight-ahead mode we could only accelerate about 1.1 mA of beam current so the FEL power was limited to 500 W. The laser was quite stable and we could operate for long periods of time at about 30 times the previous world record power.

Once lasing had been achieved, we had to learn how to energy recover. The basic principles of this were described in the previous section. Though we knew the basic ideas we had to fill in the operational details of running this very new type of machine. We had to learn how to cope with all the problems described in the last section. Once energy recovery had been achieved, we started to ramp up the current and power and met a different limitation.

The theory of high power FEL operation with absorptive mirrors had been worked out during the design stage [25]. As the mirrors heat up their surfaces bulge out and change the waist size in the optical cavity. This distortion reduces the FEL gain and ends up clamping the power at some level. Once this power is reached it cannot be exceeded even if the electron beam current is greatly increased. The extent of the distortion for a given output power can be used to define a thermal figure-of-merit for a given mirror set. The achievable output power is proportional to the figure of merit. Our initial mirrors were made of calcium fluoride, which has a relatively poor thermal figure- of-merit. As expected, the output power with these mirrors was clamped at 500 W. Using a silicon mirror for the high reflector allowed us to reach 710 W by March 11, 1999. Finally we switched to a pair of sapphire mirrors. Sapphire has a much higher thermal figure of merit than calcium fluoride. Once we installed sapphire mirrors coated for 3 micron operation we were able to ramp the power up to 1720 W with 4.4 mA of beam current on July 15. The limitation now was just the efficiency of the laser and the available current [26].

Once we had exceeded our goal of 1000 W from the IR Demo we switched over to using the beam for user experiments. This is described in the next section. We also continued to study the machine and optimize its performance. We found that the electron beam quality was sufficient to lase at not only the third harmonic, already demonstrated on several FELs, but also at the second and fifth harmonic, which had never been accomplished [27]. We also used doubling crystals to double and triple the laser when operated at the third harmonic. This produced up to 56 W of green light and 12 W of UV light (see Fig. 6). After running the accelerator for a couple of years we were able to take advantage of the improved laser operation and energy recovery and provide 2.1 kW of laser power at 3 microns, more than a factor of two over the design power.

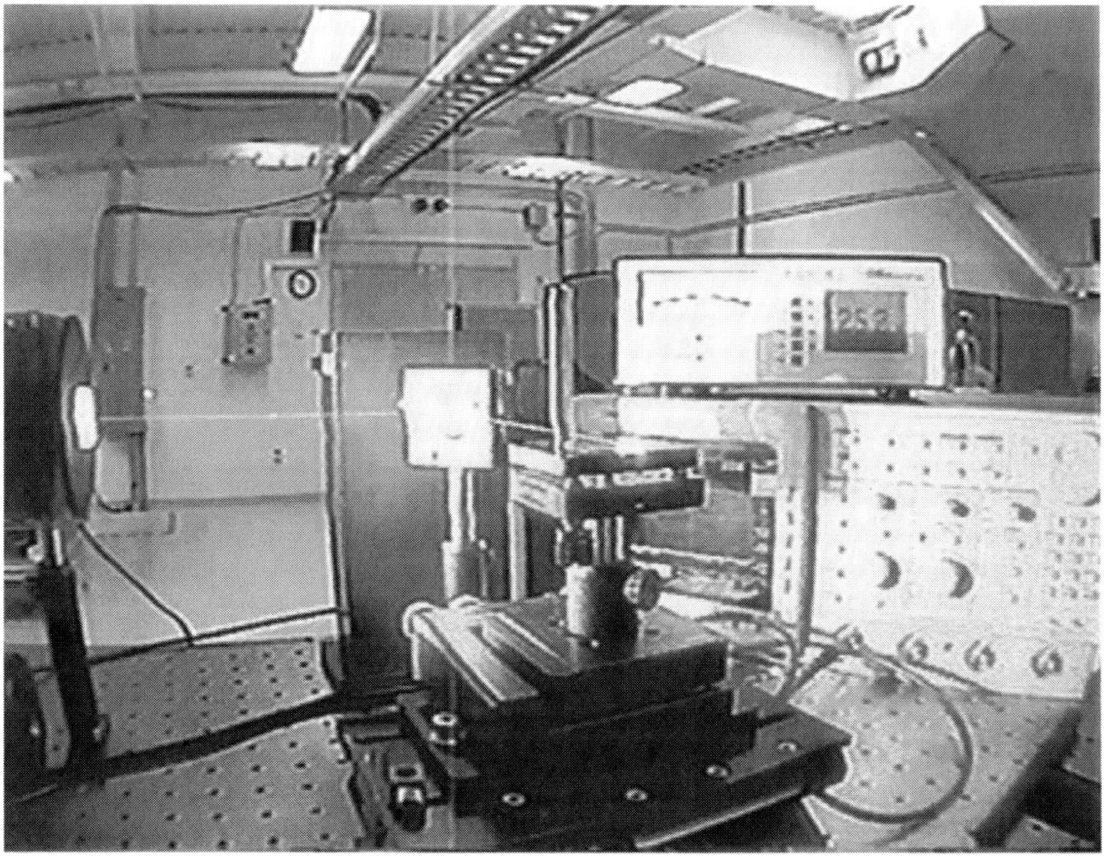

Figure 6. Frequency doubled light from the FEL. The laser was operated at 1.05 microns and doubled in a crystal to a power level up to 56 W. In this photo the 25 W beam is powerful enough to be seen going through the air.

While we delivered infrared light to users we started to look at the possibility of parasitically producing light at other wavelengths. The electron beam bunches as they enter the straight section with the FEL are very short. The synchrotron radiation produced in the last bending magnet is greatly enhanced at all wavelengths longer than the bunch length, in this case about 100 microns rms. This coherent synchrotron light is emitted in the THz part of the electromagnetic spectrum. Researchers had been working for years to produce THz radiation in a similar manner by using short pulsed lasers to produce very short electron pulses. The radiation in the 48 MeV beam was enhanced by the cube of the ratio of the electron energy to the electron rest mass energy, which is about 100 for this case. Using this simple scaling it is easy to show that we could increase the 100 μW of THz power in the laser based sources a million times to the 100 W level. To dramatically demonstrate this huge leap in power we used the THz beam to light a match [28].

The other parasitic radiation source is Thompson backscattering. The circulating light in the optical cavity collides with the electrons at the waist of the cavity where both beams are very small. The infrared photons can then scatter off the electrons and produce X-rays along the direction of the electron beam. The IR Demo produced copious amounts of these Thompson scattered X-rays and we made careful measurements of them while running the laser for other users [29].

4.2. IR Upgrade

The Navy was now interested in the next scaling step over 10 kW. If we could just triple the electron beam energy, double the electron current and keep the FEL efficiency as high as it was with the IR Demo we could get over 12 kW. This sounded like a simple, straightforward project. It turned out to be anything but. We decided at first to use an optical klystron, developed in collaboration with Advanced Energy Systems (see Fig. 7), to allow lasing in the mid-infrared with very broad tunability. This meant that we could not go to very short wavelengths unless the energy could be greatly increased. The optical klystron was also an oddity. Optical klystrons are usually used in storage rings, which have very small energy spread. An optical klystron on a storage ring has very high gain and very low efficiency. The IR Upgrade accelerator had a large energy spread and high peak current so we wanted to operate the optical klystron in a mode that had almost never been used. We used the dispersion section to just phase match between the two wigglers, creating a wiggler effectively about the same length as the actual optical klystron. This meant that the optical klystron was not the best choice since it provided less gain for the same effective number of periods compared to a simple wiggler.

Figure 7. Optical klystron used in initial lasing with the IR Upgrade. The two wigglers in the front and back have 12 periods each. The light green dispersion section in the middle phase matches the two wigglers to make the equivalent of a 25 period wiggler. The first linac cryomodule is in the background.

Initially our energy was also not initially a factor of 3 higher than the IR Demo. The third module used to raise the energy up to over 150 MeV was not yet available. With only two cryomodules we could only operate at 80 MeV. With 10 mA of beam current that gave us 800 kW of available electron beam power. If we could achieve at least 1.25% efficiency we had a

Journal of Physics: Conference Series **299** (2011) 012014 doi:10.1088/1742-6596/299/1/012014

chance of getting to our 10 kW goal. We had achieved up to 1.5% efficiency with the IR Demo so we felt that this was an achievable goal.

One new feature we had to deal with in the IR Upgrade was bending the electron beam by 180 degrees before the FEL. This was both a blessing and a curse. It was a blessing because we could now correct for the curvature in the longitudinal phase space and produce a very short bunch. It was a curse because we had to deal with huge amounts of CSR in the final bend magnets and the resulting beam degradation. With the optical klystron operated at low energy and small dispersion, the option of operation at half charge was no longer possible. According to the FEL modeling, the Upgrade wiggler needed the full charge of 135 pC to lase effectively.

As with the IR Demo, initial laser commissioning proceeded with pulsed beam at low duty cycle. Once the cavity length had been set properly the laser lased easily. We lased at about 6 microns using zinc selenide mirrors on June 17, 2003. This system was far more flexible than the IR Demo. Up to four mirror sets could be installed at one time and the wiggler could be tuned in real-time [30]. We quickly demonstrated lasing over the full reflectivity range of the mirrors but the gain seemed much smaller than expected. This was due to the longitudinal emittance being much larger than expected. Remember that, in the IR Demo, the emittances were fairly close to simulations for a 67 pC bunch. For the 135 pC beam the transverse emittance was fairly close to the simulations but the longitudinal emittance was about four times the simulation [31]. This meant that the bunch length was much larger than we had planned. The low energy exacerbated this problem. With this poor a beam quality we could not get the 1.25% efficiency we needed.

Despite this once the second recirculation arc had been installed and CW beam was achieved we expected to ramp up quickly to many kilowatts but were frustrated to find a limit of less than a kilowatt. The problem, familiar from the IR Demo days, was mirror heating. The next 6 months were spent trying out new mirror sets to try to get mirrors with less loss. One source of loss was new, however. With the shorter bunches and higher energies possible on the IR Upgrade, the downstream mirror was bombarded with THz coherent synchrotron radiation. This was absorbed in the mirror and limited the laser power. Finally, ten months after first light we achieved 4.1 kW at 5.75 microns but we realized that we could not get to 10 kW at 80 MeV. We therefore installed the third cryomodule that provided up to 80 MeV of energy gain. This now allowed us to operate at over 150 MeV. At the higher energy the longitudinal emittance was not so much of a problem so the laser gain was quite high. Unfortunately the coherent synchrotron radiation was also enhanced so mirror heating was still a problem. We therefore reversed the optical cavity so that the CSR would land on a back-plane cooled silicon mirror instead of the zinc selenide output coupler. With this change and the larger available electron beam power we could finally get up to 8.5 kW of CW laser power. We realized that we could get even higher power if we pulsed the electron beam to avoid some mirror mount thermal issues. We obtained over 10 kW in 1 second pulses with a 1/4 Hz repetition rate on July 21, 2004. Note that we had never intended the accelerator to operate pulsed so this was quite a departure from the design operation.

Since the Navy was more interested in short wavelengths and CW operation we replaced the optical klystron first with an electromagnetic wiggler that permitted operation in the 1–2 micron wavelength range and finally a tunable permanent magnet wiggler optimized for operation in that wavelength range. The latter wiggler had both high gain and high efficiency at a wavelength of great interest to the Navy: 1.6 microns. It also allowed us to tune over a very large range with ease. With a broadband resonator using hole coupling it was possible to tune while lasing between 650 nm and 5.1 microns, more than three octaves of tuning. We could also use dielectric mirrors to easily switch between fundamental, third harmonic, and fifth harmonic lasing. The fifth harmonic lasing, achieved on the IR Demo with two high reflectance mirrors, could now be easily achieved with a 4% output coupler.

Unfortunately, at short wavelengths the mirrors are even more sensitive to mirror heating so the power at 1.6 microns was still limited to about 6 kW in CW operation. We did greatly reduce the THz heating by installing a debunching chicane and THz traps after the wiggler but there was a new problem – coherent harmonics. Most lasers produce a single wavelength that is very pure, like a flute playing a single note. The FEL naturally produces many harmonics of the fundamental lasing wavelength, similar to a violin, which produces a rich spectrum of harmonics. A prism is used to separate the 5th, 6th, and 7th, harmonics when operating at 3 microns. These harmonics allow the operator to see the beam even when the laser is operated at invisible wavelengths. When operated in the near infrared, however, significant amounts of ultraviolet radiation are produced. These UV harmonics enhanced the absorption of the fundamental light. The center of the mirror heats up by more than 10 degrees Celsius compared to the edge. The solution was to enhance the thermal figure of merit by cooling the mirrors to cryogenic temperatures. The figure of merit at liquid nitrogen temperature is at least a factor of 200 larger than at room temperature. With these mirrors the efficiency of the laser, which could routinely exceed 1.6% at low power, was now independent of power or current. We could then lase at up to 14.3 kW with 8.5 mA of current at 115 MeV.

In parallel with the IR Upgrade operations we installed an optical transport line to one of the upstairs labs for the THz radiation emitted from the last bend before the wiggler. As noted above this radiation can be well over 100 W, a unique radiation source for users. With the higher energy and current of the IR Upgrade we increased THz radiation a factor of ten more than in the IR Demo. This was used to make THz movies using THz detector arrays and cameras.

4.3. UV Upgrade

Along with the IR Upgrade we were funded to build another electron beam transport to add an ultraviolet FEL. This design benefited from all the lessons learned from the IR machine plus knowledge of the detailed machine parameter space. We used a APS undulator A prototype wiggler borrowed from Cornell that matched our design energy well. The electron beam requirements for the UVFEL include a smaller transverse emittance and a smaller energy spread. This was achieved by operating the gun at half the previous charge, 67 pC, and the highest energy available, 135 MeV. Initial operation was chosen to be at 700 nm to take advantage of relaxed tolerances and high gain. At this wavelength we achieved 165 W of output power at only 0.32 mA of average beam current. A detuning curve of more than 11 microns indicated a small signal gain in excess of 100% [32]. Since that report the lasing range has been extended down into the 438 to 362 nm range with a second mirror set. Measured gains in excess of 100% are larger than predicted by one-dimensional models and efficiencies of at least 1/2N are seen. Operation at high power is much more of a challenge than in the infrared both because the optical coating absorption is much higher and because the allowed distortion is proportional to the wavelength. Average powers in excess of order 100 W will require the use of cryogenic mirrors. Work is presently underway to fully characterize the performance, reconcile the performance with models, and utilize the harmonics of the FEL light in the VUV for photonic studies of materials.

5. FEL applications and user program

5.1. Introduction

In the early 1990s, concurrent with the technical efforts leading towards a Conceptual Design Report (CDR), advances were made to industry to drum up interest in this unique laser. It became apparent early on that there was more interest in the UV region of the spectrum than the IR, so the machine design was modified to allow future production of UV light. As it was possible to also lase in the IR, applications were also contemplated for this wavelength region. Given that the cost per photon (in units of $/kJ) for an FEL becomes comparable to more conventional laser systems once the output power is in the 50 kW range [33], applications that were initially summarized [34, 35] tended to emphasize large area processing of materials in the polymer, automotive, and electronics industries.

Why the interest in these processing applications? By the late 1980s it was clear that the use of lasers for material processing could cut costs and allow some processing steps to be done that were otherwise impossible and companies making industrial lasers begin having reasonable sales volumes of order several \$100M [36]. The decade of the 1990s saw sales of all lasers used for material processing (except those based on diode lasers) quadruple, with sales exceeding \$1B by the end of the decade, and continuing to grow. The majority of these lasers are CO2 or Nd:YAG technologies with CW or long-pulsed (ms long pulses at 100s of Hz rates) temporal formats, although excimer lasers and high pulse repetition frequency (PRF), Q-switched solid state lasers with ns pulsewidths have gained in popularity. Basic physics constraints on the stored energy per unit volume and gain sets the output of single-rod Nd:YAG lasers at about 1 kW, and at a few hundred watts for excimer lasers. For the processing steps performed, such as welding, cutting, and drilling, these lasers are adequate. However, for large area processing, one requires a laser with a short pulse (fs to ps) and high PRF (100 kHz to MHz) output with average power above 10 kW [37]. This is a parameter space well- matched to the specifications of an SRF ERL-based FEL. The applications can be categorized into two categories: thermal processing or ablative processing. The thermal processing can be used for surface texturing; surface amorphization; laser glazing and annealing; adhesive bond pretreatment; crystallizing amorphous silicon; laser annealing, deposition, and cutting for photovoltaics; and solvent-free cleaning. The ablative processing can be used for micromachining; cutting and slitting; and deposition of large area thin films.

For many of these processes, the benefits of using a laser were proven (and often patented) using lower powered lasers at sub-industrial scales. However, these processes haven?t become prevalent because the power of the lasers used can't be scaled to the levels (> 10 kW) required. For example, despite a large number of publications showing the benefits of various short-pulsed (particularly ultrafast subpicosecond) lasers in materials processing [37, 38], their presence in a commercial environment has been limited. Ultrafast lasers have not achieved high average power status, and based on the nature of the amplification process, aren't likely to. And yet, there are compelling reasons to use ultrafast lasers; such as (1) a lower threshold for ablation, (2) more deterministic damage, (3) ablation with minimal heat-affected zone in metals, and no cracking or melting in insulators and ceramics. Along with a short-pulse time structure and the other desirable properties mentioned earlier is wavelength agility, so absorption bands (if present) in the material can be accessed.

5.2. Background

Nearly simultaneous with the start of construction of the IR Demo FEL was the appearance of two papers showing the efficacy of materials processing, specifically pulsed laser deposition (PLD) using short pulse, high pulse repetition rate (PRF) lasers [39, 40]. The idea is that the total power deposited to the surface is the same as with low (100s of Hz) laser systems, but because the energy/pulse is lower, the resulting ablation is less likely to produce droplets of material, plasma above the surface, *etc.* Experiments with carbon targets to produce diamond-like thin films, reported in [40], showed this to be the case. Nevertheless, while the use of a very short (1-20 ps) to ultrashort (< 1 ps) pulse minimizes the deposition of heat to the surrounding material during the pulse, the high PRF of these lasers ensures that a fresh pulse at the same position will deposit more energy raising the temperature further before the previous pulse's energy has had time to dissipate. Work to model this effect, now called accumulation, appeared in [39] and elsewhere.

A simple spreadsheet model demonstrates this effect. Treating the laser-target interaction in the thermal equilibrium approximation, the thermal conduction problem reduces to the idealized case of a semi-infinite plane and is more amenable to a simple analytical model. This is valid for laser pulse lengths greater than about 10 ps if the optical penetration depth [given by $l_o = 2(K\tau_l)^{0.5}$, where K is the diffusivity of the material and τ_l is the pulse length] is much less than the thickness of the material. One such formulation was published by Bechtel [41] and has the solutions:

$$T(t) = \frac{2I(1-R)\sqrt{K\tau_f}}{\kappa\sqrt{\pi}}, \tag{1}$$

$$T(t) = \frac{2I(1-R)\sqrt{K\tau_f}}{\kappa\sqrt{\pi}}\left(\sqrt{\frac{t}{\tau_l}} - \sqrt{\frac{t}{\tau_l} - 1},\right) \tag{2}$$

where I is the irradiance (in W/cm^2), R is the reflectivity of the material at the laser wavelength, and the other constants have the same meaning as above. Figure 8 shows the result of applying Eq. (1) and (2), to a target of 304 stainless steel, with a wavelength of 3 microns, a pulse repetition rate (PRF) of 20 MHz, and an irradiance of $\sim 3 \times 1011$ W/cm^2, this being a typical irradiance achieved with the IR Demo FEL.

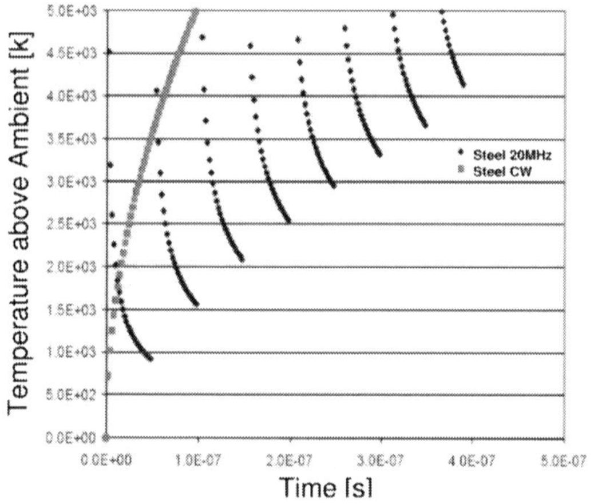

Figure 8. Modeled surface temperature of stainless steel irradiated with a high rep rate train of short pulses (blue dotted line). For comparison, the temperature profile for a CW laser with the same irradiance is shown (magenta dotted line). The parameters are given in the text.

Also shown in Fig. 8 is the surface temperature profile for CW irradiation with the same high irradiance, using the expressions published by Cohen [42]. What one sees is that, even for moderately high-reflecting metals, the first pulse takes the temperature well above the melting point. In fact, it drives the melt to vaporize, which then ionizes to form a plasma that reduces the effectiveness of subsequent pulses. Between pulses there is insufficient time for the region being irradiated to cool, so a CW-like temperature profile underlies each transient temperature excursion. This leads to the development of processing artifacts common with CW lasers, such as recast material and heat-affected zone (HAZ) on the periphery of the irradiated area. It should be noted that this was found to be true even for ablation in metals with an ultrafast laser operating at 1 kHz [43], and the same sort of techniques for mitigating the bulk heating, *e.g.*, trepanning, work equally well.

Of course, modeling is one thing, application is another. In a series of experiments, we measured the ablation rates for 304L stainless steel using the tightly focused output of the IR Demo FEL [44], this rate, as a function of plate thickness is shown in Fig. 9. The rate using an ultrafast Ti:sapphire laser operating at 1 kHz is also shown for comparison [45] in Fig. 9. What one sees is that for thin (< 100 micron) steel plates, the ablation rate for the Ti:sapphire laser is higher, but once the plate becomes sufficiently thick, the deeper hole traps plasma and the ablation rates become comparable. Due to the difference in PRFs (1 kHz for the tabletop laser *vs.* 18.7 MHz for the FEL), the processing rates are much higher for the FEL and are, in fact, comparable to CW lasers.

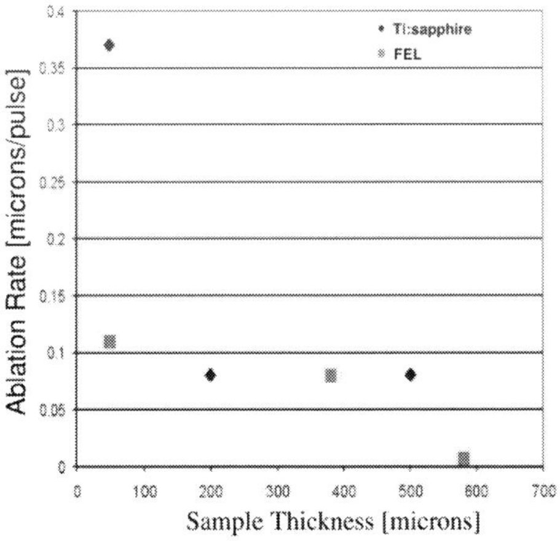

Figure 9. Comparison of ablation rates of stainless steel plates using two different ultrafast laser sources: diamonds – the FEL (wavelength = 3 microns, PRF = 18.7 MHz, F = 1 J/cm2) and squares – a Ti:sapphire laser (wavelength = 0.79 microns, PRF = 1 kHz, F = 80 J/cm2).

5.3. Applications

Carbon nanotubes are increasingly utilized in various technologies. Demand for the single-wall variety is such that the price for high-quality material is currently about $100/gm, with pure material priced at over $200/gm [46]. In 2003, current techniques produced about 0.2 gm/hr [47]. In comparison, worldwide demand is of order of thousands of kilograms annually [48]. Using the IR Demo FEL and delivering an average power of about 300 W onto the target, researchers obtained yields of 1.5 gm/hr, far higher than competing techniques [49]. As shown in Fig. 10, TEM (tunneling electron microscope) images show the single-wall nature of the tubes, and Raman characterization showed that the tubes were < 2 nm in diameter and had bundle sizes below 15 nm. Work to optimize yield and dimensions have continued using the IR Upgrade FEL delivering about 1 kW of 1.6 micron light into the reactor.

The processing of nonmetallic materials was also studied with the IR Demo FEL. One study [50], showed the benefits of wavelength tunability. In this investigation polyimide (DuPont Kapton HN100) was irradiated at two wavelengths, 3.1 microns (off-resonance with the amide I absorption) and at 5.8 microns (on-resonance with the amide absorption). Processing at the shorter wavelength, where the material was transparent, resulted in blackening of the material, a sign that the polymer has been thermally degraded. Tuning the laser to the longer wavelength resulted in cleanly-drilled holes, indicating the processing was nonthermal. This "cold-cutting" mode exhibited by IR FELs has been noted in earlier studies [51], enabling pulsed laser deposition (PLD) of polymers.

A comparative study of the PLD of NiFe, using either the FEL or a Ti:sapphire ultrafast laser, showed that the FEL produced superior quality films [52]. As shown in Fig. 11, the surface

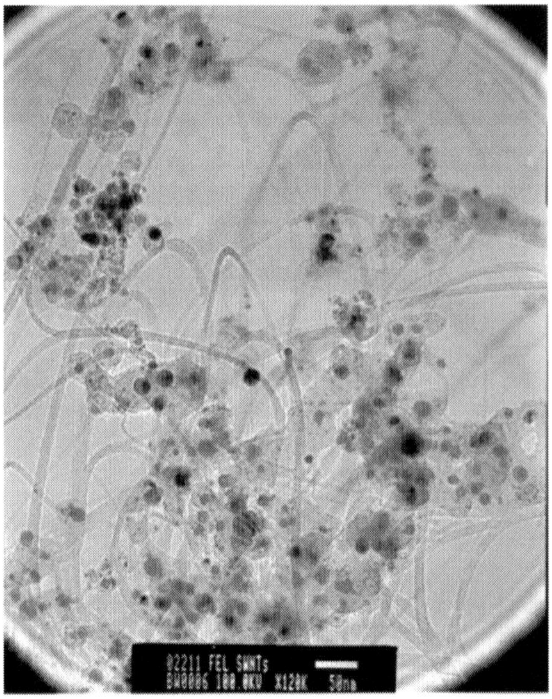

Figure 10. TEM images of single-wall carbon nanotubes produced using the IR Demo FEL operating at 3 microns.

of the films produced using the FEL is far smoother than the one made using the tabletop laser. The FEL-produced film also had a much lower coercivity. And, as one might expect from the previous discussion and data shown in Fig. 9, the deposition rate was higher for the FEL than for the tabletop laser by about 20x.

5.3.1. Surface processing Besides ablative processes, thermal (or physiochemical) processes can be employed to perform surface modifications. No material is lost, merely melted and then resolidified, or while quite hot, transformed by oxidation, nitriding, or carburization. An example which is familiar to many is the application of titanium nitride (TiN) on tool bits. This gives the bit a distinctive gold color. TiN is also used to improve the biocompatibility and wear characteristics of replacement joints [53]. In titanium nitriding studies the FEL was operated at 3 microns, usually in a burst mode (also known as a macropulse). Compared to other types of laser nitrided titanium, the FEL-produced material had a thicker and harder coating [53]. This appears to be due to the formation of oriented (200) dendrites of δ-TiN_x, as shown in Fig. 12. While one might think that this would result in a rough surface, in fact it is fairly smooth [54].

5.3.2. Medical applications With the wavelength and timing flexibility of the IR Upgrade FEL, it is natural to use it in medical applications. One such application is known as selective photothermolysis, the selective heating of tissues with light. By carefully choosing a wavelength in the near infrared, researchers targeted lipid-rich cells and heated them preferentially without heating the surrounding tissue. This study [55] paves the way for a laser treatment of acne, a condition where a sebaceous gland is producing too much lipid. Only the over-active glands, located a few millimeters under the surface of the skin are killed by the absorption of light, leaving

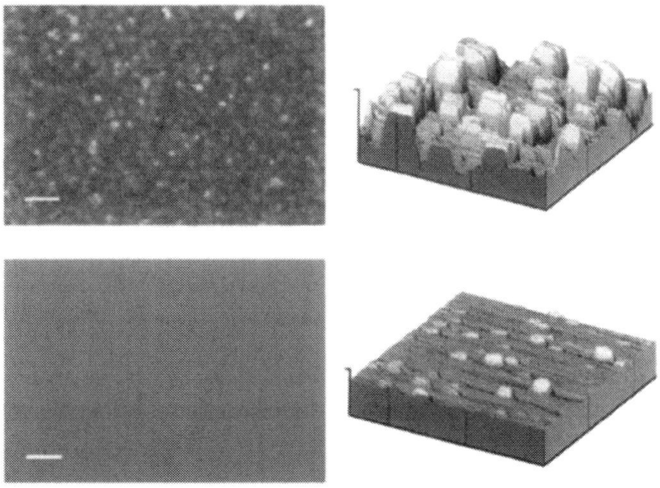

Figure 11. Scanning Electron Microscope (SEM) (left) and Atomic Force Microscope (AFM) (right) scans for thin films grown with (a) the amplified Ti:sapphire system (top pictures) and (b) the IR Upgrade FEL (bottom pictures). For the SEM scans, magnification ∼13,000x (the white line indicates 1 μmq). For the AFM scans, dimensions are 5 μm^2 and the line indicates 1500 Å height. (Figure and caption from [52].)

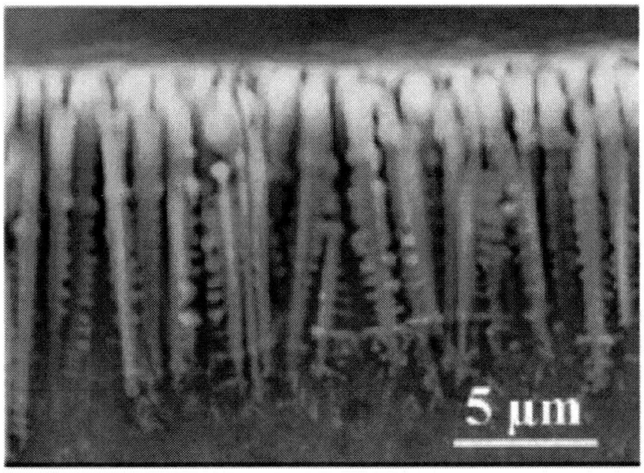

Figure 12. Formation of dendrites of δ-TiN_x after bursts of FEL irradiation at 3.15 microns [54].

the other cells unaffected. This technique also shows promise in the treatment of atherosclerosis, which if left untreated leads to heart disease and stroke.

5.3.3. Scientific applications Along with the applied research described above, the FELs ultrashort pulsewidth and wide tunability made it an excellent light source for several scientific studies. A team at the College of William and Mary used the IR Demo FEL to measure the temperature-dependent nonradiative relaxation rate of hydrogen atoms in crystalline silicon in order to determine what vibrational dynamics (*i.e.*, the coupling of the hydrogen impurity to local and lattice modes) come into play when this impurity absorbs light [56]. The high PRF, along with the relatively high average power (of order 10 W) produced data with a high SN ratio, as shown in Fig. 13.

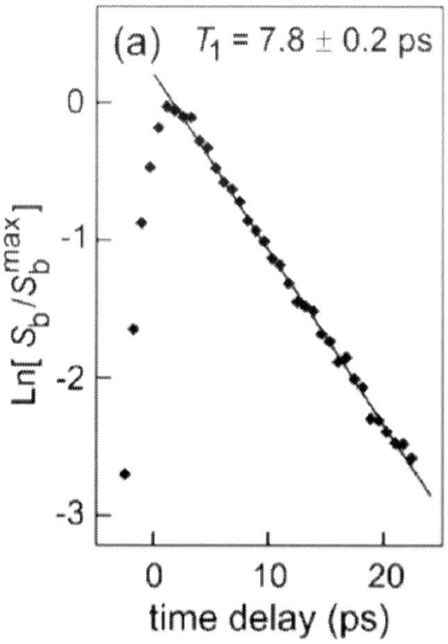

Figure 13. Decay of the transient bleaching signal from the stretch mode of the bond-centered hydrogen defect in silicon, measured at 20 K [56].

Another more recent example involves the use of the Upgrade FEL as a laser source in an effort to detect the existence of a light (sub eV) neutral boson that is a proposed dark matter candidate particle. The experiment was of the "light shining through a wall" variety, schematically diagrammed in Fig. 14. This particle, of a hypothetical family called weakly-interacting sub-eV particles (WISP), is created through an interaction between light and high magnetic field. A small fraction of photons become particles. The particles conserve momentum, so they travel in the same direction as the laser beam. Because, like dark matter, the particles couple very weakly to EM fields, they pass through the beam block and traverse another magnetic field region where they become photons once more. These are detected with an ultra low noise camera, developed for the astronomical community [57, 58, 59].

The experiment performed at Jefferson Lab was a collaborative effort, with Hampton and Yale Universities taking the lead role in scientific and results, and Jefferson Lab providing the major hardware, including the FEL. That collaboration, named the Light Pseudoscalar and Scalar Particle Search (LIPSS), took 17 hours of FEL beam, operating at a nominal wavelength of 930 nm, but detected no photons above background. Nevertheless, this result in itself set one of the most sensitive determinations of an upper bound on the scalar coupling constant (where the polarization of the FEL light is parallel to the magnetic field) for the photon-WISP conversion process [60].

5.3.4. THz applications The spectral region covered by THz radiation, roughly 1 mm to 0.1 mm in wavelength, is absorbed by transitions between electronic states and/or vibrational states in matter. Because these states are specific to a particular molecule, or ion, spectroscopy, either in

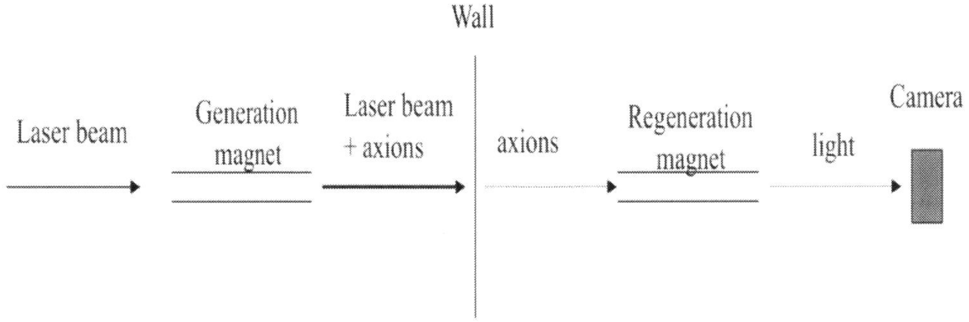

Figure 14. Schematic diagram of the dark matter candidate particle detection experiment.

the time or frequency domain can be used for location and identification. This makes the use of the THz region of the spectrum attractive for applications such as medical imaging or hazardous material detection and nondestructive testing. While the list of applications is long, progress from the lab to the field has been hampered by the lack of high average power (several watts) sources. The THz source at Jefferson Lab, is capable of producing high power, broadband, ultrashort pulsed radiation to evaluate the utility of some of these applications. It was recently used to make full-field video-frame rate images of a moving object [61]. We anticipate exciting results from the research continuing in this lab.

5.4. Conclusions

The applications discussed above are only a sampling of those pursued over the past ten years, as space does not permit us to discuss them all. The hole drilling experiments and modeling show that the FEL has a pulse format that is better suited to large area processing, whether it be via photochemical or thermal means. In addition, given the pulse format and tunablility, the FEL can be used to optimize the parameters that a tabletop laser may be configured to operate at, but don't know *a priori*. That in itself is enabling to a variety of applications that would not be readily explored, because the parameter space is too large, and the capabilities of readily available lasers are often too limited. At the time of this writing, with the addition of the UV FEL, experiments are either in the planning stages, or ready to be tested. Indeed, the prevalent view is that the best application for a FEL has yet to be found. Excitement in the field of FELs has never been higher though because of the advances into shorter and shorter wavelengths. In the past the extension of storage ring X-ray performance from first to now third generation sources has enabled an extensive set of crucial developments in material science and biology to determine static structures. However, storage rings are nearly at the apex of their possible capability. They will never be able to achieve the short pulses needed to follow chemical interactions or configuration changes of molecules. Nor will they be able to achieve the brightness required to do in vivo imaging of single proteins. The collective gain effect in FELs is expected to achieve the brightness levels needed and the application of linac rather than ring based electron sources is expected to enable the ultrashort pulses needed for dynamical studies. The developments discussed in this paper may play a major role in future light source activity by allowing high repetition rate generation of light to support many simultaneous users or achieve the sensitivity needed for examining rare events. We also hope to eventually achieve our early goal of practical use of industrial light at an affordable price for applications that cannot be achieved using conventional processing or existing lasers.

References

[1] C. Leemann, D. Douglas and G. Krafft, Ann. Rev. Nucl. Part. Sci. **51**, 413 (2001).
[2] G. A. Krafft and J. J. Bisognano, Proceedings of the 1989 IEEE Particle Accelerator Conference, 1256 (1989).
[3] R. A. Warren, Star Wars and the FEL, unpublished (1985).
[4] Laser Processing Consortium (1995) available from the FEL Dept. Office, Jefferson Lab, Newport News, Virginia 23606.
[5] M. A. Tigner, Nuovo Cim. **37**, 1228 (1965).
[6] D. A. G. Deacon, L. R. Elias, J. M. J. Madey, G. J. Ramian, H. A. Schwettman and T. I. Smith, Phys. Rev. Lett. **38**, 892 (1977).
[7] H. Motz, W. Thon and R. N. Whitehurst, Appl. Phys. **24**, 826 (1953).
[8] H. P. Freund and G. R. Neil, *Free electron generators of microwave radiation Electron Beam Generators of Microwave Radiation*, Proc. 87 Number 5 (Santa Fe, NM: IEEE) pp. 782-803 (1999).
[9] G. R. Neil and L. Merminga, Rev. Mod. Phys. **74**, 685 (2002).
[10] T. I. Smith, H. A. Schwettmann, K. W. Berryman and R. L. Swent, *Facilities at the Stanford Picosecond FEL Center*, Proc. SPIE 1854 (San Diego, CA) pp. 23-33 (1993).
[11] C. Brau, Nucl. Instrum. Meth. A **318**, 38 (1992).
[12] S. V. Benson, B. Richman and L. Vintro, Nucl. Instrum. Meth. A **296**, 110 (1990).
[13] H. Hortike *et al.*, Proceeding of the 2004 FEL Conference, THPOS17, 251 (2004).
[14] D. Oepts, R. J. Bakker, D. A. Jaroszynski, A. F. G. van der Meer and P. W. van Amersfoort, Nucl. Instrum. Meth. A **341**, 28 (1994).
[15] *Free Electron Lasers and Other Advanced Sources of Light*, National Academy Press, Washington, DC (1994).
[16] C. K. Sinclair, Nucl. Inst. and Meth. A **318**, 410 (1992).
[17] D. H. Dowell *et al.*, *Commissioning Results of the LCLS Injector*, Proc. of the 2007 FEL Conference (Novosibirsk, Russia) (2007).
[18] C. Hernandez-Garcia, P. G. O'Shea and M. L. Stutzman, Physics Today **61**, 44 (2008).
[19] J. B. Flanz and C. P. Sargent, Nucl. Inst. and Meth. A **241**, 325 (1985).
[20] T. I. Smith *et al.*, Nucl. Inst. and Meth. A **259**, 1 (1987).
[21] C. Levi-Strauss, *The Savage Mind* (Chicago, IL: University of Chicago Press) (1996).
[22] L. Merminga *et al.*, *High current effects in energy recovery linacs*, Proc. 2001 IEEE Particle Accelerator Conf. (Chicago, IL) (2001).
[23] C. Tennant C *et al.*, Phys. Rev. ST Accel. Beams **8**, 074403 (2005).
[24] S. V. Benson *et al.*, Nucl. Inst. and Meth. A **429**, 27 (1999).
[25] S. V. Benson *et al.*, Nucl. Inst. and Meth. A **407**, 401 (1998).
[26] G. R. Neil *et al.*, Phys. Rev. Let. **84** 662 (2000).
[27] G. R. Neil *et al.*, Phys. Rev. Lett. **87**, 084801 (2001).
[28] G. L. Carr *et al.*, Nature **420**, 153 (2002).
[29] J. R. Boyce, D. R. Douglas, H. Toyokawa, W. J. Brown and F. Hartemann, *Sub-picosecond, high flux, Thomson X-ray sources at Jefferson Lab's high power FEL*, Proc. 2003 Particle Accelerator Conf. (Portland, OR), 938 (2003).
[30] M. D. Shinn *et al.*, Nucl. Inst. and Meth. A **507**, 196 (2003).
[31] C. Hernandez-Garcia C *et al.*, *Longitudinal space charge effects in the Jefferson Lab IR FEL SRF Linac*, Proc. of the 2004 FEL Conference (Trieste, Italy) p. 363 (2004).
[32] R. Benson, *First lasing of the Jefferson Lab UV Demo laser*, Preprint paper MOOAI8, to appear in Proc. of the 2010 FEL Conference (Malmo, Sweden) (2010).
[33] G. R. Neil, *A cost estimation model for high power FELs*, Proc. of the 14th Particle Accelerator Conf. (Dallas, TX), 137 (1995).
[34] *High Power Ultraviolet and Infrared Free-Electron Laser for Industrial Processing Proposal* (Laser Processing Consortium) (1994) available from the FEL Dept. Office, Jefferson Lab, Newport News, Virginia 23606.
[35] *Free-Electron Lasers for Industry Vol. 1 Commercial and Technological Rationale* (Laser Processing Consortium) (1995), available from the FEL Dept. Office, Jefferson Lab, Newport News, Virginia 23606.
[36] W. M. Steen, *Laser Material Processing* 2nd ed. (New York, NY: Springer) (1998).
[37] M. J. Kelley, Proc. of SPIE **3888**, 598 (2000).
[38] M. D. Perry *et al.*, J. Appl. Phys. **85**, 6803 (1999).
[39] E. G. Gamaly, A. V. Rode and B. Luther-Davies, J. Appl. Phys. **85**, 4213 (1999).
[40] A. V. Rode, B. Luther-Davies and E. G. Gamaly, J. Appl. Phys. **85**, 4222 (1999).
[41] J. H. Bechtel, J. Appl. Phys. **46**, 1585 (1975).
[42] M. Cohen, *Material Processing Laser Handbook* (Amsterdam:North-Holland) 1578 (1972).
[43] F. Dausinger, H. Hugel and V. Konov, Proc. of SPIE **5147**, 106 (2003).
[44] M. D. Shinn, *Metal processing experiments with a high rep rate laser – the Jefferson Lab FEL*, presented at

the 5th Intl. Symp. on Laser Precision Microfabrication Conference (Williamsburg, Virginia) (2005).
[45] S. M. Klimetov *et al.*, Proc. of SPIE **4830**, 515 (2003).
[46] CheapTubes.com web site: *http://www.cheaptubesinc.com/*.
[47] M. W. Smith, private communication (2003).
[48] J. Ouellette, The Industrial Physicist **8**(6), 18 (2002).
[49] P. C. Elkund *et al.*, Nano. Lett. **2**, 561 (2002).
[50] M. J. Kelley, *Surface processing and micromachining of polyimide driven by a high average power infra-red free electron laser*, Mat. Res. Soc. Symp. (San Francisco, CA: MRS), 617 (2000).
[51] D. Bubb *et al.*, J. Vac. Sci. Tech A **19**, 2698 (2001).
[52] A. Reilly *et al.*, J. Appl. Phys. **93**, 3098 (2003).
[53] E. Carpene, P. Schaaf, M. Han, P. K. Lieb and M. D. Shinn, Applied Surface Science **186**, 195 (2002).
[54] E. Carpene, M. Shinn and P. Schaaf, Appl. Phys. A **247**, 307 (2005).
[55] R. Rox Anderson *et al.*, Lasers in Surgery and Medicine **38**, 913 (2006).
[56] M. Budde, G. Lüpke, C. Parks Cheney, N. H. Tolk and L. C. Feldman, Phys. Rev. Lett. **85**, 1452 (2000).
[57] R. Cameron *et al.*, Phys. Rev. D **47**, 3707 (1993).
[58] A. S. Chou *et al.*, Phys. Rev. Lett. **100**, 080402 (2008).
[59] C. Robilliard *et al.*, Phys. Rev. Lett. **99**, 190403 (2007).
[60] A. Afanasev *et al.*, Phys. Rev. Lett. **101**, 120401 (2008).
[61] J. M. Klopf *et al.*, Bull. Amer. Phys. Soc. **52**, X39.00013 (2007).

Journal of Physics: Conference Series **299** (2011) 012015 doi:10.1088/1742-6596/299/1/012015

CEBAF Accelerator Achievements

Y. C. Chao, M. Drury, C. Hovater, A. Hutton, G. A. Krafft, M. Poelker, C. Reece, and M. Tiefenback
Jefferson Lab, Newport News, Virginia 23606, USA

Abstract. In the past decade, nuclear physics users of Jefferson Lab's Continuous Electron Beam Accelerator Facility (CEBAF) have benefited from accelerator physics advances and machine improvements. As of early 2011, CEBAF operates routinely at 6 GeV, with a 12 GeV upgrade underway. This article reports highlights of CEBAF's scientific and technological evolution in the areas of cryomodule refurbishment, RF control, polarized source development, beam transport for parity experiments, magnets and hysteresis handling, beam breakup, and helium refrigerator operational optimization.

1. Overview

Reference [1] describes Jefferson Lab's Continuous Electron Beam Accelerator Facility (CEBAF) circa 2000, when operation for experiments was well established, but when much remained to be learned about improving and optimizing the accelerator. As of early 2011, CEBAF operates routinely at 6 GeV, with substantially improved reliability, and with a 12 GeV upgrade underway [2, 3, 4]. The following sections report highlights of CEBAF's scientific and technological evolution over the past decade, beginning with the cryomodule refurbishment program (Section 2) that has enable robust five-pass CEBAF operation at 6 GeV, 1.5 times the original design for final energy. A new radiofrequency (RF) control system (Section 3) has been implemented using a digital feedback system that eliminated the components most susceptible to temperature-induced phase drifts and that greatly improved injector stability. A polarized-source development process (Section 4), complemented by the optimization of beam transport for parity experiments (Section 5), has enabled the parity-violation and polarization-transfer experiments that are now yielding essential new information on nucleon quark-gluon structure. Section 6, on magnets and hysteresis handling, describes the control mechanisms that have been devised to satisfy reproducibility and stability requirements of CEBAF magnets. Section 7 reports improvements in beam breakup simulations that have enabled confirmation of the higher-order-mode (HOM) damping specification for the 12 GeV upgrade. Section 8 describes the Ganni Cycle, an optimizing reconfiguration of CEBAF's helium refrigeration system that nearly doubles the lifetime of some refrigerator components and that saves more than $1000 dollars/day in power costs.

2. Cryomodule refurbishment project

Jefferson Lab has recently completed the C50 refurbishment project, so called because it sought to increase the peak effective voltage of the refurbished cryomodules to a maximum of 50 MV. (The original specification was 20 MV, although several provide up to 35 MV.) The ultimate goal was to enable robust five-pass CEBAF operation at 6 GeV, 1.5 times the original requirement.

Ten of the weakest of the accelerator's 42 cryomodules were refurbished. This activity included complete disassembly to enable the reprocessing of the SRF cavities with improved protocols (Figure 1). The chief objective was to extend the operating range of each cavity from as low as 4 MV/m to an average of 12.5 MV/m, the performance needed to yield 50 MV for a cryomodule.

Figure 1. Reassembly of a reprocessed C50 cavity pair into a quarter-cryomodule.

While two of the cryomodules were selected for rework due to vacuum leaks, the principal performance-limiting phenomenon addressed was parasitic electron field emission due to residual particulate contamination on cavity surfaces. Such field emission produces two ill effects: increased cryogenic load and electrostatic charging of the ceramic RF input window which sits at 2 K in the original cryomodule design. This charging in turn induces the occurrence of periodic arc discharges that require both RF and beam interruption. As a result, the available gradient in many cases is governed not by cavity limitations but by arc rate management. Operationally, it is the sum of these trips over hundreds of cavities that has constrained the maximum useful energy of CEBAF. Elimination of the arc rate limitation was a C50 goal.

Two measures were taken in the C50 project to address this issue. First, to eliminate almost all of the offending particulate contamination, the niobium SRF cavities each received a fresh acid etch followed by a high-pressure rinse with ultra-pure water, both within a cleanroom environment. Neither of these procedures was available during the initial construction of CEBAF. Second, the RF window and waveguide attached to each cavity were replaced with a new "dog-leg" configuration that significantly reduces the risk of charging and trips, in the event that field emission is yet present. These measures have proven to be quite effective. For almost all of the cavities, field-emission-induced radiation was eliminated in the operating range of interest. Cavity performance, if limited, was constrained only by quench conditions derived from original fabrication flaws. While window arc-related trips do occasionally occur, they are greatly reduced and occur at much higher operational conditions. None of the C50 cavities have gradient limitations imposed by arc rate management.

During refurbishment prior to receiving their fresh etch, the cavities were subjected to a 600°C bake in a vacuum furnace. Such a bake reduces the hydrogen content dissolved in the niobium bulk. When that concentration is too high, NbH will precipitate on the surface at temperatures around 100 K, given sufficient opportunity. NbH is a normal conducting metal, so its presence can dramatically increase cryogenic losses in what has come to be known as Q-disease. During CEBAF construction this phenomenon was only just being identified and no remedial vacuum furnace was available. To manage the situation, it became necessary to enforce rather fast cooldowns for the cryomodules, especially through the vulnerable hydride

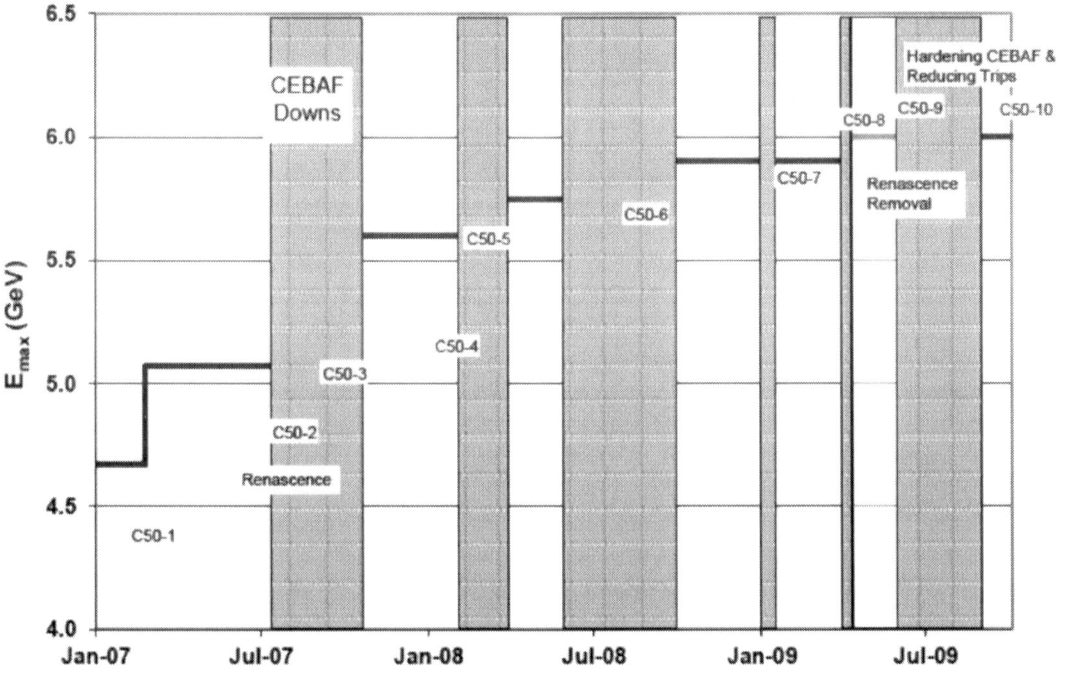

Figure 2. Staggered C50 cryomodule reinstallation and CEBAF's associated performance rise.

precipitation range. With the C50 cavities now baked, we can allow for slower cryomodule cooldown. This in turn reduces the likelihood of mechanical damage, such as the creation of vacuum leaks, during thermal cycles.

Other C50 improvements included replacement of the external polyethylene RF windows with a new ceramic model and modifications to the mechanical cavity tuners to prevent backlash. Damaged or worn parts were also replaced.

The C50 cryomodules were refurbished at a rate of three per year. To maximally support CEBAF operation during this time, cryomodules were rotated in and out during machine downs, all the while maintaining a rough balance between the voltage capacity of the north and south linacs. Figure 2 shows the progression of C50 cryomodule reinstallation and the associated rise in CEBAF's maximum five-pass final energy for experiments. All ten refurbished cryomodules are now once again fully operational in CEBAF.

The raw cavity performance after reprocessing was excellent. Field emission was dramatically improved and the superconducting RF dissipation was essentially textbook performance. After incorporation into finished cryomodules, however, the results were mixed. Of the eighty reprocessed cavities, 75% met or exceeded the goal of 12.5 MV/m. Under operational conditions, the C50 cryomodules have been able to sustain an average voltage of 47 MV, with three of them meeting or exceeding the 50 MV goal. Unfortunately, the installed C50 cavities were unable to meet the improved Q specification. During cavity vertical tests, the reprocessed cavities routinely demonstrated Q's of 1×10^{10} or higher. However, assembly into a cryomodule resulted in significant Q degradation. The RF dissipation losses increased by roughly a factor of two, essentially returning to their previous operational values, less the contribution from field emission. The degradation is now believed to be due to unrecognized magnetized components built into the original cryomodules, such as the helium vessels or the tuners. (The presence of magnetic field during cooldown creates "frozen-in" fluxoids which produce dissipation in response to RF fields.) The low Q's had been previously attributed to Q-disease. Unfortunately, due to schedule pressure, investigations into the specific causes along with possible fixes were not completed before the end of the C50 project. C50 cryomodules operating at or near 50 MV are generating heat loads for the 2 K cryogenic system in excess of 150 W. Thus, while the C50 project has been quite successful in enabling robust 6 GeV operation, the cost in terms of refrigeration load has been much higher than expected.

3. RF control

A new RF control system has been implemented using a digital feedback system that eliminated the components most susceptible to temperature-induced phase drifts. The new design directly digitizes the intermediate frequency, eliminating all of the analog signal processing. The new system incorporates a field programmable gate array for all of the digital signal processing. The feedback controller uses a proportional-integral controller to maintain cavity field, removing the finite gain offset that was seen in the analog proportional controller. The new controller reduced the phase drift susceptibility from 1°/°C to 0.1°/°C. Starting with the photogun drive laser and up through the buncher cavity, the injector RF was converted to the new design. The new RF control system greatly improved the stability of the injector.

4. Polarized source development

Approximately 80% of the experiments conducted at CEBAF require spin-polarized electron beams, and even those experiments that do not require polarized beam receive it, since all electrons originate from a common photogun. Today's CEBAF polarized source is a load-locked gun (more below) that provides 85% polarization and up to 180 μA delivered to a single hall, with combined current to all three halls in excess of 200 μA. The operating lifetime of the polarized source is limited primarily by the photogun vacuum level, which is difficult to measure because it is so low (low 10^{-12} Torr) and commercial vacuum gauges are not reliable in this pressure range. The photogun can deliver ~500 C before steps must be taken to restore photocathode yield, or QE. This represents approximately six weeks of beam delivery without interruption under high-current operation, and longer time periods at lower current.

The CEBAF/Jefferson Lab load-locked gun is shown in Figure 3 [5]. It consists of four vacuum chambers: the high-voltage chamber, the photocathode preparation chamber, a "suitcase" chamber used for replacing photocathode samples, and an intermediate chamber that must be evacuated and baked each time the suitcase is attached. The suitcase is normally detached from the photogun and stored elsewhere. This approach helps to reduce the overall footprint of the photogun when in operation. Desirable features incorporated into this design include the ability to store multiple photocathode samples, reliable sample transport from one chamber to the next without dropping, and rapid heating and cooling of samples for fast turn-around at activation.

With a load-locked gun, new photocathode samples are introduced via the suitcase, and then moved from one chamber to the next, with the best vacuum obtained in the high-voltage chamber. In this manner, lengthy bakeouts of the entire photogun are avoided. In addition, since photocathode activation takes place inside a dedicated preparation chamber, cesium is not inadvertently applied to the gun high-voltage electrode as was the case with the previous CEBAF photogun design. Cesium reduces the work function of materials and leads to catastrophic field emission from the cathode electrode. Another benefit of the load-locked photogun is the use of a compact tapered insulator that extends into the gun high-voltage vacuum chamber, which greatly reduces the amount of metal biased at high voltage, and consequently minimizes the likelihood of field emission.

Today's state-of-the-art high-polarization photocathode is the GaAs/GaAsP strained-superlattice structure [6], which consists of a very thin GaAs surface layer (~5 nm) grown atop

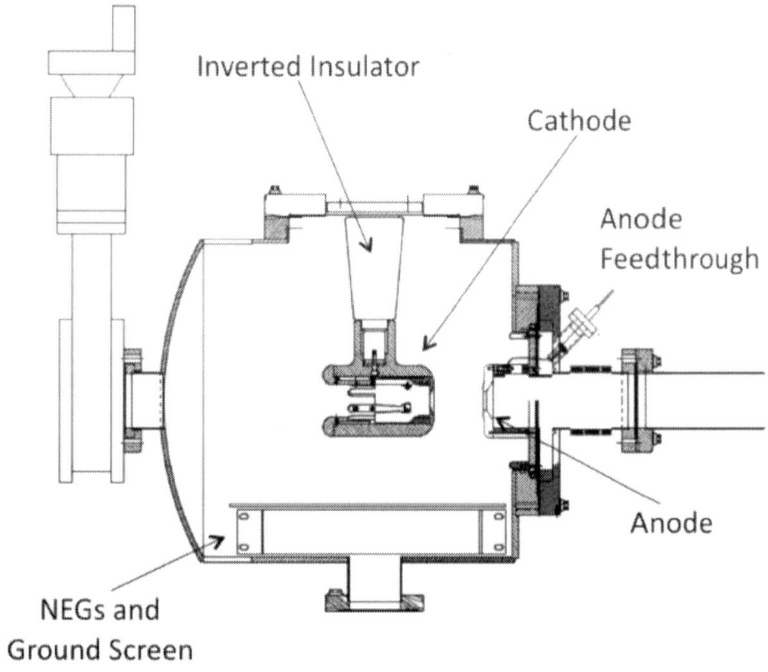

Figure 3. The high-voltage vacuum chamber of the latest polarized electron source, with compact inverted geometry high-voltage insulator.

10 to 20 pairs of thin, alternating layers of GaAsP and GaAs. By growing very thin GaAs layers, the strain can be maintained, which improves polarization, and electrons in subsurface layers efficiently tunnel through the GaAsP layers. And by using many thin layers of GaAs/GaAsP, the QE can be considerably higher than obtained from a single (thicker) layer of strained GaAs. The net result is polarization ~85% and QE ~1%.

Originally, we expected to use a DC laser to generate a DC electron beam which would be given the appropriate RF structure required for acceleration using the photoinjector RF chopper system. However, this would have been very inefficient, with most of the extracted electrons dumped on a copper plate within the photoinjector RF chopper. Instead, CEBAF drive lasers emit short pulses of light, to directly produce an electron beam with the appropriate RF structure required for acceleration. This approach, termed *synchronous photoinjection*, has now been adopted by other accelerator facilities. CEBAF drive lasers rely on the accelerator-friendly technique gain-switching to generate RF-pulsed laser light, and exploit recent advances from the fiber-based telecommunications industry [7]. Light from a fiber-coupled, gain-switched diode laser is delivered to an ErYb-doped fiber amplifier to obtain ~5 W average power at 1560 nm. This light is converted to a useful wavelength via second harmonic generation to obtain over 2 W average power at 780 nm with ~40 ps pulses and pulse repetition rate of 499 MHz.

4.1. Spin manipulation and polarimetry

Polarized-beam experiments require a specific orientation of the electron spin direction at the target, typically parallel to the direction of beam motion. Moreover, the spin direction must flip sign at some specified frequency. Spin flipping is accomplished by reversing the polarity of voltage applied to an electro-optical element called a Pockels cell, which is located on the drive laser table at the photoinjector. Typically, the polarization direction of the electron beam flips at 30 Hz, and more recently a technique was developed to flip polarization much faster, up to ~1 kHz. Fast flipping was implemented to help reduce the effects of target boiling for the Qweak experiment and has significantly improved the experimenters' ability to make a precise measurement of an extremely small physics asymmetry.

Electrons leave the photocathode with spin direction pointing parallel/antiparallel to the direction of beam motion, depending on the helicity of the laser circular polarization (right or left circular) created by the Pockels cell. But the spin direction precesses in the horizontal plane as the beam passes through the arcs and transport lines to the halls, and this net spin precession must be cancelled out by orienting the spin direction at the injector by the opposite amount using a spin manipulator.

The original Z-style spin manipulator was overly complicated, and quickly replaced with a Wien filter spin manipulator [8]. A Wien filter is a device with static electric and magnetic fields perpendicular to each other and to the velocity of charged particles passing through it. Unit charged particles with a velocity of $\beta c = E/B$ are undeflected in passing through the Wien filter, while the spin is rotated in the plane of the electric field. Most recently, a second Wien filter was added to the injector beamline, but with field forces oriented in the vertical plane. Two Wien filters and intervening solenoids are used to flip the sign of the beam polarization weekly, to assist parity violation experiments cancel out unwanted helicity-correlated spot size variations.

There are five polarimeters at CEBAF: one Compton polarimeter, three Møller polarimeters, and the 5 MeV Mott polarimeter at the CEBAF photoinjector. The term *spin dance* was coined to describe the process of sending identical beam to multiple polarimeters, while using the photoinjector Wien-filter spin manipulator to vary the direction of the electron spin. Each polarimeter receives polarization with the same magnitude, but the portion of the polarization pointing in the longitudinal direction follows a sinusoidal trend as the spin direction is changed using the Wien filter (see Fig. 4). The spin dance provided the first precise comparison of the analyzing powers of Compton, Møller, and Mott polarimeters [9]. It revealed disagreements among the polarimeters that were associated with errors in the analyzing power of some polarimeters and an incomplete appreciation for all systematic effects.

To understand the different curves in Figure 4, it must be noted that beams delivered to the three experimental halls follow different paths after extraction from the accelerator, and are often derived from different numbers of recirculation passes, so the total spin precession from the polarized source to each experimental hall is in general quite different. The Wien angle that provides longitudinal polarization at one hall does not necessarily provide longitudinal polarization at neighboring halls. In fact, there are approximately 400 two-hall energy combinations that give maximum polarization in two halls [10] and only two beam energies (2.1 GeV and 4.2 GeV) can provide maximum longitudinal polarization to all three halls simultaneously. Recently it was realized that by redistributing the energy gain of the two linacs, while keeping total energy gain constant, there is significantly greater flexibility in maximizing the degree of longitudinal polarization at multiple halls [11].

Finally, it's worth mentioning another beneficial outcome of the spin dance, namely, evaluation of the value of the spin orientation exiting the injector that provides maximum longitudinal polarization in each experimental hall leads to an independent and very precise (better than 10^{-4}) absolute measurement of the final electron beam energy.

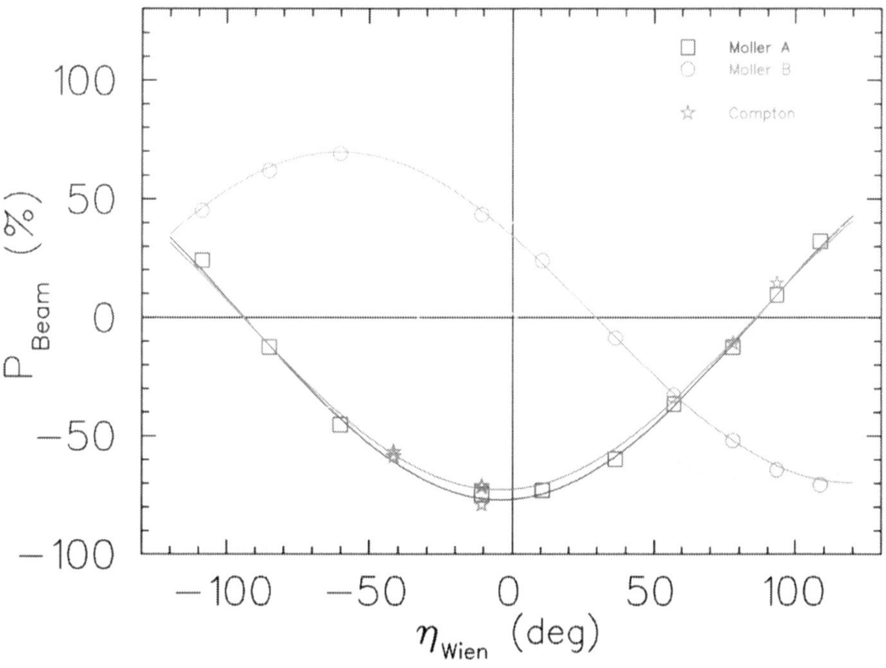

Figure 4. The results of a spin dance, where the longitudinal component of the beam polarization is measured as a function of Wien spin manipulator angle, at multiple polarimeters.

5. Optimizing injector-to-target beam transport for parity violation experiments

Here we discuss efforts to identify anomalous beam envelope mismatches (transport anomalies) across different energy sections of CEBAF and the correction techniques we developed that resulted in an unprecedented low level of helicity-correlated orbit variation, partly contributing to the first Jefferson Lab parity violation experiment to achieve precision better than 100 parts per billion (ppb).

An important and demanding class of experiments conducted at CEBAF is illistrated by the parity violation (PV) experiment [12] devoted to the study of extremely small asymmetries in elastic electron-nucleon scattering generated by electromagnetic and weak interactions. The electron beam is polarized in either positive or negative handedness (helicity state), and the asymmetry, A_{PV}, is measured in terms of left- and right-handed helicity state cross sections, $\sigma_{L,R}$:

$$A_{PV} = (\sigma_R - \sigma_L)/(\sigma_R + \sigma_L) \approx Q^2/M_Z^2 \approx 10^{-6} - 10^{-4}, \quad (1)$$

where Q^2 is the 4-momentum transfer and M_Z the mass of the Z-boson. It is noted that the typical magnitude of such asymmetry is in the range of 1 to 100 parts per million (ppm). To confidently measure such a small asymmetry, the systematic variations of the electron beam must be controlled to a few percent of the asymmetry itself, or to the sub-100 ppb level. This translates into very tight specifications on beam quality, especially the position, angle or intensity of the beam at the target. These are referred to as helicity-correlated (HC) parameters. Table 1 gives an overview of the intrinsic physics asymmetries and requirements on HC beam properties of various PV experiments at Jefferson Lab. The tight constraints on the HC orbit translate into exacting demands on the performance of the accelerator, with over 6 km of beam transport and four decades of momentum gain from the injector to the target location. The typical beam quality measurements performed at CEBAF for non-parity experiments, while sufficient for loss-

Table 1. Helicity-correlated beam specifications for parity violation experiments at CEBAF.

Experiment	Physics Asym. (ppm)	Intensity (ppm)	Position on Target (nm)	Angle on Target (nRad)
HAPPEX-I	13	1.0	10	10
G0	2-50	1.0	20	2
HAPPEX-He	8	0.6	3	3
HAPPEX-II	1.3	0.6	2	2
QWeak	0.3	0.1	2	10
PREX	<1	0.1	1	1

free delivery on target and efficient event generation, are not enough for controlling HC orbit to the sub-100 ppb level. The following describes the mechanism of runaway HC orbit variation caused by transport anomaly and the methods implemented to combat it.

5.1. Helicity-correlated orbit and transport anomaly

Origin of Helicity-Correlated Orbits and Their Consequence — In simple terms, PV experiments rely on comparing scattering cross sections between electron bunch trains in positive and negative helicity states, which in turn are alternatingly generated at 30 Hz from the CEBAF photoinjector through a switching system in the laser path before the laser impinges on the photocathode to produce electron beam. The switching system, an electro-optic element called a Pockels cell, unavoidably introduces systematic steering on the laser spot that develops into fluctuation in the electron beam orbit correlated to the helicity state. The true asymmetry A_{PV} is extracted from raw experimental data by applying experimentally determined correction factors. The measurement of correction factors, however, can be contaminated by HC orbit variations, resulting in deficient correction, and thus compromised precision in A_{PV}.

Consequence of Transport Anomaly — In an ideal world, the phenomenon known as adiabatic damping would ensure reduction of the HC orbit variation by a factor of almost 100 from the injector at 100 keV to 3 GeV, a typical energy for parity experiments. In reality, however, deviations from ideal beam transport can not only prevent the damping from occuring, but can, in extreme cases, actually make the HC orbit grow. This transport anomaly can potentially have detrimental effects on measuring the asymmetry A_{PV}. Transport anomaly can be categorized into cross-plane coupling (coupling) and near-singular transport due to grossly mismatched optics (mismatch). These two effects can compound each other, with consequences not possible by either one alone. Despite having detrimental effects on measuring the asymmetry, their small absolute magnitude may be barely discernible in terms of beam loss or beam size on target, and thus they are very difficult to detect or track. A more subtle effect is the evolution of the mismatch between the beam ellipse and beam orbit (such as the HC orbit itself), even if they start in full congruence [13]. This means in a situation with compounded mismatch and coupling, we can quickly lose the ability to simultaneously correct the beam shape and the orbit as they fall out of coherence.

Methods for Eliminating Transport Anomaly — In 2004 the decision was made to characterize the transport throughout the five-pass CEBAF. A series of beam-based measurements followed. The technique relied on making intentional helicity-correlated orbits by moving the laser beam at the photocathode using a mirror attached to a piezoelectric transducer (PZT). Analysis of the outcome soon pointed to the low-energy front end as the main source of strong beam phase-space blowup. This was primarily attributed to near singular transport across the SRF cavities (cryounit and cryomodules) because effective modeling of beam transport through these cavities

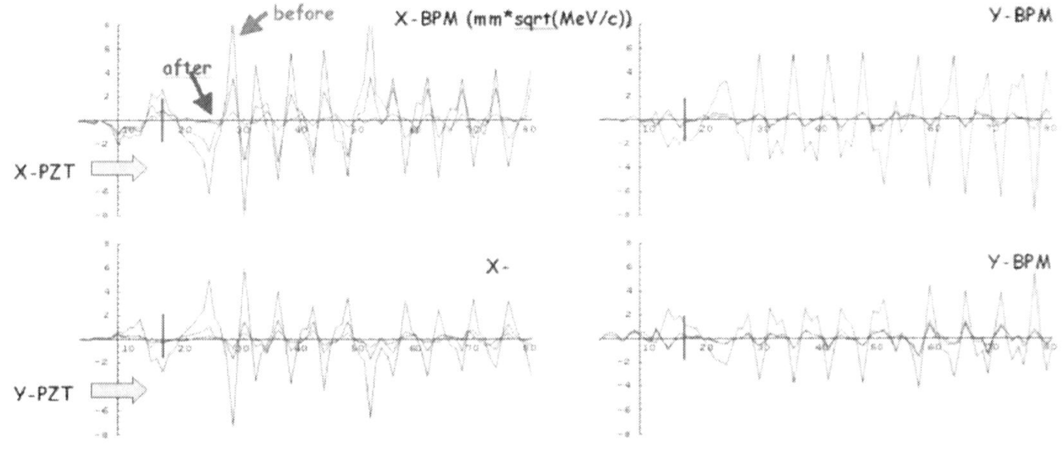

Figure 5. The PZT response before and after injector matching from 100 keV to 800 MeV.

at such low energy was very difficult. This was likely exacerbated by coupling present in the HOM couplers of these cavities, leading to phase-space blowup described above. Coincidence between jumps in PZT amplitude and locations of the cryo-cavities was unmistakable. Similar blowup signatures were also observed in the area between the end of the injector (60 MeV) and the first two linacs, although at a much lower level. With the above observations, a strategy was formulated in two stages to address transport anomaly and reclaim expected adiabatic damping: first, for 100 keV to 60 MeV, with accurately measured transport and sufficient correction elements, a model-based solution is possible, and second, for 60 MeV to 3 GeV, lacking an accurate long-range model, an empirical approach appears more practical.

100 keV to 60 MeV — Four-dimensional (X, X', Y, Y') transfer matrices were measured using difference-orbit methods across all injector cryo-cavities. The result displayed high degrees of near-singularity in 4D transport, confirming initial suspicion. Furthermore all measured transfer matrices satisfy the linear 4D symplectic condition, implying that proper transport can be restored with linear elements only. An optimization program was developed to use existing quadrupoles and skew quadrupoles to eliminate XY coupling and minimize singularity from 100 keV to 60 MeV while complying with multiple constraints dictated by operation and hardware considerations. The solution, once implemented, showed significant improvement in all transport characteristics. The solution eliminated XY coupling (off-diagonal determinant as percent of total), and brought singularity (SVD condition number) in the original transport down to near-ideal level.

60 MeV to 3 GeV — The empirical approach adopted for correcting mismatch from 60 MeV to 3 GeV is betatron matching under the guidance of the X and Y PZT signals, as they define a subspace in the 4D phase space shown to be representative of the characteristics of true HC orbits. This has proved effective except when PZT orbits are not sufficiently congruent with the beam spot, in which case a compromise must be empirically reached.

Outcome of Injector Matching — Figure 5 shows the BPM response, from 100 keV to 800 MeV, to the X and Y PZTs before and after anomalous transport in the Injector was corrected. Figure 6 shows similar signals at 3 GeV in experimental Hall A (HAPPEX). The HC orbit reduction from cathode to target was significant. This translated into unprecedented HC orbit reported by HAPPEX 2005 at 3 GeV and by G0 at a much lower energy of 650 MeV.

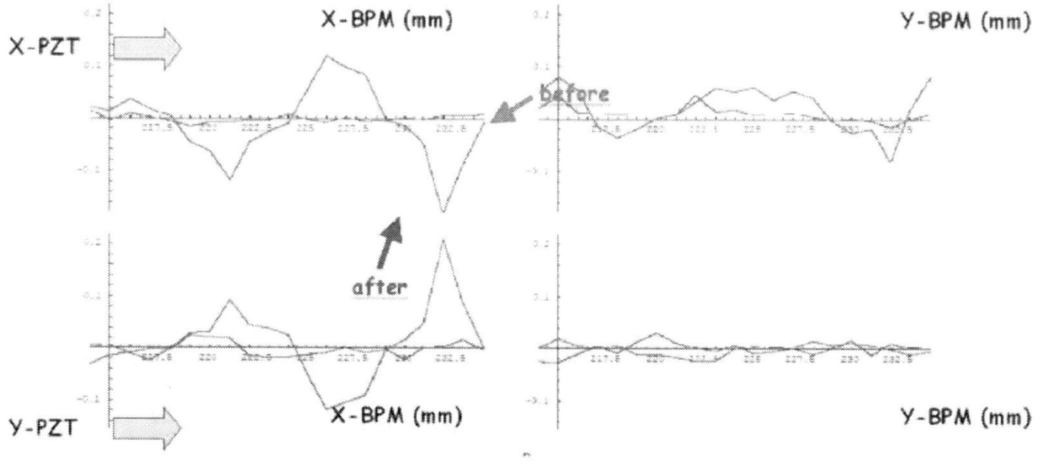

Figure 6. The PZT response before and after injector matching with 3 GeV into Hall A.

6. Magnets and hysteresis handling

The CEBAF accelerator magnets, all normal conducting, are typically static for extended periods, often for weeks with no change of set point. Most magnets in the system have iron flux cores, reducing the total amp-turn requirements to establish the desired magnetic field. For such magnets, providing the desired magnetic field requires control not only of the current in the magnet at the time of interest, but also of the history of the current. This section describes the control mechanisms which have been devised to satisfy reproducibility and stability requirements of CEBAF magnets. It also describes some vulnerabilities of earlier protocols and provisions which have been made to identify soft failures in the magnet power supplies. We do not attempt here to discuss issues such as ripple control, but specialize to hysteresis handling and fault detection.

6.1. Operational philosophy

The magnets are set up during tune-up operations, and hysteresis cycled so that the magnetic configuration can be reliably restored at later times. Stability and reproducibility of the magnetic fields are the principal concerns for magnet operation. The accelerator changes subtly over time scales of a week due to such effects as ground motion driven changes in the machine circumference and changes in the accelerating gradient profile along the linacs as various accelerating cavities are taken out of or restored to service. These influence the detailed optics of the accelerator, and are compensated at setup (and such other times as required) by manual tuning operations which establish certain global transfer function properties. The quadrupoles of the accelerator are almost all independently powered, with independent power supply calibration errors which change when the power supplies are replaced. The procedures used in accelerator operation are, of necessity, tolerant of these variables.

6.2. Magnet power supply characteristics

The dipole bending magnets are powered in groups on common power supplies. The subset of dipoles on a given power supply bus which were required to be independently adjustable were sized to provide slightly greater bending field integral at any given current than their companion magnets. Individual control was provided through a parallel electronic load, a controllable shunt to reduce the individual magnet current without affecting the bus current. The primary dipole

Journal of Physics: Conference Series **299** (2011) 012015 doi:10.1088/1742-6596/299/1/012015

power supplies are physically large units called box supplies. The box supplies are controlled through an 18-bit DAC in the power supply with full-scale currents ranging from 270 to 640 A. The associated set-point granularity is at the mA level. This provides resolution comparable to the intrinsic energy spread of the beam, and is sufficient and convenient for operational purposes.

Almost all of the quadrupole and steering corrector magnets in the CEBAF accelerator are controlled by the trim magnet system. This system is implemented as 32-channel blocks of rack-mounted electronic loads fed from common bus sources at ± 35 V in the rack. Each of the channels provides ground-referenced power up to ± 30 V and (typically) up to ± 10 A. Specialized versions of the trim cards are used for low-current injector magnets (± 1 A) and in a few locations for marginally higher currents (± 12 A). The trim system uses 16-bit DACs with consequent current resolution of 0.3 mA for the usual ± 10 A full range. The sampled read-back current from the trim cards also has 16-bit resolution. While bench calibration of the trim cards is attempted only at the ± 5 mA level, the bit noise on the read-back current is used to trigger a noise counter which is very helpful in identifying marginal stability failures even before the card performance violates calibration thresholds.

6.3. Hysteresis handling

For many magnets in the injector and linacs, as well as a few in scattered service elsewhere, iron cores were not beneficial. These magnets have hysteresis handling disabled and are controlled by simply setting the coil current. However, for the iron-bearing magnets in the system, the history of the current must be controlled in order to ensure that the magnetic fields seen by the beam are as desired. Field integral versus current curves are shown in Figure 7 for unipolar-operated, 3 m length BA dipole magnets and in Figure 8 for bipolar-operated, 0.3 m length QA quadrupoles. The figures' left-hand sides show the field integrals versus current, and the right-hand sides show the nonlinear, hysteretic residue after subtracting the dominant linear contribution in the magnet current. (The current proportionality constant was selected for graphical purposes rather than for having any special physical meaning.) These curves illustrate the need to control the extremes of current during hysteresis cycling.

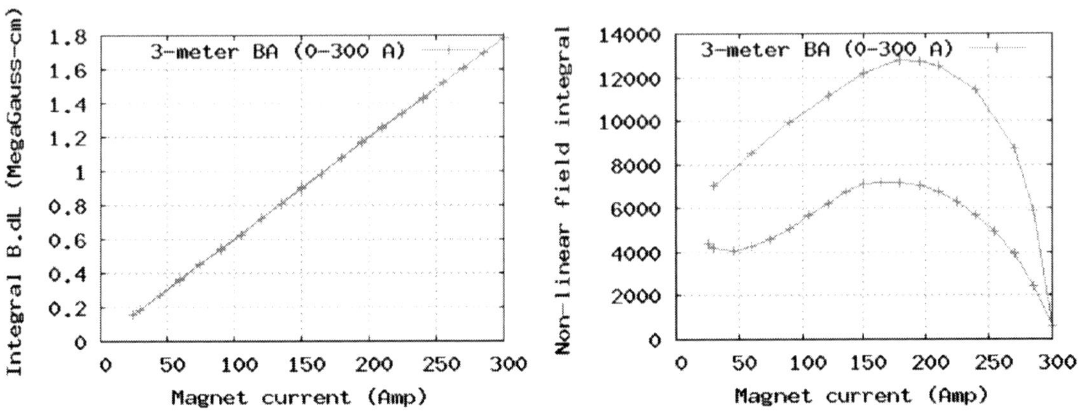

Figure 7. On the left, the field integral (Gauss-cm) through a 3 m dipole magnet as a function of current, and on the right, the field integral with dominant linear portion subtracted, showing saturation and hysteresis.

During commissioning in the early 1990s, the quadrupole magnet controls were configured to use the disconnected parts of the hysteresis curve for decreasing magnitude of current. Positive current set points were approached from the positive extreme current, negative set points

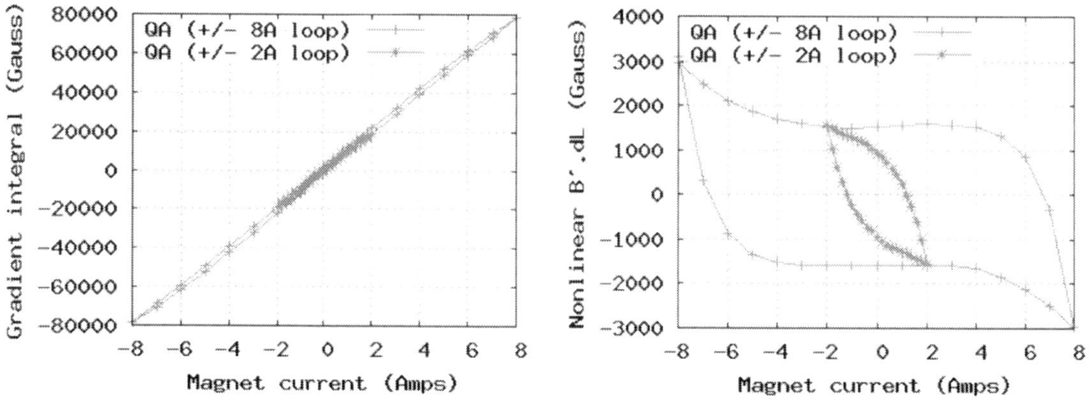

Figure 8. On the left, the gradient integral for a 30 cm QA quadrupole with ± 10 A and ± 3 A hysteresis cycles, and on the right, the nonlinear residue after subtracting the (same) linear component from each.

from the negative extreme. This had the side effect of rendering very low field integral set points unreachable "on-loop". Disabling hysteresis control for these magnets caused confusion concerning the magnetic field configuration. To avoid this difficulty, all bipolar magnets were redesignated to use the portion of the curve starting at the positive extreme and moving in the direction of algebraically decreasing current. This portion of the hysteresis curve remains in use at Jefferson Lab for the (bipolar) trim magnets.

The dipoles also initially used the field versus current relation from the upper part of the hysteresis curve, relaxing from the extreme current. Hysteresis cycling was done by 1) raising the bus power supply to peak current, 2) then shunting the desired current around the magnet, and 3) reducing the bus current to the desired operating point. As the beam energy was raised toward 6 GeV, some of the shunted magnets were more saturated than their partners, making supplementary power supplies necessary. Retaining the initial protocol would have resulted in adding supplementary current at the peak bus current, which would alter the hysteretic contribution to the magnetic field. Operating the magnets by setting them on the rising current curve was a better fit for a unified protocol using both shunts and supplementary power. The dipole magnets are now cycled from peak to minimum current. The supplementary power supplies and shunt controls are activated at the hysteresis minimum, and the bus current is then raised to the desired set point. The shunts, being fundamentally passive devices, will never reverse the magnet currents. They only begin shunting current after the bus voltage begins to rise, and each magnet experiences its intended peak and minimum values of current.

The nonzero lower hysteresis bound for the (unipolar) dipoles is driven by a hardware constraint: the power supplies trip off when run too close to zero. For the BA dipoles, for set points in the upward-concave region of the lower curve in Figure 7 (currents less than about 30 A), varying the lower extreme current alters the field from the expected value. During operation at very low beam energy, this has caused confusion over energy calibration at the parts per thousand level.

The rate of change of current during hysteresis cycling is also important. For the solid-core dipoles, there are eddy current effects which should be limited. We have settled on dI/dt values of approximately 5 A/sec, implemented in the power supply hardware, to provide reproducible fields. The maximum and minimum set points are issued by the control system, which waits for the power supply to achieve the set current before issuing the next current set point. The cycle from maximum current to minimum current is executed twice, and then with all shunt

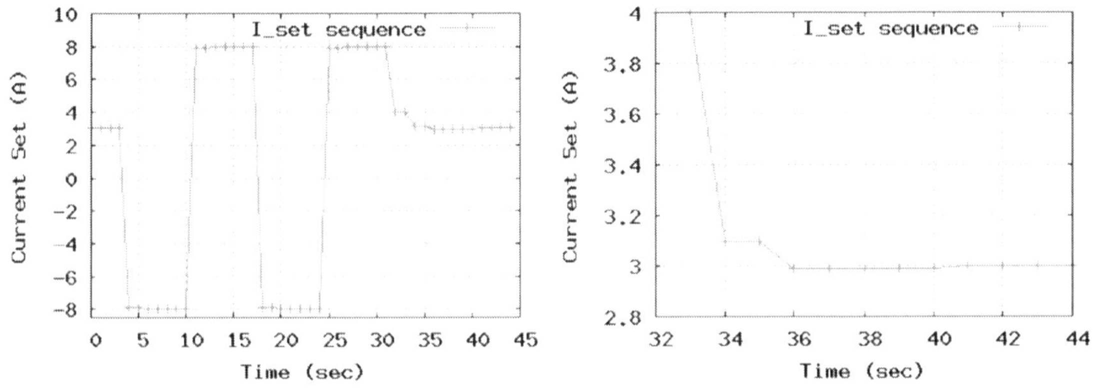

Figure 9. The left figure shows the time course of hysteresis for QA magnet (± 8 A) and the right figure shows the deliberate undershoot by 10 mA.

and supplementary power enabled, the ramp to final set point is done.

CEBAF quadrupoles are laminated, rather than solid, iron. While testing and improving our hysteresis protocols, we uncovered surprisingly large variability due to latency in execution on the local I/O controllers both in EPICS, the Experimental Physics and Industrial Control System (http://www.aps.anl.gov/epics/EPICS), and in the previous TACL systems. We obtain measurably more reproducible magnet behavior by relying upon internal power supply time constants to control fine-grained dI/dt, rather than by trying to specify the I(t) curve explicitly. The detailed set-point sequence now in use avoids vulnerability to latency effects and communication errors by issuing an arguably minimal number of set points, and avoids vulnerability to the variable characteristics of power supply and magnet combinations by controlling the size of the steps issued.

The time course specified for the trim magnet hysteresis protocol is shown in Figure 9. These illustrate the protocol applied to a QA magnet with ± 8 A current extremes. The current is taken from the negative extreme to the positive extreme twice, preventing the underdamped power supply from exceeding the ± 8 A bounds by using initial targets of ± 7.9 A. After the final positive extreme is attained, the current is stepped to 1 A above and then to 100 mA above the final current, and then to the final value. These steps are selected to avoid undershoot of the final current value and consequent variation in the remanent field contribution to the magnets.

The remanent field in the laminated iron quadrupoles relaxes by a small but detectable amount during the few hours after cycling. This relaxation shifts the field of all of the quadrupoles slightly more negative (vertically focusing) by about 50 Gauss of gradient integral. The 100 mA undershoot in current was selected to target this property of the QA quadrupoles. Reducing the current by 100 mA reduces the gradient integral by 100 Gauss, after which restoring the current to nominal set point restores only 50 Gauss. This tactic has been demonstrated to improve the stability of the large-scale optics of the CEBAF accelerator, rendering systematic drifts undetectable.

6.4. Fault detection

During hysteresis cycling, all magnet power supplies are verified to achieve each current set point to within the relevant calibration standards for each type. During operation, the measured currents are also required to remain within a calibration window. However, fluctuations of the magnet current even within the calibration window can cause unacceptable fluctuation of the beam position in the accelerator or at the experimental target. The 16-bit resolution of measured current in the trim magnets is much finer than the calibration window (0.3 mA versus ± 5 mA for 10 A cards). A cumulative noise alarm counter was implemented to allow identification of magnets showing sample-to-sample variation in current. The noise threshold is selectable, but presently generates a count for steps of three bits. The noise alarm technique was also applied to the measured voltage, anticipating that magnet faults may occur even when the current is held constant. These counters have been instrumental in identifying incipient wiring faults, resistive connections, and card failures and preventing accelerator down time.

7. Beam breakup

In superconducting recirculating linacs, multipass beam breakup (BBU) instability can create a significant limitation on average current [14, 15]. In this instability, roughly equivalent to multibunch instabilities in storage rings, inadequately damped transverse deflecting higher-order modes (HOMs) are excited by beam displacement on subsequent beam passes causing the beam to become deflected transversely and eventually lost, should the HOM excitation become large enough. The instability grows exponentially in time as long as the beam average current exceeds a threshold current, given by the condition that the energy deposited by the displaced beam can no longer be dissipated by the usual HOM damping in the HOM absorbers.

The CEBAF accelerator was constructed with cavities developed for storage ring applications at Cornell University. These cavities had been designed for very high HOM damping to mitigate ring multibunch instability, and applying these cavities in CEBAF was natural, given the previous difficulties with multibunch BBU. Early in the CEBAF project, it was shown both theoretically [16] and experimentally [17] that the actual threshold current for multipass BBU was at least two orders of magnitude beyond the maximum operating current of the accelerator. It was known at the time that CEBAF was overdesigned in HOM damping. Because a recirculating linac at the CEBAF scale and cost had not been built previously, and because at the time the theoretical codes were not deemed to be very reliable, completely eliminating this potential source of trouble to the project was thought to be wise.

Because of possibilities afforded by the Cornell/CEBAF cavities, other projects were keen to apply these cavities to yield high average current. In particular, the series of high-average-current energy-recovering linacs built as free-electron laser drivers at Jefferson Lab at first used essentially the same cavity design, but later used cavity designs with worse HOM damping incorporated from the beginning as cost-saving design choices. These devices ran at beam average currents up to 10 mA, roughly two orders of magnitude above that at CEBAF. The free-electron lasers recirculated beam at lower energies than at CEBAF, enhancing the possibilities for instability. Multipass BBU instability was indeed predicted [18] observed [19] and eliminated [20] by a variety of means. Indeed, one significant conclusion of the work during this period was that BBU calculations and HOM damping measurements had become good enough that reliable predictions of the instability threshold current could be obtained. Predictions good to much better than 10% were possible on machines of the small scale of the free-electron lasers, consisting of perhaps two dozen accelerating cavities.

For a brief period, multipass BBU was observed in the full CEBAF accelerator [21]. The reasons for this situation are instructive, and provide compelling evidence about the accuracy of present simulation methods. The HOM damping requirements for the new type of cavities to be used in the 12 GeV upgrade project were relaxed by about two orders of magnitude compared to the original CEBAF cavities. The 12 GeV HOM damping requirements were obtained after thorough calculations taking into account the most unfavorable operations scenario supported by the 12 GeV accelerator and by using a statistical analysis including the fact that in a large machine the HOMs will not necessarily be degenerate in frequency [22]. The original 12 GeV HOM damping specifications were promulgated in 2004 [23], and have a safety factor of only 2 compared to two orders of magnitude previously. This situation reflects our greater confidence in the results of thorough, modern simulation calculations. Since soon after the specification calculations were completed, new cavities installed in the CEBAF accelerator have been nominally required to meet only the 12 GeV requirements, as they should allow for stable operation of the present 6 GeV accelerator.

In 2007, a new cryomodule incorporating cavity prototypes of differently shaped cell profiles — high gradient (HG) and low loss (LL) — was installed in the fourth slot of CEBAF's north linac (NL04) as part of ongoing activity supporting increasing the final energy of the baseline CEBAF accelerator to 6 GeV. The LL cavity shape (see Fig. 10) was chosen as the most suitable for the 12 GeV era because of its better cryogenic efficiency. Soon thereafter, multipass BBU instability was observed, including the interesting observation that the operating current could be tuned precisely enough to the threshold current so the HOM excitation level could be made static in time. The instability manifested itself initially as a vertically extended spot at the end of the linac that eventually becomes large enough that the machine protection system trips the beam off.

Figure 10. Image of a new high gradient and low loss cavity used to increase the CEBAF base energy to 6 GeV as well as prepare for the 12 GeV upgrade.

Evidence was gathered confirming the effect was multipass BBU. Voltages on beam position monitors were Fourier analyzed when the instability was present indicating transverse movement at an aliased frequency of 1920 MHz. This information, coupled with the fact that HOM frequencies were recorded with a spectrum analyzer from coaxial HOM coupler ports of cavities in the newly installed NL04, led to direct observations of the excited HOMs in the cavities in question when BBU instabilities were triggered. These observations unambiguously confirmed that multipass BBU was present. Particularly, RF measurements detected loaded Q-values of two BBU-corresponding vertically polarized dipole modes in one HG cavity (HG002) residing in the second passband (TM_{110}-like modes) as high as 10^8. The nominal phase advances of the two modes ($4\pi/7$ and $5\pi/7$ respectively) correspond to phase velocities highly synchronized with the electron velocities yielding comparably large geometry-dependent beam coupling factors R/Q. Together with the unexpectedly high Q's, the HOM impedances in fact exceeded the 12 GeV requirement by a factor of 3. Any BBU-driven instability was cured during the period the module was installed in the north linac by judicious changes in the recirculation optics. Also, the experimental observation was made that in some situations of beam recirculation optics, three passes through the accelerator can be more stable (have higher threshold current) than two passes, even though there is greater average current in the former case. In 2009, the module was eventually removed from the north linac.

Investigations were launched during the period 2007–2009 to understand the lack of HOM damping. With NL04 still in CEBAF, extensive numerical investigations have been carried out to predict the cavity shape of the BBU culprit HG002 merely from frequency and external Q measurements at HOM couplers using today's high-end supercomputing capabilities combined with an elaborated cavity-shape algorithm tool [24]. This effort has forecast that both BBU-causing modes are tilted unfavorably away from the HOM couplers. The findings could not be proven until HG002 became available for field profile measurements after removing NL04 from CEBAF and disassembly in 2009. In fact, the tilted modes could be experimentally verified, though profiles deviated somewhat from the numerical results. For a more thorough understanding, the helium vessel of HG002 was removed to map external cell contours with a coordinate-measuring machine (CMM). This enabled reconstruction of the cavity interior with a parametric CAD model assuming uniform wall thickness [25]. Utilizing this model, HOMs have been found by numerical eigenmode analysis that differed significantly from expected fields in an undeformed cavity. When compared with field profile measurements, the BBU-generating HOMs could eventually be identified unambiguously (see Figure 11). Residual deviations from experimental results can be explained by the fact that the wall thickness is not uniform as assumed and that cells are not fully concentric, as confirmed by CMM measurements. The analysis also revealed that the culprit of the BBU scenario was a single deformed cell that required extensive tuning to flatten the accelerating field. The source of this deformation was traced to shrinkage due to a single extra weld pass on this cell during fabrication. This eventually caused modes to tilt severely, leaving practically no field at the HOM coupler positions, preventing adequate coupling of the HOMs and hence couplers, and leading to the observed elevated Q-values. It was also determined that the corresponding R/Q-factors would be adjusted to a value very close to the value deduced experimentally from the observed BBU thresholds. Another factor conspired to maximize vulnerability of this cavity to BBU; it so happened that the input waveguide window on this cavity had developed a leak, requiring installation of a shorting cap and auxiliary vacuum pump. This configuration not only eliminated opportunity to operate HG002, but also eliminated the input waveguide as a means for damping the TM_{110}-like modes now tilted toward it, thus contributing to raising their external Q values and lowering the BBU threshold.

To summarize, recent studies of multipass BBU theory and experiment show that BBU simulations are now good at predicting the instability threshold for small machines at the sub 10/of HOM damping Q-values and accurate calculation of beam coupling factors are needed for the simulation results to be correct, that the reasonableness and adequacy of the present 12 GeV HOM damping specifications have been confirmed. Such a level of accuracy is far beyond that believed achieved during the original CEBAF project.

8. Floating-pressure Ganni Cycle helium refrigeration

In the CEBAF Central Helium Liquefier, or CHL, the compression of helium to between 16 and 21 atm means substantial electrical power costs. Operating such a system at maximum capacity, even at the frequent times when maximum capacity isn't needed, is analogous to driving a car by flooring the accelerator while controlling the speed by strong use of the brake. However, by reconfiguring the CHL control structure, the Ganni Cycle enables the efficient scale-back of compressor energy consumption by reducing the operating pressures when full capacity isn't needed. Again figuratively speaking, it's like controlling automobile speed with varying acceleration and without use of the brake to offset over-acceleration. The process is seen in terms of floating pressure because it involves regulating the compressor discharge pressure to match the refrigeration load. The reconfigured control system senses when more refrigeration is needed by monitoring the amount of helium being used for cooling. The pressure floats to

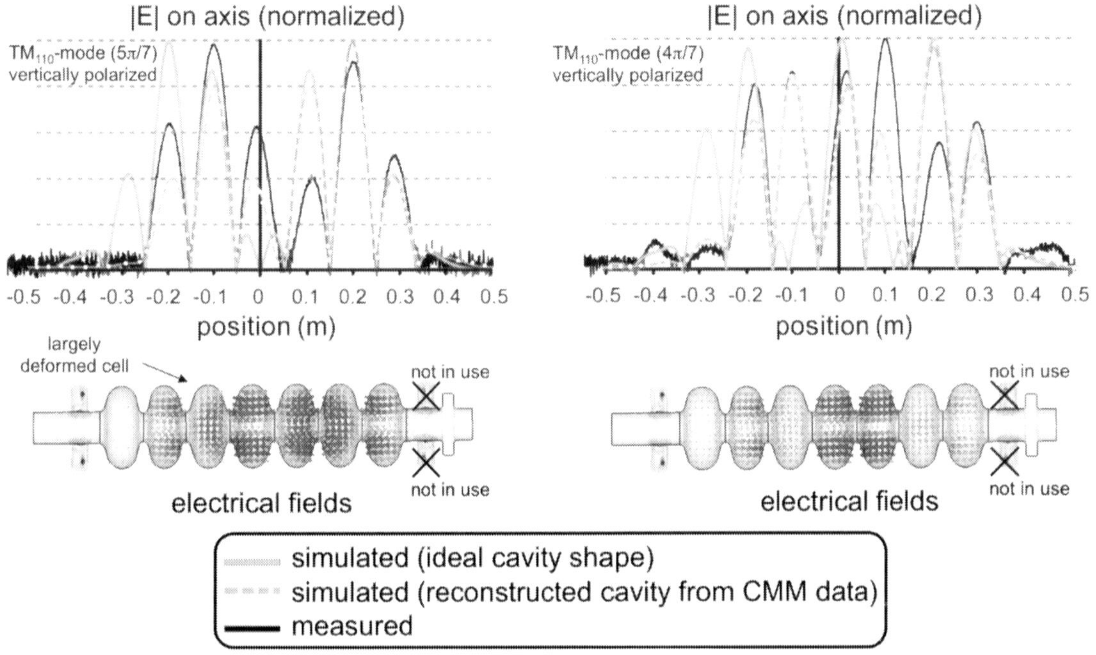

Figure 11. Calculated and measured high-order modes for a prototype high-gain low-loss cavity.

meet the load. If there is a reduction in load, there is a reduction in pressure, in power for compression, and consequently in electricity consumption.

The Ganni Cycle provides a basis for an optimal design at maximum load, turn-down cases and mixed modes, addressing the compressor system as the major input power loss contributor. It provides a solution to implement on as-built systems (existing or new) to improve system efficiency, reliability, availability and load stability under actual loads and help to improve the experimental envelope. It invalidates the philosophy that operating as-built systems at the presumed optimum of the manufacturer's theoretical design is optimal by properly identifying the fundamental process system parameters for control. It is a constant-pressure-ratio process cycle and maintains the compressor efficiency for varying loads. The Ganni Cycle maintains a constant volume flow (and thus velocity) at any point in the system and preserves the expander efficiency and the oil-removal effectiveness during the turn-down cases.

The Ganni Cycle nearly doubles the lifetime of some refrigerator components. Implementation for CEBAF, which mostly involved reconfiguration, did not require extensive capital expenses. For several years, this approach to CHL operation has been saving more than \$1000/day in power costs, having lowered the CHL's electrical demand from 6 MW to 4.2 MW. The process has also been adopted at Brookhaven's Relativistic Heavy Ion Collider and at the Spallation Neutron Source at Oak Ridge. It has been licensed by Jefferson Lab to Linde Cryogenics, Division of Linde Process Plants, Inc., and Linde Kryotechnik AG for worldwide commercialization. NASA is using it in a cryogenics plant supporting the preflight, Earth-bound testing of components of the James Webb Space Telescope, Hubble's envisioned replacement, in ultra-cold operational conditions.

References

[1] C. Leemann, D. Douglas, and G. Krafft, Ann. Rev. Nucl. Part. Sci. **51**, 413 (2001).
[2] L. S. Cardman and L. Harwood, Proceedings of PAC07, Albuquerque, New Mexico, USA page 58 (2007).
[3] A. Lung, Proceedings of the XVI International Workshop on Deep-Inelastic Scattering page 230 (2008).
[4] R. McKeown, AIP Conf. Proc. (2011).
[5] P. Adderley, J. Clark, J. Grames, J. Hansknecht, K. Surles-Law *et al.*, Phys. Rev. ST Accel.Beams **13**, 010101 (2010).
[6] T. Nakanishi *et al.*, AIP Conf. Proc. **421**, 300 (1998).
[7] J. Hansknecht and M. Poelker, Phys. Rev. ST Accel. Beams **9**, 063501 (2006).
[8] C. Sinclair *et al.*, Phys. Rev. ST Accel. Beams **10**, 023501 (2007).
[9] J. Grames *et al.*, Phys. Rev. ST Accel. Beams **7**, 042802 (2004).
[10] C. K. Sinclair, Jefferson Lab Technical Note CEBAF-TN-96-032 (1996).
[11] D. Higinbotham, AIP Conf.Proc. **1149**, 751 (2009).
[12] K. Paschke, Eur.Phys.J. **A32**, 549 (2007).
[13] Y.-C. Chao, Proceedings of PAC07, Albuquergue, New Mexico, USA page 780 (2007).
[14] P. Axel *et al.*, IEEE Trans. Nucl. Sci. **24**, 1133 (1977).
[15] C. Lyneis, M. Mcashan, R. Rand, H. Schwettman, T. Smith *et al.*, IEEE Trans. Nucl. Sci. **28**, 3445 (1981).
[16] G. Krafft and J. Bisognano, Proceedings of the 1987 IEEE Particle Accelerator Conference page 1356 (1987).
[17] N. Sereno, L. Cardman, G. Kraft, C. Sinclair, and J. Bisognano, Proceedings of 1993 Particle Accelerator Conference page 3246 (1993).
[18] K. Beard, Jefferson Lab Technical Note JLAB-TN-05-060 (2005).
[19] C. Tennant *et al.*, Phys. Rev. ST Accel. Beams **9**, 074403 (2005).
[20] D. Douglas, K. Jordan, L. Merminga, E. Pozdeev, C. Tennant *et al.*, Phys. Rev. ST Accel. Beams **9**, 064403 (2006).
[21] R. Kazimi, A. P. Freyberger, C. Hovater, G. Krafft, F. Marhauser *et al.*, Proceedings of the 2008 Europeon Paricle Accelerator Conference page 2722 (2008).
[22] I. Shin and B. Yuun, Jefferson Lab Technical Note JLAB-TN-09-004 (2009).
[23] B. Yunn, Jefferson Lab Technical Note JLAB-TN-04-035 (2004).
[24] Z. Li *et al.*, SLAC-PUB-13266 (2008).
[25] F. Marhauser *et al.*, To appear in Proceedings of the XXV Linear Accelerator Conference (2010).

AUTHOR INDEX

Institute of Physics
Dirac House, Temple Back
Bristol BS1 6BE UK

proceedings
.com